Accessing the E-book edition

Using the VitalSource® ebook

Access to the VitalBook™ ebook accompanying this book is via VitalSource® Bookshelf – an ebook reader which allows you to make and share notes and highlights on your ebooks and search across all of the ebooks that you hold on your VitalSource Bookshelf. You can access the ebook online or offline on your smartphone, tablet or PC/Mac and your notes and highlights will automatically stay in sync no matter where you make them.

Create a VitalSource Bookshelf account at ***https://online.vitalsource.com/user/new*** or log into your existing account if you already have one.

Redeem the code provided in the panel below to get online access to the ebook. Log in to Bookshelf and click the **Account** menu at the top right of the screen. Select **Redeem** and enter the redemption code shown on the scratch-off panel below in the **Code To Redeem** box. Press **Redeem**. Once the code has been redeemed your ebook will download and appear in your library.

DOWNLOAD AND READ OFFLINE

To use your ebook offline, download BookShelf to your PC, Mac, iOS device, Android device or Kindle Fire, and log in to your Bookshelf account to access your ebook:

On your PC/Mac

Go to ***http://bookshelf.vitalsource.com/*** and follow the instructions to download the free **VitalSource Bookshelf** app to your PC or Mac and log into your Bookshelf account.

On your iPhone/iPod Touch/iPad

Download the free **VitalSource Bookshelf** App available via the iTunes App Store and log into your Bookshelf account. You can find more information at ***https://support.vitalsource.com/hc/en-us/categories/200134217-Bookshelf-for-iOS***

On your Android™ smartphone or tablet

Download the free **VitalSource Bookshelf** App available via Google Play and log into your Bookshelf account. You can find more information at ***https://support.vitalsource.com/hc/en-us/categories/200139976-Bookshelf-for-Android-and-Kindle-Fire***

On your Kindle Fire

Download the free **VitalSource Bookshelf** App available from Amazon and log into your Bookshelf account. You can find more information at ***https://support.vitalsource.com/hc/en-us/categories/200139976-Bookshelf-for-Android-and-Kindle-Fire***

N.B. The code in the scratch-off panel can only be used once. When you have created a Bookshelf account and redeemed the code you will be able to access the ebook online or offline on your smartphone, tablet or PC/Mac.

SUPPORT

If you have any questions about downloading Bookshelf, creating your account, or accessing and using your ebook edition, please visit ***http://support.vitalsource.com/***

Electromagnetics for Electrical Machines

Electromagnetics for Electrical Machines

Saurabh Kumar Mukerji
Ahmad Shahid Khan
Yatendra Pal Singh

CRC Press
Taylor & Francis Group
Boca Raton London New York

CRC Press is an imprint of the
Taylor & Francis Group, an **informa** business

Reprint 2019

CRC Press
Taylor & Francis Group
6000 Broken Sound Parkway NW, Suite 300
Boca Raton, FL 33487-2742

ISBN: 978-0-367-24049-3 (hbk)

Digitally Printed at Replika Press Pvt. Ltd.

Library of Congress Cataloging-in-Publication Data

Mukerji, Saurabh Kumar.
Electromagnetics for electrical machines / Saurabh Kumar Mukerji, Ahmad Shahid Khan, Yatendra Pal Singh.
pages cm
Includes bibliographical references and index.
ISBN 978-1-4987-0913-2
1. Electric machinery--Mathematical models. 2. Electromagnetism--Mathematics. I. Khan, Ahmad Shahid II. Singh, Yatendra Pal. III. Title.

TK2211.M85 2015
621.31'042--dc23 2014040868

Visit the Taylor & Francis Web site at
http://www.taylorandfrancis.com

and the CRC Press Web site at
http://www.crcpress.com

For sale in India, Pakistan, Nepal, Bhutan, Bangladesh and Sri Lanka only.

Contents

Foreword xiii
Preface xv
Authors xvii

1. Introduction 1
1.1 Introduction 1
1.2 Field Approach 1
1.3 Domain of Machines 2
1.4 Review of Field Theory 2
1.5 Field Theorems 3
1.5.1 Uniqueness Theorem 4
1.5.2 Poynting Theorem 4
1.5.3 Approximation Theorem 4
1.6 Problem of Slotting 4
1.7 Eddy Current Phenomena 5
1.8 Polyphase Induction Machines 5
1.8.1 Laminated Iron Cores 6
1.8.2 Unlaminated Iron Cores 6
1.8.3 Simulation of Armature Winding 6
1.9 Case Studies 7
1.10 Numerical Techniques 7
1.10.1 Finite Element Method 7
1.10.2 Analytical Techniques 7
References 8

2. Review of Field Equations 11
2.1 Introduction 11
2.2 Maxwell's Equations in Integral Form 11
2.3 Maxwell's Equations in Point Form 13
2.4 General Equations for One Type of Field 14
2.5 Maxwell's Equations for Fields in Moving Media 16
2.6 Scalar Electric and Magnetic Potentials 17
2.7 Vector Magnetic Potential 18
2.8 Periodic Fields, Field Equations in Phasor Form 21
2.9 Retarded Potentials 23
2.10 Continuity Equation and Relaxation Time 25
2.11 A Rear Window View 28
Presentation Problems 29
References 29

3. Theorems, Revisited 31
3.1 Introduction 31
3.2 Uniqueness Theorem 31
3.2.1 Uniqueness Theorem for Laplace and Poisson Equations 31
3.2.1.1 Example of a Cuboid 33
3.2.1.2 Example of a Rectangular Region 37
3.2.2 Uniqueness Theorem for Vector Magnetic Potentials 41
3.2.3 Uniqueness Theorem for Maxwell's Equations 43
3.3 Helmholtz Theorem 45
3.4 Generalised Poynting Theorem 49
3.4.1 Components of Power Flow 54
3.4.2 Components of Force 55
3.5 Approximation Theorems 57
3.5.1 Approximation Theorem for Laplacian Field 57
3.5.2 Approximation Theorem for Vector Magnetic Potential 63
3.5.3 Approximation Theorem for Maxwell's Equations 68
Project Problems 73
Disclaimer 73
References 74

4. Laplacian Fields 75
4.1 Introduction 75
4.2 Potential Distribution for Rectangular Double-Slotting 76
4.2.1 Tooth-Opposite-Tooth Orientation 76
4.2.1.1 Evaluation of Arbitrary Constants 77
4.2.2 Tooth-Opposite-Slot Orientation 81
4.2.2.1 Evaluation of Arbitrary Constants 82
4.2.3 Arbitrary Orientation of Tooth and Slot 86
4.2.3.1 Evaluation of Arbitrary Constants 88
4.2.4 Air-Gap Permeance 92
4.3 Modelling for Aperiodical Field Distributions 94
4.3.1 Tooth-Opposite-Tooth Orientation 94
4.3.1.1 Evaluation of Unknown Functions 95
4.3.2 Tooth-Opposite-Slot Orientation 97
4.3.2.1 Evaluation of Unknown Functions 98
4.3.3 Arbitrary Orientation of Two Teeth 99
4.3.3.1 Evaluation of Unknown Functions 102
4.4 Fringing Flux for Tooth-Opposite-Tooth Orientation with Small Air Gap 107
4.4.1 Schwarz–Christoffel Transformation 108
4.4.1.1 Transformation from z Plane to w Plane 108
4.4.1.2 Transformation from χ Plane to w Plane 110

4.5 Air-Gap Field of a Conductor Deep inside an Open Slot 111
4.5.1 Schwarz–Christoffel Transformation 112
4.5.1.1 Transformation from *z*-Plane to *w*-Plane 112
4.5.1.2 Transformation from χ-Plane to *w*-Plane 114
4.6 Magnetic Field near Armature Winding Overhang 116
4.6.1 Surface Current Density 117
4.6.2 Magnetic Field Intensity 118
4.6.2.1 Field in Region 1 118
4.6.2.2 Field in Region 2 119
4.6.2.3 Field in Region 3 119
4.6.2.4 Field in Region 4 120
4.6.3 Boundary Conditions 121
4.6.3.1 Selection of Field Expressions 121
4.6.3.2 Evaluation of Arbitrary Constants 121
Project Problems 124
References 125

5. Eddy Currents in Magnetic Cores 127
5.1 Introduction 127
5.2 Eddy Current Machines (Solid Rotor Induction Machines) 127
5.2.1 Two-Dimensional Model 128
5.2.1.1 Field Components 129
5.2.1.2 Eddy Current Density 130
5.2.1.3 Eddy Current Loss 130
5.2.1.4 Force Density 131
5.2.1.5 Mechanical Power Developed 131
5.2.1.6 Rotor Power Input 131
5.3 Eddy Currents in Large Plates due to Alternating Excitation Current 132
5.3.1 Single-Phase Excitation 133
5.3.2 Polyphase Excitation 134
5.4 Eddy Currents in Cores with Rectangular Cross-Sections 137
5.5 Eddy Currents in Cores with Triangular Cross-Sections 139
5.6 Eddy Currents in Cores with Regular Polygonal Cross-Sections 143
5.6.1 Cores with Triangular Cross-Sections 144
5.6.2 Cores with Hexagonal Cross-Sections 149
5.6.3 Cores with Octagonal Cross-Sections 152
5.7 Eddy Currents in Circular Cores 157
5.8 Distribution of Current Density in Circular Conductors 159
5.9 Eddy Currents in Laminated Rectangular Cores 161
Project Problems 168
References 169

6. Laminated-Rotor Polyphase Induction Machines 171
6.1 Introduction 171
6.2 Two-Dimensional Fields in Anisotropic Media 172
6.3 Cage or Wound Rotor Induction Machines 175
6.3.1 Rotor Parameters 183
6.4 Induction Machines with Skewed Rotor Slots 184
6.4.1 Air-Gap Field 186
6.4.2 Fields in the Anisotropic Rotor Region 189
6.4.3 Determination of Arbitrary Constants 194
Project Problems 197
References 197

7. Unlaminated Rotor Polyphase Induction Machines 199
7.1 Introduction 199
7.2 Tooth-Ripple Harmonics in Solid-Rotor Induction Machines 199
7.2.1 Physical Description 199
7.2.1.1 Slip/Torque Characteristics 200
7.2.1.2 Idealised Configuration 200
7.2.2 Field Distribution in Stator Slots 203
7.2.2.1 Vector Magnetic Potential 203
7.2.2.2 Magnetic Field Intensity 204
7.2.3 Field Distribution in the Air Gap 206
7.2.4 Field Distribution in the Solid Rotor 217
7.2.5 Machine Performances 227
7.2.5.1 Eddy Current Loss 228
7.2.5.2 Force Density 229
7.2.5.3 Mechanical Power Developed 229
7.2.5.4 Rotor Input Power 230
7.2.5.5 Torque 231
7.3 Three-Dimensional Fields in Solid-Rotor Induction Machines 231
7.3.1 Idealised Model 231
7.3.2 Field Distributions 233
7.3.3 Effects of Finite Machine Length 248
7.3.4 Effect of Different Rotor and Stator Lengths 248
7.3.5 Performance Parameters 248
7.3.5.1 Eddy Current Loss 252
7.3.5.2 Force Density 257
7.3.5.3 Mechanical Power Developed 258
7.3.5.4 Rotor Input Power 259
7.3.5.5 Slip-Power Relation 259
Project Problems 259
References 259

8. Case Studies 261
8.1 Introduction 261
8.2 Slot Leakage Inductance for Conductors in Open Slots 261
8.2.1 Physical Configuration 261
8.2.2 Current Density Distribution 261
8.2.3 Vector Magnetic Potential 263
8.2.4 Flux Density 263
8.2.5 Inductance 264
8.3 Leakage Inductance of Transformers 265
8.3.1 Physical Configuration 266
8.3.2 Current Density Distribution 266
8.3.3 Vector Magnetic Potential 267
8.3.4 Magnetic Flux Density 269
8.3.5 Arbitrary Constants 270
8.3.6 Leakage Inductance 271
8.4 Field Theory of Hysteresis Machines 273
8.4.1 Simplifying Assumptions 274
8.4.2 Field Distributions 274
8.4.3 Induction Machine Action 278
8.4.3.1 Eddy Current Density 278
8.4.3.2 Eddy Current Loss 278
8.4.3.3 Force 279
8.4.3.4 Mechanical Power 280
8.4.3.5 Slip-Power Relation 280
8.4.4 Hysteresis Machine Action 281
8.4.4.1 Power Components 282
8.4.4.2 Slip-Power Relation 283
8.4.5 Impact of Different Parameters 283
8.5 Single-Phase Induction Motors with Composite Poles 286
8.5.1 Simplifying Assumptions 287
8.5.2 Idealised Machine Structure 287
8.5.3 Field Distribution 288
8.6 Transient Fields in Plates due to Type 2 Impact Excitations 298
8.6.1 Current Impact Excitation 300
8.6.2 Voltage Impact Excitation 305
Project Problems 310
References 310

9. Numerical Computation 313
9.1 Introduction 313
9.2 Numerical Analysis 313
9.2.1 Computational Errors 314
9.2.2 Numerical Stability 314
9.3 Domain of Numerical Analysis 314

9.3.1 Values of Functions.......... 314
9.3.2 Equations and Systems of Equations.......... 315
9.3.2.1 Linear Algebraic Equations.......... 315
9.3.3 Eigen-Value or Singular Value Problems.......... 316
9.3.4 Optimisation Problem.......... 316
9.3.5 Differential Equations.......... 317
9.3.5.1 Finite Difference Method.......... 317
9.3.5.2 Finite Element Method.......... 317
9.3.6 Numerical Integration.......... 318
9.3.6.1 Integration over Bounded Intervals.......... 318
9.3.6.2 Integration over Unbounded Intervals.......... 319
9.3.7 Software.......... 322
9.4 Types of Equations.......... 323
9.4.1 Relations without Summation or Integration.......... 324
9.4.1.1 Fringing Flux.......... 324
9.4.1.2 Air-Gap Field of a Conductor Deep inside an Open Slot.......... 324
9.4.1.3 Analysis of Eddy Current Induction Machines.......... 325
9.4.1.4 Eddy Currents in a Conducting Plate.......... 325
9.4.1.5 Eddy Currents within a Circular Core.......... 326
9.4.1.6 Distribution of Current Density in a Circular Conductor.......... 326
9.4.1.7 Two-Dimensional Fields in Anisotropic Media.......... 326
9.4.1.8 Field in the Cage or Wound Rotor Machine.......... 327
9.4.1.9 Induction Machine with Skewed Rotor Slots.......... 327
9.4.1.10 Field Theory of Hysteresis Machines.......... 328
9.4.2 Relations Involving Simple Summations.......... 329
9.4.2.1 Eddy Currents in Rectangular and Square Cores.......... 329
9.4.2.2 Eddy Currents in Triangular Core.......... 329
9.4.2.3 Slot Leakage Inductance.......... 330
9.4.2.4 Leakage Inductance of Transformers.......... 330
9.4.2.5 Transient Fields in Plates.......... 330
9.4.3 Summations Leading to Simultaneous Linear Algebraic Equations.......... 331
9.4.3.1 Potential Distributions in Tooth-Opposite-Tooth Orientation Case.......... 331
9.4.3.2 Potential Distributions in Tooth-Opposite-Slot Orientation.......... 334
9.4.3.3 Potential Distributions in Arbitrary Tooth Orientation.......... 336
9.4.3.4 Air-Gap Permeance.......... 338

9.4.3.5 Magnetic Field near Armature Winding Overhang 338
9.4.3.6 Eddy Currents in Regular Polygonal Cross-Section Cores 339
9.4.3.7 Eddy Currents in Laminated Rectangular Cores 341
9.4.3.8 Tooth-Ripple Harmonics in Solid Rotor Induction Machines 341
9.4.3.9 Three-Dimensional Fields in Solid Rotor Induction Machines 344
9.4.3.10 Single-Phase Induction Motors with Composite Poles Being Considered 346
9.4.4 Integrations Leading to Simultaneous Algebraic Equations 347
9.4.4.1 Tooth-Opposite-Tooth Orientation Case 348
9.4.4.2 Tooth-Opposite-Slot Orientation 352
9.4.4.3 Arbitrary Orientation of Teeth 358
Further Reading 363

Appendix 1: Hilbert Transform 365

Appendix 2: Evaluation of Integrals Involved in Section 4.3.1 367

Appendix 3: Evaluation of Integrals Involved in Section 4.3.2 375

Appendix 4: Evaluation of Integrals Involved in Section 4.3.3 381

Appendix 5: Evaluation of Arbitrary Constants Involved in Section 5.9 395

Appendix 6: Current Sheet Simulation of Stator Winding 399

Index 409

Foreword

Electromagnetics for Electrical Machines explains the intricacies of its subject in a simple and systematic manner. I thoroughly enjoyed poring over it. It is one of the few books that covers a difficult subject through investigation and by using programmed concepts for learning. The authors have spent considerable time in formulating the structure of the book and its contents. I believe they have been successful in their attempt. There have been several books on electromagnetics, each having its own facets. However, this book is unique in that it attempts to teach the concept of EMFT and its application to the theory and design of electrical machines.

The contributions of the authors of this book in various research and scientific areas have been outstanding. They are academicians who have devoted themselves to the task of educating young minds and inculcating scientific temper in them.

I must heartily congratulate the authors for the magnificent job they have done.

Brig. (Dr.) Surjit Pabla
Vice Chancellor
Mangalayatan University
Beswan, Aligarh

Preface

The contents of this book are based on the syllabi of MTech courses on electrical machines in some of the Indian universities and technical institutes. It is basically designed to serve for a one-semester course at the postgraduate level for electrical engineering students. Its contents are mainly confined to the linear theory of electromagnetics. Since this book lays more emphasis on concepts than on developing problem-solving skills, no numerical methods (viz. an iterative solution using relaxation techniques, finite difference methods, finite-element methods, the method of moment, etc.) are included. The first few introductory chapters of this book may also be used by students of physics, electronics and communication engineering, as well as those research scholars who are concerned with the problems involving electromagnetics.

The contents of this book are divided into the following nine chapters.

Chapter 1 is an introductory chapter that highlights the essence of field theory and its correlation with the electrical machines.

Chapter 2 includes a review of Maxwell's equations and scalar and vector potentials. It briefly describes the special cases leading to the Laplace, Poisson's, eddy current and wave equations.

Chapter 3 includes the uniqueness theorems, generalised Poynting theorem and a brief treatment of the Helmholtz theorem. It also deals with the approximation theorems developed to enhance the acceptability of approximate solutions of the field equations obtained under certain specified boundary conditions.

Chapter 4 is devoted to the solution of Laplace equation encountered in the design of electrical machines. It also introduces the Schwarz–Christoffel transformation and its applications and includes the determination of airgap permeance.

Chapter 5 outlines the solutions for eddy current equations. The skin effects in circular conductors, eddy currents in solid and laminated iron cores are also discussed.

Chapter 6 is devoted to the analyses of electromagnetic fields in laminated-rotor induction machines. Field theory for anisotropic media and its application in laminated-rotor induction machines is briefly presented. The effects of skewed rotor slots in laminated-rotor induction machines are discussed.

Chapter 7 presents three-dimensional field analyses for three-phase solid rotor induction machines. It also describes the end effects in solid rotor induction machines. Field analysis for harmonic fields, in solid rotor induction machines due to tooth ripples, is also discussed.

Chapter 8 includes the examples relating to the slot leakage inductance of rotating electrical machines, transformer leakage inductance and theory of hysteresis machines. It presents the modelling of fields for a potentially new type of single-phase induction motor with composite poles. It also includes the electromagnetic transients in solid-conducting plates. An extension of this treatment may lead to the study of electromagnetic transients in electrical machines.

Chapter 9 briefly describes the common techniques employed in numerical analysis. It also classifies all the cases described in Chapters 4 through 8 in accordance with the equations involved therein. This chapter also indicates the methods through which these equations can be solved to yield the desired results.

In order to enhance the subject knowledge of the reader, this book includes some presentation/project problems at the end of each chapter. Some of the mathematical material, for example, evaluation of some lengthy integrals, is included in the appendices. One of these appendices includes a brief note on the Hilbert transform. One appendix is fully devoted to the current sheet simulation of a balanced three-phase armature winding carrying a balanced three-phase alternating current. Throughout this book, standard symbols are used without defining them. Any nonstandard symbol is defined at the point it is first introduced; the meaning continues as long as it is not redefined.

The authors thankfully acknowledge the help and encouragement received from Professor S. C. Jain, former vice-chancellor; Professor A. B. Datye, former director of engineering and technology; Ms. Nidhi Singh, assistant professor, Department of Mechanical Engineering; Mr. Manoj Kumar Bhardwaj, assistant professor, Department of Electrical Engineering, all from Mangalayatan University, Aligarh. The authors are indebted to Professor Satnam Prasad Mathur, former pro vice-chancellor, Mangalayatan University, who has meticulously gone through second and third chapters and provided many valuable suggestions. The authors are thankful to Ms. Shivani Sharma, lecturer and Mr. Ashok Kumar, both from the Electrical Engineering Department, Mangalayatan University, for drawing figures included in this book. Thanks are also due to Mr. Anuj Verma, Gaurav Kumar Sharma and other MTech students who helped us in one way or the other in the preparation of this book.

Saurabh Kumar Mukerji, Ahmad Shahid Khan and Yatendra Pal Singh

Authors

Saurabh Kumar Mukerji obtained his BSc (Engg) degree in electrical engineering from Aligarh Muslim University (AMU), Aligarh, in 1958, and MTech and PhD degrees in electrical engineering from IIT Bombay, in 1963 and 1968, respectively. He has more than 50 years of teaching experience. During this period, he served Madhav Engineering College, Gwalior (India), AMU, Aligarh (India), SRMS College of Engineering and Technology, Bareilly (India), Multimedia University (MMU), Melaka (Malaysia) and Alfatha University (Air Academy), Misurata (Libya), in various capacities, including as a professor. Having served for five years as senior professor and head of the Electrical Engineering Department at Mangalayatan University, Aligarh (India), he is presently teaching as guest professor in the same university. He has served as a professor and chairman, Department of Electrical Engineering, AMU, Aligarh (India). He has published more than 40 research papers in various national and international journals and conference proceedings. He taught electrical machines and electromagnetic fields to graduate and postgraduate classes for more than 20 years. He has also taught network theory, optimization techniques, electrical machine design, antenna and wave propagation, microwave engineering, electrical measurements and measuring instruments and so on. His main area of research includes electromagnetic field applications to electrical machines.

Professor Mukerji has been on the review board for the *Journal of the Institution of Engineers* (India), *Progress in Electromagnetic Research* (PIER) and *Journal of Electromagnetic Waves and Applications* (JEMWA) (MIT, USA) and the *Journal of Engineering Science and Technology* (JESTEC) (Malaysia). He has served as an executive editor with Thomson George Publishing House, Malaysia.

Ahmad Shahid Khan has more than 42 years of teaching, research and administrative experience. He obtained his BSc (Engg), MSc (Engg) and PhD degrees from Aligarh Muslim University (AMU), Aligarh (India), in 1968, 1971 and 1980, respectively. He served at AMU, Aligarh in various capacities, including as a professor and chairman, Department of Electronics Engineering; registrar, AMU, Aligarh; and as estate officer (gazetted). After retirement in December 2006 from AMU, Aligarh, he served as the director, International Institute of Management and Technology, Meerut; director, Vivekananda College of Technology and Management, Aligarh; and director, Jauhar College of Engineering and Technology, Rampur. He also served as a professor in Krishna Institute of Engineering and Technology, Ghaziabad, and Institute of Management Studies, Ghaziabad. All these engineering institutes are affiliated with UP Technical University, Lucknow (India). He is currently a visiting professor in Mewat Engineering College (Wakf) Palla, Nuh affiliated to Mehrishi Dayanand University Rohtak, Haryana (India).

Dr. Khan is a fellow of the Institution of Electronics and Telecommunication Engineers (India) and a life member of the Institution of Engineers (India), Indian Society for Technical Education and Systems Society of India. He has attended many international and national conferences and refresher/orientation courses in the emerging areas of electronics and telecommunication. His area of interest is mainly related to electromagnetics, antennas and wave propagation, microwaves and radar systems. He has published 23 papers mainly related to electromagnetics. He is a recipient of Pandit Madan Mohan Malviya Memorial Gold Medal for one of his research papers published in the *Journal of the Institution of Engineers*, India, in 1978.

Dr. Khan edited *A Guide to Laboratory Practice in Electronics and Communication Engineering* in 1998, which was republished in 2002 by the Department of Electronics Engineering. His name was added as a coauthor in the Indian-adopted third edition of *Antennas for All Applications* published by Tata McGraw-Hill, New Delhi, in 2006. The fourth edition of this book was published in April 2010 with a new title, *Antennas and Wave Propagation* published by Tata McGraw-Hill, New Delhi. Dr. Khan added six new chapters in this edition. Its 13th printing appeared in October 2014. His latest book, *Microwave Engineering: Concepts and Fundamentals*, was published by CRC Press in March 2014.

Yatendra Pal Singh has more than 7 years of teaching experience. He obtained his graduate and postgraduate degrees in physics from Aligarh Muslim University (AMU), Aligarh, in 1999 and 2001, respectively. He obtained his PhD degree in the area of solar physics from AMU, Aligarh, in 2008. He is currently working as an assistant professor in the Department of Applied Physics, Institute of Engineering and Technology, Mangalayatan University, Aligarh, India. He has taught applied physics, electromagnetic field theory, mechanics, waves and oscillations to graduate students and quantum field theory, mathematical physics to postgraduate students for more than 5 years. He has so far published 15 papers in peer-reviewed journals, including *Journal of Geophysical Research* (JGR), *Astronomy & Astrophysics* (A&A), *Solar Physics* and in Elsevier journals such as *Journal of Atmospheric and Solar Terrestrial Physics* and *Planetary and Space Science* and 10 papers in proceedings of international conferences. Dr. Singh is a reviewer of research papers for the *Progress in Electromagnetic Research* (PIER), MIT, USA, and *Astrophysics and Space Science*, Europe. His major areas of interest include solar physics, heliospheric magnetic fields and magnetosphere.

1

Introduction

1.1 Introduction

Electromagnetic field theory has innumerable practical applications. The terms resistance, inductance, capacitance, conductance, potential, power, energy, force and torque emanate from the field concepts. The involvement of its concepts is imminent in all electrical and electronic devices and systems and the literature about most of these is available in abundance. The area of electrical machines is also deeply related to the electromagnetic field theory but has not yet been addressed to the level of satisfaction. One of the likely reasons for this neglect is that the field theory is presumed to be very conceptual with the general notion that it cannot be easily grasped without deep insight. It is further presumed that the field theory leads to complicated mathematical expressions, which create a sort of repulsive effect in the mind of the reader. In fact these myths are more psychological than real. The involvement of such mathematical expressions is quite common in other fields of science and engineering. Besides, the easy availability of design software based on numerical techniques has provided an excuse to those who want to avoid field concepts, mathematical complexities, and significant inaccuracies.

1.2 Field Approach

As of today, most of electrical machines are designed by employing software based on numerical techniques. Some of such readily available programs rely on over simplification of the problems, whereas some others are overly dependent on the data curves or empirical relations. These programs are neither generic nor optimal in nature. Besides, in implementing these techniques, computers normally execute voluminous calculations and even then, the ultimate outcome remains approximate. This scenario, therefore, calls for more accurate approaches. In this book, this task has been taken as a challenge. In essence, the basic aim of this book is to systematically correlate electrical machines to the field theory.

1.3 Domain of Machines

At this stage, it is pertinent to mention that the domain of electrical machines in itself is quite wide. As far their types are concerned, these contain devices with stationary and rotating (or moving) parts. Except transformers, which belong to the first category, all other machines fall to the second. The transformer in itself may have different types of cores and windings. The rotating (or linear) electrical machines may include direct current machines, induction machines, synchronous machines, eddy current machines, hysteresis machines, actuators and so forth. Many of these can further be classified as generators and motors. Besides, their rotors, stators, windings and modes of connections may also vary in shapes and sizes. These machines pose problems of electrical and thermal insulation. The shapes, sizes and types of slots also differ greatly. In particular, induction machines use semi-closed or fully closed slots. These machines contain rims and shafts and their presence is bound to influence the distribution of fields. There are problems of the induced eddies in the conductor cross sections, magnetic cores and other such artefacts. There are problems of teeth saturation, end effects and the nonlinearity of the core materials. All these aspects and factors play pivotal roles in electrical machines. Thus, accounting of all the above factors and aspects in one go is a horrendous task. It calls for tremendous effort and energy not usually seen in books of this scope.

At first sight the situation appears to be quite discouraging if not hopeless. As a way out, this book deals only with the general aspects of the involvement of field theory, in some of the commonly used machines. In order to provide for its extension to other machines, an effort has been made to develop the theory in a generalised way. Thus this book discusses only the simplified models of machines in regard to field theory and leaves the complicated cases to others to tread in. In general, this book deals with two-dimensional fields, but in some cases the variations in third dimension are also accounted for.

1.4 Review of Field Theory[1–5]

The book begins with the review of field theory as the effective operation of rotating electrical machines, transformers, inductors and other devices rely on proper magnetic field arrangement. In micro-electro-mechanical systems (MEMS) wherein the size reduces to the order of nanometres, both magnetic and electric fields are used for motion control. The electric motors, generators and actuators are referred to as energy conversion devices. This conversion between electrical and mechanical energy takes place in the coupling fields. Thus even a cursory look at any electrical machine reveals that the field phenomenon is deeply involved in its analysis and operation. It, therefore, appears to be proper to first understand the various aspects of the electromagnetic field and then to understand its role in the functioning of electrical machines.

The study of the electromagnetic field revolves around some basic laws. These include Coulomb's, Gauss's, Biot–Savart's, Ampere's, Faraday's laws and so on. These laws lead to certain quantities referred to as field quantities, which include electric field intensity ($\boldsymbol{E}$), electric flux density ($\boldsymbol{D}$), magnetic field intensity ($\boldsymbol{H}$) and magnetic flux density ($\boldsymbol{B}$). The integrated effect of these gives rise to the concept of current (I), surface current density (K), (volumetric) current density (J), scalar electric potential (V), scalar magnetic potential ($\mathcal{V}$), vector magnetic potential ($\boldsymbol{A}$), force (F), torque (T), power (P), energy ($\hat{E}$) and so on. Some of these quantities are interrelated by a set of relations called Maxwell's equations, whereas some others are obtained by manipulating this set. Maxwell's equations can be expressed either in integral or in differential (or point) form. These equations may take different forms for time-variant and time-invariant conditions, in static and moving media, and in accordance with the presence or absence of charge and current densities. These laws and relations are deeply related to the operation of rotating electrical machines. In the transient performance for many of these machines the concept of the retarded potentials, continuity equation and relaxation time is also deep rooted. Thus these laws, relations and concepts need to be reviewed for ready reference.

1.5 Field Theorems[6–8]

The four magical Maxwell's equations form the basis of design and analysis of rotating electrical machines. The manipulation of Maxwell's equations along with other field relations ultimately leads to another set of four equations known as Laplace's, Poisson's, wave and diffusion (or eddy current) equations. All of these are involved in the analysis of electrical machines in one way or the other. In some cases, the Laplace's and Poisson's equations have the lion's share, whereas in others the eddy current equation plays a pivotal role. In some situations, the wave equations too play a role and lead to the concept of retarded potentials.

1.5.1 Uniqueness Theorem

The solutions of these equations deserve careful scrutiny vis-à-vis their correctness and accuracy. The uniqueness theorem provides a mean through which the solutions obtained can be scrutinised. It is, therefore, necessary to explore the utility of this theorem in relation to Laplace and Poisson's equations, vector magnetic potentials and Maxwell's equations.

1.5.2 Poynting Theorem

Beside the uniqueness theorem, the Helmholtz theorem and the generalized Poynting theorem play a significant role in the analysis of electrical machines. The Helmholtz theorem assumes significance as and where the fields are of vector nature, whereas the Poynting theorem takes cognizance of electromechanical energy transfer.

1.5.3 Approximation Theorem

Since in most of the cases the solutions of field equations satisfy boundary conditions only approximately, for such cases, the approximation theorems are developed to review the acceptability of the solution.

1.6 Problem of Slotting[9–12]

In most electrical machines the rotor and/or stator are slotted, which affects the overall performance of the machine. These effects can be noticed while the analysis is carried out for the magnetic flux density distribution in the air gap, the magneto-motive-force (mmf) resulting due to the field current, the torque exerted on the rotor and the induced electro-motive-force (emf). The intensity of these effects depends on the types, shapes, sizes, relative displacements of slots and the relative dimensions of teeth and air gap. While designing an electrical machine, due consideration needs to be given to the effect of slotting to predict its exact behaviour. In most of such problems, the slots are considered to be source free. In cases wherein the sources are present, these sources are presumed to be located deep inside the slots. Thus, the problem reduces to the solution of the Laplace equation. The fields involved in such problems can be referred to as Laplacian fields. In view of the significance of such problems, due emphasis is to be given to the Laplacian fields. Such problems can be solved by using the Schwarz–Christoffel transformation method or separation of variables method. These methods lead to the determination of air gap permeance.

1.7 Eddy Current Phenomena[13,14]

The flow of eddy currents is another important aspect of field phenomena, which is deeply involved in some of the rotating electrical machines. The genesis of this phenomenon relates to the presence of a time-varying magnetic flux in a conducting medium, which generates an electromotive force. This force lies in the plane perpendicular to the direction of flux change. This electromotive force causes flow of current in the material, which is referred to as eddy current. These currents depend on the geometry of the medium, the rate of alternation of the flux and the electric and magnetic properties of the materials involved. The flow of these currents is always in such a direction so as to oppose the change in the flux that produces them. The net effect of such a flow is to prevent immediate penetration into the interior of the matter. As a result, for continuously varying applied field the magnetising force in the interior of the material never exceeds a small fraction of the magnetising force at the surface. When these eddy currents react with the inducing field, they create a mechanical torque. Such a phenomenon is noticed in case of a polyphase induction machine with a solid rotor acting either as a motor or as a break. The flow of eddy currents also produces loss of power, generating heat in a material. Such a loss is referred to as eddy current loss. Induction heating is an example of using this power.

The study of eddy current phenomena gains ground particularly when one studies the distribution of field and current in a conducting medium, which is subjected to a time-varying field. Such a situation arises in electrical machines and devices wherein magnetic cores are used. In the design of electromechanical clutches, eddy current brakes, electromagnetic couplings, solid iron rotor-induction machines and self-starting synchronous motors, the penetration of alternating flux into the solid iron needs proper estimation and thus requires a thorough understanding of eddy current phenomena. Such a study must encompass magnetic cores of different shapes and types, induction machines with solid rotors, large plates subjected to ac single-phase and polyphase excitations.

1.8 Polyphase Induction Machines[15–20]

The polyphase induction machines can be classified into two broad groups. These can be referred to as the laminated-rotor-induction machines and the unlaminated rotor-induction machines. The squirrel cage induction machines and slip-ring-type induction machines belong to the first category, while solid rotor-induction machines and drag cup-type induction machines belong to the second group. Some of their salient aspects are briefly described in the following sections.

1.8.1 Laminated Iron Cores

Laminated iron cores are used for cage-rotor and wound-rotor-induction machines. The slotted region of the rotor is isotropic but inhomogeneous. This region can be roughly represented as an equivalent anisotropic homogeneous region extending radially from the rotor air gap surface to the base of the rotor slots. In a linear treatment, a finite constant value for the saturated rotor teeth permeability may be chosen. The rotor core beyond the slotted region is rather magnetically unsaturated and the permeability for this region, being large, can be taken as infinite.

In view of the above description, the analyses of electromagnetic fields in laminated-rotor-induction machines need thorough consideration. Field theory for anisotropic media and its application to laminated-rotor-induction machines also invites due attention. Besides, the effects of skewed rotor slots in laminated-rotor-induction machines need to be properly addressed in terms of the field theory.

In a highly simplified treatment for the eddy current loss in laminated cores, the inhomogeneous laminated core region is often simulated by an anisotropic homogeneous region. The slotted region of laminated-rotor-induction machines can likewise be represented by an equivalent anisotropic homogeneous region. An exhaustive treatment for the electromagnetic field theory of squirrel cage induction machines was published by Mishkin, as early as 1954, wherein he simulated the heterogeneous isotropic slotted regions by homogeneous anisotropic regions.

1.8.2 Unlaminated Iron Cores

As noted above, the solid-rotor-induction machines fall into the second group of polyphase induction machines. In view of its importance, the proper field analyses for harmonic fields in solid-rotor-induction machines due to tooth ripples become more than a necessity. The study of end effects in the solid-rotor-induction machines requiring three-dimensional field analyses for three phase solid-rotor-induction machines also deserves due consideration.

1.8.3 Simulation of Armature Winding

For the field analyses of rotating electrical machines, the three-phase armature winding is often simulated by a suitable surface current sheet. Due attention is, therefore, required for simulating a balanced three-phase armature winding carrying a balanced three-phase alternating current.

1.9 Case Studies

The study of field analysis of rotating electrical machines can be better understood if some assorted examples covering slot leakage inductance of rotating electrical machines, transformer leakage inductance and theory of hysteresis machines are included. For the completion, this book also presents modelling of steady-state fields for a potentially new type of single-phase induction motor with composite poles and electromagnetic transients in solid conducting plates.

1.10 Numerical Techniques

In Section 1.2, it was noted that the design of most of the electrical machines rely on software based on numerical techniques, particularly the software based on the finite element methods that are dominating the scene. It was further noted that these applications are neither generic nor optimal in nature and the computers that execute such tasks have to perform numerous calculations to yield approximate results. A very brief comparison of the finite element method and the analysis presented in this book is given in the following subsections.

1.10.1 Finite Element Method

The genesis of the finite element method lies in dividing a two- or three-dimensional space into a large (finite) number of segments referred to as elements. This space, with field distribution, is enclosed by some real or imaginary boundaries. At these boundaries some field parameters are specified that are referred to as boundary conditions. With this technique it is expected that by taking more numbers of elements the resulting distribution will be closer to the real ones. However, in view of different types of computational errors (see Chapter 9), this goal is not always achieved. Moreover, in the case of numerical techniques a picture depicting a phenomenon may emerge only after obtaining a large number of curves or table.

1.10.2 Analytical Techniques

The analysis presented in this book is based on analytical techniques. More specifically the solutions are obtained by using the powerful method of separation of variables. The resulting mathematical expressions can readily be used to evaluate the required parameters. The analysis, in all the subsequent chapters, involves only the Cartesian system wherein X, Y and Z are used to represent the coordinates and x, y and z to represent the values of locations in these coordinates.

The analysis presented in Sections 4.4, 4.5, 5.2, 5.3, 5.7, 5.8, 6.2 through 6.4 and 8.40 has no involvement of any summation or integration, whereas in Sections 5.4, 5.5, 8.2, 8.3 and 8.6 summations are needed only to evaluate some of the field quantities. Thus, out of a total of 23 sections, the analysis of 15 sections is quite simplified. In Sections 4.2, 4.6, 5.6, 5.9, 7.2, 7.3 and 8.5, the obtained relations involve some arbitrary constants that are to be evaluated by solving some simultaneous linear algebraic equations. Similarly, Section 4.3 involves a number of improper integrals and some arbitrary functions. The evaluation of these functions also requires solutions of simultaneous linear algebraic equations. Thus, in the last two categories once the arbitrary constants/functions are evaluated, the field parameters can easily be estimated. This estimation, however, will further require summation of involved series or evaluation of involved improper integrals. In view of these resulting expressions, it is relatively easier to visualise the effects of different involved parameters even by looking at the mathematical expressions.

This analysis too will require numerical computations to obtain the field distributions. In this case both the infinite series and the integrals involved can be converted to the form of linear simultaneous algebraic equations. These equations can be tackled by using matrix operations with much ease and with much less numerical computations. It needs to be mentioned that during the process of analysis a large number of summations and integration are already evaluated. Thus, the situation is much more relaxed in comparison to the numerical methods. As will be evident later in some of the cases, the ultimate results are quite close to the closed-form solutions.

References

1. Maxwell, C., *A Treatise on Electricity and Magnetism*, Vols. I and II, Clarendon Press, England, 1904.
2. Stratton, J. A., *Electromagnetic Theory*, McGraw-Hill Book Company, Inc., New York, 1941.
3. Hague, B., *The Principles of Electromagnetism (Applied to Electrical Machines)*, Dover Publications, New York, 1962.
4. Hayt, W. H. Jr., *Engineering Electromagnetics*, Tata McGraw-Hill, New Delhi, India, 1997.
5. Sadiku, M. N. O., *Principles of Electromagnetics*, 4th Edn., Oxford University Press, New York, 2007.
6. Mukerji, S. K., Goel, S. K., Bhooshan, S., and Basu, K. P., Electromagnetic fields theory of electrical machines Part II: Uniqueness theorem for time-varying electromagnetic fields in hysteretic media, *International Journal of Electrical Engineering Education*, 42(2), 203–208, April 2005.
7. Mukerji, S. K., Goel, S. K., Bhooshan, S., and Basu, K. P., Electromagnetic fields theory of electrical machines Part I: Poynting theorem for electromechanical

energy conversion, *International Journal of Electrical Engineering Education*, 41(2), 137–145, April 2004.

8. George B. A. and Weber, H. J., *Mathematical Methods for Physicists*, Elsevier Academic Press, San Diego, California, 2005.
9. Mukherji, K. C. and Neville, S., Magnetic permeance of identical double slotting, *Proceedings of the IEE*, 118(9), 1257–1268, 1971.
10. Gibbs, W. J., *Conformal Transformation in Electrical Engineering*, Chapman & Hall Ltd., London, 1958.
11. Khan, A. S. and Mukerji, S. K., Field between two unequal opposite and displaced slots, *IEEE Transactions*, 90IC561-I, EC, 1–7, 1990.
12. Khan, A. S., Distribution of magnetic field between slotted equipotential surfaces, PhD thesis, Department of Electrical Engineering, Aligarh Muslim University, Aligarh, India, 1981.
13. Subbarao, V., *Eddy Currents in Linear and Non-Linear Media*, Omega Scientific Publishers, New Delhi, India, 1991.
14. Kesavamurthy, N. and Bedford, R. E., *Fields and Circuits in Electrical Machines*, Thacker Spike & Co. (1933) Private Ltd., Calcutta, India, pp. 223–228, 1966.
15. Mishkin, E., Theory of the squirrel-cage induction machine derived directly from Maxwell's field equations, *Quarterly Journal of Mechanics and Applied Mathematics*, VII(Pt. 4), 473–487, 1954.
16. Mukerji, S. K., Srivastava, D. S., Singh, Y. P., and Avasthi, D. V., Eddy current phenomena in laminated structures due to travelling electromagnetic fields, *Progress in Electromagnetics Research M*, 18, 159–169, 2011.
17. Mukerji, S. K., Linear electromagnetic field analysis of the solid rotor induction machine (tooth-ripple phenomena and end-effects), PhD thesis, IIT Bombay, India, 1967.
18. Mukerji, S. K., George, M., Ramamurthy, M. B., and Asaduzzaman, K., Eddy currents in solid rectangular cores, *Progress in Electromagnetics Research*, 7, 117–131, 2008.
19. Mukerji, S. K., George, M., Ramamurthy, M. B., and Asaduzzaman, K., Eddy currents in laminated rectangular cores, *Progress in Electromagnetics Research*, 83, 435–445, 2008.
20. Sharma, N. D. and Bedford, R. E., *Hysteresis Machines*, Perfect Prints, Thane (w), India, 2003.

2

Review of Field Equations

2.1 Introduction

The electromagnetic field theory has the capability of uniting quite diverse areas such as propagation of electromagnetic waves and functioning of dynamos. The electromagnetic field theory basically revolves around four simple equations called Maxwell's equations. These equations have made it possible to derive laws that govern the emission and reception of radio waves, propagation of current through wire grids and the operation of electric motors and transformers.

It is no exaggeration to state that the whole physics of electromagnetism is contained, though deeply buried, in Maxwell's four magical equations. Thus, it is proper to have a brief review of field theory and to explore the simplicity and versatility of Maxwell's equations. The current chapter presents this concisely. This chapter first describes Maxwell's equations in the integral and in point forms and later discusses their application in the fields in moving media. It also includes descriptions of scalar electric, scalar magnetic and vector magnetic potentials, periodic fields and field equations in phasor form. The text further spells the meanings of retarded potentials, continuity equation and the relaxation time.

2.2 Maxwell's Equations in Integral Form

Electric lines of force, also called flux lines, emanate from positive electric charge and terminate on negative electric charge. Likewise, in the case of permanent magnets, magnetic flux lines outside the magnet emanate from its north pole and terminate on the south pole. Inside the magnet, the direction of flux lines are reversed, that is, these lines emanate from south pole and terminate on the north pole. Therefore, magnetic flux lines are closed curves. Clearly, the electric flux lines and the magnetic flux lines behave differently. This is because the positive and negative charges exist independently while north and south magnetic poles are not. Therefore, it is possible that only one type of electric charge may reside inside a closed surface. However, no closed surface can enclose only one type of magnetic pole. The total electric flux emanating from a closed surface is defined as the total charge inside the surface. If both positive and negative charges are present inside the surface, the net charge will be the algebraic sum of discrete charges. In the case of distributed charge, the net charge will be the integrated value of the charge density. The magnetic poles being inseparable (magnetic monopoles do not exist in nature even though we may sometimes postulate their existence purely as a mathematical construct), the total magnetic flux emanating from any closed surface is always zero. This leads to the Gauss's law, or the integral form of Maxwell's[1] equations for magnetic and electric fields

$$\oiint_s \boldsymbol{B} \cdot \boldsymbol{ds} = 0 \tag{2.1}$$

$$\oiint_s \boldsymbol{D} \cdot \boldsymbol{ds} = Q \tag{2.2}$$

In these equations, **B** and **D**, respectively, indicate magnetic and electric flux density on the closed surface *s*, and *Q* indicates net charge inside this surface.

Faraday's law of electromagnetic induction states that electro-motive-force (e.m.f.), in a closed contour (i.e. closed path), is equal to the rate of decrease of magnetic flux φ, linking (or passing through) the closed contour. The e.m.f. is defined as the integrated value of the electric field intensity along the closed contour. Therefore,

$$e.m.f. = -\frac{d\varphi}{dt} \tag{2.3}$$

or

$$\oint_c \boldsymbol{E} \cdot \boldsymbol{dl} = -\frac{d}{dt}\iint_s B \cdot ds \tag{2.4}$$

In these equations, the electric field intensity **E** is on the closed contour *c* and the magnetic flux density **B** is on the surface *s*, while the contour *c* is the edge of the open surface *s*. Equation 2.4 is identified as Maxwell's third equation in integral form.

The fourth Maxwell's equation in integral form is based on Ampere's law. This law states that magneto-motive-force (m.m.f.), in a closed contour *c*, is equal to the total current linking the closed contour. The m.m.f. is defined as the integrated value of the magnetic field intensity along the closed contour. Therefore,

$$m.m.f. = I \tag{2.5}$$

or

$$\oint_c \boldsymbol{H} \cdot \boldsymbol{dl} = I \tag{2.6}$$

In these equations I indicates the total current linking the closed contour c. The value of I is the algebraic sum of discrete currents. In the case of distributed current, the net current will be the integrated value of the current density over the open surface s. The contour c is again the edge of the open surface s. Therefore, Equation 2.6 can be rewritten as

$$\oint_c \boldsymbol{H} \cdot \boldsymbol{dl} = \iint_s \boldsymbol{J} \cdot \boldsymbol{ds} \tag{2.7}$$

Ampere's experimental law was based on dc current I, leading to time-invariant magnetic field. Maxwell[1], without performing any experiment, generalised this equation for time-varying fields. Thus, Maxwell's fourth equation in the integral form is

$$\oint_c \boldsymbol{H} \cdot \boldsymbol{dl} = \iint_s \left(\boldsymbol{J} + \frac{\partial \boldsymbol{D}}{\partial t} \right) \cdot \boldsymbol{ds} \tag{2.8}$$

where $\boldsymbol{J}$ is usually called the conduction current density (not necessarily dc). It may include source current density, if there is any. The vector $\boldsymbol{D}$ can be defined as the electric flux density. The term $\partial \boldsymbol{D}/\partial t$ is said to be the displacement current density. In free space conduction current is zero, though there can be nonzero value for the displacement current. Many authors opine that the introduction of this displacement current is the major contribution of Maxwell with far-reaching consequences (e.g. wireless communications).

2.3 Maxwell's Equations in Point Form

Maxwell's equations in point (or differential) form can be readily obtained by applying the divergence theorem to Equations 2.1 and 2.2, and using Stokes theorem to Equations 2.4 and 2.8. The resulting equations are

$$\nabla \cdot \boldsymbol{B} = 0 \tag{2.9a}$$

$$\nabla \cdot \boldsymbol{D} = \rho \tag{2.9b}$$

$$\nabla \times \boldsymbol{E} = -\frac{\partial \boldsymbol{B}}{\partial t} \tag{2.9c}$$

$$\nabla \times \boldsymbol{H} = \boldsymbol{J} + \frac{\partial \boldsymbol{D}}{\partial t} \tag{2.9d}$$

In Equation 2.9b, the volume charge density ρ, at a given point, is a source for electric field. In Equation 2.9d, the current density $\boldsymbol{J}$ has two components

$$\boldsymbol{J} = \boldsymbol{J}_o + \sigma \boldsymbol{E} \tag{2.10}$$

where $\boldsymbol{J}_o$ indicates current source density, while the second term ($\sigma\boldsymbol{E}$) indicates induced current density in a conductor with conductivity σ.

2.4 General Equations for One Type of Field

Field vectors also depend on electromagnetic properties of the media where they exist. For linear homogeneous bilateral isotropic media, the following constitutive relations are satisfied:

$$\boldsymbol{B} = \mu \boldsymbol{H} \tag{2.11a}$$

$$\boldsymbol{D} = \varepsilon \boldsymbol{E} \tag{2.11b}$$

$$\boldsymbol{J} = \sigma \boldsymbol{E} \tag{2.11c}$$

where permeability μ, permittivity ε and the conductivity σ are constant scalar quantities for a given medium.

Considering the vector identity

$$\nabla \times \nabla \times \boldsymbol{E} \equiv \nabla(\nabla \cdot \boldsymbol{E}) - \nabla^2 \boldsymbol{E} \tag{2.12}$$

In view of Equations 2.9b through 2.9d and 2.10 the left-hand side (LHS) of Equation 2.12 becomes

$$-\nabla \times \frac{\partial \boldsymbol{B}}{\partial t} = -\mu \frac{\partial}{\partial t}(\nabla \times \boldsymbol{H}) = -\mu \frac{\partial}{\partial t}\left(\boldsymbol{J} + \frac{\partial \boldsymbol{D}}{\partial t}\right) = -\mu \frac{\partial \boldsymbol{J}_o}{\partial t} - \mu\sigma \frac{\partial \boldsymbol{E}}{\partial t} - \mu\varepsilon \frac{\partial^2 \boldsymbol{E}}{\partial t^2} \tag{2.13a}$$

Also, the right-hand side (RHS) of Equation 2.12 gives

$$\frac{1}{\varepsilon}\nabla(\nabla \cdot \boldsymbol{D}) - \nabla^2\boldsymbol{E} = \frac{1}{\varepsilon}\nabla\rho - \nabla^2\boldsymbol{E} \tag{2.13b}$$

On equating Equations 2.13a and 2.13b, we get

$$\nabla^2\boldsymbol{E} = \mu\sigma\frac{\partial \boldsymbol{E}}{\partial t} + \mu\varepsilon\frac{\partial^2 \boldsymbol{E}}{\partial t^2} + \left(\mu\frac{\partial \boldsymbol{J}_o}{\partial t} + \frac{1}{\varepsilon}\nabla\rho\right) \tag{2.13c}$$

Similarly, since

$$\nabla \times \nabla \times \boldsymbol{H} \equiv \nabla(\nabla \cdot \boldsymbol{H}) - \nabla^2\boldsymbol{H} \tag{2.14}$$

In view of Equations 2.9a, 2.9c, 2.9d and 2.10, we get, for the LHS of this equation.

$$\begin{aligned}\nabla \times \left(\boldsymbol{J}_o + \sigma\boldsymbol{E} + \varepsilon\frac{\partial \boldsymbol{E}}{\partial t}\right) &= \nabla \times \boldsymbol{J}_o + \sigma(\nabla \times \boldsymbol{E}) + \varepsilon\frac{\partial}{\partial t}(\nabla \times \boldsymbol{E}) \\ &= \nabla \times \boldsymbol{J}_o - \mu\sigma\frac{\partial \boldsymbol{H}}{\partial t} - \mu\varepsilon\frac{\partial^2 \boldsymbol{H}}{\partial t^2}\end{aligned} \tag{2.15a}$$

while the RHS of Equation 2.14 gives

$$\nabla(0) - \nabla^2\boldsymbol{H} = -\nabla^2\boldsymbol{H} \tag{2.15b}$$

On equating Equations 2.15a and 2.15b, we get

$$\nabla^2\boldsymbol{H} = \mu\sigma\frac{\partial \boldsymbol{H}}{\partial t} + \mu\varepsilon\frac{\partial^2 \boldsymbol{H}}{\partial t^2} + (-\nabla \times \boldsymbol{J}_o) \tag{2.16}$$

Expressions for vectors **D** and **B** obtained from Equations 2.13c and 2.16, respectively, are

$$\nabla^2\boldsymbol{D} = \mu\sigma\frac{\partial \boldsymbol{D}}{\partial t} + \mu\varepsilon\frac{\partial^2 \boldsymbol{D}}{\partial t^2} + \left(\mu\varepsilon\frac{\partial \boldsymbol{J}_o}{\partial t} + \nabla\rho\right) \tag{2.17}$$

$$\nabla^2\boldsymbol{B} = \mu\sigma\frac{\partial \boldsymbol{B}}{\partial t} + \mu\varepsilon\frac{\partial^2 \boldsymbol{B}}{\partial t^2} + (-\mu\nabla \times \boldsymbol{J}_o) \tag{2.18}$$

The terms inside the parentheses in Equations 2.13c, 2.16 through 2.18 indicate sources. The presence of sources renders these equations inhomogeneous. In the absence of sources, field equations are identified as damped wave equations. Time-varying electromagnetic fields in a conductor produce eddy currents opposing the original field. This results in damping. In nonconducting media, these equations reduce to wave equations. While for quasi-stationary fields in good conductors, these equations reduce to eddy current equation (also called diffusion equation). In case of time-invariant fields, these equations reduce to Poisson equations if sources are present, and Laplace equations if there is no source distribution.

2.5 Maxwell's Equations for Fields in Moving Media

In textbooks, two different rules are often described for induced e.m.f. One is based on 'flux linkage' and the other is based on 'flux cutting'. The former gives *'transformer e.m.f.'* and the latter termed as *'motional e.m.f.'* gives induced e.m.f. in conductors of rotating electrical machines. This second type of induced e.m.f. is not apparent from Maxwell's equations cited above.

Consider a reference system R' fixed on a medium, say the rotor of a moving (rotating) electrical machine, which moves with the constant velocity v relative to the system R. Let the latter system be at rest with respect to the stator of the machine. Electromagnetic fields in the reference system R' satisfy the following Maxwell's equations:

$$\nabla' \cdot \boldsymbol{B}' = 0 \tag{2.19}$$

$$\nabla' \cdot \boldsymbol{D}' = \rho' \tag{2.20}$$

$$\nabla' \times \boldsymbol{E}' = -\frac{\partial \boldsymbol{B}'}{\partial t'} \tag{2.21}$$

$$\nabla' \times \boldsymbol{H}' = \boldsymbol{J}' + \frac{\partial \boldsymbol{D}'}{\partial t'} \tag{2.22}$$

Since the rotor velocity is very small compared to the velocity of light, that is, $v \ll c$, the relativistic effects[2] can be ignored. Thus, $\nabla' = \nabla$ and $t' = t$. Also, $\mu' = \mu$, $\varepsilon' = \varepsilon$ and $\sigma' = \sigma$. The components of field vectors in the direction of motion and those that are normal to this direction are indicated using subscripts M and N, respectively. Therefore[2],

$$\boldsymbol{B}'_N = \boldsymbol{B}_N, \quad \boldsymbol{D}'_N = \boldsymbol{D}_N, \quad \boldsymbol{B}'_M = \boldsymbol{B}_M, \quad \boldsymbol{D}'_M = \boldsymbol{D}_M \tag{2.23a}$$

$$E'_N = E_N + (v \times B)_N, \quad H'_N = H_N - (v \times D)_N, \quad E'_M = E_M, \quad H'_M = H_M \tag{2.23b}$$

$$J'_N = J_N, \quad J'_M = J_M - v\rho', \quad \text{and} \quad \rho' = \rho \tag{2.23c}$$

Hence, Equations 2.19 through 2.22 can be rewritten as

$$\nabla \cdot \boldsymbol{B} = 0 \tag{2.24}$$

$$\nabla \cdot \boldsymbol{D} = \rho \tag{2.25}$$

$$\nabla \times (\boldsymbol{E} + v \times \boldsymbol{B}) = -\frac{\partial \boldsymbol{B}}{\partial t} \tag{2.26}$$

$$\nabla \times (\boldsymbol{H} - v \times \boldsymbol{D}) = (\boldsymbol{J} - v\rho) + \frac{\partial \boldsymbol{D}}{\partial t} \tag{2.27}$$

In the theory of rotating electrical machines if the curvature of air-gap surfaces is neglected, the angular velocity is transformed into linear velocity v, and above equations can be applied. It may be noted that Equation 2.26 includes electric field due to *'magnetic flux cutting'* or motional e.m.f., whereas Equation 2.27 includes magnetic field due to *'electric flux cutting'* or motional m.m.f.

2.6 Scalar Electric and Magnetic Potentials

For time-invariant fields, Maxwell's equations reduce to

$$\nabla \cdot \boldsymbol{B} = 0 \tag{2.28}$$

$$\nabla \cdot \boldsymbol{D} = \rho \tag{2.29}$$

$$\nabla \times \boldsymbol{E} = 0 \tag{2.30}$$

$$\nabla \times \boldsymbol{H} = \boldsymbol{J} \tag{2.31a}$$

Further, for current-free regions, Equation 2.31a reduces to

$$\nabla \times \boldsymbol{H} = 0 \tag{2.31b}$$

Now, since curl of the gradient of any scalar function is identically zero, in view of Equations 2.30 and 2.31b we may define

$$E \stackrel{def}{=} -\nabla V \tag{2.32a}$$

$$H \stackrel{def}{=} -\nabla \mathcal{V} \tag{2.32b}$$

where the scalar electric potential V for electrostatic field in the charge-free homogeneous regions and the scalar magnetic potential $\mathcal{V}$ for magneto-static field in the current-free homogeneous regions satisfy Laplace equations given as

$$\nabla^2 V = 0 \tag{2.33a}$$

$$\nabla^2 \mathcal{V} = 0 \tag{2.33b}$$

The scalar electric potentials for homogeneous regions with charge distribution satisfy the Poisson equation, instead of the Laplace equation, that is,

$$\nabla^2 V = -\frac{\rho}{\varepsilon} \tag{2.34}$$

2.7 Vector Magnetic Potential

In current carrying regions, instead of scalar magnetic potential, vector magnetic potential is defined. Since divergence of curl of a vector is identically zero, the vector B in Equation 2.28 can be expressed as curl of the vector magnetic potential A. Thus,

$$B \stackrel{def}{=} \nabla \times A \tag{2.35}$$

Now, since

$$\nabla \times \nabla \times A \equiv \nabla(\nabla \cdot A) - \nabla^2 A \tag{2.36}$$

Therefore, in view of Equation 2.35, for homogeneous regions we have

$$\begin{aligned} \nabla^2 A &= \nabla(\nabla \cdot A) - \nabla \times B = \nabla(\nabla \cdot A) - \mu \nabla \times H \\ &= \nabla(\nabla \cdot A) - \mu(J + \varepsilon(\partial E/\partial t)) \end{aligned} \tag{2.37a}$$

For slowly time-varying (or quasi-stationary) fields, it is assumed that

$$\varepsilon(\partial E/\partial t) \ll J \tag{2.37b}$$

Therefore, for homogeneous regions

$$\nabla^2 A \cong \nabla(\nabla \cdot A) - \mu J \tag{2.37c}$$

If gradient of an arbitrary scalar function of space coordinate is added to vector A, Equation 2.28 will remain unaltered. In this sense, vector A is not uniquely defined by Equation 2.35. According to the Helmholtz theorem (see Section 3.2), the vector potential A can be uniquely defined if its divergence is also defined. This is usually done in a way that simplifies Equation 2.37c.

An obvious choice is

$$\nabla \cdot A = 0 \tag{2.38}$$

The relation spelled by Equation 2.38 is known as Coulomb's gauge condition.

It may be noted that for high-frequency fields Equation 2.37b is not valid. As a result, Equation 2.38 is to be suitably modified[3,4] (see Section 2.8).

In view of Equation 2.38, the simplified form of Equation 2.37c valid for quasi-stationary field is given as

$$\nabla^2 A = -\mu J \tag{2.39}$$

For nonconducting regions, Equation 2.10 gives

$$\nabla^2 A = -\mu J_o \tag{2.40}$$

For known values of the source current density J_o, this equation is identified as the Poisson equation for the vector magnetic potential A.

It may be noted that the Laplacian operator operating on a scalar function is defined as the divergence of gradient of the scalar function. However, if the operand is a vector, its Laplacian must be defined from Equation 2.36. In the Cartesian system of space coordinates, Equation 2.39 results

$$\nabla^2(A_x a_x + A_y a_y + A_z a_z) = -\mu(J_x a_x + J_y a_y + J_z a_z) \tag{2.41a}$$

In the Cartesian system of coordinates, the three unit vectors (a_x, a_y and a_z, are constants, that is, invariant of coordinates of a point. Therefore, by equating the coefficients of respective unit vectors, Equation 2.41a can be resolved into the following three scalar equations:

$$\nabla^2 A_x = -\mu J_x \tag{2.41b}$$

$$\nabla^2 A_y = -\mu J_y \tag{2.41c}$$

$$\nabla^2 A_z = -\mu J_z \tag{2.41d}$$

In these equations the Laplacian is operating on scalar functions, thus it can be defined as divergence of gradient. Similar conclusion is reached by considering Equation 2.40 in the Cartesian system of space coordinates.

Magnetic flux, φ, through a surface s is given as

$$\varphi = \iint_S \boldsymbol{B} \cdot \boldsymbol{ds} \tag{2.42a}$$

Using Equation 2.35, we get

$$\varphi = \iint_S (\nabla \times \boldsymbol{A}) \cdot \boldsymbol{ds} \tag{2.42b}$$

Now, in view of the Stokes theorem, one obtains

$$\varphi = \oint_c \boldsymbol{A} \cdot \boldsymbol{dl} \tag{2.42c}$$

where the contour c is the edge (or border) of the surface s.

In two-dimensional field problems, Equation 2.42c reduces to a simpler form as elaborated through the following example.

Let the vectors $\boldsymbol{A}$ and $\boldsymbol{B}$ be independent of the z coordinate. Consider a rectangular contour abcda shown in Figure 2.1, with two sides, bc and da, parallel to the Z-axis. Let the vector potential along the side bc be $\boldsymbol{A}_1$ and along ad be $\boldsymbol{A}_2$. Along the side ab the potential varies from $\boldsymbol{A}_2$ to $\boldsymbol{A}_1$, while, along cd, it varies from $\boldsymbol{A}_1$ to $\boldsymbol{A}_2$. The magnetic flux through this contour, found from Equation 2.42c, is

$$\varphi = \ell(\boldsymbol{A}_{1z} - \boldsymbol{A}_{2z}) \tag{2.43}$$

where ℓ indicates the length of the side bc (and ad).

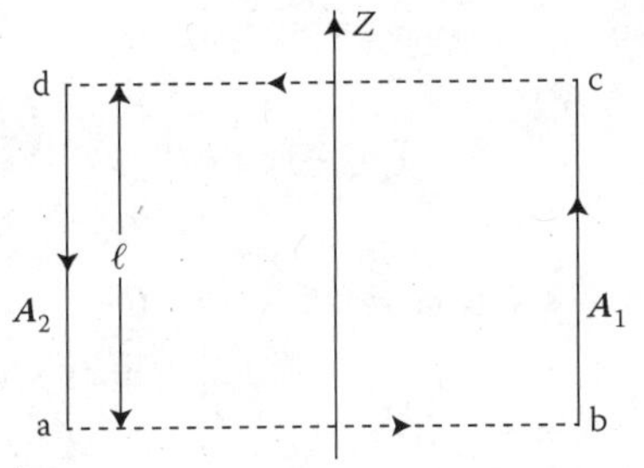

FIGURE 2.1
Rectangular contour abcda.

2.8 Periodic Fields, Field Equations in Phasor Form

Maxwell's equations in point (or differential) form for harmonic fields can be readily obtained from Equations 2.9 and 2.10. We define field vectors as complex phasors instead of instantaneous sinusoidal quantities. The moduli of these complex phasors give root mean square values. If these complex phasors are multiplied with $\sqrt{2}$ times the exponential factor $\exp(j\omega t)$, the real part gives the instantaneous expressions for harmonic field of frequency ω. The time derivative of field is simply the phasor field multiplied by $j\omega$. Thus, Maxwell's equations for harmonic fields are

$$\nabla \cdot \tilde{\boldsymbol{B}} = 0 \tag{2.44}$$

$$\nabla \cdot \tilde{\boldsymbol{D}} = \tilde{\rho} \tag{2.45}$$

$$\nabla \times \tilde{\boldsymbol{E}} = -j\omega \cdot \tilde{B} \tag{2.46}$$

$$\nabla \times \tilde{\boldsymbol{H}} = \tilde{J} + j\omega \cdot \tilde{D} \tag{2.47}$$

In the above equations, phasor quantities are distinguished from instantaneous quantities by placing a cap on each symbol. In subsequent treatment, however, no such distinction is made.

Maxwell's equations for harmonic fields in homogeneous source-free regions are given as

$$\nabla \cdot \boldsymbol{H} = 0 \tag{2.48}$$

$$\nabla \cdot \boldsymbol{E} = 0 \tag{2.49}$$

$$\nabla \times \boldsymbol{E} = -j\omega\mu \cdot \boldsymbol{H} \tag{2.50}$$

$$\nabla \times \boldsymbol{H} = \sigma \boldsymbol{E} + j\omega\varepsilon \cdot \boldsymbol{E} \tag{2.51a}$$

Using Equations 2.50 and 2.51a

$$\nabla \times \nabla \times \boldsymbol{E} = -j\omega\mu \cdot \nabla \times \boldsymbol{H} = -j\omega\mu \cdot (\sigma + j\omega\varepsilon)\boldsymbol{E} \tag{2.51b}$$

Also,

$$\nabla \times \nabla \times \boldsymbol{E} \equiv \nabla(\nabla \cdot \boldsymbol{E}) - \nabla^2 \boldsymbol{E} \tag{2.51c}$$

Therefore, in view of Equations 2.49, 2.51a and 2.51b

$$\nabla^2 \boldsymbol{E} = j\omega\mu\tilde{\sigma}\boldsymbol{E} \tag{2.52a}$$

where the complex conductivity $\tilde{\sigma}$ is defined as

$$\tilde{\sigma} \overset{def}{=} \sigma + j\omega\varepsilon \tag{2.52b}$$

Equation 2.52a reduces to the eddy current equation for harmonic field if $\sigma \gg \omega\varepsilon$, giving

$$\nabla^2 \boldsymbol{E} \cong \delta^2 \boldsymbol{E} \tag{2.53a}$$

where

$$\delta = (1 + j)/d \tag{2.53b}$$

The classical depth of penetration (d) is given by

$$d = \sqrt{\frac{2}{\omega\mu\sigma}} \tag{2.53c}$$

2.9 Retarded Potentials

Earlier, Maxwell's equations for time-varying fields were given by Equations 2.9a through 2.9d and the current density by Equation 2.10. Also **B** was given by Equation 2.35. These are reproduced below:

$$\nabla \cdot \boldsymbol{B} = 0 \tag{2.9a}$$

$$\nabla \cdot \boldsymbol{D} = \rho \tag{2.9b}$$

$$\nabla \times \boldsymbol{E} = -\frac{\partial \boldsymbol{B}}{\partial t} \tag{2.9c}$$

$$\nabla \times \boldsymbol{H} = \boldsymbol{J} + \frac{\partial \boldsymbol{D}}{\partial t} \tag{2.9d}$$

$$\boldsymbol{J} = \boldsymbol{J}_o + \sigma \boldsymbol{E} \tag{2.10}$$

$$\boldsymbol{B} \stackrel{def}{=} \nabla \times A \tag{2.35}$$

From Equations 2.9c and 2.35, we get

$$\nabla \times \boldsymbol{E} = -\frac{\partial}{\partial t}(\nabla \times \boldsymbol{A}) \tag{2.54a}$$

For stationary medium, space and time coordinates being independent of each other, we may rewrite this equation as

$$\nabla \times \boldsymbol{E} = -\nabla \times \left(\frac{\partial \boldsymbol{A}}{\partial t}\right) \tag{2.54b}$$

Therefore,

$$\nabla \times \left(\boldsymbol{E} + \frac{\partial \boldsymbol{A}}{\partial t}\right) = 0 \tag{2.54c}$$

In view of Equation 2.54c, we may define

$$\boldsymbol{E} + \frac{\partial \boldsymbol{A}}{\partial t} \stackrel{def}{=} -\nabla V \tag{2.55a}$$

or

$$\boldsymbol{E} = -\nabla V - \frac{\partial \boldsymbol{A}}{\partial t} \tag{2.55b}$$

In view of Equations 2.9d, 2.10, 2.35 and 2.55b, we can proceed as below:

$$\nabla \times \boldsymbol{B} = \nabla \times (\nabla \times \boldsymbol{A}) \equiv \nabla(\nabla \cdot \boldsymbol{A}) - \nabla^2 \boldsymbol{A} \tag{2.56a}$$

Also,

$$\nabla \times \boldsymbol{B} = \mu \nabla \times \boldsymbol{H} = \mu\left(\boldsymbol{J}_o + \sigma \boldsymbol{E} + \varepsilon \frac{\partial \boldsymbol{E}}{\partial t}\right) \tag{2.56b}$$

or

$$\nabla \times \boldsymbol{B} = \mu \boldsymbol{J}_o - \mu\sigma\left(\nabla V + \frac{\partial \boldsymbol{A}}{\partial t}\right) - \mu\varepsilon\left(\nabla \frac{\partial V}{\partial t} + \frac{\partial^2 \boldsymbol{A}}{\partial t^2}\right) \tag{2.56c}$$

Equating RHS of Equations 2.56a and 2.56c, we get

$$\nabla^2 \boldsymbol{A} = \nabla\left(\nabla \cdot \boldsymbol{A} + \mu\sigma V + \mu\varepsilon \frac{\partial V}{\partial t}\right) + \mu\sigma \frac{\partial \boldsymbol{A}}{\partial t} + \mu\varepsilon \frac{\partial^2 \boldsymbol{A}}{\partial t^2} - \mu \boldsymbol{J}_o \tag{2.57}$$

Next, consider Equations 2.9b and 2.55b

$$\nabla \cdot \boldsymbol{E} = \rho/\varepsilon \tag{2.58a}$$

Also,

$$\nabla \cdot \boldsymbol{E} = -\nabla \cdot \left(\nabla V + \frac{\partial \boldsymbol{A}}{\partial t}\right) = -\nabla^2 V - \frac{\partial}{\partial t}(\nabla \cdot \boldsymbol{A}) \tag{2.58b}$$

Equating RHS of Equations 2.58a and 2.58b, we get

$$\nabla^2 V = -\frac{\partial}{\partial t}(\nabla \cdot \boldsymbol{A}) - \frac{\rho}{\varepsilon} \tag{2.59}$$

To define the vector potential $\boldsymbol{A}$ uniquely, it is necessary to define its divergence as well (see Helmholtz theorem, Section 3.2).

Consider Equations 2.57 and 2.59; if we choose

$$\nabla \cdot \boldsymbol{A} \overset{def}{=} -\left(\mu\sigma V + \mu\varepsilon\frac{\partial V}{\partial t}\right) \tag{2.60}$$

we get

$$\nabla^2 \boldsymbol{A} = \mu\sigma\frac{\partial \boldsymbol{A}}{\partial t} + \mu\varepsilon\frac{\partial^2 \boldsymbol{A}}{\partial t^2} - \mu \boldsymbol{J}_o \tag{2.61}$$

and

$$\nabla^2 V = \mu\sigma\frac{\partial V}{\partial t} + \mu\varepsilon\frac{\partial^2 V}{\partial t^2} - \frac{\rho}{\varepsilon} \tag{2.62}$$

The vector $\boldsymbol{A}$ and the scalar V are called retarded potentials. These are related through Equation 2.60 and satisfy Equations 2.61 and 2.62, respectively. If the source densities are absent, these potentials satisfy damped wave equations in conducting regions and wave equations in nonconducting regions. Though Equation 2.60 has been chosen rather arbitrarily, it has the advantage of generating similar equations for both potentials. The name *retarded potentials* are given because the value of these potentials at a given instant t depends on the value of sources at an earlier instant since these potentials obey wave equation with a finite wave velocity ($v = 1/\sqrt{\mu\varepsilon}$).

2.10 Continuity Equation and Relaxation Time

Consider a region with volume v bounded by its surface s. Let the total current coming out of this surface be I. Outflow of this current causes a progressive reduction of any charge Q, that may be present in the volume, provided that the electric charge is neither created nor annihilated. Therefore,

$$I = -\frac{dQ}{dt} \tag{2.63}$$

This is called the *continuity equation*. Since the current I is given by integrating the current density over the closed surface s, and the charge Q is obtained by integrating the charge density in the volume v, the *integral form* of the continuity equation can be given as

$$\oint\!\!\!\oint_s \boldsymbol{J} \cdot d\boldsymbol{s} = -\frac{d}{dt}\iiint_v \rho\, dv \tag{2.64}$$

To obtain the *point form* of the continuity equation, apply Stokes' theorem to the LHS of this equation. Noting that the time and space coordinates are independent of each other, the sequence of operations on the RHS of Equation 2.64 can be interchanged. Since the resulting equation is valid for any arbitrary volume v, integrands on the two sides can be equated. Therefore, we get

$$\nabla \cdot \boldsymbol{J} = -\frac{\partial \rho}{\partial t} \tag{2.65}$$

The treatment presented above is valid for both homogeneous as well as inhomogeneous regions. For homogeneous regions and for some special cases of inhomogeneous regions where (σ/ε) is constant, the LHS of Equation 2.65 can be expressed, in view of Maxwell's second equation, in terms of the volume charge density ρ as

$$\nabla \cdot \boldsymbol{J} = \nabla \cdot \left(\frac{\sigma}{\varepsilon}\boldsymbol{D}\right) = \frac{\sigma}{\varepsilon}\nabla \cdot \boldsymbol{D} = \frac{\sigma}{\varepsilon} \cdot \rho$$

Therefore, Equation 2.65 can be rewritten as

$$\frac{\partial \rho}{\partial t} + \frac{\sigma}{\varepsilon} \cdot \rho = 0 \tag{2.66}$$

This equation can be readily solved to get

$$\rho = k \cdot e^{-(t/\tau)} \tag{2.67a}$$

where the relaxation time τ is

$$\tau \stackrel{def}{=} \varepsilon/\sigma \tag{2.67b}$$

The parameter k indicates an arbitrary constant.

The relaxation time τ is the time taken by charge density to reduce to e^{-1} of its original value, that is, 36.8% of initial value. Some typical values of τ are given below:

1. For copper with $\sigma = 5.9 \times 10^7$ S/m and $\varepsilon = \varepsilon_o = 8.854 \times 10^{-12}$ F/m; $\tau \cong 1.5 \times 10^{-19}$ s.
2. For water having $\sigma = 10^{-4}$ S/m, $\varepsilon_r = 81$; $\tau \cong 717 \times 10^{-8} \cong 7$ μs.

The arbitrary constant k is to be found from the initial condition. Let the charge density distribution at the instant $t = t_o$ be given as ρ_o, a function of space coordinates only. Therefore, from Equation 2.67a

$$k = \rho_o \cdot e^{(t_o/\tau)} \tag{2.67c}$$

Thus, the solution for Equation 2.66 is given as

$$\rho = \rho_o \cdot e^{-(t-t_o)/\tau} \quad (\text{for } t \geq t_o) \tag{2.68}$$

This shows that in a region with a constant value of relaxation time τ, if there is any arbitrary distribution of charge, it decays uniformly with time.

Consider Maxwell's equations for a region free from current sources, with constant value for the relaxation time:

$$\nabla \cdot \boldsymbol{D} = \rho \tag{2.69a}$$

$$\nabla \times \boldsymbol{H} = \boldsymbol{J} + \frac{\partial \boldsymbol{D}}{\partial t} = \frac{1}{\tau}\boldsymbol{D} + \frac{\partial \boldsymbol{D}}{\partial t} \tag{2.69b}$$

From the first equation, for ρ as given by Equation 2.68, we can infer that

$$\boldsymbol{D} = \boldsymbol{D}_o e^{-(t-t_o)/\tau} \tag{2.70a}$$

where $\boldsymbol{D}_o$ is a function of space coordinates only, such that

$$\nabla \cdot \boldsymbol{D}_o = \rho_o \tag{2.70b}$$

While, using Equation 2.70a, we have

$$\frac{\partial \boldsymbol{D}}{\partial t} = -\frac{1}{\tau}\boldsymbol{D}$$

Therefore, from Equation 2.69b, one gets

$$\nabla \times \boldsymbol{H} = 0 \tag{2.71}$$

If there is no external source for the magnetic field, the value of $\boldsymbol{H}$ will be zero. This example shows that even a time-varying electric field may not produce any magnetic field. It may be noted that due to conduction currents energy is dissipated. This energy is released from the electric field due to the rearrangement of charge density distribution. Since the magnetic field is zero in the conducting volume, it is zero outside as well. Thus, at every point on a closed surface bounding the conducting volume, the Poynting vector is zero. Therefore, there is no radiation of power into or out of this volume.

2.11 A Rear Window View

In our journey through the kingdom of Maxwell's equations, we have now covered some distance. It is now time to look back. Maxwell's equations are the experimental results described in mathematical language. These equations are based on two axioms, namely,

1. Nonexistence of magnetic monopoles
2. Conservation of electric charges

Consider Maxwell's third equation, that is, $\nabla \times \boldsymbol{E} = -(\partial \boldsymbol{B}/\partial t)$. If we take divergence on both sides of this equation, and interchange the order of differentiations with respect to time and space coordinates, we get

$$\frac{\partial}{\partial t}(\nabla \cdot \boldsymbol{B}) = 0 \tag{2.72}$$

It follows from the above equation that at every point in the field the divergence of $\boldsymbol{B}$ is time invariant. Thus, if $\nabla \cdot \boldsymbol{B}$ is ever zero, it remains at the zero value. Therefore, Maxwell's first equation, that is, $\nabla \cdot \boldsymbol{B} = 0$, is included in (consistent with) Maxwell's third equation.

Next, consider Maxwell's fourth equation, that is, $\nabla \times \boldsymbol{H} = \boldsymbol{J} + (\partial \boldsymbol{D}/\partial t)$. Again, if we take divergence on both sides of this equation, and interchange the order of differentiations with respect to time and space coordinates, we get

$$\nabla \cdot \boldsymbol{J} = -\frac{\partial}{\partial t}(\nabla \cdot \boldsymbol{D}) \tag{2.73}$$

Now, consider the continuity equation in point form, that is, $\nabla \cdot \boldsymbol{J} = -(\partial \rho/\partial t)$. On comparing this equation with Equation 2.73, we obtain $\nabla \cdot \boldsymbol{D} - \rho = \rho'$, where ρ' is time invariant. Thus, if ρ' is ever zero, it remains at the zero value. Therefore, Maxwell's second equation, that is, $\nabla \cdot \boldsymbol{D} = \rho$, is included in (consistent with) Maxwell's fourth equation.

Maxwell's equations describe large-scale electromagnetic phenomena. Those in atomic and subatomic distances are in the domain of *quantum mechanics*. It has been said that '*Whatever form the equations of quantum electrodynamics ultimately assume, their statistical average over large numbers of atoms must lead to Maxwell's equations'*[5].

PRESENTATION PROBLEMS

1. The generalised Coulomb law is given by

$$E(r,t) = e\left\{\left[\frac{n}{KR^2}\right] + \frac{1}{c}\cdot\frac{\partial}{\partial t}\left[\frac{n}{KR}\right] - \frac{1}{c}\cdot\frac{\partial}{\partial t}\left[\frac{\beta}{KR}\right]\right\}$$

2. The generalised Biot–Savart's law is given by

$$B(r,t) = e\left\{\left[\frac{\beta \times n}{KR^2}\right] + \frac{1}{c}\cdot\frac{\partial}{\partial t}\left[\frac{\beta \times n}{KR}\right]\right\}$$

3. The expressions of generalised Lienard–Wiechert fields are given by

$$E(r,t) = e\left\{\left[\frac{(n-\beta)\cdot(1-\beta^2)}{K^3R^2}\right] + \left[\frac{n \times ((n-\beta)\times a)}{c^2K^3R}\right]\right\}$$

and

$$B(r,t) = e\left\{\left[\frac{(\beta \times n)\cdot(1-\beta^2)}{K^3R^2}\right] + \left[\frac{(a\cdot n)((\beta\times n)\times a)}{c^2K^3R}\right] + \frac{a\times n}{c^2K^2R}\right\}$$

4. The generalised expressions for scalar and vector potentials are $\varphi(r,t) = \dfrac{e}{[R - \beta\cdot R]}$ and $A(r,t) = \dfrac{e[\beta]}{[R - \beta\cdot R]}$

In all above problems $n = R/R$; $\beta = u/c$; $K = 1 - \beta\cdot n$ and in Problem 3 $a = du(tr)/dtr$ (*Hint for all problems*: Chapter 8 of Reference 6).

References

1. Maxwell, C., *A Treatise on Electricity and Magnetism,* Vols. I and II, Clarendon Press, England, 1904.
2. Stratton, J. A., *Electromagnetic Theory,* McGraw-Hill Book Company, Inc., New York, pp. 78–81, 1941.
3. Hayt, W. H. Jr., *Engineering Electromagnetics,* Tata McGraw-Hill, New Delhi, India, pp. 326–331, 1997.

4. Matthew, N. O. S., *Principles of Electromagnetics*, 4th Edn., Oxford University Press, New York, pp. 345–347, 2007.
5. Stratton, J. A., *Electromagnetic Theory*, McGraw-Hill Book Company, Inc., New York, pp. vii, 1941.
6. Heald, M. A. and Marion, J. B., *Classical Electromagnetic Radiation*, 3rd Edn., Brooks/Cole, Thomson Learning, Dover, New York, pp. 256–288, 1995.

3

Theorems, Revisited

3.1 Introduction

The majority of field problems reduce either to the solution of Laplace's equation or that of Poisson's equation. Once a solution is found, it becomes a bone of contention whether the obtained solution is the only possible solution. The answer to this question is given by the uniqueness theorem, which is considered as a litmus test in all such cases. This chapter begins with the uniqueness theorem and explores its applicability and compatibility for the solutions of Laplace's and Poisson's equations. The domain of the uniqueness theorem is further extended to encompass the vector magnetic potential and Maxwell's equations. This chapter also includes the discussion on Helmholtz's theorem and the generalised Poynting theorem. A new concept of approximation theorems is also introduced and their usefulness for the Laplace equation, vector magnetic potential and eddy current equation is explored.

3.2 Uniqueness Theorem

The uniqueness theorem spells that if a solution exists for a given equation under certain specified boundary conditions it is the only possible solution and may be referred to as unique. In the following subsections this theorem is discussed vis-a-vis the Laplace and Poisson equations, vector magnetic potential and Maxwell's equations.

3.2.1 Uniqueness Theorem for Laplace and Poisson Equations

In general, there are infinite solutions for the Laplace and Poisson equations. However, under certain boundary conditions unique solutions for these equations can be found. The uniqueness theorem[1] reviewed here describes these conditions that are equally valid for both Laplace and Poisson equations.

Let V_1 and V_2 be any two solutions for the Laplace equation for potential distribution in a given volume v. The difference potential V_o is defined as

$$V_o \stackrel{def}{=} V_1 - V_2 \tag{3.1}$$

It may be noted that V_o also satisfies the Laplace equation

$$\nabla^2 V_o = 0 \tag{3.2}$$

Consider the identity

$$\nabla \cdot (V_o \nabla V_o) \equiv |\nabla V_o|^2 + \nabla^2 V_o \tag{3.3}$$

In view of Equation 3.2, Equation 3.3 reduces to

$$\nabla \cdot (V_o \nabla V_o) = |\nabla V_o|^2 \tag{3.4}$$

The integration of both sides of Equation 3.4 over the volume v, and the application of the divergence theorem gives

$$\oiint_s (V_o \nabla V_o) \cdot ds = \iiint_v |\nabla V_o|^2 \, dv \tag{3.5}$$

where s indicates the surface bounding the volume v, internally as well as externally.

Under certain boundary conditions satisfied by the two solutions, the left-hand side (LHS) of this equation reduces to zero. The zero value for the volume integration on the right-hand side (RHS) of this equation implies that since the integrand $|\nabla V_o|^2$ is nowhere negative, ∇V_o must be zero at every point in the volume v. Therefore, the zero value for the LHS implies that the difference potential must be a constant, that is, the two solutions, at every point in the volume as well as at every point on the bounding surface of this volume, must only differ by a constant.

The conditions for the LHS of Equation 3.5 to be zero are satisfied if V_o is zero on all or parts of the bounding surface s, and the normal derivative of V_o is zero on its remaining parts. Therefore, to obtain the unique solution, potential distribution must be defined over parts of the boundary surface and its normal derivative over the remaining parts. Once the unique expression for potential distribution in a region is found, its normal derivative over the entire boundary can be uniquely obtained using this expression. Therefore, at no part on the boundary surface, both potential and its normal derivative can be specified independently. It is important to note that one or the other (but not both) boundary conditions must be specified over the entire bounding surface and the solution that satisfies the given boundary conditions over this closed surface s is the unique solution. It is further to be noted that for a unique solution, potential must be defined at least at one point on the boundary and the solution must satisfy this boundary condition.

The uniqueness theorem[1] presented above provides for a set of boundary conditions over *closed surfaces*. Any solution of the Laplace equation that satisfies these conditions is unique. A question that arises is, are these conditions *'necessary'* or *'sufficient'* or both? As shown in the following examples, unique solutions can be achieved even with undefined boundary conditions for a part of the closed boundary surface, provided that sufficient conditions elsewhere within the region where the field distribution is being obtained are specified and satisfied by the solution. Therefore, the theorem provides *'sufficient conditions'* and not the *'necessary conditions'* for uniqueness of the solution of Laplace or Poisson equations.

3.2.1.1 Example of a Cuboid

The cuboid shown in Figure 3.1 has boundary conditions specified on all surfaces except the top surface. Since neither the potential distribution nor its normal derivative is specified on the top surface, the region inside the cuboid may be called a dark region with unknown field distribution. To determine the field distribution, the region must be illuminated with the torch function suitably distributed on the top surface. Synthesis of the torch function is done using certain boundary conditions given elsewhere on the boundary. In this example, the boundary condition is provided on the bottom surface of the cuboid, vide Equation 3.8.[1]

The potentials on four side surfaces of the cuboid are given as

$$V\big|_{x=\pm a/2} = V\big|_{y=\pm b/2} = 0 \tag{3.6}$$

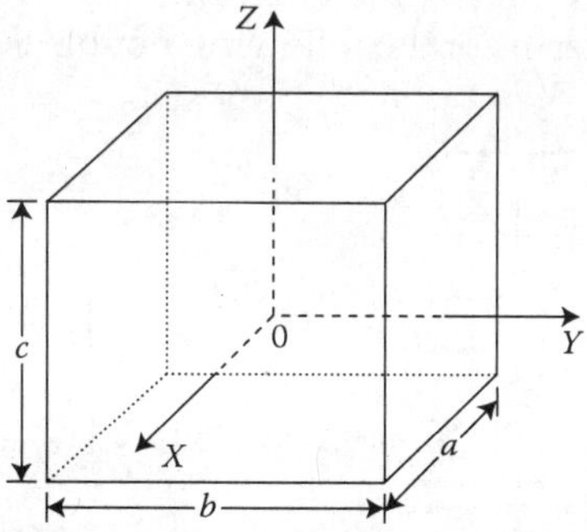

FIGURE 3.1
A cuboid with Cartesian system of space coordinates.

On the bottom surface, the distributions of potential and its normal derivative are given as

$$\begin{aligned} V\Big|_{z=-c/2} &= \sum_{m-odd}^{\infty}\sum_{n-odd}^{\infty} A_{mn}^{(1)} \cdot \cos\left(\frac{m\pi}{a}\cdot x\right)\cdot \cos\left(\frac{n\pi}{b}\cdot y\right) \\ &+ \sum_{m-odd}^{\infty}\sum_{n-even}^{\infty} A_{mn}^{(2)} \cdot \cos\left(\frac{m\pi}{a}\cdot x\right)\cdot \sin\left(\frac{n\pi}{b}\cdot y\right) \\ &+ \sum_{m-even}^{\infty}\sum_{n-odd}^{\infty} A_{mn}^{(3)} \cdot \sin\left(\frac{m\pi}{a}\cdot x\right)\cdot \cos\left(\frac{n\pi}{b}\cdot y\right) \\ &+ \sum_{m-even}^{\infty}\sum_{n-even}^{\infty} A_{mn}^{(4)} \cdot \sin\left(\frac{m\pi}{a}\cdot x\right)\cdot \sin\left(\frac{n\pi}{b}\cdot y\right) \end{aligned} \tag{3.7}$$

and

$$\begin{aligned} \frac{\partial V}{\partial z}\bigg|_{z=-c/2} &= \sum_{m-odd}^{\infty}\sum_{n-odd}^{\infty} B_{mn}^{(1)} \cdot \cos\left(\frac{m\pi}{a}\cdot x\right)\cdot \cos\left(\frac{n\pi}{b}\cdot y\right) \\ &+ \sum_{m-odd}^{\infty}\sum_{n-even}^{\infty} B_{mn}^{(2)} \cdot \cos\left(\frac{m\pi}{a}\cdot x\right)\cdot \sin\left(\frac{n\pi}{b}\cdot y\right) \\ &+ \sum_{m-even}^{\infty}\sum_{n-odd}^{\infty} B_{mn}^{(3)} \cdot \sin\left(\frac{m\pi}{a}\cdot x\right)\cdot \cos\left(\frac{n\pi}{b}\cdot y\right) \\ &+ \sum_{m-even}^{\infty}\sum_{n-odd}^{\infty} B_{mn}^{(4)} \cdot \sin\left(\frac{m\pi}{a}\cdot x\right)\cdot \sin\left(\frac{n\pi}{b}\cdot y\right) \end{aligned} \tag{3.8}$$

where $A_{mn}^{(1)-(4)}$ and $B_{mn}^{(1)-(4)}$ are known coefficients of the double Fourier series.

On the top surface of this cuboid neither the potential nor its normal derivative is specified. Thus thereat, the torch function consistent with Equation 3.6, can be defined in terms of the following double Fourier series:

$$\begin{aligned} V\Big|_{z=c/2} &\stackrel{def}{=} \sum_{m-odd}^{\infty}\sum_{n-odd}^{\infty} T_{mn}^{(1)} \cdot \cos\left(\frac{m\pi}{a}\cdot x\right)\cdot \cos\left(\frac{n\pi}{b}\cdot y\right) \\ &+ \sum_{m-odd}^{\infty}\sum_{n-even}^{\infty} T_{mn}^{(2)} \cdot \cos\left(\frac{m\pi}{a}\cdot x\right)\cdot \sin\left(\frac{n\pi}{b}\cdot y\right) \\ &+ \sum_{m-even}^{\infty}\sum_{n-odd}^{\infty} T_{mn}^{(3)} \cdot \sin\left(\frac{m\pi}{a}\cdot x\right)\cdot \cos\left(\frac{n\pi}{b}\cdot y\right) \\ &+ \sum_{m-even}^{\infty}\sum_{n-even}^{\infty} T_{mn}^{(4)} \cdot \sin\left(\frac{m\pi}{a}\cdot x\right)\cdot \sin\left(\frac{n\pi}{b}\cdot y\right) \end{aligned} \tag{3.9}$$

Synthesis of the torch function will be completed with the determination of coefficients in the four double Fourier series, involved in the above equation.

In view of Equations 3.6 through 3.9, the potential distribution inside the cuboid can be written as

$$V = \sum_{m-odd}^{\infty}\sum_{n-odd}^{\infty}\cos\left(\frac{m\pi}{a}\cdot x\right)\cdot\cos\left(\frac{n\pi}{b}\cdot y\right)\cdot\left\{C_{mn}^{(1)}\cdot\frac{\cosh\tau_{mn}\cdot(z+c/2)}{\cosh(\tau_{mn}\cdot c)}+D_{mn}^{(1)}\right.$$
$$\left.\times\frac{\sinh\tau_{mn}\cdot(z+c/2)}{\sinh(\tau_{mn}\cdot c)}\right\}+\sum_{m-odd}^{\infty}\sum_{n-even}^{\infty}\cos\left(\frac{m\pi}{a}\cdot x\right)\cdot\sin\left(\frac{n\pi}{b}\cdot y\right)$$
$$\cdot\left\{C_{mn}^{(2)}\cdot\frac{\cosh\tau_{mn}\cdot(z+c/2)}{\cosh(\tau_{mn}\cdot c)}+D_{mn}^{(2)}\times\frac{\sinh\tau_{mn}\cdot(z+c/2)}{\sinh(\tau_{mn}\cdot c)}\right\}$$
$$+\sum_{m-even}^{\infty}\sum_{n-odd}^{\infty}\sin\left(\frac{m\pi}{a}\cdot x\right)\cdot\cos\left(\frac{n\pi}{b}\cdot y\right)\cdot\left\{C_{mn}^{(3)}\cdot\frac{\cosh\tau_{mn}\cdot(z+c/2)}{\cosh(\tau_{mn}\cdot c)}+D_{mn}^{(3)}\right.$$
$$\left.\times\frac{\sinh\tau_{mn}\cdot(z+c/2)}{\sinh(\tau_{mn}\cdot c)}\right\}+\sum_{m-even}^{\infty}\sum_{n-even}^{\infty}\sin\left(\frac{m\pi}{a}\cdot x\right)\cdot\sin\left(\frac{n\pi}{b}\cdot y\right)$$
$$\cdot\left\{C_{mn}^{(4)}\cdot\frac{\cosh\tau_{mn}\cdot(z+c/2)}{\cosh(\tau_{mn}\cdot c)}+D_{mn}^{(4)}\times\frac{\sinh\tau_{mn}\cdot(z+c/2)}{\sinh(\tau_{mn}\cdot c)}\right\} \quad (3.10a)$$

where

$$C_{mn}^{(1)}+D_{mn}^{(1)}=T_{mn}^{(1)} \quad (3.10b)$$

$$C_{mn}^{(2)}+D_{mn}^{(2)}=T_{mn}^{(2)} \quad (3.10c)$$

$$C_{mn}^{(3)}+D_{mn}^{(3)}=T_{mn}^{(3)} \quad (3.10d)$$

$$C_{mn}^{(4)}+D_{mn}^{(4)}=T_{mn}^{(4)} \quad (3.10e)$$

and

$$\tau_{mn}=\sqrt{\left(\frac{m\pi}{a}\right)^2+\left(\frac{n\pi}{b}\right)^2} \quad (3.10f)$$

Therefore, in view of Equation 3.7

$$C_{mn}^{(1)} = A_{mn}^{(1)} \cdot \cosh(\tau_{mn} \cdot c) \tag{3.11a}$$

$$C_{mn}^{(2)} = A_{mn}^{(2)} \cdot \cosh(\tau_{mn} \cdot c) \tag{3.11b}$$

$$C_{mn}^{(3)} = A_{mn}^{(3)} \cdot \cosh(\tau_{mn} \cdot c) \tag{3.11c}$$

and

$$C_{mn}^{(4)} = A_{mn}^{(4)} \cdot \cosh(\tau_{mn} \cdot c) \tag{3.11d}$$

Similarly, in view of Equation 3.8

$$D_{mn}^{(1)} = B_{mn}^{(1)} \cdot \frac{\sinh(\tau_{mn} \cdot c)}{\tau_{mn}} \tag{3.12a}$$

$$D_{mn}^{(2)} = B_{mn}^{(2)} \cdot \frac{\sinh(\tau_{mn} \cdot c)}{\tau_{mn}} \tag{3.12b}$$

$$D_{mn}^{(3)} = B_{mn}^{(3)} \cdot \frac{\sinh(\tau_{mn} \cdot c)}{\tau_{mn}} \tag{3.12c}$$

and

$$D_{mn}^{(4)} = B_{mn}^{(4)} \cdot \frac{\sinh(\tau_{mn} \cdot c)}{\tau_{mn}} \tag{3.12d}$$

From Equations 3.10a through 3.12, the coefficients of torch function can be given as

$$T_{mn}^{(1)} = A_{mn}^{(1)} \cdot \cosh(\tau_{mn} \cdot c) + B_{mn}^{(1)} \cdot \frac{\sinh(\tau_{mn} \cdot c)}{\tau_{mn}} \tag{3.13a}$$

$$T_{mn}^{(2)} = A_{mn}^{(2)} \cdot \cosh(\tau_{mn} \cdot c) + B_{mn}^{(2)} \cdot \frac{\sinh(\tau_{mn} \cdot c)}{\tau_{mn}} \tag{3.13b}$$

$$T_{mn}^{(3)} = A_{mn}^{(3)} \cdot \cosh(\tau_{mn} \cdot c) + B_{mn}^{(3)} \cdot \frac{\sinh(\tau_{mn} \cdot c)}{\tau_{mn}} \tag{3.13c}$$

$$T_{mn}^{(4)} = A_{mn}^{(4)} \cdot \cosh(\tau_{mn} \cdot c) + B_{mn}^{(4)} \cdot \frac{\sinh(\tau_{mn} \cdot c)}{\tau_{mn}} \tag{3.13d}$$

In view of Equations 3.10a through 3.12, potential distribution inside the cuboid is found as

$$\begin{aligned} V = & \sum_{m-odd}^{\infty} \sum_{n-odd}^{\infty} \cos\left(\frac{m\pi}{a} \cdot x\right) \cdot \cos\left(\frac{n\pi}{b} \cdot y\right) \\ & \times \left\{ A_{mn}^{(1)} \cdot \cosh\tau_{mn} \cdot (z + c/2) + B_{mn}^{(1)} \cdot \frac{\sinh\tau_{mn} \cdot (z + c/2)}{\tau_{mn}} \right\} \\ & + \sum_{m-odd}^{\infty} \sum_{n-even}^{\infty} \cos\left(\frac{m\pi}{a} \cdot x\right) \cdot \sin\left(\frac{n\pi}{b} \cdot y\right) \\ & \times \left\{ A_{mn}^{(2)} \cdot \cosh\tau_{mn} \cdot (z + c/2) + B_{mn}^{(2)} \cdot \frac{\sinh\tau_{mn} \cdot (z + c/2)}{\tau_{mn}} \right\} \\ & + \sum_{m-even}^{\infty} \sum_{n-odd}^{\infty} \sin\left(\frac{m\pi}{a} \cdot x\right) \cdot \cos\left(\frac{n\pi}{b} \cdot y\right) \\ & \times \left\{ A_{mn}^{(3)} \cdot \cosh\tau_{mn} \cdot (z + c/2) + B_{mn}^{(3)} \cdot \frac{\sinh\tau_{mn} \cdot (z + c/2)}{\tau_{mn}} \right\} \\ & + \sum_{m-even}^{\infty} \sum_{n-even}^{\infty} \sin\left(\frac{m\pi}{a} \cdot x\right) \cdot \sin\left(\frac{n\pi}{b} \cdot y\right) \\ & \times \left\{ A_{mn}^{(4)} \cdot \cosh\tau_{mn} \cdot (z + c/2) + B_{mn}^{(4)} \cdot \frac{\sinh\tau_{mn} \cdot (z + c/2)}{\tau_{mn}} \right\} \end{aligned} \tag{3.14}$$

3.2.1.2 Example of a Rectangular Region

Consider the rectangular region ($-W/2 \le x \le W/2$, $0 \le y \le h$) shown in Figure 3.2. The region is piecewise homogeneous, with permittivity ε_1 for ($0 < y < g$) and ε_2 for ($g < y < h$). The boundary conditions specified are

$$V\big|_{x=\pm W/2} = 0 \tag{3.15a}$$

$$V\big|_{y=0} = 0 \tag{3.15b}$$

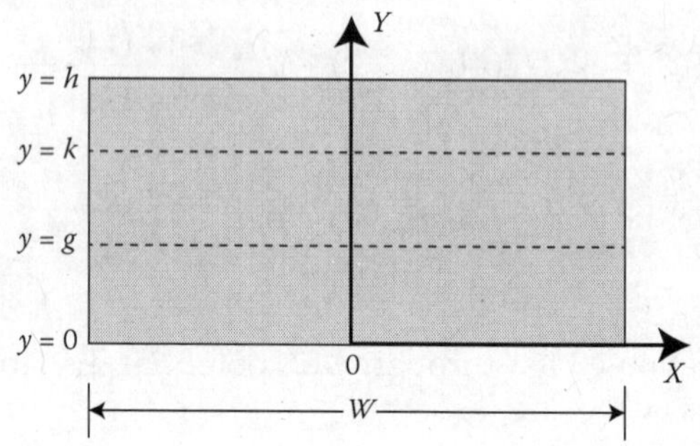

FIGURE 3.2
A rectangular region with Cartesian system of space coordinates.

No boundary condition is specified on the top surface, that is, at $y = h$. However, instead of another boundary condition, it is given that

$$V\Big|_{y=k} = \sum_{m-odd}^{\infty} a_m \cos\left(\frac{m\pi}{W}\cdot x\right) + \sum_{n=1}^{\infty} b_n \sin\left(\frac{n2\pi}{W}\cdot x\right) \tag{3.15c}$$

where $g < k < h$.

Let us divide this rectangular region into two subregions, such that the subregion 1 extends over $(-W/2 \le x \le W/2,\ 0 \le y \le k)$ and the subregion 2 extends over $(-W/2 \le x \le W/2,\ k \le y \le h)$. For the subregion 1, the potential distributions along all the four boundary sides are specified; therefore, it should be possible to obtain a unique solution for the potential distribution had it been a homogeneous region. Since subregion 1 is piecewise homogeneous with $y = g$ serving as a boundary between two homogeneous zones, viz. 1a stretching over $0 < y < g$, and 1b stretching over $g < y < k$ the potential distributions in these zones can be readily obtained if the potential distribution is known along the boundary $y = g$. Let us assume a pseudo-torch function

$$V\Big|_{y=g} = \sum_{m-odd}^{\infty} c_m \cos\left(\frac{m\pi}{W}\cdot x\right) + \sum_{n=1}^{\infty} d_n \sin\left(\frac{n2\pi}{W}\cdot x\right) \tag{3.16a}$$

where c_m and d_n indicate two sets of unknown constants.

The solution of the Laplace equation for the two zones is

$$V_{1a} = \sum_{m-odd}^{\infty} c_m \cos\left(\frac{m\pi}{W}\cdot x\right)\cdot\frac{\sinh((m\pi/W)\cdot y)}{\sinh((m\pi/W)\cdot g)} + \sum_{n=1}^{\infty} d_n \sin\left(\frac{n2\pi}{W}\cdot x\right) \times \frac{\sinh\left((n2\pi/W)\cdot y\right)}{\sinh\left((n2\pi/W)\cdot g\right)} \tag{3.16b}$$

$$V_{1b} = \sum_{m-odd}^{\infty} \cos\left(\frac{m\pi}{W} \cdot x\right) \cdot \left[\frac{\sinh\{(m\pi/W)\cdot(y-g)\}}{\sinh\{(m\pi/W)\cdot(k-g)\}} \cdot a_m - \frac{\sinh\{(m\pi/W)\cdot(y-k)\}}{\sinh\{(m\pi/W)\cdot(k-g)\}} \cdot c_m\right] + \sum_{n=1}^{\infty} \sin\left(\frac{n2\pi}{W} \cdot x\right) \cdot \left[\frac{\sinh\{(n2\pi/W)\cdot(y-g)\}}{\sinh\{(n2\pi/W)\cdot(k-g)\}} \cdot b_m - \frac{\sinh\{(n2\pi/W)\cdot(y-k)\}}{\sinh\{(n2\pi/W)\cdot(k-g)\}} \cdot d_m\right] \quad (3.16c)$$

To synthesise the pseudo-torch function defined at the boundary $y = g$, we need to consider the following boundary condition at the same boundary, that is, at $y = g$. This boundary condition is obtained in view of the requirement that the normal component of the displacement vector $\boldsymbol{D}$ must be continuous at the boundary in the absence of any surface charge distribution there.

$$\varepsilon_1 \frac{V_{1a}}{\partial y}\bigg|_{y=g} = \varepsilon_2 \frac{V_{1b}}{\partial y}\bigg|_{y=g} \quad (3.17a)$$

Therefore, using Equations 3.16a and 3.16b, one obtains

$$c_m = a_m \cdot \left[\frac{\varepsilon_2 \cdot \mathrm{cosech}\{(m\pi/W)\cdot(k-g)\}}{\varepsilon_1 \cdot \coth((m\pi/W)\cdot g) + \varepsilon_2 \cdot \coth\{(m\pi/W)\cdot(k-g)\}}\right] \quad (3.17b)$$

$$d_n = b_n \cdot \left[\frac{\varepsilon_2 \cdot \mathrm{cosech}\{(n2\pi/W)\cdot(k-g)\}}{\varepsilon_1 \cdot \coth((n2\pi/W)\cdot g) + \varepsilon_2 \cdot \coth\{(n2\pi/W)\cdot(k-g)\}}\right] \quad (3.17c)$$

Next, consider the subregion 2 extending over ($-W/2 \le x \le W/2$, $k \le y \le h$). The potential distributions on its three sides, viz. $x = \pm W/2$ and $y = k$, are given. For the fourth side, viz. at $y = h$, let us define torch function as

$$V_2\big|_{y=h} \overset{def}{=} \sum_{m-odd}^{\infty} T_{1m} \cos\left(\frac{m\pi}{W} \cdot x\right) + \sum_{n=1}^{\infty} T_{2n} \sin\left(\frac{n2\pi}{W} \cdot x\right) \quad (3.18)$$

where T_{1m} and T_{2n} indicate two sets of unknown coefficients in the expression of the torch function.

For the determination of these coefficients, we shall make use of a continuity condition at a different place, viz. at $y = k$, instead of at $y = h$.

In view of Equation 3.18, the potential distribution in subregion 2 is given as

$$V_2 = \sum_{m-odd}^{\infty} \cos\left(\frac{m\pi}{W}\cdot x\right)\cdot\left[\frac{\sinh\{(m\pi/W)\cdot(y-h)\}}{\sinh\{(m\pi/W)\cdot(k-h)\}}\right.$$
$$\left.\times a_m - \frac{\sinh\{(m\pi/W)\cdot(y-k)\}}{\sinh\{(m\pi/W)\cdot(k-h)\}}\cdot T_{1m}\right]$$
$$+\sum_{n=1}^{\infty} \sin\left(\frac{n2\pi}{W}\cdot x\right)\cdot\left[\frac{\sinh\{(n2\pi/W)\cdot(y-h)\}}{\sinh\{(n2\pi/W)\cdot(k-h)\}}\right.$$
$$\left.\times b_m - \frac{\sinh\{(n2\pi/W)\cdot(y-k)\}}{\sinh\{(n2\pi/W)\cdot(k-h)\}}\cdot T_{2n}\right] \quad (3.19)$$

$$\left.\frac{V_2}{\partial y}\right|_{y=k} = \left.\frac{V_{1b}}{\partial y}\right|_{y=k} \quad (3.20)$$

Therefore, in view of Equations 3.16c and 3.19, we get

$$T_{1m} = \frac{a_m\cdot\coth\{(m\pi/W)(k-g)\} + c_m\cdot\begin{bmatrix}\coth\{(m\pi/W)(h-k)\}\\ -\operatorname{cosech}\{(m\pi/W)(k-g)\}\end{bmatrix}}{\operatorname{cosech}\{(m\pi/W)(h-k)\}} \quad (3.21a)$$

$$T_{2n} = \frac{b_n\cdot\coth\{(n2\pi/W)(k-g)\} + d_n\cdot\begin{bmatrix}\coth\{(n2\pi/W)(h-k)\}\\ -\operatorname{cosech}\{(n2\pi/W)(k-g)\}\end{bmatrix}}{\operatorname{cosech}\{(n2\pi/W)(h-k)\}} \quad (3.21b)$$

Having obtained all the unknown coefficients, vide Equations 3.17b, 3.17c, 3.21a and 3.21b, the distributions of scalar electric potentials in rectangular region are found from Equations 3.16b, 3.16c and 3.19.

The determination of unique expression for torch function provides potential distribution on that part of the boundary where no boundary condition was originally specified. Now, according to the classical uniqueness theorem, the corresponding field distribution inside the region is unique. It may, therefore, be concluded that the classical uniqueness theorem provides *'sufficient conditions'* and not *'necessary conditions'* for unique solution of the Laplace equation.[2] Uniqueness theorems presented in the two subsequent sections, likewise, provide *'sufficient conditions'* for unique solutions. Applications for the *torch function* synthesis method for the solution of eddy current equation are discussed in Chapter 5.

3.2.2 Uniqueness Theorem for Vector Magnetic Potentials

In general, there are infinite solutions of Laplace and Poisson equations for the vector magnetic potential A. The uniqueness theorem[2] reviewed here describes boundary conditions to be satisfied for a unique solution of these equations.

Let A_1 and A_2 be any two solutions for Equation 2.40, given the distribution of magnetic potential in a volume v bounded by the closed surface s. The difference potential A_o is

$$A_o = A_1 - A_2 \tag{3.22}$$

It may be noted that the difference potential A_o satisfies the Laplace equation

$$\nabla^2 A_o = 0 \tag{3.23}$$

In view of Equation 2.38

$$\nabla \cdot A_1 = \nabla \cdot A_2 = 0 \tag{3.24a}$$

Therefore,

$$\nabla \cdot A_o = 0 \tag{3.24b}$$

Consider the identity[3]

$$\iiint_v [(\nabla \times P) \cdot (\nabla \times Q) - P \cdot (\nabla \times \nabla \times Q)] dv \equiv \oiint_s [P \times (\nabla \times Q)] \cdot ds \tag{3.25a}$$

Let us take

$$P \triangleq Q \triangleq A_o \tag{3.25b}$$

In view of Equations 3.23 and 3.24b, we get

$$\nabla \times \nabla \times A_o = 0 \tag{3.25c}$$

Therefore, Equation 3.25a results

$$\iiint_v |\nabla \times A_o|^2 \, dv \equiv \oiint_s [A_o \times (\nabla \times A_o)] \cdot ds \tag{3.26a}$$

or

$$\iiint_v |B_o|^2 \, dv \equiv \oiint_s [A_o \times B_o] \cdot ds \tag{3.26b}$$

If the RHS of this equation is zero, since the integrand on the LHS is nowhere negative, it follows that

$$\nabla \times A_o \overset{def}{=} B_o = 0 \tag{3.27a}$$

Therefore, the difference of the two solutions, that is, A_o must be a constant at every point in the volume v, including on its surface s. This constant will be zero if at least at one point in the volume v or on the bounding surface s, the value of the vector potential A is defined and the solution satisfies this value.

With reference to Equation 3.26a, it may be noted that on the bounding surface s, if either the vector A_o or its curl is normal to the surface ds, in that case the vector $[A_o \times (\nabla \times A_o)]$ will be entirely in the tangential direction. Let us replace $\nabla \times A_o$ by B_o, that is, the difference of the two flux density vectors, and resolve each of these vectors into two vector components, one in the tangential direction, (A_o^T and B_o^T) and the other in the normal direction, (A_o^N and B_o^N) to the surface. Thus,

$$[A_o \times (\nabla \times A_o)] = [A_o \times B_o] = (A_o^T + A_o^N) \times (B_o^T + B_o^N) \tag{3.27b}$$

or

$$[A_o \times (\nabla \times A_o)] = A_o^T \times B_o^T + (A_o^T \times B_o^N + A_o^N \times B_o^T) \tag{3.27c}$$

The vector quantity within parenthesis on the RHS of this equation is in tangential direction therefore,

$$[A_o \times (\nabla \times A_o)] \cdot ds = (A_o^T \times B_o^T) \cdot ds \tag{3.27d}$$

Now, we arbitrarily chose two distinct tangential directions, say, $T1$ and $T2$. Let each of the two vectors, A_o^T and B_o^T, be expressed as sum of vector components in these directions, that is,

$$A_o^T = A_o^{T1} + A_o^{T2} \tag{3.28a}$$

and

$$B_o^T = B_o^{T1} + B_o^{T2} \tag{3.28b}$$

thus,

$$\left[A_o \times (\nabla \times A_o)\right] \cdot ds = \left[(A_o^{T1} + A_o^{T2}) \times (B_o^{T1} + B_o^{T2})\right] \cdot ds \tag{3.28c}$$

or

$$\left[A_o \times (\nabla \times A_o)\right] \cdot ds = \left[A_o^{T1} \times B_o^{T2} - B_o^{T1} \times A_o^{T2}\right] \cdot ds \tag{3.28d}$$

This shows that the RHS of the above equation will be zero, if in an arbitrary tangential direction, $T1$ or $T2$, the components of both vectors, A_o and B_o, are specified and the two solutions, (A_1, B_1) and (A_2, B_2), satisfy this boundary condition.

We, therefore, conclude that the solution for the magnetic flux density B will be unique provided that the tangential component of vector A is specified on parts of the bounding surface, the tangential component of vector B is specified on some other parts and on the remaining parts of the closed boundary surface, the component of A as well as of B, in an arbitrary tangential direction, are specified. This direction may vary, in any arbitrary manner, from point to point on the boundary surface. For the unique determination of the vector potential A, its value must be specified at least at one point on the bounding surface. This is in addition to the boundary conditions stated above.

3.2.3 Uniqueness Theorem for Maxwell's Equations

Consider an arbitrary volume v, bounded by a closed surface s. Let $(H_1, E_1, B_1, D_1, J_1)$ and $(H_2, E_2, B_2, D_2, J_2)$ indicate two sets of solutions of Maxwell's equations for this region. The difference fields $H = H_1 - H_2$, $E = E_1 - E_2$, $B = B_1 - B_2$, $D = D_1 - D_2$ and $J = J_1 - J_2$ satisfy Maxwell's equations as well as all constitutive equations. This field, however, takes no cognizance of any source that might be present in the volume. Thus,

$$\nabla \times H = J + \frac{\partial D}{\partial t} \tag{3.29}$$

$$\nabla \times E = -\frac{\partial B}{\partial t} \tag{3.30}$$

$$B = \mu H \tag{3.31}$$

$$D = \varepsilon E \tag{3.32}$$

and

$$J = \sigma E \tag{3.33}$$

where μ, ε and σ are scalar functions of position with positive real values, while the difference fields *E*, *H*, *B*, *D* and *J* are vector functions of position as well as time. Now, consider the identity

$$-\nabla \cdot (\boldsymbol{E} \times \boldsymbol{H}) = \boldsymbol{E} \cdot (\nabla \times \boldsymbol{H}) - \boldsymbol{H} \cdot (\nabla \times \boldsymbol{E}) \tag{3.34}$$

Therefore, in view of Equations 3.29 and 3.30, we get

$$-\nabla \cdot (\boldsymbol{E} \times \boldsymbol{H}) = \boldsymbol{E} \cdot \left(\boldsymbol{J} + \frac{\partial \boldsymbol{D}}{\partial t} \right) + \boldsymbol{H} \cdot \frac{\partial \boldsymbol{B}}{\partial t} \tag{3.35}$$

or

$$-\nabla \cdot (\boldsymbol{E} \times \boldsymbol{H}) = \boldsymbol{E} \cdot \boldsymbol{J} + \frac{\partial}{\partial t}\left(\frac{1}{2} \cdot \mu H^2 + \frac{1}{2} \cdot \varepsilon E^2 \right) \tag{3.36}$$

where

$$H^2 = \boldsymbol{H} \cdot \boldsymbol{H} \tag{3.37a}$$

and

$$E^2 = \boldsymbol{E} \cdot \boldsymbol{E} \tag{3.37b}$$

are scalar functions of position and time with zero or positive values.

On integrating both sides of Equation 3.36 over the volume *v*, which is bounded by surface *s*, and then using divergence theorem on the LHS, we get in view of Equation 3.33

$$-\oiint_s (\boldsymbol{E} \times \boldsymbol{H}) \cdot d\boldsymbol{s} \iiint_v \left[\sigma E^2 + \frac{\partial}{\partial t}\left(\frac{1}{2} \cdot \mu H^2 + \frac{1}{2} \cdot \varepsilon E^2 \right) \right] dv \tag{3.38}$$

For certain boundary conditions on *s*, the LHS of this equation reduces to zero. Then, the integral on the RHS must vanish. Let the fields distributed in the volume *v* be specified at the initial time t_o and this initial field is satisfied by the two solutions. For $t > t_o$, if the two solutions deviate from each other, then the integrand on the RHS of Equation 3.38 will be more than zero and the integral will cease to be zero. Therefore, any solution of Maxwell's equations that satisfies initial condition and certain boundary conditions that ensure that the LHS of Equation 3.38 is zero, is the unique solution.

Boundary conditions that result in a zero value for the LHS of Equation 3.38 can be readily found by proceeding in a manner similar to that presented in the preceding section. We may conclude that the solution for Maxwell's equations will be unique provided that on parts of the bounding surface the tangential component of vector **E** is specified, on some other parts the tangential component of vector **H** is specified and on the remaining parts of the closed boundary surface, the component of **E** as well as of **H**, in an arbitrary tangential direction, are specified. This direction may vary, in any arbitrary manner, from point to point on the boundary surface.

In this treatment field distributions, only in bounded regions, are considered. Stratton[4] has discussed the uniqueness theorem for field distributions in externally unbounded regions. The present treatment assumes fields in hysteresis free media. A linearised treatment for hysteretic media is found in literature.[5,6]

3.3 Helmholtz Theorem

If scalar and vector source densities of a differentiable vector field $\mathcal{F}$, that vanishes at infinity faster than $1/r$, are given, the vector field is uniquely specified provided that the source densities are extending over finite distances. Let the source densities be given as

$$\nabla \cdot \mathcal{F} = \mathcal{S} \tag{3.39}$$

and

$$\nabla \times \mathcal{F} = \mathcal{V} \tag{3.40}$$

Now, if the vector field is not uniquely specified, these equations will be satisfied by, say, $\mathcal{F}_1$ and $\mathcal{F}_2$. Let the difference field be

$$\mathcal{F}_0 = \mathcal{F}_1 - \mathcal{F}_2 \tag{3.41}$$

such that

$$\nabla \cdot \mathcal{F}_0 = 0 \tag{3.42}$$

and

$$\nabla \times \mathcal{F}_0 = 0 \tag{3.43}$$

Therefore, we may define

$$\mathcal{F}_0 \overset{def}{=} \nabla\varphi \tag{3.44}$$

Hence, in view of Equation 3.42,

$$\nabla^2\varphi = 0 \tag{3.45}$$

Now

$$\nabla \cdot (\varphi\nabla\varphi) \equiv \varphi\nabla^2\varphi + (\nabla\varphi)^2 = (\nabla\varphi)^2 = (\mathcal{F}_0)^2 \tag{3.46}$$

On integrating over a volume and then applying the divergence theorem, we get

$$\iiint_v \nabla \cdot (\varphi\nabla\varphi)dv = \oiint_s (\varphi\nabla\varphi) \cdot ds = \oiint_s (\varphi\mathcal{F}_0) \cdot ds = \iiint_v (\mathcal{F}_0)^2 dv \tag{3.47}$$

If the normal component of $\mathcal{F}$ on the boundary surface s is specified, the normal component of $\mathcal{F}_0$ will be zero. Thus

$$\iiint_v (\mathcal{F}_0)^2 dv = 0 \tag{3.48}$$

Since $(\mathcal{F}_0)^2$ is nowhere negative, we get $\mathcal{F}_0$ equals to zero. Thus, the vector field $\mathcal{F}$ is uniquely defined. In the case of infinite region, the vector $\mathcal{F}$ is uniquely specified provided that $\mathcal{F}$ vanishes at infinity as $1/r^2$.

An alternative presentation of the Helmholtz theorem[7] states that a vector field $\mathcal{F}$ can be uniquely expressed as a sum of the gradient of a scalar potential and curl of a vector potential, that is,

$$\mathcal{F} = -\nabla\varphi + \nabla \times \Psi \tag{3.49}$$

The scalar and vector potentials at $\boldsymbol{r}$ are related to scalar and vector source densities at $\boldsymbol{r}'$ as given by Equations 3.39 and 3.40

$$d\varphi(\boldsymbol{r}) = \frac{\mathcal{S}(\boldsymbol{r}')\,dv'}{4\pi|\boldsymbol{r} - \boldsymbol{r}'|}$$

Thus,

$$\varphi(\boldsymbol{r}) = \iiint\limits_{v'} \frac{\mathcal{S}(\boldsymbol{r}')\,dv'}{4\pi|\boldsymbol{r}-\boldsymbol{r}'|} \tag{3.50}$$

and

$$d\Psi(\boldsymbol{r}) = \frac{\mathcal{V}(\boldsymbol{r}')\,dv'}{4\pi|\boldsymbol{r}-\boldsymbol{r}'|}$$

giving

$$\Psi(\boldsymbol{r}) = \iiint\limits_{dv'} \frac{\mathcal{V}(\boldsymbol{r}')\,dv'}{4\pi|\boldsymbol{r}-\boldsymbol{r}'|}. \tag{3.51}$$

On taking divergence on both sides of Equation 3.49, in view of Equation 3.50 one gets

$$\nabla\cdot\mathcal{F}(\boldsymbol{r}) = -\nabla^2\varphi(\boldsymbol{r}) = -\iiint\limits_{v'} \frac{\mathcal{S}(\boldsymbol{r}')}{4\pi}\nabla^2\left(\frac{1}{|\boldsymbol{r}-\boldsymbol{r}'|}\right)dv' \tag{3.52a}$$

or

$$\nabla\cdot\mathcal{F}(\boldsymbol{r}) = \iiint\limits_{v'} \mathcal{S}(\boldsymbol{r}')\delta^3(\boldsymbol{r}-\boldsymbol{r}')dv' = \mathcal{S}(\boldsymbol{r}) \tag{3.52b}$$

This is consistent with Equation 3.39. Similarly, on taking curl on both sides of Equation 3.49, we get

$$\nabla\times\mathcal{F}(\boldsymbol{r}) = \nabla\times\{\nabla\times\Psi(\boldsymbol{r})\} \equiv \nabla\{\nabla\cdot\Psi(\boldsymbol{r})\} - \nabla^2\Psi(\boldsymbol{r}) \tag{3.53}$$

Now, using Equation 3.51

$$\nabla\cdot\Psi(\boldsymbol{r}) = \iiint\limits_{dv'} \frac{\mathcal{V}(\boldsymbol{r}')}{4\pi}\cdot\nabla\left(\frac{1}{|\boldsymbol{r}-\boldsymbol{r}'|}\right)dv' \tag{3.54a}$$

or

$$\nabla\cdot\Psi(\boldsymbol{r}) = -\iiint\limits_{dv'} \frac{\mathcal{V}(\boldsymbol{r}')}{4\pi}\cdot\nabla'\left(\frac{1}{|\boldsymbol{r}-\boldsymbol{r}'|}\right)dv' \tag{3.54b}$$

Since,

$$\nabla' \cdot \left[\left(\frac{1}{|\boldsymbol{r}-\boldsymbol{r}'|}\right)\mathcal{V}(\boldsymbol{r}')\right] = \left(\frac{1}{|\boldsymbol{r}-\boldsymbol{r}'|}\right)\nabla' \cdot \mathcal{V}(\boldsymbol{r}') + \mathcal{V}(\boldsymbol{r}') \cdot \nabla'\left(\frac{1}{|\boldsymbol{r}-\boldsymbol{r}'|}\right) \tag{3.55a}$$

or

$$\mathcal{V}(\boldsymbol{r}') \cdot \nabla'\left(\frac{1}{|\boldsymbol{r}-\boldsymbol{r}'|}\right) = \nabla' \cdot \left[\left(\frac{1}{|\boldsymbol{r}-\boldsymbol{r}'|}\right)\mathcal{V}(\boldsymbol{r}')\right] - \left(\frac{1}{|\boldsymbol{r}-\boldsymbol{r}'|}\right)\nabla' \cdot \mathcal{V}(\boldsymbol{r}') \tag{3.55b}$$

In view of Equation 3.40, the second term on the RHS being zero, we have from Equation 3.54b

$$\nabla \cdot \Psi(\boldsymbol{r}) = -\iiint_{dv'} \nabla' \cdot \left[\frac{1}{4\pi} \cdot \left(\frac{1}{|\boldsymbol{r}-\boldsymbol{r}'|}\right)\mathcal{V}(\boldsymbol{r}')\right] dv' \tag{3.56a}$$

or

$$\nabla \cdot \Psi(\boldsymbol{r}) = -\oiint_{s'} \left[\frac{1}{4\pi} \cdot \left(\frac{1}{|\boldsymbol{r}-\boldsymbol{r}'|}\right)\mathcal{V}(\boldsymbol{r}')\right] \cdot ds' \tag{3.56b}$$

Since $\mathcal{V}(\boldsymbol{r}')$ extends over finite distances, the RHS of this equation tends to zero as s' tends to infinity. Therefore, Equation 3.53 reduces to

$$\nabla \times \mathcal{F}(\boldsymbol{r}) = -\nabla^2\Psi(\boldsymbol{r}) \tag{3.57}$$

Now, using Equation 3.41, we get

$$\nabla \times \mathcal{F}(\boldsymbol{r}) = -\iiint_{dv'} \frac{\mathcal{V}(\boldsymbol{r}')}{4\pi} \nabla^2\left(\frac{1}{|\boldsymbol{r}-\boldsymbol{r}'|}\right) dv' \tag{3.58a}$$

or

$$\nabla \times \mathcal{F}(\boldsymbol{r}) = \iiint_{dv'} \mathcal{V}(\boldsymbol{r}')\delta^3(\boldsymbol{r}-\boldsymbol{r}')dv' = \mathcal{V}(\boldsymbol{r}) \tag{3.58b}$$

This is consistent with Equation 3.40.

3.4 Generalised Poynting Theorem

Electrical motors consist of two parts. The part that remains stationary is called the *stator*, and the part that is movable is called *rotor*. Let the electric power be supplied to the stator armature winding. A small part of this input power is dissipated as eddy-current loss and hysteresis loss in the stator iron core. A fraction of the input power is also dissipated as ohmic loss in the armature winding. The remaining power is transferred across the air gap into the rotor. The Poynting theorem and its application to radiation from antennas are discussed in most textbooks on field theory.[7] The treatment is generalised by considering Maxwell's equations for electromagnetic fields in moving media. Generalised Poynting theorem[8] finds its application in electromechanical energy conversion. Electromechanical energy conversion takes place only in moving media. Consider a medium, say, the rotor of an electrical machine, moving with a velocity v, relative to a stationary reference frame, fixed on the stator of this machine. The field equations in the stationary reference frame in terms of fields seen in the moving reference frame are given as

$$\nabla \cdot \boldsymbol{B}' = 0 \tag{3.59}$$

$$\nabla \cdot \boldsymbol{D}' = \rho' \tag{3.60}$$

$$\nabla \times (\boldsymbol{E}' - \boldsymbol{v} \times \boldsymbol{B}') = -\frac{\partial \boldsymbol{B}'}{\partial t} \tag{3.61}$$

$$\nabla \times (\boldsymbol{H}' + \boldsymbol{v} \times \boldsymbol{D}') = (\boldsymbol{J}' + \boldsymbol{v}\rho') + \frac{\partial \boldsymbol{D}'}{\partial t} \tag{3.62}$$

For a linearised treatment of hysteresis effects, let the constitutive equations be given as

$$\boldsymbol{B}' = \mu \boldsymbol{H}' + \mu' \widehat{\boldsymbol{H}'} \tag{3.63}$$

and

$$\boldsymbol{D}' = \varepsilon \boldsymbol{E}' + \varepsilon' \widehat{\boldsymbol{E}'} \tag{3.64}$$

where μ, μ', ε and ε' are positive real scalar functions of space coordinates, while $\widehat{\boldsymbol{H}'}$ and $\widehat{\boldsymbol{E}'}$ indicate *Hilbert transform* (Appendix 1) of $\boldsymbol{H}'$ and $\boldsymbol{E}'$, respectively. For steady-state sinusoidal time-varying electromagnetic fields, these relations result in complex permeability and complex permittivity, if field vectors are given in phasor forms.[9] The resulting hysteresis loop takes elliptic shape.

The Poynting vector at the instant t can be given as

$$\boldsymbol{E} \times \boldsymbol{H} = (\boldsymbol{E}' - \boldsymbol{v} \times \boldsymbol{B}') \times (\boldsymbol{H}' + \boldsymbol{v} \times \boldsymbol{D}') \tag{3.65}$$

Therefore,

$$\nabla \cdot (\boldsymbol{E} \times \boldsymbol{H}) = \nabla \cdot \left[(\boldsymbol{E}' - \boldsymbol{v} \times \boldsymbol{B}') \times (\boldsymbol{H}' + \boldsymbol{v} \times \boldsymbol{D}') \right] \tag{3.66a}$$

or

$$\begin{aligned} \nabla \cdot (\boldsymbol{E} \times \boldsymbol{H}) \equiv\; & (\boldsymbol{H}' + \boldsymbol{v} \times \boldsymbol{D}') \cdot \left[\nabla \times (\boldsymbol{E}' - \boldsymbol{v} \times \boldsymbol{B}') \right] \\ & - (\boldsymbol{E}' - \boldsymbol{v} \times \boldsymbol{B}') \cdot \left[\nabla \times (\boldsymbol{H}' + \boldsymbol{v} \times \boldsymbol{D}') \right] \end{aligned} \tag{3.66b}$$

Using Maxwell's equations for moving media as given by Equations 3.61 and 3.62

$$\nabla \cdot (\boldsymbol{E} \times \boldsymbol{H}) = -(\boldsymbol{H}' + \boldsymbol{v} \times \boldsymbol{D}') \cdot \frac{\partial \boldsymbol{B}'}{\partial t} - (\boldsymbol{E}' - \boldsymbol{v} \times \boldsymbol{B}') \cdot \left[(\boldsymbol{J}' + \boldsymbol{v}\rho') + \frac{\partial \boldsymbol{D}'}{\partial t} \right] \tag{3.66c}$$

On simplification, we get

$$-\nabla \cdot (\boldsymbol{E} \times \boldsymbol{H}) = \left\{ \boldsymbol{H}' \cdot \frac{\partial \boldsymbol{B}'}{\partial t} + \boldsymbol{E}' \cdot \frac{\partial \boldsymbol{D}'}{\partial t} \right\} + \boldsymbol{E}' \cdot \boldsymbol{J}' + \boldsymbol{v} \cdot \left\{ \rho' \boldsymbol{E}' + \boldsymbol{J}' \times \boldsymbol{B}' + \frac{\partial (\boldsymbol{D}' \times \boldsymbol{B}')}{\partial t} \right\} \tag{3.66d}$$

Now, we have

$$\boldsymbol{H}' \cdot \frac{\partial \boldsymbol{B}'}{\partial t} = \frac{\partial}{\partial t} \left(\frac{1}{2} \boldsymbol{H}' \cdot \boldsymbol{B}' \right) + \frac{1}{2} \left(\boldsymbol{H}' \cdot \frac{\partial \boldsymbol{B}'}{\partial t} - \boldsymbol{B}' \cdot \frac{\partial \boldsymbol{H}'}{\partial t} \right) \tag{3.67a}$$

Similarly,

$$\boldsymbol{E}' \cdot \frac{\partial \boldsymbol{D}'}{\partial t} = \frac{\partial}{\partial t} \left(\frac{1}{2} \boldsymbol{E}' \cdot \boldsymbol{D}' \right) + \frac{1}{2} \left(\boldsymbol{E}' \cdot \frac{\partial \boldsymbol{D}'}{\partial t} - \boldsymbol{D}' \cdot \frac{\partial \boldsymbol{E}'}{\partial t} \right) \tag{3.67b}$$

In order to account for the time derivatives of various quantities appearing in Equations 3.67a and 3.67b, consider the inhomogeneous medium moving with a velocity v relative to a point fixed on the stationary reference frame. Medium parameters at this point are seen by an observer in this reference frame to vary with time, that is,

$$\frac{d\mu}{dt} = \frac{\partial \mu}{\partial t} \tag{3.68}$$

However, for an observer moving with the medium, this point is moving relative to him with a velocity $(-v)$, and in doing so it moves across the inhomogeneous medium. Thus,

$$\frac{d\mu}{dt} = \frac{\partial \mu}{\partial x} \cdot \frac{dx}{dt} + \frac{\partial \mu}{\partial y} \cdot \frac{dy}{dt} + \frac{\partial \mu}{\partial z} \cdot \frac{dz}{dt} = -v \cdot \nabla \mu \tag{3.69}$$

Therefore, from Equations 3.68 and 3.69

$$\frac{\partial \mu}{\partial t} = -v \cdot \nabla \mu \tag{3.70}$$

Similarly,

$$\frac{\partial \mu'}{\partial t} = -v \cdot \nabla \mu' \tag{3.71}$$

$$\frac{\partial \varepsilon}{\partial t} = -v \cdot \nabla \varepsilon \tag{3.72}$$

$$\frac{\partial \varepsilon'}{\partial t} = -v \cdot \nabla \varepsilon' \tag{3.73}$$

Therefore, in view of Equations 3.63, 3.70 and 3.71

$$\frac{\partial \boldsymbol{B}'}{\partial t} = \mu \frac{\partial \boldsymbol{H}'}{\partial t} + \mu' \frac{\partial \widehat{\boldsymbol{H}'}}{\partial t} + (-v \cdot \nabla \mu)\boldsymbol{H}' + (-v \cdot \nabla \mu')\widehat{\boldsymbol{H}'} \tag{3.74a}$$

and from Equations 3.64, 3.72 and 3.73, we get

$$\frac{\partial \boldsymbol{D}'}{\partial t} = \varepsilon \frac{\partial \boldsymbol{E}'}{\partial t} + \varepsilon \frac{\partial \widehat{\boldsymbol{E}'}}{\partial t} + (-v \cdot \nabla \varepsilon)\boldsymbol{E}' + (-v \cdot \nabla \varepsilon')\widehat{\boldsymbol{E}'} \tag{3.74b}$$

Further, for the observer moving with the medium, since fields are functions of time as well as the time-varying coordinates of the point moving with a velocity ($-v$),

$$\frac{d\boldsymbol{H}'}{dt} = \frac{\partial \boldsymbol{H}'}{\partial x}\cdot\frac{dx}{dt} + \frac{\partial \boldsymbol{H}'}{\partial y}\cdot\frac{dy}{dt} + \frac{\partial \boldsymbol{H}'}{\partial z}\cdot\frac{dz}{dt} + \frac{\partial \boldsymbol{H}'}{\partial T} \tag{3.75}$$

Therefore,

$$\frac{\partial \boldsymbol{H}'}{\partial t} = \frac{d\boldsymbol{H}'}{dt} - (-\boldsymbol{v}\cdot\nabla)\boldsymbol{H}' \tag{3.76a}$$

Similarly,

$$\frac{\partial \widehat{\boldsymbol{H}'}}{\partial t} = \frac{d\widehat{\boldsymbol{H}'}}{dt} - (-\boldsymbol{v}\cdot\nabla)\widehat{\boldsymbol{H}'} \tag{3.76b}$$

$$\frac{\partial \boldsymbol{E}'}{\partial t} = \frac{d\boldsymbol{E}''}{dt} - (-\boldsymbol{v}\cdot\nabla)\boldsymbol{E}' \tag{3.76c}$$

and

$$\frac{\partial \widehat{\boldsymbol{E}'}}{\partial t} = \frac{d\widehat{\boldsymbol{E}'}}{dt} - (-\boldsymbol{v}\cdot\nabla)\widehat{\boldsymbol{E}'} \tag{3.76d}$$

Thus, in view of Equations 3.63, 3.70, 3.71 and 3.74a, one gets

$$\begin{aligned}\boldsymbol{H}'\cdot\frac{\partial \boldsymbol{B}'}{\partial t} &= \frac{\partial}{\partial t}\left(\frac{1}{2}\boldsymbol{H}'\cdot\boldsymbol{B}'\right) + \frac{1}{2}\mu'\left(\boldsymbol{H}'\cdot\frac{\partial \widehat{\boldsymbol{H}'}}{\partial t} - \widehat{\boldsymbol{H}'}\cdot\frac{\partial \boldsymbol{H}'}{\partial t}\right)\\ &\quad + \frac{1}{2}(-\boldsymbol{v}\cdot\nabla\mu)H'^2 + \frac{1}{2}(-\boldsymbol{v}\cdot\nabla\mu')(\boldsymbol{H}'\cdot\widehat{\boldsymbol{H}'})\end{aligned} \tag{3.77a}$$

Similarly,

$$\begin{aligned}\boldsymbol{E}'\cdot\frac{\partial \boldsymbol{D}'}{\partial t} &= \frac{\partial}{\partial t}\left(\frac{1}{2}\boldsymbol{E}'\cdot\boldsymbol{D}'\right) + \frac{1}{2}\varepsilon'\left(\boldsymbol{E}'\cdot\frac{\partial \widehat{\boldsymbol{E}'}}{\partial t} - \widehat{\boldsymbol{E}'}\cdot\frac{\partial \boldsymbol{E}'}{\partial t}\right)\\ &\quad + \frac{1}{2}(-\boldsymbol{v}\cdot\nabla\varepsilon)E'^2 + \frac{1}{2}(-\boldsymbol{v}\cdot\nabla\varepsilon')(\boldsymbol{E}'\cdot\widehat{\boldsymbol{E}'})\end{aligned} \tag{3.77b}$$

where

$$H'^2 = \boldsymbol{H}'\cdot\boldsymbol{H}' \tag{3.78a}$$

and

$$E'^2 = \boldsymbol{E}' \cdot \boldsymbol{E}' \tag{3.78b}$$

Let the velocity vector $\boldsymbol{v}$ be defined as

$$\boldsymbol{v} \overset{def}{=} v\boldsymbol{a}_v \tag{3.79}$$

where $\boldsymbol{a}_v$ indicates a unit vector in the direction of the velocity vector $\boldsymbol{v}$.

Equations 3.77a and 3.77b can, therefore, be rewritten as

$$\begin{aligned} \boldsymbol{H}' \cdot \frac{\partial \boldsymbol{B}'}{\partial t} &= \frac{\partial}{\partial t}\left(\frac{1}{2}\boldsymbol{H}' \cdot \boldsymbol{B}'\right) + \frac{1}{2}\mu'\left(\boldsymbol{H}' \cdot \frac{\partial \widehat{\boldsymbol{H}'}}{\partial t} - \widehat{\boldsymbol{H}'} \cdot \frac{\partial \boldsymbol{H}'}{\partial t}\right) \\ &\quad + v\left\{\frac{1}{2}(-\boldsymbol{a}_v \cdot \nabla\mu)H'^2 + \frac{1}{2}(-\boldsymbol{a}_v \cdot \nabla\mu')(\boldsymbol{H}' \cdot \widehat{\boldsymbol{H}'})\right\} \end{aligned} \tag{3.80a}$$

and

$$\begin{aligned} \boldsymbol{E}' \cdot \frac{\partial \boldsymbol{D}'}{\partial t} &= \frac{\partial}{\partial t}\left(\frac{1}{2}\boldsymbol{E}' \cdot \boldsymbol{D}'\right) + \frac{1}{2}\varepsilon'\left(\boldsymbol{E}' \cdot \frac{\partial \widehat{\boldsymbol{E}'}}{\partial t} - \widehat{\boldsymbol{E}'} \cdot \frac{\partial \boldsymbol{E}'}{\partial t}\right) \\ &\quad + v\left\{\frac{1}{2}(-\boldsymbol{a}_v \cdot \nabla\varepsilon)E'^2 + \frac{1}{2}(-\boldsymbol{a}_v \cdot \nabla\varepsilon')(\boldsymbol{E}' \cdot \widehat{\boldsymbol{E}'})\right\} \end{aligned} \tag{3.80b}$$

Now, in view of Equations 3.64a, 3.64b, 3.80a and 3.80b, we get

$$\begin{aligned} \boldsymbol{H}' \cdot \frac{\partial \boldsymbol{B}'}{\partial t} &= \frac{\partial}{\partial t}\left(\frac{1}{2}\boldsymbol{H}' \cdot \boldsymbol{B}'\right) + \frac{1}{2}\mu'\left(\boldsymbol{H}' \cdot \frac{d\widehat{\boldsymbol{H}'}}{dt} - \widehat{\boldsymbol{H}'} \cdot \frac{d\boldsymbol{H}'}{dt}\right) \\ &\quad + v\left[\frac{1}{2}\mu'\left\{\widehat{\boldsymbol{H}'} \cdot (-\boldsymbol{a}_v \cdot \nabla)\boldsymbol{H}' - \boldsymbol{H}' \cdot (-\boldsymbol{a}_v \cdot \nabla)\widehat{\boldsymbol{H}'}\right\}\right. \\ &\quad \left. + \left\{\frac{1}{2}(-\boldsymbol{a}_v \cdot \nabla\mu)H'^2 + \frac{1}{2}(-\boldsymbol{a}_v \cdot \nabla\mu')(\boldsymbol{H}' \cdot \widehat{\boldsymbol{H}'})\right\}\right] \end{aligned} \tag{3.81a}$$

and

$$\begin{aligned} \boldsymbol{E}' \cdot \frac{\partial \boldsymbol{D}'}{\partial t} &= \frac{\partial}{\partial t}\left(\frac{1}{2}\boldsymbol{E}' \cdot \boldsymbol{D}'\right) + \frac{1}{2}\varepsilon'\left(\boldsymbol{E}' \cdot \frac{d\widehat{\boldsymbol{E}'}}{dt} - \widehat{\boldsymbol{E}'} \cdot \frac{d\boldsymbol{E}'}{dt}\right) + v\left[\frac{1}{2}\varepsilon'\left\{\widehat{\boldsymbol{E}'} \cdot (-\boldsymbol{a}_v \cdot \nabla)\boldsymbol{E}'\right.\right. \\ &\quad \left.\left. - \boldsymbol{E}' \cdot (-\boldsymbol{a}_v \cdot \nabla)\widehat{\boldsymbol{E}'}\right\} + \left\{\frac{1}{2}(-\boldsymbol{a}_v \cdot \nabla\varepsilon)E'^2 + \frac{1}{2}(-\boldsymbol{a}_v \cdot \nabla\varepsilon')(\boldsymbol{E}' \cdot \widehat{\boldsymbol{E}'})\right\}\right] \end{aligned} \tag{3.81b}$$

3.4.1 Components of Power Flow

Let the rotor of a common induction or synchronous machine be located inside the stator bore. Therefore, on integrating both sides of Equation 3.66d over the rotor volume V bounded by its surface s, one gets using divergence theorem on its LHS

$$
\begin{aligned}
-\oint_{S}(\boldsymbol{E}\times\boldsymbol{H})\cdot d\boldsymbol{S} = {} & \overset{(1)}{\iiint_{V}}\left[\frac{\partial}{\partial t}\left(\frac{1}{2}\boldsymbol{H}'\cdot\boldsymbol{B}'\right)+\frac{\partial}{\partial t}\left(\frac{1}{2}\boldsymbol{E}'\cdot\boldsymbol{D}'\right)\right]dv+\overset{(1)}{\iiint_{V}}(\sigma E'^{2})dv \\
& +\overset{(3)}{\iiint_{V}}\left[+\frac{1}{2}\mu'\left(\boldsymbol{H}'\cdot\frac{d\widehat{\boldsymbol{H}'}}{dt}-\widehat{\boldsymbol{H}'}\cdot\frac{d\boldsymbol{H}'}{dt}\right)+\frac{1}{2}\varepsilon'\left(\boldsymbol{E}'\cdot\frac{d\widehat{\boldsymbol{E}'}}{dt}-\widehat{\boldsymbol{E}'}\cdot\frac{d\boldsymbol{E}'}{dt}\right)\right]dv \\
& +\overset{(4)}{\iiint_{V}}\left[v\cdot\left\{\rho\boldsymbol{E}'+\boldsymbol{J}'\times\boldsymbol{B}'+\frac{\partial(\boldsymbol{D}'\times\boldsymbol{B}')}{\partial t}\right\}\right]dv \\
& +\overset{(5)}{\iiint_{V}}\left[v\frac{1}{2}\mu'\left\{\widehat{\boldsymbol{H}'}\cdot(-\boldsymbol{a}_{v}\cdot\nabla)\boldsymbol{H}'-\boldsymbol{H}'\cdot(-\boldsymbol{a}_{v}\cdot\nabla)\widehat{\boldsymbol{H}'}\right\}\right]dv \\
& +\overset{(6)}{\iiint_{V}}\left[v\frac{1}{2}\varepsilon'\left\{\widehat{\boldsymbol{E}'}\cdot(-\boldsymbol{a}_{v}\cdot\nabla)\boldsymbol{E}'-\boldsymbol{E}'\cdot(-\boldsymbol{a}_{v}\cdot\nabla)\widehat{\boldsymbol{E}'}\right\}\right]dv \\
& +\overset{(7)}{\iiint_{V}}\left[v\frac{1}{2}\left\{(-\boldsymbol{a}_{v}\cdot\nabla\mu)H'^{2}+(-\boldsymbol{a}_{v}\cdot\nabla\mu')(\boldsymbol{H}'\cdot\widehat{\boldsymbol{H}'})\right\}\right]dv \\
& +\overset{(8)}{\iiint_{V}}\left[v\frac{1}{2}\left\{(-\boldsymbol{a}_{v}\cdot\nabla\varepsilon)E'^{2}+(-\boldsymbol{a}_{v}\cdot\nabla\varepsilon')(\boldsymbol{E}'\cdot\widehat{\boldsymbol{E}'})\right\}\right]dv
\end{aligned}
\tag{3.82}
$$

The generalised Poynting theorem as given above states that the total power flowing into a closed surface is equal to the sum of the following components:

1. The rate of increase in the energy stored in the electromagnetic field in the region bounded by the closed surface
2. Eddy current loss in conducting regions
3. Hysteresis loss due to magnetic and electric hysteresis
4. A combination of mechanical power developed containing mechanical power (i) for moving electric charge in electric field, (ii) for moving current carrying conductors in magnetic field, (iii) for transfer of electromagnetic momentum
5. Mechanical power developed due to magnetic hysteresis

6. Mechanical power developed due to dielectric hysteresis
7. Mechanical power developed due to space variations of magnetic properties of media
8. Mechanical power developed due to space variations of dielectric properties of media in the volume

3.4.2 Components of Force

In Equation 3.82, the coefficients of velocity v can be considered as force densities. Thus,

$$\mathcal{F} = \sum_{n=1}^{7} \mathcal{F}_n \tag{3.83a}$$

where

$$\mathcal{F}_1 = \rho E' \tag{3.83b}$$

$$\mathcal{F}_2 = J' \times B' \tag{3.83c}$$

$$\mathcal{F}_3 = \frac{\partial(D' \times B')}{\partial t} \tag{3.83d}$$

$$\mathcal{F}_4 = \frac{1}{2}\mu'\left\{\widehat{H'}\cdot(-a_v\cdot\nabla)H' - H'\cdot(-a_v\cdot\nabla)\widehat{H'}\right\} \tag{3.83e}$$

$$\mathcal{F}_5 = \frac{1}{2}\varepsilon'\left\{\widehat{E'}\cdot(-a_v\cdot\nabla)E' - E'\cdot(-a_v\cdot\nabla)\widehat{E'}\right\} \tag{3.83f}$$

$$\mathcal{F}_6 = \frac{1}{2}\left\{(-a_v\cdot\nabla\mu)H'^2 + (-a_v\cdot\nabla\mu')(H'\cdot\widehat{H'})\right\} \tag{3.83g}$$

$$\mathcal{F}_7 = \frac{1}{2}\left\{(-a_v\cdot\nabla\varepsilon)E'^2 + (-a_v\cdot\nabla\varepsilon')(E'\cdot\widehat{E'})\right\} \tag{3.83h}$$

The RHS of Equation 3.83a or of Equations 3.83b to h indicates the following forces:

1. First term gives Lorentz's force acting on electric charge due to electric field.
2. Second term gives Lorentz's force on current carrying conductor placed in magnetic field. It governs the principle of operation of most

general-purpose linear as well as rotating electrical machines, such as induction machines or synchronous machines.

3. Third term indicates the rate of increase of electromagnetic momentum per unit volume.
4. Fourth term gives the forces developed due to magnetic hysteresis.
5. Fifth term gives the forces developed due to dielectric hysteresis.
6. Sixth term gives force densities due to the space variation of magnetic properties of the medium.
7. Seventh term gives force densities due to the space variation of dielectric properties of the medium.

It may be mentioned that terms 4 and 5 explain the operation of magnetic and dielectric hysteresis machines, whereas terms 6 and 7 govern the principle of operation of magnetic and dielectric reluctance motors.

In many electrical machines, for instance dc machines, the electrical power is given to the rotating part, that is, rotor of the machine. The rotor is placed inside the stator bore. Now, if both sides of Equation 3.82 are multiplied by minus one, and noting that the stator is moving with a velocity '$-v$', relative to the rotor, then the resulting equation becomes

$$\begin{aligned}
\oint_S (\boldsymbol{E}\times\boldsymbol{H})\cdot d\boldsymbol{S} = &-\overset{(1)}{\iiint_V}\left[\frac{\partial}{\partial t}\left(\frac{1}{2}\boldsymbol{H}'\cdot\boldsymbol{B}'\right)+\frac{\partial}{\partial t}\left(\frac{1}{2}\boldsymbol{E}'\cdot\boldsymbol{D}'\right)\right]dV-\overset{(2)}{\iiint_V}(\sigma E'^2)dv\\
&-\overset{(3)}{\iiint_V}\left[\frac{1}{2}\mu'\left(\boldsymbol{H}'\cdot\frac{d\widehat{\boldsymbol{H}'}}{dt}-\widehat{\boldsymbol{H}'}\cdot\frac{d\boldsymbol{H}'}{dt}\right)+\frac{1}{2}\varepsilon'\left(\boldsymbol{E}'\cdot\frac{d\widehat{\boldsymbol{E}'}}{dt}-\widehat{\boldsymbol{E}'}\cdot\frac{d\boldsymbol{E}'}{dt}\right)\right]dv\\
&+\overset{(4)}{\iiint_V}\left[v\cdot\left\{\rho\boldsymbol{E}'+\boldsymbol{J}'\times\boldsymbol{B}'+\frac{\partial(\boldsymbol{D}'\times\boldsymbol{B}')}{\partial t}\right\}\right]dv\\
&+\overset{(5)}{\iiint_V}\left[v\frac{1}{2}\mu'\left\{\widehat{\boldsymbol{H}'}\cdot(-\boldsymbol{a}_v\cdot\nabla)\boldsymbol{H}'-\boldsymbol{H}'\cdot(-\boldsymbol{a}_v\cdot\nabla)\widehat{\boldsymbol{H}'}\right\}\right]dv\\
&+\overset{(6)}{\iiint_V}\left[v\frac{1}{2}\varepsilon'\left\{\widehat{\boldsymbol{E}'}\cdot(-\boldsymbol{a}_v\cdot\nabla)\boldsymbol{E}'-\boldsymbol{E}'\cdot(-\boldsymbol{a}_v\cdot\nabla)\widehat{\boldsymbol{E}'}\right\}\right]dv\\
&+\overset{(7)}{\iiint_V}\left[v\frac{1}{2}\left\{(-\boldsymbol{a}_v\cdot\nabla\mu)H'^2+(-\boldsymbol{a}_v\cdot\nabla\mu')(\boldsymbol{H}'\cdot\widehat{\boldsymbol{H}'})\right\}\right]dv\\
&+\overset{(8)}{\iiint_V}\left[v\frac{1}{2}\left\{(-\boldsymbol{a}_v\cdot\nabla\varepsilon)E'^2+(-\boldsymbol{a}_v\cdot\nabla\varepsilon')(\boldsymbol{E}'\cdot\widehat{\boldsymbol{E}'})\right\}\right]dv
\end{aligned}\tag{3.84}$$

The LHS of Equation 3.84 shows the power flowing out of the closed rotor surface and its RHS gives, in the rotor region, the rate of decrease in the energy stored in the magnetic and electric field; minus ohmic and hysteresis losses, plus mechanical power developed.

In certain machines, such as the common air conditioner, ceiling fans and so on the electric power is supplied to the stator, which is housed inside the rotor bore. For such cases the LHS of Equation 3.84 gives the power flowing out from the closed stator surface into the rotor. Consequently, the RHS of this equation gives, in the stator region, the rate of decrease in the energy stored in the magnetic and electric field; minus ohmic and hysteresis losses, plus the mechanical power developed.

3.5 Approximation Theorems

For the unique solution of boundary value problems, it is necessary that the solution exactly satisfies certain boundary conditions. If the boundary conditions are satisfied only approximately, the resulting field distribution in the region inside the boundary will be erroneous. The error may vary from point to point in the bounded region. Therefore, for an acceptable solution of a boundary value problem, a tolerance limit must be prescribed. This can be done variously. For instance, it could be (1) the maximum acceptable error(s) at one or more points in the region, (2) the maximum acceptable error anywhere in the region, (3) the mean square error over the entire region and so on. The theorems presented below consider mean square error over the bounded region. It is assumed that though the boundary conditions are only approximately satisfied, the solution of the differential equation giving the field distribution is otherwise exact. In the treatment given below, the term *'approximate solution'* implies a field distribution that approximately satisfies the boundary conditions.

3.5.1 Approximation Theorem for Laplacian Field

To obtain a unique solution for the Laplace equation giving the distribution of potential field in a region, certain boundary conditions are required to be satisfied. Often, it may not be possible to exactly satisfy boundary conditions. An exact solution of the Laplace equation that approximately satisfies boundary conditions results in erroneous distribution of potential field in the region.

Consider a solution V, for a region of volume v, that satisfies boundary conditions approximately. We define absolute error α as

$$\alpha \overset{def}{=} V - V_o \tag{3.85}$$

where V_o is the solution of the Laplace equation for this region that accurately satisfies the boundary conditions on its bounding surface *s*. The error α is a function of space coordinates of any point in *v*. This error for a point on the bounding surface, *s*, is defined as α^s,

$$\alpha^s \overset{def}{=} (V - V_o)\big|_s \tag{3.86}$$

Another error function β^s is defined as the normal derivative of α, at the bounding surface *s*

$$\beta^s \overset{def}{=} \frac{\partial \alpha}{\partial n} \boldsymbol{a}_n \bigg|_s \tag{3.87}$$

Both α^s and β^s are functions of coordinates of points on *s*, and $\boldsymbol{a}_n$ is a unit vector normal to the bounding surface *s*.

Let the cumulative error ϵ (the symbol ϵ is different from ε which stands for permittivity) be defined as

$$\epsilon \overset{def}{=} \oiint_s (\alpha^s \cdot \beta^s) \cdot ds \tag{3.88}$$

After defining various error functions, the theorem can be point wise stated as

1. The cumulative error ϵ cannot be negative
2. The cumulative error ϵ may not be zero, unless the absolute error α is a constant
3. The mean square value of the error in the volume is the cumulative error per unit volume

In order to prove the theorem, consider the following identity:

$$\nabla \cdot (\alpha \nabla \alpha) \equiv \alpha \nabla^2 \alpha + |\nabla \alpha|^2 \tag{3.89a}$$

In view of Equation 3.85, the Laplacian of absolute error α is zero. Therefore, the first term on the RHS of this equation is zero, resulting in

$$\nabla \cdot (\alpha \nabla \alpha) = |\nabla \alpha|^2 \tag{3.89b}$$

Now, on integrating both the sides of this equation over the volume of the region *v*, and on applying divergence theorem we get

$$\oiint_s (\alpha \nabla \alpha) \cdot ds = \oiint_s (\alpha^s \cdot \beta^s) \cdot ds \stackrel{def}{=} \epsilon$$
$$= \iiint_v |\nabla \alpha|^2 \, dv \stackrel{def}{=} v \cdot |\nabla \alpha|^2 \Big|_{\text{average}} \quad (3.90)$$

From this equation we note that its RHS cannot be negative, thus the cumulative error ϵ cannot be negative. Further, ϵ will be zero provided that $\nabla\alpha$ is zero at every point in v, that is, if α is a constant. The mean square error being the average value of $|\nabla\alpha|^2$ in v, is the cumulative error ϵ divided by the volume v. Thus, the root mean square (rms) value of the error in the volume is

$$|\nabla \alpha|_{RMS} = \sqrt{\epsilon / v} \quad (3.91)$$

Lastly, it may be seen that at every point on the boundary surface, the absolute error, α^s and its normal derivative, $(\partial\alpha/\partial n)|_s$, must be of the same sign. For if it is not so, Equation 3.90 could be violated as we cannot exclude the possibility of zero or even negative values for the cumulative error ϵ.

As a corollary to this theorem, we define relative error α_r between two solutions of the Laplace equation, both approximately satisfying boundary conditions on the bounding surface s

$$\alpha_r \stackrel{def}{=} V_1 - V_2 \quad (3.92)$$

Similarly,

$$\alpha_r^s \stackrel{def}{=} (V_1 - V_2)\big|_s \quad (3.93)$$

$$\beta_r^s \stackrel{def}{=} \frac{\partial \alpha_r}{\partial n} a_n \Big|_s \quad (3.94)$$

and

$$\epsilon_r \stackrel{def}{=} \oiint_s (\alpha_r^s \cdot \beta_r^s) \cdot ds = \iiint_v |\nabla \alpha|^2 \, dv \stackrel{def}{=} v \cdot |\nabla \alpha|^2 \Big|_{\text{average}} \quad (3.95)$$

From Equation 3.90, we get

$$\epsilon_r = \iiint_v |\nabla \alpha_r|^2 \, dv \quad (3.96)$$

The rms value of the relative error in v is given as

$$|\nabla \alpha_r|_{\text{RMS}} = \sqrt{\epsilon_r / v} \tag{3.97}$$

Having defined various relative error functions, the statement of the theorem follows:

1. The relative cumulative error ϵ_r cannot be negative
2. The relative cumulative error ϵ_r may not be zero, unless the relative error α_r is a constant
3. In volume v, the rms value for the modulus of gradient of relative error α_r is equal to the square root of the relative cumulative error per unit volume

It may be noted that for an *approximate solution* of the Laplace equation, together with the boundary conditions (the potential and its normal derivative, both on the entire bounding surface), the tolerance limit for the mean square (or the rms) error in the volume must also be prescribed. It is to be ensured that the cumulative error per unit volume must be less than or equal to the prescribed tolerance limit for the *mean square* error in the volume.

It may also be noted that from the point of view of the uniqueness theorem, on any part of the bounding surface only one type of boundary condition, either the potential or its normal derivative, may be specified. The other boundary condition follows from the solution, as these two boundary conditions are related to each other. In the approximate solution, this relationship between the two types of boundary conditions becomes rather hazy. Therefore, let us make a hypothesis. Noting that as the approximate solution approaches the exact one, the ratio $(1/V) \cdot (\partial V/\partial n)\boldsymbol{a}_n|_s$ approaches $(1/V_o) \cdot (\partial V_o/\partial n)\boldsymbol{a}_n|_s$, thus as a first approximation one may take

$$\frac{1}{V_o} \cdot \frac{\partial V_o}{\partial n}\boldsymbol{a}_n\bigg|_s \cong \frac{1}{V} \cdot \frac{\partial V}{\partial n}\boldsymbol{a}_n\bigg|_s \tag{3.98}$$

or

$$\frac{\partial V_o}{\partial n}\boldsymbol{a}_n\bigg|_s \cong \frac{V_o}{V} \cdot \frac{\partial V}{\partial n}\boldsymbol{a}_n\bigg|_s \tag{3.99}$$

Therefore,

$$\beta^s \stackrel{def}{=} \frac{\partial(V - V_o)}{\partial n}\boldsymbol{a}_n\bigg|_s = \frac{\partial V}{\partial n}\boldsymbol{a}_n\bigg|_s - \frac{\partial V_o}{\partial n}\boldsymbol{a}_n\bigg|_s = \left[\frac{\alpha}{V} \cdot \frac{\partial V}{\partial n}\boldsymbol{a}_n\right]_s = \frac{\alpha^s}{V} \cdot \left[\frac{\partial V}{\partial n}\boldsymbol{a}_n\bigg|_s\right] \tag{3.100}$$

If the closed bounding surface is divided into a number of open surfaces, one or the other (but not both) type of boundary conditions can be specified to each of these open surfaces. In a problem where only the distribution of potential $V_o|_{s_i}$ on the open boundary surface, s_i, is given,

$$\alpha^{s_i} \overset{def}{=} (V - V_o)\Big|_{s_i} \tag{3.101a}$$

where V spells the approximate solution.

Thus,

$$\beta^{s_i} = \left[\frac{\alpha}{V} \cdot \frac{\partial V}{\partial n} \boldsymbol{a}_n\right]\Bigg|_{s_i} \tag{3.101b}$$

where s_i stands for all such surfaces with specified distributions of potential $V_o|_{s_i}$.

Further, on the open boundary surface, s_j, where only the distributions of normal derivative of potentials $(\partial V_o/\partial n)\boldsymbol{a}_n|_{s_j}$ is given, and $(\partial V/\partial n)\boldsymbol{a}_n|_{s_j}$ is found from the approximate solution, we have, in view of Equation 3.100

$$\beta^{s_j} \overset{def}{=} \frac{\partial (V - V_o)}{\partial n} \boldsymbol{a}_n\Bigg|_{s_j} = \frac{\alpha^{s_j}}{V} \cdot \left[\frac{\partial V}{\partial n} \boldsymbol{a}_n\Bigg|_{s_j}\right] \tag{3.102a}$$

It gives

$$\alpha^{s_j} = \left[\frac{\beta^{s_j}}{(\partial V/\partial n)}\right] \cdot V\Big|_{s_j} \tag{3.102b}$$

The cumulative error ϵ can now be given as

$$\epsilon \overset{def}{=} \oiint_s (\alpha^s \cdot \beta^s) \cdot ds = \iint_{s_i} (\alpha^{s_i} \cdot \beta^{s_i}) \cdot ds + \iint_{s_j} (\alpha^{s_j} \cdot \beta^{s_j}) \cdot ds \tag{3.103}$$

The approximate solution needs to be improved if the cumulative error per unit volume fails to satisfy the following relation:

$$\sqrt{\epsilon/v} \le \left[\nabla \alpha_r|_{\text{RMS}}\right]_{\text{prescribed}} \tag{3.104a}$$

or

$$\epsilon/v \le |\nabla \alpha|^2\Big|_{\text{prescribed mean square value}} \tag{3.104b}$$

It is often desired that error may be expressed as a fraction or percentage of some base value. We may choose, for instance, a base value ϵ_b as defined below

$$\epsilon_b \stackrel{def}{=} \oiint_s (V_o \nabla V_o) \cdot ds = \oiint_s \left(V_o \frac{\partial V_o}{\partial n} \right)\Bigg|_s ds \tag{3.105}$$

where V_o and $(\partial V_o/\partial n)$ are, respectively, the given potential and its normal derivative on the bounding surface s.

Let us split the closed surface s into two sets of open surfaces, s_i and s_j, such that the potential distributions are specified on s_i, and normal derivative of potentials on s_j. We assume

$$\frac{1}{V_o} \cdot \frac{\partial V_o}{\partial n} \cong \frac{1}{V} \cdot \frac{\partial V}{\partial n} \tag{3.106a}$$

where V is a solution that approximately satisfies boundary conditions. Therefore,

$$\frac{\partial V_o}{\partial n}\Bigg|_{s_i} \cong \frac{V_o}{V} \cdot \frac{\partial V}{\partial n}\Bigg|_{s_i} \tag{3.106b}$$

and

$$V_o\Big|_{s_j} \cong V \cdot \frac{\partial V_o/\partial n}{\partial V/\partial n}\Bigg|_{s_j} \tag{3.106c}$$

Thus, from Equation 3.105, we get

$$\epsilon_b \stackrel{def}{=} \iint_{s_i} \left[V_o \cdot \frac{V_o}{V} \cdot \frac{\partial V}{\partial n} \right]_{s_i} ds + \iint_{s_j} \left[V \cdot \frac{\partial V_o/\partial n}{\partial V/\partial n} \cdot \frac{\partial V_o}{\partial n} \right]_{s_j} ds \tag{3.107}$$

Since V and V_o are solutions of the Laplace equation in volume v, bounded by the surface s, in view of Equations 3.89b and 3.105 it may be seen that

$$\epsilon_b = \iiint_v |\nabla V_o|^2 \, dv \tag{3.108}$$

The definition of the base value ϵ_b is rather arbitrary. A more convenient definition could be as follows:

$$\epsilon_b \stackrel{def}{=} \oiint_s (V\nabla V) \cdot ds = \oiint_s \left(V \frac{\partial V}{\partial n} \right)\Big|_s ds \tag{3.109}$$

The advantage of using this definition is that the bounding surface is not required to be divided into open surfaces. If only one type of boundary condition (V_o or ($\partial V_o/\partial n$)) is specified over the entire boundary, the other type follows from the approximate solution V, directly.

Now, the error as percentage of base value is defined, in view of Equation 3.104b, as

$$\epsilon_{p.c.} = \frac{\epsilon/v}{\epsilon_b/v} \times 100 = \frac{\epsilon}{\epsilon_b} \times 100 \tag{3.110}$$

3.5.2 Approximation Theorem for Vector Magnetic Potential

Consider Poisson's equation for the vector magnetic potential. Let a solution of this equation, for a region of volume v, that satisfies boundary conditions approximately be $\boldsymbol{A}$. We define absolute error α as

$$\alpha \stackrel{def}{=} \boldsymbol{A} - \boldsymbol{A}_o \tag{3.111}$$

where $\boldsymbol{A}_o$ is the solution of Poisson's equation for this region that accurately satisfies the boundary conditions on its bounding surface s. The vector error α is a function of space coordinates of any point in v. This error for a point on the bounding surface is defined as α^s,

$$\alpha^s \stackrel{def}{=} (\boldsymbol{A} - \boldsymbol{A}_o)\big|_s \tag{3.112}$$

Since

$$\nabla^2 \boldsymbol{A} = -\mu \boldsymbol{J} \tag{3.113a}$$

and

$$\nabla^2 \boldsymbol{A}_o = -\mu \boldsymbol{J} \tag{3.113b}$$

Thus,

$$\nabla^2 \alpha = 0 \tag{3.114}$$

Further, we have

$$\nabla \cdot \alpha = \nabla \cdot \boldsymbol{A} - \nabla \cdot \boldsymbol{A}_o = 0 - 0 = 0 \tag{3.115}$$

Setting

$$\beta \overset{def}{=} \nabla \times \alpha \tag{3.116}$$

Let us consider the identity

$$\begin{aligned} \nabla \cdot (\alpha \times \beta) &\equiv \beta \cdot (\nabla \times \alpha) - \alpha \cdot (\nabla \times \beta) \\ &= |\beta|^2 - \alpha \cdot [\nabla \times \nabla \times \alpha] \\ &\equiv |\beta|^2 - \alpha \cdot \left[\nabla(\nabla \cdot \alpha) - \nabla^2 \alpha\right] \end{aligned}$$

Since the divergence as well as the Laplacian of the vector error α is zero,

$$\nabla \cdot (\alpha \times \beta) = |\beta|^2 \tag{3.117}$$

Now, on integrating both sides of this equation over the volume of the region v, and then using the divergence theorem, we get

$$\oiint_s (\alpha \times \beta) \cdot ds \equiv \iiint_v |\beta|^2 \, dv \tag{3.118}$$

The cumulative error ϵ can be defined as

$$\epsilon \overset{def}{=} \oiint_s (\alpha \times \beta) \cdot ds \tag{3.119}$$

Let us define the average value for the modulus squared curl of absolute error α as the mean square error. Thus, the approximation theorem for vector magnetic potential can be stated as

1. The cumulative error ϵ cannot be negative
2. The cumulative error ϵ may not be zero, unless the absolute error α is a constant
3. In the volume v, the mean square error is equal to the cumulative error per unit volume

Equation 3.119 can be simplified as

$$\epsilon \overset{def}{=} \oiint_s (\alpha_T^s \times \beta_T^s) \cdot ds \tag{3.120}$$

The suffix *T* indicates vector component tangential to the bounding surface *s*.

Let these error functions be expressed as sum of two tangential vectors on the bounding surface such that

$$\alpha_T^s = \alpha_{T1}^s + \alpha_{T2}^s \tag{3.121a}$$

and

$$\beta_T^s = \beta_{T1}^s + \beta_{T2}^s \tag{3.121b}$$

where *T*1 and *T*2 are two orthogonal tangential directions on *s*, then

$$\alpha_T^s \times \beta_T^s = \alpha_{T1}^s \times \beta_{T2}^s + \alpha_{T2}^s \times \beta_{T1}^s \tag{3.122}$$

Thus,

$$\epsilon \overset{def}{=} \oiint_s (\alpha_{T1}^s \cdot \beta_{T2}^s + \alpha_{T2}^s \cdot \beta_{T1}^s) \cdot ds \tag{3.123}$$

where

$$\alpha_{T1}^s = (A_{T1} - A_{oT1})\big|_s \tag{3.124a}$$

$$\alpha_{T2}^s = (A_{T2} - A_{oT2})\big|_s \tag{3.124b}$$

$$\beta_{T1}^s = \nabla \times (A_{T1} - A_{oT1})\big|_s \tag{3.124c}$$

and

$$\beta_{T2}^s = \nabla \times (A_{T2} - A_{oT2})\big|_s \tag{3.124d}$$

Let the closed surface *s* be divided into three sets of open surfaces s_i, s_j, and s_k, such that A_{oT} is defined on s_i, $\nabla \times A_{oT}$ on s_j and (A_{oT1} and $\nabla \times A_{oT1}$) on s_k. Let the solution *A* that approximately satisfies boundary condition on the bounding surfaces s be known and so also $\nabla \times \boldsymbol{A}$ be approximately known on s (i.e. on s_i, s_j and s_k). This implies that $\boldsymbol{A}_T$, $\boldsymbol{A}_{T1}$, $\boldsymbol{A}_{T2}$, $\nabla \times \boldsymbol{A}_T$, $\nabla \times \boldsymbol{A}_{T1}$ and $\nabla \times \boldsymbol{A}_{T2}$ are approximately known on each open surface s_i, s_j and s_k.

For the surface s_i, we have

$$\alpha^{s_i} \stackrel{def}{=} (\boldsymbol{A}_T - \boldsymbol{A}_{oT})\Big|_{s_i} \tag{3.125}$$

We assume that

$$\frac{\nabla \times \boldsymbol{A}_{oT}}{|\boldsymbol{A}_{oT}|} \cong \frac{\nabla \times \boldsymbol{A}_T}{|\boldsymbol{A}_T|} \tag{3.126a}$$

or

$$\nabla \times \boldsymbol{A}_{oT} = \left\{ \frac{|\boldsymbol{A}_{oT}|}{|\boldsymbol{A}_T|} \right\} \cdot (\nabla \times \boldsymbol{A}_T) \tag{3.126b}$$

Thus,

$$\beta^{s_i} \stackrel{def}{=} (\nabla \times \boldsymbol{A}_T - \nabla \times \boldsymbol{A}_{oT})\Big|_{s_i} = \left\{ \frac{|\boldsymbol{A}_T| - |\boldsymbol{A}_{oT}|}{|\boldsymbol{A}_T|} \cdot (\nabla \times \boldsymbol{A}_T) \right\}\Bigg|_{s_i} \tag{3.127}$$

For the surface s_j, we have

$$\beta^{s_j} \stackrel{def}{=} (\nabla \times \boldsymbol{A}_T - \nabla \times \boldsymbol{A}_{oT})\Big|_{s_j} \tag{3.128}$$

Now, we assume that

$$\frac{\boldsymbol{A}_{oT}}{|\nabla \times \boldsymbol{A}_{oT}|} \cong \frac{\boldsymbol{A}_T}{|\nabla \times \boldsymbol{A}_T|} \tag{3.129a}$$

Giving,

$$\boldsymbol{A}_{oT} = \left\{ \frac{|\nabla \times \boldsymbol{A}_{oT}|}{|\nabla \times \boldsymbol{A}_T|} \right\} \cdot \boldsymbol{A}_T \tag{3.129b}$$

Thus,

$$\alpha^{s_j} \stackrel{def}{=} (\boldsymbol{A}_T - \boldsymbol{A}_{oT})\Big|_{s_j} = \left\{ \frac{|\nabla \times \boldsymbol{A}_T| - |\nabla \times \boldsymbol{A}_{oT}|}{|\nabla \times \boldsymbol{A}_T|} \cdot \boldsymbol{A}_T \right\}\Bigg|_{s_j} \tag{3.130}$$

and for the surface s_k, let $T1$ and $T2$ be the two orthogonal tangential directions with the direction $T2$ leading $T1$ by 90°. We define an operator j that rotates a space vector by 90° in anticlockwise direction. This operator is similar to the j operator that rotates a time phasor by 90° in anticlockwise direction (the j operator is extensively used in the study of ac circuits). Therefore,

$$\mathbf{T2} \stackrel{def}{=} j\,\mathbf{T1} \tag{3.131}$$

Now, let us assume

$$\frac{\mathbf{A}_{oT2}}{|\mathbf{A}_{T2}|} \cong j\,\frac{\mathbf{A}_{oT1}}{|\mathbf{A}_{T1}|} \tag{3.132a}$$

and

$$\frac{\nabla \times \mathbf{A}_{oT2}}{|\nabla \times \mathbf{A}_{T2}|} \cong j\,\frac{\nabla \times \mathbf{A}_{oT1}}{|\nabla \times \mathbf{A}_{T1}|} \tag{3.132b}$$

Therefore,

$$\mathbf{A}_{oT2} = j\,\frac{|\mathbf{A}_{T2}|}{|\mathbf{A}_{T1}|}\mathbf{A}_{oT1} \tag{3.133}$$

and

$$\nabla \times \mathbf{A}_{oT2} = j\,\frac{|\nabla \times \mathbf{A}_{T2}|}{|\nabla \times \mathbf{A}_{T1}|}(\nabla \times \mathbf{A}_{oT1}) \tag{3.134}$$

Thus,

$$\alpha^{s_k} = \left\{\mathbf{A}_T - \mathbf{A}_{oT1} - \frac{|\mathbf{A}_{T2}|}{|\mathbf{A}_{T1}|} \cdot j\,\mathbf{A}_{oT1}\right\} \tag{3.135}$$

and

$$\beta^{s_k} \stackrel{def}{=} \nabla \times \left\{\mathbf{A}_T - \mathbf{A}_{oT1} - \frac{|\nabla \times \mathbf{A}_{T2}|}{|\nabla \times \mathbf{A}_{T1}|} \cdot j\,\mathbf{A}_{oT1}\right\}\Bigg|_{s_k} \tag{3.136}$$

Therefore, Equation 3.119 can be rewritten as

$$\epsilon \cong \iint_{s_i} (\alpha^{s_i} \times \beta^{s_i}) \cdot ds + \iint_{s_j} (\alpha^{s_j} \times \beta^{s_j}) \cdot ds + \iint_{s_k} (\alpha^{s_k} \times \beta^{s_k}) \cdot ds \tag{3.137}$$

On substituting expressions for α's and β's from Equations 3.125, 3.127, 3.128, 3.130, 3.133 and 3.136 in the above equation, cumulative error ϵ can be found. As indicated in the preceding section, this error can also be expressed as percentage of a chosen base value.

3.5.3 Approximation Theorem for Maxwell's Equations

Consider Maxwell's two curl equations in phasor form for harmonic fields characterised by the factor $e^{-j\omega t}$

$$\nabla \times \boldsymbol{E} = j\omega\mu\boldsymbol{H} \tag{3.138a}$$

$$\nabla \times \boldsymbol{H} = \boldsymbol{J} - j\omega\varepsilon\boldsymbol{E} \tag{3.139a}$$

Let the field vectors involved in these equations only approximately satisfy the prescribed boundary conditions. Maxwell's equations for field vectors exactly satisfying the given boundary conditions are

$$\nabla \times \boldsymbol{E}_o = j\omega\mu\boldsymbol{H}_o \tag{3.138b}$$

$$\nabla \times \boldsymbol{H}_o = \boldsymbol{J}_o - j\omega\varepsilon\boldsymbol{E}_o \tag{3.139b}$$

Therefore,

$$\nabla \times \boldsymbol{e} = j\omega\mu\boldsymbol{h} \tag{3.138c}$$

$$\nabla \times \boldsymbol{h} = \boldsymbol{j} - j\omega\varepsilon\boldsymbol{e} \tag{3.139c}$$

where

$$\boldsymbol{e} = \boldsymbol{E} - \boldsymbol{E}_o \tag{3.140a}$$

$$\boldsymbol{h} = \boldsymbol{H} - \boldsymbol{H}_o \tag{3.140b}$$

$$\boldsymbol{j} = \boldsymbol{J} - \boldsymbol{J}_o \tag{3.140c}$$

and

$$j = \sigma e \tag{3.140d}$$

Now, since

$$-\nabla \cdot (e \times h^*) \equiv -h^* \cdot (\nabla \times e) + e \cdot (\nabla \times h^*) \tag{3.141a}$$

Therefore, in view of Equations 3.138c and 3.139c, we get

$$-\nabla \cdot (\mathbf{e} \times h^*) = \sigma e^2 + j\omega(\varepsilon e^2 - \mu h^2) \tag{3.141b}$$

where

$$e^2 \overset{def}{=} e \cdot e^* \tag{3.141c}$$

$$h^2 \overset{def}{=} h \cdot h^* \tag{3.141d}$$

Consider a conducting region of volume v bounded by its surface s. If both sides of Equation 3.141b are integrated over this volume, on applying divergence theorem to the LHS of the resulting equation, one gets on equating the real parts on the two sides:

$$\oiint_s \Re e\left[-(e \times h^*)\right] \cdot ds = \iiint_v (\sigma e^2) dv \tag{3.142}$$

The LHS of this equation can be simplified to

$$\oiint_s \Re e\left[-(e_T \times h_T^*)\right] \cdot ds = \iiint_v (\sigma e^2) dv \tag{3.143}$$

The subscript T indicates the tangential component of field vector on the bounding surface. Thus, in view of Equations 3.140, 3.142 and 3.143, the cumulative error ϵ is defined as

$$\epsilon \overset{def}{=} \oiint_s \Re e\left[-\left\{(E_T - E_{oT}) \times (H_T^* - H_{oT}^*)\right\}\right] \cdot ds \overset{def}{=} \oiint_s \Re e\left[-\left\{e_T \times h_T^*\right\}\right] \cdot ds \tag{3.144a}$$

where

$$e_T \stackrel{def}{=} E_T - E_{oT} \tag{3.144b}$$

and

$$h_T \stackrel{def}{=} H_T - H_{oT} \tag{3.144c}$$

while (E_{oT}, H_{oT}) indicate specified value of the tangential components of fields on the bounding surface s, and (E_T, H_T) indicate tangential components of fields on the bounding surface, approximately satisfying boundary conditions. The symbol e_T and h_T indicate the tangential components of error vectors on the surface s.

The mean square error in the volume v is given, in view of Equation 3.144a, as

$$\left|\Re e\left[e \cdot (\nabla \times h^*)\right]\right|_{\text{mean square}} = \left[\sigma(e \cdot e^*)\right]_{\text{mean square}} = \frac{\epsilon}{v} \tag{3.145}$$

Having defined various error functions through Equations 3.144a through 3.144c, in view of Equation 3.145 it is concluded that

1. The cumulative error ϵ cannot be negative
2. The cumulative error ϵ may not be zero, unless both solutions are identical and error free
3. In the volume v, the mean square value for the error is equal to the cumulative error on the bounding surface s, per unit volume

From the uniqueness theorem it can be seen that as many as three types of boundary conditions can be specified and on a given part of the boundary surface only one of these three types of the boundary condition can be independently specified. The other boundary conditions follow from the solution. Let the bounding surface s be divided into three sets of open surfaces, such that on s_i the tangential component of electric field E is specified, on s_j the tangential component of magnetic field H is specified, and on s_k, the component of both E and H in an arbitrarily chosen tangential direction is specified. In the approximate solution, this relationship between the three types of boundary conditions becomes rather hazy. Let us assume that on s_i, the tangential component E_{oT} is specified and it is satisfied exactly by the unknown solution E_o, while the known solution E and H with field components E_T and H_T on s_i provides approximate field values on the boundary surface on s_i. With known expressions for E_T and E_{oT}, e_T on s_i is obtained from Equation 3.144b as

$$e_T^{s_i} \stackrel{def}{=} (\boldsymbol{E_T} - \boldsymbol{E_{oT}})\big|_{s_i} \tag{3.146}$$

Since $\boldsymbol{H_{oT}}$ is not known on s_i, to find $\boldsymbol{h_T}$, we assume that

$$\frac{E_{oT}}{E_T} \cong \frac{H_{oT}}{H_T} \tag{3.147a}$$

Therefore,

$$H_{oT} \cong H_T \cdot \left(\frac{E_{oT}}{E_T}\right) \tag{3.147b}$$

or

$$\boldsymbol{H_{oT}} \cong \boldsymbol{H_T} \cdot \left(\frac{E_{oT}}{E_T}\right) \tag{3.148}$$

Giving,

$$h_T^{s_i} \stackrel{def}{=} (\boldsymbol{H_T} - \boldsymbol{H_{oT}})\big|_{s_i} = \left(\frac{e_T}{E_T}\boldsymbol{H_T}\right)\Bigg|_{s_i} \tag{3.149}$$

Next, let us assume that on s_j, the tangential component $\boldsymbol{H_{oT}}$ is specified and it is satisfied exactly by the unknown solution $\boldsymbol{H_o}$, while the known solution $\boldsymbol{E}$ and $\boldsymbol{H}$ with field components $\boldsymbol{E_T}$ and $\boldsymbol{H_T}$ on s_j provides approximate field values on the boundary surface on s_j. With known expressions for $\boldsymbol{H_T}$ and $\boldsymbol{H_{oT}}$, $\boldsymbol{h_T}$ is obtained from Equation 3.144c as

$$\boldsymbol{h_T^{s_j}} \stackrel{def}{=} (\boldsymbol{H_T} - \boldsymbol{H_{oT}})\big|_{s_j} = \left(\frac{e_T}{H_T}\boldsymbol{H_T}\right)\Bigg|_{s_j} \tag{3.150}$$

Since $\boldsymbol{E_{oT}}$ is not known on s_j, to find $\boldsymbol{e_T}$ consider Equation 3.146

$$E_{oT} \cong E_T \cdot \left(\frac{H_{oT}}{H_T}\right) \tag{3.151a}$$

or

$$\boldsymbol{E_{oT}} \cong \boldsymbol{E_T} \cdot \left(\frac{H_{oT}}{H_T}\right) \tag{3.151b}$$

Giving,

$$e_T^{s_j} \stackrel{def}{=} (\boldsymbol{E}_T - \boldsymbol{E}_{oT})\Big|_{s_j} = \left(\frac{h_T}{H_T}\boldsymbol{E}_T\right)\Bigg|_{s_j} \tag{3.152}$$

Next, consider the boundary conditions on open surface s_k. Note that $\boldsymbol{E}_T$, $\boldsymbol{E}_{T1}$ (parallel to $\boldsymbol{E}_{oT1}$), $\boldsymbol{E}_{T2}$ (= $\boldsymbol{E}_T - \boldsymbol{E}_{T1}$ and parallel to $\boldsymbol{E}_{oT2}$), and also $\boldsymbol{H}_T$, $\boldsymbol{H}_{T1}$ (parallel to $\boldsymbol{H}_{oT1}$) and $\boldsymbol{H}_{T2}$ (= $\boldsymbol{H}_T - \boldsymbol{H}_{T1}$ and parallel to $\boldsymbol{H}_{oT2}$) are approximately known. Further, let ($\boldsymbol{E}_{oT1}$ and $\boldsymbol{H}_{oT1}$) on s_k be given as boundary conditions.

We assume that

$$\frac{\boldsymbol{E}_{oT2}}{|\boldsymbol{E}_{T2}|} \cong j\,\frac{\boldsymbol{E}_{oT1}}{|\boldsymbol{E}_{T1}|} \tag{3.153a}$$

and

$$\frac{\boldsymbol{H}_{oT2}}{|\boldsymbol{H}_{T2}|} \cong j\,\frac{\boldsymbol{H}_{oT1}}{|\boldsymbol{H}_{T1}|} \tag{3.153b}$$

$$\boldsymbol{E}_{oT} = \boldsymbol{E}_{To1} + \boldsymbol{E}_{To2} = \boldsymbol{E}_{To1}\frac{|\boldsymbol{E}_{T2}|}{|\boldsymbol{E}_{T1}|} \cdot j\,\boldsymbol{E}_{oT1} \tag{3.154}$$

$$\boldsymbol{H}_{oT} = \boldsymbol{H}_{To1} + \boldsymbol{H}_{To2} = \boldsymbol{H}_{To1} + \frac{|\boldsymbol{H}_{T2}|}{|\boldsymbol{H}_{T1}|} \bullet j\,\boldsymbol{H}_{oT1} \tag{3.155}$$

Giving,

$$e_T^{s_k} \stackrel{def}{=} (\boldsymbol{E}_T - \boldsymbol{E}_{oT})\Big|_{s_k} = \left(\boldsymbol{E}_T - \boldsymbol{E}_{To1} - \frac{|\boldsymbol{E}_{T2}|}{|\boldsymbol{E}_{T1}|} \cdot j\,\boldsymbol{E}_{oT1}\right)\Bigg|_{s_k} \tag{3.156}$$

and

$$h_T^{s_k} \stackrel{def}{=} (\boldsymbol{H}_T - \boldsymbol{H}_{oT})\Big|_{s_k} = \left(\boldsymbol{H}_T - \boldsymbol{H}_{To1} - \frac{|\boldsymbol{H}_{T2}|}{|\boldsymbol{H}_{T1}|} \cdot j\,\boldsymbol{H}_{oT1}\right)\Bigg|_{s_k} \tag{3.157}$$

Therefore, Equation 3.144a can be rewritten as

$$\epsilon \stackrel{def}{=} \oiint_s \Re e\left[-\left\{e_T \times h_T^*\right\}\right] \cdot ds \tag{3.158}$$

or

$$\epsilon \cong \iint_{s_i} \Re e\left[-\left\{e_T^{s_i} \times h_T^{s_i*}\right\}\right] \cdot ds + \iint_{s_j} \Re e\left[-\left\{e_T^{s_j} \times h_T^{s_j*}\right\}\right] \cdot ds$$

$$+ \iint_{s_k} \Re e\left[-\left\{e_T^{s_k} \times h_T^{s_k*}\right\}\right] \cdot ds \tag{3.159}$$

On substituting expressions for e_T e_T and h_T^* s from Equations 3.144a, 3.149, 3.150, 3.152, 3.156 and 3.157 in the above equation, cumulative error ϵ can be found. As indicated in the preceding section, this error can also be expressed as a percentage of a chosen base value.

For approximate solution of Maxwell's equations for harmonic fields, together with the tangential component of either $\boldsymbol{E}$ or $\boldsymbol{H}$ on the bounding surface, the tolerance limit for the mean square error in the volume must also be specified. It is to be ensured that the cumulative error per unit volume of the region must be less than or equal to the prescribed tolerance limit for the mean square error in the volume, that is,

$$\frac{\epsilon}{v} \le \left|\Re e\left[e \cdot (\nabla \times h^*)\right]\right|_{\text{specified}} \tag{3.160}$$

PROJECT PROBLEMS

1. Extend the generalised Poynting theorem to take cognizance of moving charged particles present in MHD generators and some microwave devices.
2. Develop a uniqueness theorem for electromagnetic fields in anisotropic media.
3. State and prove the approximation theorem for the analysis of electromagnetic fields in anisotropic media.
4. In Example 2 of Section 3.2.1.2, assume that the value of the potential at $y = k$ is unity, that is, $V|_{y=k} = 1$ over $-w/2 < x < w/2$, where $g < k < h$. Prepare a computer program to determine the potential distribution in the rectangular region for a mean square error not to exceed x%. The 'x' may be assigned a suitable value (preferably < 1) before solving the problem.

DISCLAIMER

A portion of this chapter is based on reference 8 with permission from Sage Publications.

References

1. Mukerji, S. K., Sharma, S. and Shukla, S., Solution of Laplace equation using torch function synthesis technique, *National Conference on Recent Advances in Technology and Engineering*, Mangalayatan University, Aligarh, India, pp. 1–9, 2012.
2. Stratton, J. A., *Electromagnetic Theory*, McGraw-Hill Book Company, Inc., New York, pp. 256–257, 1941.
3. Stratton, J. A., *Electromagnetic Theory*, McGraw-Hill Book Company, Inc., New York, pp. 250, 1941.
4. Stratton, J. A., *Electromagnetic Theory*, McGraw-Hill Book Company, Inc., New York, pp. 486–488, 1941.
5. Arfken, G. B. and Weber, H. J., *Mathematical Methods for Physicists*, Elsevier Academic Press, San Diego, California, pp. 95–100, 2005.
6. Mukerji, S. K., Goel, S. K., Bhooshan, S. and Basu, K. P., Electromagnetic fields theory of electrical machines, Part II: Uniqueness theorem for time-varying electromagnetic fields in hysteretic media, *International Journal of Electrical Engineering Education*, 42(2), 203–208, April 2005.
7. Griffiths, D. J., *Introduction to Electrodynamics*, Pearson Education, New Delhi, India, pp. 555–557, 2003.
8. Mukerji, S. K., Goel, S. K., Bhooshan, S. and Basu, K. P., Electromagnetic fields theory of electrical machines, Part I: Poynting theorem for electromechanical energy conversion, *International Journal of Electrical Engineering Education*, 41(2), 137–145, April 2004.
9. Macfadyen, K. A., Vector permeability, *Journal of Institution of Electrical Engineers*, 94(Pt.III), 407, 1947.

4

Laplacian Fields

4.1 Introduction

The rotors and stators in most electrical machines are provided with slots to accommodate the current-carrying conductors. The overall performance of a machine is greatly affected due to the presence of these slots in accordance with their types, shapes, sizes and the relative displacements and the relative dimensions of teeth and air gap. These effects can be readily visualised when analysis is carried out for the magnetic flux density distribution in the air gap, the mmf due to the field current, the torque exerted on the rotor and the induced emf. Thus, the effect of slotting needs to be given due weightage to predict the behaviour of an electrical machine.

While designing an electrical machine, in most cases, the slots are assumed to be source free. In the cases wherein the sources are present these are presumed to be located deep inside the slots. In view of these assumptions, the problem reduces to the solution of the Laplace equation. The fields involved in these problems can be referred to as Laplacian fields. Such problems can readily be solved by using the Schwarz–Christoffel transformation or the separation of variables method. The solution so obtained ultimately leads to the determination of air-gap permeance. This chapter, therefore, describes the potential distribution between identical double slotting. These distributions are obtained for the cases of (i) tooth-opposite-tooth, (ii) tooth-opposite-slot and (iii) an arbitrary orientation of tooth-slot configuration. The relations for air-gap permeance are also obtained. This chapter also discusses the modelling for aperiodic field distributions for the three above-mentioned cases. Besides, an effort has also been made to describe the method of determination of (i) fringing flux for tooth-opposite-tooth orientation with small air gap, (ii) air-gap field of a conductor deep inside an open slot and (iii) the magnetic field near armature winding overhang.

4.2 Potential Distribution for Rectangular Double-Slotting

Magnetic permeance, between two equipotential, slotted iron surfaces, can be readily determined if the magnetic field distribution between the two surfaces is known. In his classical paper, Carter[1] presented an analytical solution for the distribution of field between two identically slotted surfaces oriented tooth-opposite-tooth. Subsequently, Coe and Taylor[2] developed this work for the purposes of numerical treatment. In an unpublished work, Carter[3] dealt with the problem of two identically slotted iron surfaces oriented tooth-opposite-slot. This latter work of Carter[3] was described by Mukherji and Neville[3], who also presented numerical developments. The analysis presented here differs basically from that of Carter[1,3] in the method of approach. Carter used the method of Schwarz–Christoffel transformation[4], whereas the present treatment is based on the method of separation of variables. Using Fourier series and Fourier integrals, field distribution, between two rectangular slots was obtained by Khan and Mukerji.[5] The field distribution between two rectangular teeth was also formulated by Khan.[6]

The double-slotting comprises an infinite iron surface, with a sequence of identical rectangular teeth and slots separated by a constant air gap of length g, from another identically slotted surface. These two parallel surfaces can be oriented in any arbitrary relative position. The subsequent subsections describe the field distribution for the three possible cases of orientations.

4.2.1 Tooth-Opposite-Tooth Orientation

Figure 4.1 illustrates two slotted surfaces for tooth-opposite-tooth orientation. The two surfaces are identical with deep slots ($d \approx \infty$). It is assumed that slot width is s, tooth width is t and tooth pitch is λ for both the surfaces. The distribution of scalar magnetic potential satisfies the Laplace equation.

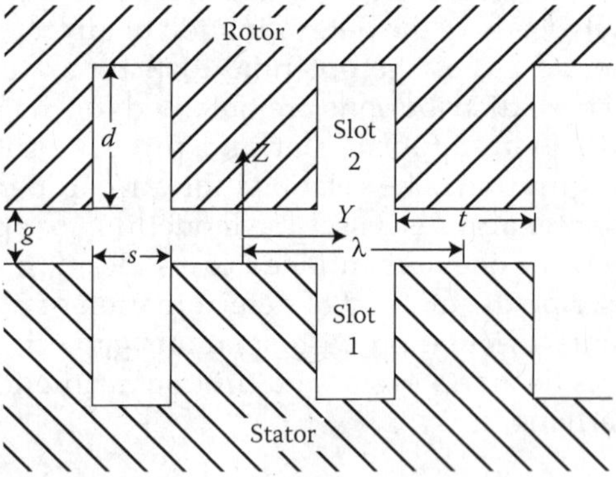

FIGURE 4.1
Tooth-opposite-tooth orientation.

The stator- and the rotor-iron surfaces are assumed to be maintained at magnetic potentials +1/2 and –1/2 m.k.s. units. For this configuration the distribution of scalar magnetic potential ($\mathcal{V}_1$) in the stator slot 1 can be given as

$$\mathcal{V}_1 = \frac{1}{2} + \sum_{m-odd}^{\infty} a_m \cdot \cos\left\{\frac{m\pi}{s}(y - \lambda/2)\right\} \cdot e^{+(m\pi/s)(z+g/z)} \tag{4.1}$$

Similarly, in the rotor slot 2, the magnetic potential ($\mathcal{V}_2$) can be given as

$$\mathcal{V}_2 = -\frac{1}{2} - \sum_{m-odd}^{\infty} a_m \cdot \cos\left\{\frac{m\pi}{s}(y - \lambda/2)\right\} \cdot e^{-(m\pi/s)(z-g/2)} \tag{4.2}$$

In Equations 4.1 and 4.2, a_m indicates a set of arbitrary constants.

The distribution of scalar magnetic potential in the air gap ($\mathcal{V}_o$) is an even periodic function of y with a period length λ, and an odd function of z. It can thus be written as

$$\mathcal{V}_o = -b_o \cdot \frac{2}{g} \cdot z - \sum_{n=1}^{\infty} b_n \cdot \cos\left(\frac{2n\pi}{\lambda} y\right) \cdot \frac{\sinh((2n\pi/\lambda) \cdot z)}{\sinh((2n\pi/\lambda) \cdot (g/2))} \tag{4.3}$$

where b_n indicates a set of arbitrary constants.

4.2.1.1 Evaluation of Arbitrary Constants

The arbitrary constants a_m and b_n involved in Equations 4.1 through 4.3 can be evaluated by using the following boundary conditions:

$$\mathcal{V}_o\Big|_{z=-g/2} = \mathcal{V}_1\Big|_{z=-g/2} \quad \text{over } t/2 \le y \le (\lambda - t/2) \tag{4.4a}$$

$$= \frac{1}{2} \quad \text{over } 0 \le y \le t/2 \text{ and } (\lambda - t/2) \le y \le \lambda \tag{4.4b}$$

$$\left.\frac{\partial \mathcal{V}_o}{\partial z}\right|_{z=-g/2} = \left.\frac{\partial \mathcal{V}_1}{\partial z}\right|_{z=-g/2} \quad \text{over } \frac{t}{2} < y < (\lambda - t/2) \tag{4.5}$$

Application of boundary conditions given by Equations 4.4a and 4.4b on Equations 4.1 and 4.3 gives

$$b_o + \sum_{n=1}^{\infty} b_n \cdot \cos\left(\frac{2n\pi}{\lambda} y\right) = \frac{1}{2} + \sum_{m-odd}^{\infty} a_m \cdot \cos\left\{\frac{m\pi}{s}(y - \lambda/2)\right\}$$
$$\text{over } t/2 \le y \le (\lambda - t/2) \tag{4.6a}$$

$$= \frac{1}{2} \quad \text{over } 0 \le y \le t/2 \text{ and } (\lambda - t/2) \le y \le \lambda \tag{4.6b}$$

Therefore,

$$b_o = \frac{1}{\lambda} \cdot \left[\int_0^{\lambda} \frac{1}{2} dy + \sum_{m-odd}^{\infty} a_m \cdot \int_{t/2}^{(\lambda - t/2)} \cos\left\{ \frac{m\pi}{s} (y - \lambda/2) \right\} \cdot dy \right] \tag{4.7a}$$

or

$$b_o = \left[\frac{1}{2} + \frac{s}{\lambda} \cdot \sum_{M-odd}^{\infty} a_M \cdot \frac{\sin(M\pi/2)}{(M\pi/2)} \right] \tag{4.7b}$$

and

$$b_n = \frac{2}{\lambda} . \sum_{m-odd}^{\infty} a_m . \int_{t/2}^{(\lambda - t/2)} \cos\left\{ \frac{m\pi}{s} (y - \lambda/2) \right\} \cos\left(\frac{2n\pi}{\lambda} y \right) \cdot dy \tag{4.8a}$$

or,

$$b_n = \left[-\frac{1}{\pi} \cdot \frac{\lambda}{s} \cdot \sum_{M-odd}^{\infty} a_M \cdot (M) \cdot \sin\left(\frac{M\pi}{2} \right) \cdot \left\{ \frac{\cos((n\pi t/\lambda))}{(n)^2 - ((M/2) \cdot (\lambda/s))^2} \right\} \right] \tag{4.8b}$$

In Equations 4.8a and 4.8b: $n = 1, 2, 3, \ldots$.

Further, in view of Equations 4.1, 4.3 and 4.5, we get

$$\sum_{m-odd}^{\infty} a_m \cdot \frac{m\pi}{s} \cdot \cos\left\{ \frac{m\pi}{s} (y - \lambda/2) \right\}$$
$$= -b_o \cdot \frac{2}{g} - \sum_{n=1}^{\infty} b_n \cdot \frac{2n\pi}{\lambda} \cdot \cos\left(\frac{2n\pi}{\lambda} y \right) \cdot \coth\left(\frac{2n\pi}{\lambda} \cdot \frac{g}{2} \right)$$
$$\text{over } t/2 < y < (\lambda - t/2) \tag{4.9}$$

On multiplying both sides of Equation 4.9 by $\cos\{(p\pi/s)(y - \lambda/2)\}$ with an odd integer p, and then integrating the resulting expressions over $t/2 < y < (\lambda - t/2)$, one gets on replacing p by another odd integer m

$$a_m \cdot \frac{m\pi}{2} = -b_o \cdot \frac{s}{g} \cdot \frac{2}{m\pi} \cdot \sin\left(\frac{m\pi}{2}\right) + \frac{\lambda}{s} \cdot \sum_{n=1}^{\infty} b_n \cdot (n) \coth\left(\frac{2n\pi}{\lambda} \cdot \frac{g}{2}\right) \cdot (m) \cdot \sin\left(\frac{m\pi}{2}\right) \times \left\{\frac{\cos(n\pi t/\lambda)}{(n)^2 - ((m/2)\cdot(\lambda/s))^2}\right\} \tag{4.10}$$

for, $m = 1, 3, 5, \ldots$.

In Equation 4.10 if the expressions for b_o and b_n given by Equations 4.7b and 4.8b are substituted, we get

$$a_m \cdot \frac{m\pi}{2} = -\left[\frac{1}{2} + \frac{s}{\lambda} \cdot \sum_{M-odd}^{\infty} a_M \cdot \frac{\sin(M\pi/2)}{(M\pi/2)}\right] \cdot \frac{s}{g} \cdot \frac{2}{m\pi} \cdot \sin\left(\frac{m\pi}{2}\right) - \frac{1}{\pi} \cdot \left(\frac{\lambda}{S}\right)^2 \cdot (m) \cdot \sin\left(\frac{m\pi}{2}\right) \cdot \sum_{M-odd}^{\infty} \left[a_M \cdot (M) \cdot \sin\left(\frac{M\pi}{2}\right) \cdot \mathcal{S}(m, M)\right] \tag{4.11}$$

for $m = 1, 3, 5, \ldots$.

where

$$\mathcal{S}(m, M) = \sum_{n=1}^{\infty} \left\{\frac{\cos(n\pi t/\lambda)}{(n)^2 - ((M/2)\cdot(\lambda/s))^2}\right\} \cdot \left\{\frac{\cos(n\pi t/\lambda)}{(n)^2 - ((m/2)\cdot(\lambda/s))^2}\right\} \times (n) \cdot \coth\left(\frac{2n\pi}{\lambda} \cdot \frac{g}{2}\right) \tag{4.12}$$

for $m, M = 1, 3, 5, \ldots$.

Since the expression inside the summation sign is an even function of n, we also have

$$\mathcal{S}(m, M) = \sum_{n=-\infty}^{-1} \left\{\frac{\cos(n\pi t/\lambda)}{(n)^2 - ((m/2)\cdot(\lambda/s))^2}\right\} \cdot \left\{\frac{\cos(n\pi t/\lambda)}{(n)^2 - ((M/2)\cdot(\lambda/s))^2}\right\} \times (n) \cdot \coth\left(\frac{2n\pi}{\lambda} \cdot \frac{g}{2}\right) \tag{4.13}$$

for $m, M = 1, 3, 5, \ldots$.

Now, since

$$(n)\cdot\coth\left(\frac{2n\pi}{\lambda}\cdot\frac{g}{2}\right)\cong|n|,\quad \text{for } n\neq 0 \tag{4.14}$$

for $n=\pm(1,2,3,\ldots)$.

Thus, Equations 4.12 and 4.13 can be rewritten, respectively, as

$$S(m,M)\cong\sum_{n=1}^{\infty}\left\{\frac{\cos(n\pi t/\lambda)}{(n)^2-((m/2)\cdot(\lambda/s))^2}\right\}\cdot\left\{\frac{\cos(n\pi t/\lambda)}{(n)^2-((M/2)\cdot(\lambda/s))^2}\right\}\cdot|n| \tag{4.15}$$

for $m, M = 1, 3, 5, \ldots$.

And

$$S(m,M)\cong\sum_{n=-\infty}^{-1}\left\{\frac{\cos(n\pi t/\lambda)}{(n)^2-((m/2)\cdot(\lambda/s))^2}\right\}\cdot\left\{\frac{\cos(n\pi t/\lambda)}{(n)^2-((M/2)\cdot(\lambda/s))^2}\right\}\cdot|n| \tag{4.16}$$

for $m, M = 1, 3, 5, \ldots$.

Since both expressions give zero value for $n = 0$, we may combine them resulting:

$$S(m,M)\cong\sum_{n=-\infty}^{\infty}\left\{\frac{\cos(n\pi t/\lambda)}{(n)^2-((m/2)\cdot(\lambda/s))^2}\right\}\cdot\left\{\frac{\cos(n\pi t/\lambda)}{(n)^2-((M/2)\cdot(\lambda/s))^2}\right\}\cdot\frac{|n|}{2} \tag{4.17}$$

for $m, M = 1, 3, 5, \ldots$.

This infinite series can be readily summed up using Poisson's summation formula:

$$\sum_{n=-\infty}^{n=\infty}f(n)=\sum \text{Residues of}\,[-\pi\cdot f(z)\cdot\cot(\pi z)],\quad \text{at the poles of } f(z) \tag{4.18a}$$

Therefore,

$$\begin{aligned}S(m,M)&\cong\sum_{n=-\infty}^{\infty}\left\{\frac{\cos(n\pi t/\lambda)}{(n)^2-((m/2)\cdot(\lambda/s))^2}\right\}\cdot\left\{\frac{\cos(n\pi t/\lambda)}{(n)^2-((M/2)\cdot(\lambda/s))^2}\right\}\cdot\frac{|n|}{2}\\&=-\frac{\pi}{2}\cdot\left[\frac{\cos^2(M\pi t/2s)\cdot\cot(M\pi\lambda/2s)-\cos^2(m\pi t/2s)\cdot\cot(m\pi\lambda/2s)}{(M\lambda/2s)^2-(m\lambda/2s)^2}\right]\end{aligned} \tag{4.18b}$$

for $m, M = 1, 3, 5, \ldots$.

In view of Equations 4.11 and 4.18b, if the infinite series in m and M are terminated after retaining first $\mathcal{M}$-terms, one obtains $\mathcal{M}$ linearly independent algebraic simultaneous equations between equal numbers of unknowns. Numerical solution of these equations determines the values of unknowns a_M (i.e., $a_1, a_3, a_5, \ldots, a_{2\mathcal{M}-1}$). The substitution of these values of a_M in Equations 4.7b and 4.8b gives, respectively, the values for b_o and b_n.

4.2.2 Tooth-Opposite-Slot Orientation

For tooth-opposite-slot orientation shown in Figure 4.2, the distribution of scalar magnetic potential ($\mathcal{V}_1$) in the stator slot 1 can, vide Equation 4.1, still be given as:

$$\mathcal{V}_1 = \frac{1}{2} + \sum_{m-odd}^{\infty} c_m \cdot \cos\left\{\frac{m\pi}{s}(y - \lambda/2)\right\} \cdot e^{+(m\pi/s)(z+g/2)} \tag{4.19}$$

while the potential distribution in the rotor slot 2 ($\mathcal{V}_2$) can be expressed as

$$\mathcal{V}_2 = -\frac{1}{2} - \sum_{m-odd}^{\infty} c_m \cdot \cos\left(\frac{m\pi}{s} \cdot y\right) \cdot e^{-(m\pi/s)(z-g/2)} \tag{4.20}$$

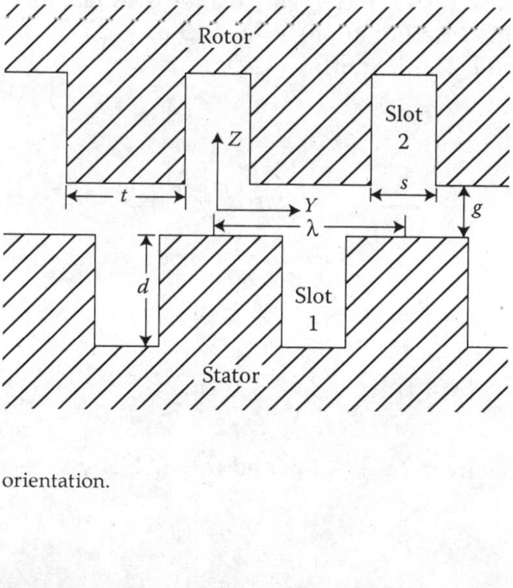

FIGURE 4.2
Tooth-opposite-slot orientation.

The potential distribution on the air-gap surfaces, being an even periodic function of y with a period λ, can be written as

$$\mathcal{V}_o\Big|_{z=-g/2} = d_o + \sum_{n=1}^{\infty} d_n \cdot \cos\left(\frac{2n\pi}{\lambda}y\right) \tag{4.21a}$$

$$\mathcal{V}_o\Big|_{z=g/2} = -d_o - \sum_{n=1}^{\infty} d_n \cdot \cos\left\{\frac{2n\pi}{\lambda}\left(y - \frac{\lambda}{2}\right)\right\} \tag{4.21b}$$

where d_n indicates a set of arbitrary constants. Therefore, the solution of the Laplace equation giving the air-gap potential distribution is

$$\begin{aligned}\mathcal{V}_o = -\frac{2}{g} \cdot z \cdot d_o - \sum_{n=1}^{\infty} d_n \cdot \Bigg[&\cos\left(\frac{2n\pi}{\lambda}y\right) \cdot \frac{\sinh\{(2n\pi/\lambda)(z - g/2)\}}{\sinh((2n\pi/\lambda) \cdot g)} \\ &+ \cos\left\{\frac{2n\pi}{\lambda}\left(y - \frac{\lambda}{2}\right)\right\} \cdot \frac{\sinh\{(2n\pi/\lambda)(z + g/2)\}}{\sinh((2n\pi/\lambda) \cdot g)} \Bigg]\end{aligned} \tag{4.22}$$

4.2.2.1 Evaluation of Arbitrary Constants

The arbitrary constants c_m and d_n can also be evaluated by using boundary conditions specified by Equations 4.4a, 4.4b and 4.5. Thus, from Equations 4.4a, 4.4b, 4.19 and 4.21a, we get

$$\begin{aligned} d_o + \sum_{n=1}^{\infty} d_n \cdot \cos\left(\frac{2n\pi}{\lambda}y\right) \\ = 1/2 + \sum_{m-odd}^{\infty} c_m \cdot \cos\left\{\frac{m\pi}{s}(y - \lambda/2)\right\} \end{aligned} \tag{4.22a}$$

$$\text{over } t/2 \le y \le (\lambda - t/2)$$
$$= 1/2$$
$$\text{over } 0 \le y \le t/2 \text{ and } (\lambda - t/2) \le y \le \lambda \tag{4.22b}$$

Therefore, the Fourier coefficients are

$$d_o = \frac{1}{\lambda}\cdot\left[\int_0^{\lambda}\frac{1}{2}dy + \sum_{m-odd}^{\infty} c_m \cdot \int_{t/2}^{(\lambda-t/2)} \cos\left\{\frac{m\pi}{s}(y-\lambda/2)\right\}\cdot dy\right]$$

or

$$d_o = \left[\frac{1}{2} + \frac{s}{\lambda}\cdot\sum_{M-odd}^{\infty} c_M \cdot \frac{\sin(M\pi/2)}{(M\pi/2)}\right] \tag{4.23a}$$

and

$$d_n = \frac{2}{\lambda}\cdot\int_{t/2}^{(\lambda-t/2)}\left[\sum_{m-odd}^{\infty} c_m \cdot \cos\left\{\frac{m\pi}{s}(y-\lambda/2)\right\}\right]\cdot\cos\left(\frac{2n\pi}{\lambda}y\right)\cdot dy$$

or,

$$d_n = \left[-\frac{1}{\pi}\cdot\frac{\lambda}{s}\cdot\sum_{M-odd}^{\infty} c_M \cdot (M)\cdot\sin\left(\frac{M\pi}{2}\right)\cdot\left\{\frac{\cos(n\pi t/\lambda)}{(n)^2 - ((M/2)\cdot(\lambda/s))^2}\right\}\right] \tag{4.23b}$$

for, $n = 1, 2, 3, \ldots$.

Using Equations 4.5, 4.19 and 4.22, we get

$$\begin{aligned}\sum_{m-odd}^{\infty} c_m \cdot \left(\frac{m\pi}{s}\right)\cdot\cos\left\{\frac{m\pi}{s}(y-\lambda/2)\right\} \\ = -\frac{2}{g}\cdot d_o - \sum_{n=1}^{\infty} d_n \cdot \left(\frac{2n\pi}{\lambda}\right)\cdot\left[\cos\left\{\frac{2n\pi}{\lambda}\left(y-\frac{\lambda}{2}\right)\right\}\cdot\operatorname{cosech}\left(\frac{2n\pi}{\lambda}\cdot g\right)\right. \\ \left. + \cos\left(\frac{2n\pi}{\lambda}y\right)\cdot\coth\left(\frac{2n\pi}{\lambda}\cdot g\right)\right] \quad \text{over } t/2 < y < (\lambda - t/2)\end{aligned} \tag{4.24}$$

Considering the following identities

$$\int_{t/2}^{(\lambda-t/2)} 2\cdot\cos\left\{\frac{M\pi}{s}(y-\lambda/2)\right\}\cdot dy = \frac{4s}{M\pi}\cdot\sin\left(\frac{M\pi}{2}\right) \tag{4.25a}$$

$$\int_{t/2}^{(\lambda-t/2)} 2\cdot\cos\left\{\frac{M\pi}{s}(y-\lambda/2)\right\}\cdot\cos\left\{\frac{m\pi}{s}(y-\lambda/2)\right\}\cdot dy = s \quad \text{for } m = M \tag{4.25b}$$

$$\int_{t/2}^{(\lambda-t/2)} 2\cdot\cos\left\{\frac{M\pi}{s}(y-\lambda/2)\right\}\cdot\cos\left\{\frac{m\pi}{s}(y-\lambda/2)\right\}\cdot dy = 0 \quad \text{for } m \neq M \tag{4.25c}$$

$$\int_{t/2}^{(\lambda-t/2)} 2\cdot\cos\left\{\frac{M\pi}{s}(y-\lambda/2)\right\}\cdot\cos\left\{\frac{2n\pi}{\lambda}(y-\lambda/2)\right\}\cdot dy$$
$$= -\frac{\lambda}{\pi}\cdot\frac{\lambda}{s}\cdot(M)\cdot\sin\left(\frac{M\pi}{2}\right)\cdot\frac{\cos(n\pi\cdot(s/\lambda))}{(n)^2-((M/2)\cdot(\lambda/s))^2} \tag{4.25d}$$

and

$$\int_{t/2}^{(\lambda-t/2)} 2\cdot\cos\left\{\frac{M\pi}{s}(y-\lambda/2)\right\}\cdot\cos\left(\frac{2n\pi}{\lambda}y\right)\cdot dy$$
$$= -\frac{\lambda}{\pi}\cdot\frac{\lambda}{s}\cdot(M)\cdot\sin\left(\frac{M\pi}{2}\right)\cdot\frac{\cos(n\pi\cdot(t/\lambda))}{(n)^2-((M/2)\cdot(\lambda/s))^2} \tag{4.25e}$$

on multiplying both sides of Equation 4.24 by $2\cdot\cos\{(M\pi/s)(y-\lambda/2)\}$, for any odd integer M, and then integrating both sides over $t/2 < y < (\lambda - t/2)$, we get on replacing M by m:

$$c_m = -2\cdot\frac{s}{g}\cdot\left(\frac{2}{m\pi}\right)^2\cdot\sin\left(\frac{m\pi}{2}\right)\cdot d_o$$
$$+\frac{\lambda}{s}\cdot\frac{2}{\pi}\cdot\sin\left(\frac{m\pi}{2}\right)\cdot\sum_{n=1}^{\infty}d_n\cdot\left[\left\{\frac{n\cdot\cos(n\pi\cdot(s/\lambda))}{(n)^2-((m/2)\cdot(\lambda/s))^2}\right\}\cdot\operatorname{cosech}\left(\frac{2n\pi}{\lambda}\cdot g\right)\right.$$
$$\left.+\left\{\frac{n\cdot\cos(n\pi\cdot(t/\lambda))}{(n)^2-((m/2)\cdot(\lambda/s))^2}\right\}\cdot\coth\left(\frac{2n\pi}{\lambda}\cdot g\right)\right] \tag{4.26}$$

for $m = 1, 3, 5, \ldots$.

The substitution of the expressions for d_o and d_n from Equations 4.23a and 4.23b in Equation 4.26 gives

$$c_m = -2\cdot\frac{s}{g}\cdot\left(\frac{2}{m\pi}\right)^2\cdot\sin\left(\frac{m\pi}{2}\right)\cdot\left[\frac{1}{2}+\frac{s}{\lambda}\cdot\sum_{M-odd}^{\infty}c_M\cdot\frac{\sin(M\pi/2)}{(M\pi/2)}\right]$$
$$-\left(\frac{\lambda}{s}\right)^2\cdot\frac{2}{\pi^2}\cdot\sin\left(\frac{m\pi}{2}\right)\cdot\sum_{M-odd}^{\infty}c_M\cdot(M)\cdot\sin\left(\frac{M\pi}{2}\right)$$
$$\times[\mathcal{S}_1(m,M)+\mathcal{S}_2(m,M)] \quad \text{for } m,\ M = 1, 3, 5, \ldots \tag{4.27}$$

where

$$\mathcal{S}_1(m,M) \stackrel{def}{=} \sum_{n=1}^{\infty} \left\{ \frac{\cos(n\pi t/\lambda)}{(n)^2 - ((M/2)\cdot(\lambda/s))^2} \right\} \cdot \left\{ \frac{\cos(n\pi\cdot(s/\lambda))}{(n)^2 - ((M/2)\cdot(\lambda/s))^2} \right\} \times (n)\cdot \operatorname{cosech}\left(\frac{2n\pi}{\lambda}\cdot g \right) \tag{4.27a}$$

for $m, M = 1, 3, 5, \ldots$.

And

$$\mathcal{S}_2(m,M) \stackrel{def}{=} \sum_{n=1}^{\infty} \left\{ \frac{\cos((n\pi t/\lambda))}{(n)^2 - ((M/2)\cdot(\lambda/s))^2} \right\} \cdot \left\{ \frac{\cos((n\pi t/\lambda))}{(n)^2 - ((m/2)\cdot(\lambda/s))^2} \right\} \times (n)\cdot \coth\left(\frac{2n\pi}{\lambda}\cdot g \right) \tag{4.27b}$$

for $m, M = 1, 3, 5, \ldots$.

Now, because of the cosec hyperbolic function, the infinite series in Equation 4.27a is fast convergent. Thus, this series may be terminated after, say, $\mathcal{N}$ number of terms. The truncated series is given as

$$\mathcal{S}_1(m,M) \cong \sum_{n=1}^{\mathcal{N}} \left\{ \frac{\cos(n\pi t/\lambda)}{(n)^2 - ((M/2)\cdot(\lambda/s))^2} \right\} \cdot \left\{ \frac{\cos(n\pi\cdot(s/\lambda))}{(n)^2 - ((m/2)\cdot(\lambda/s))^2} \right\} \times n\cdot \operatorname{cosech}\left(\frac{2n\pi}{\lambda}\cdot g \right) \tag{4.28}$$

for $m, M = 1, 3, 5, \ldots \infty$.

Next, consider the infinite series in the Equation 4.27b. The infinite series is an even function of n and since

$$(n)\cdot \coth\left(\frac{2n\pi}{\lambda}\cdot g \right) \cong |n|, \quad \text{for } n \neq 0 \tag{4.29}$$

Therefore, Equation 4.27b can be rewritten as

$$\mathcal{S}_2(m,M) \cong \sum_{n=1}^{\infty} \left\{ \frac{\cos(n\pi t/\lambda)}{(n)^2 - ((M/2)\cdot(\lambda/s))^2} \right\} \cdot \left\{ \frac{\cos(n\pi t/\lambda)}{(n)^2 - ((m/2)\cdot(\lambda/s))^2} \right\} \cdot |n| \tag{4.30a}$$

Also, as

$$\mathcal{S}_2(m,M) \cong \sum_{n=-\infty}^{-1} \left\{ \frac{\cos(n\pi t/\lambda)}{(n)^2 - ((M/2)\cdot(\lambda/s))^2} \right\} \cdot \left\{ \frac{\cos(n\pi t/\lambda)}{(n)^2 - ((m/2)\cdot(\lambda/s))^2} \right\} \cdot |n| \tag{4.30b}$$

Thus,

$$\mathcal{S}_2(m,M) \cong \sum_{n=-\infty}^{\infty} \left\{ \frac{\cos(n\pi t/\lambda)}{(n)^2 - ((M/2)\cdot(\lambda/s))^2} \right\} \cdot \left\{ \frac{\cos(n\pi t/\lambda)}{(n)^2 - ((m/2)\cdot(\lambda/s))^2} \right\} \cdot \frac{|n|}{2} \tag{4.31}$$

on comparing with Equation 4.17, it is evident that

$$\mathcal{S}_2(m,M) = \mathcal{S}(m,M) \tag{4.32a}$$

Therefore, in view of Equation 4.18b, we have

$$\begin{aligned}\mathcal{S}_2(m,M) &\cong \sum_{n=-\infty}^{\infty} \left\{ \frac{\cos(n\pi t/\lambda)}{(n)^2 - ((M/2)\cdot(\lambda/s))^2} \right\} \cdot \left\{ \frac{\cos(n\pi t/\lambda)}{(n)^2 - ((m/2)\cdot(\lambda/s))^2} \right\} \cdot \frac{|n|}{2} \\ &= -\frac{\pi}{2} \cdot \left[\frac{\cos^2(M\pi t/2s)\cdot\cot(M\pi t/2s) - \cos^2(m\pi t/2s)\cdot\cot(m\pi t/2s)}{((M\lambda/2s))^2 - ((m\lambda/2s))^2} \right]\end{aligned} \tag{4.32b}$$

for $m, M = 1, 3, 5, \ldots\infty$.

In view of Equations 4.28 and 4.32b, if the infinite series in m and M occurring in the Equations 4.23a, 4.23b and 4.27 are terminated after retaining first $\mathcal{M}$ terms, one obtains $\mathcal{M}$ linearly independent algebraic simultaneous equations between equal numbers of unknowns. Numerical solution of these equations determines the unknowns c_m (i.e., $c_1, c_3, c_5, \ldots, c_{2\mathcal{M}-1}$). On substituting these values of c_M in Equations 4.23a and 4.23b, values for d_o and d_n can be readily found.

4.2.3 Arbitrary Orientation of Tooth and Slot

Consider the identical double-slotting shown in Figure 4.3. The tooth-centres of the two equipotential surfaces are separated by a distance δ, where $0 \le \delta \le \lambda/2$. For $\delta = 0$, the orientation will be tooth-opposite-tooth and for $\delta = \lambda/2$ it will result in tooth-opposite-slot orientation. Let the rotor- and stator-iron slotted surfaces be at a magnetic potential of $-1/2$ and $+1/2$ m.k.s.

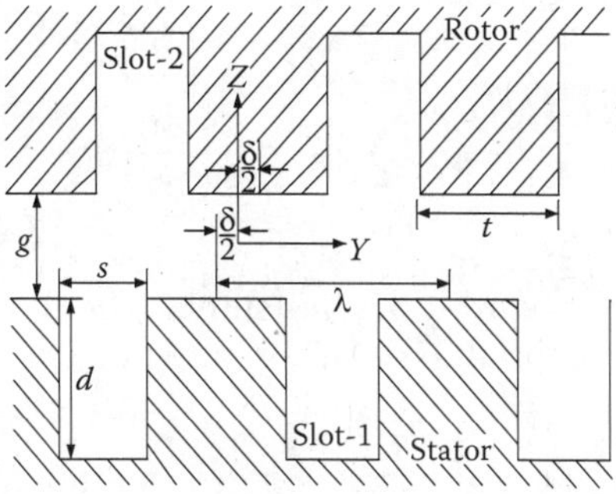

FIGURE 4.3
Rectangular double-slotting.

units respectively. The distribution of magneto-static potential in the stator slot 1 can be given as

$$\mathcal{V}_1 = \frac{1}{2} - \sum_{m=1}^{\infty} p_m \cdot \sin\left\{\frac{m\pi}{s}\left(y - \frac{t}{2} + \frac{\delta}{2}\right)\right\} \cdot e^{+(m\pi/s)(z+g/2)}$$
$$\text{over } (t/2 - \delta/2) \le y \le (s + t/2 - \delta/2) \quad (4.33)$$

And in the rotor slot 2

$$\mathcal{V}_2 = -\frac{1}{2} - \sum_{m=1}^{\infty} p_m \cdot \sin\left\{\frac{m\pi}{s}\left(y + \frac{t}{2} - \frac{\delta}{2}\right)\right\} \cdot e^{-(m\pi/s)(z-g/2)}$$
$$\text{over } (-s - t/2 + \delta/2) \le y \le (-t/2 + \delta/2) \quad (4.34)$$

Note that the same set of arbitrary constants p_m is involved in the two expressions.

The distributions of scalar magnetic potential on the two air-gap surfaces are periodic in the y direction with a period equal to the slot-pitch λ. For $\delta = 0$, these distributions are even functions of y that correspond to the tooth-opposite-tooth orientation. Further, as $\delta = \lambda/2$ corresponds to the tooth-opposite-slot orientation, the potential distributions on the air-gap surfaces will be again even functions of y provided that the origin is shifted to coincide with a tooth axis, that is, if y is replaced by $(y \pm \delta/2)$. These distributions can, therefore, be given by the following Fourier series expansions:

$$\mathcal{V}_0\big|_{z=-g/2} = q_o + \sum_{n=1}^{\infty} q_n \cdot \cos\left\{\frac{n2\pi}{\lambda}(y + \delta/2)\right\} \quad (4.35a)$$

and

$$\mathcal{V}_o\Big|_{z=g/2} = -q_o - \sum_{n=1}^{\infty} q_n \cdot \cos\left\{\frac{2\pi n}{\lambda}(y-\delta/2)\right\} \tag{4.35b}$$

where q_n indicates a set of arbitrary constants. Therefore, the distribution of the scalar magnetic potential in the air-gap region is

$$\begin{aligned}\mathcal{V}_o = -q_o \cdot \frac{2z}{g} - \sum_{n=1}^{\infty} q_n \cdot \Bigg[\cos\left\{\frac{n2\pi}{\lambda}(y-\delta/2)\right\} \cdot \frac{\sinh\{(n2\pi/\lambda)(z+g/2)\}}{\sinh((2\pi n/\lambda)\cdot g)} \\ + \cos\left\{\frac{n2\pi}{\lambda}(y+\delta/2)\right\} \cdot \frac{\sinh\{(n2\pi/\lambda)(z-g/2)\}}{\sinh((2\pi n/\lambda)\cdot g)} \Bigg]\end{aligned} \tag{4.36}$$

This expression for the air-gap potential $\mathcal{V}_o$ satisfies the requirements stated above.

4.2.3.1 Evaluation of Arbitrary Constants

The arbitrary constants p_m, q_o and q_n involved in Equations 4.33, 4.34 and 4.36 can be evaluated by using the following boundary conditions:

$$\mathcal{V}_o\Big|_{z=-g/2} = \mathcal{V}_1\Big|_{z=-g/2} \quad \text{over } (t/2-\delta/2) \le y \le (s+t/2-\delta/2) \tag{4.37}$$

$$= 1/2 \quad \text{over } (-\delta/2) \le y \le (t/2-\delta/2)$$

$$\text{and } (s+t/2-\delta/2) \le y \le (\lambda-\delta/2) \tag{4.38}$$

$$\frac{\partial \mathcal{V}_o}{\partial z}\bigg|_{z=-g/2} = \frac{\partial \mathcal{V}_1}{\partial z}\bigg|_{z=-g/2} \quad \text{over } (t/2-\delta/2) \le y \le (s+t/2-\delta/2) \tag{4.39}$$

Thus, in view of Equations 4.37, 4.38, 4.35a and 4.33, we get

$$q_o + \sum_{n=1}^{\infty} q_n \cdot \cos\left\{\frac{n2\pi}{\lambda}(y+\delta/2)\right\} = \frac{1}{2} - \sum_{m=1}^{\infty} p_m \cdot \sin\left\{\frac{m\pi}{s}\left(y-\frac{t}{2}+\delta/2\right)\right\}$$

$$\text{over } (t/2-\delta/2) \le y \le (s+t/2-\delta/2) \tag{4.40a}$$

$$\text{and} \quad q_o + \sum_{n=1}^{\infty} q_n \cdot \cos\left\{\frac{2\pi n}{\lambda}(y+\delta/2)\right\} = \frac{1}{2}$$

over $(-\delta/2) \le y \le (t/2 - \delta/2)$ and $(s + t/2 - \delta/2) \le y \le (\lambda - \delta/2)$ (4.40b)

Therefore, the Fourier coefficient q_o is found as

$$q_o = \frac{1}{\lambda}\cdot\left[\frac{1}{2}\cdot\int_{(-\delta/2)}^{(\lambda-\delta/2)} dy - \sum_{m=1}^{\infty} p_m \cdot \int_{(t/2-\delta/2)}^{(s+t/2-\delta/2)} \sin\left\{\frac{m\pi}{s}\left(y - \frac{t}{2} + \frac{\delta}{2}\right)\right\} dy\right]$$

$$\text{or,}\quad q_o = \frac{1}{2} - \frac{s}{\lambda}\cdot\sum_{M=1}^{\infty} p_M \cdot \frac{[1 - \cos(M\pi)]}{M\pi} \tag{4.41a}$$

Now, to find q_n, multiply Equations 4.40a and 4.40b, by $[(2/\lambda)\cdot\cos\{(N2\pi/\lambda)(y + \delta/2)\}]$; where N is an integer. Then integrate it over $(-\delta/2) \le y \le (\lambda - \delta/2)$. This will result in the term containing q_o on the LHS and (1/2) on the RHS to disappear, while the limits of integration for the remaining term on the RHS are modified. In the resulting expression, on replacing N by n:

$$q_n = -\frac{2}{\lambda}\cdot\sum_{m=1}^{\infty} p_m \cdot \int_{(t/2-\delta/2)}^{(s+t/2-\delta/2)} \sin\left\{\frac{m\pi}{s}\left(y - \frac{t}{2} + \frac{\delta}{2}\right)\right\}\cdot\cos\left\{\frac{N2\pi}{\lambda}(y + \delta/2)\right\} dy$$

$$\text{or,}\quad q_n = \frac{1}{2\pi}\cdot\left(\frac{\lambda}{s}\right)\cdot\sum_{M=1}^{\infty} p_M \cdot (M)\cdot\left\{\frac{\cos(n\pi\cdot(t/\lambda))}{(n)^2 - ((M/2)\cdot(\lambda/s))^2}\right\}\cdot\{1 - \cos(M\pi)\} \tag{4.41b}$$

for $n = 1,2,3, \ldots$.

In view of Equations 4.39, 4.33 and 4.36, we get

$$\sum_{m=1}^{\infty} p_m \cdot\left(\frac{m\pi}{s}\right)\cdot\sin\left\{\frac{m\pi}{s}\left(y - \frac{t}{2} + \frac{\delta}{2}\right)\right\}$$

$$= q_o \cdot \frac{2}{g} + \sum_{n=1}^{\infty} q_n \cdot\left(\frac{n2\pi}{\lambda}\right)\cdot\left[\cos\left\{\frac{n2\pi}{\lambda}\left(y - \frac{\delta}{2}\right)\right\}\cdot\operatorname{cosech}\left(\frac{2\pi n}{\lambda}\cdot g\right)\right.$$

$$\left. + \cos\left\{\frac{n2\pi}{\lambda}\left(y + \frac{\delta}{2}\right)\right\}\cdot\coth\left(\frac{2\pi n}{\lambda}\cdot g\right)\right]$$

$$\text{over}\quad (t/2 - \delta/2) \le y \le (s + t/2 - \delta/2) \tag{4.42}$$

Multiplying both sides of Equation 4.42 by $\sin\{(M\pi/s)(y - (t/2) + (\delta/2))\}$, and then integrating over $(t/2 - \delta/2) \le y \le (s + t/2 - \delta/2)$, on replacing M by m one gets on simplification:

$$p_m = q_o \cdot \left(\frac{s}{g}\right) \cdot \left(\frac{2}{m\pi}\right)^2 \cdot [1 - \cos(m\pi)] - \left(\frac{1}{\pi}\right) \cdot \left(\frac{\lambda}{s}\right) \cdot \sum_{n=1}^{\infty} q_n$$
$$\times \left[\left\{ \frac{n \cdot \operatorname{cosech}((2\pi n/\lambda) \cdot g)}{(n)^2 - ((m/2) \cdot (\lambda/s))^2} \right\} \cdot \left[\cos\left\{ n\pi \cdot \frac{(t - 2\delta)}{\lambda} \right\} - \cos(m\pi) \cdot \cos\left\{ n\pi \cdot \frac{(t + 2\delta)}{\lambda} \right\} \right] \right.$$
$$\left. + \left[\{1 - \cos(m\pi)\} \cdot \cos\left(n\pi \cdot \frac{t}{\lambda} \right) \right] \cdot \left\{ \frac{n \cdot \coth((2\pi n/\lambda) \cdot g)}{(n)^2 - ((m/2) \cdot (\lambda/s))^2} \right\} \right]$$

for $m = 1,2,3, \ldots\infty$.

If the expressions for q_o and q_n, as given respectively by Equations 4.41a and 4.41b, are substituted in this equation, on simplification we get

$$p_m = \left(\frac{s}{g}\right) \cdot \left(\frac{2}{m\pi}\right)^2 \cdot \{1 - \cos(m\pi)\} \cdot \left[\frac{1}{2} - \frac{s}{\lambda} \cdot \sum_{M=1}^{\infty} p_M \cdot \frac{[1 - \cos(M\pi)]}{M\pi} \right]$$
$$- \left(\frac{1}{\pi}\right) \cdot \left(\frac{\lambda}{s}\right) \cdot \frac{1}{2\pi} \cdot \left(\frac{\lambda}{s}\right) \cdot \sum_{M=1}^{\infty} p_M \cdot (M) \cdot \{1 - \cos(M\pi)\}$$
$$\times [\mathcal{S}_3(m, M) - \cos(m\pi) \cdot \mathcal{S}_4(m, M) + \{1 - \cos(m\pi)\} \cdot \mathcal{S}_5(m, M)] \tag{4.43}$$

for $m = 1,2,3, \ldots$.
where,

$$\mathcal{S}_3(m, M) \overset{def}{=} \sum_{n=1}^{\infty} \left\{ \frac{\cos(n\pi \cdot (t/\lambda))}{(n)^2 - ((M/2) \cdot (\lambda/s))^2} \right\} \cdot \left\{ \frac{n \cdot \operatorname{cosecth}((2\pi n/\lambda) \cdot g)}{(n)^2 - ((m/2) \cdot (\lambda/s))^2} \right\}$$
$$\times \cos\left\{ n\pi \cdot \frac{(t - 2\delta)}{\lambda} \right\} \tag{4.43a}$$

$$\mathcal{S}_4(m, M) \overset{def}{=} \sum_{n=1}^{\infty} \left\{ \frac{\cos(n\pi \cdot (t/\lambda))}{(n)^2 - ((M/2) \cdot (\lambda/s))^2} \right\} \cdot \left\{ \frac{n \cdot \operatorname{cosecth}((2\pi n/\lambda) \cdot g)}{(n)^2 - ((m/2) \cdot (\lambda/s))^2} \right\}$$
$$\times \cos\left\{ n\pi \cdot \frac{(t + 2\delta)}{\lambda} \right\} \tag{4.43b}$$

and

$$\mathcal{S}_5(m,M) \stackrel{def}{=} \sum_{n=1}^{\infty} \left\{ \frac{\cos^2(n\pi \cdot (t/\lambda))}{(n)^2 - ((M/2)\cdot(\lambda/s))^2} \right\} \cdot \left\{ \frac{n \cdot \coth((2\pi n/\lambda)\cdot g)}{(n)^2 - ((m/2)\cdot(\lambda/s))^2} \right\} \tag{4.43c}$$

The first two infinite series are fast convergent because of the cosec hyperbolic function in the numerator. These series can be truncated having retained, say the first $\mathcal{N}$ number of terms. The truncated series are given as

$$\mathcal{S}_3(m,M) \stackrel{def}{=} \sum_{n=1}^{\mathcal{N}} \left\{ \frac{\cos(n\pi \cdot (t/\lambda))}{(n)^2 - ((M/2)\cdot(\lambda/s))^2} \right\} \cdot \left\{ \frac{n \cdot \operatorname{cosech}((2\pi n/\lambda)\cdot g)}{(n)^2 - ((m/2)\cdot(\lambda/s))^2} \right\} \times \cos\left\{ n\pi \cdot \frac{(t - 2\delta)}{\lambda} \right\} \tag{4.43d}$$

and

$$\mathcal{S}_4(m,M) \stackrel{def}{=} \sum_{n=1}^{\mathcal{N}} \left\{ \frac{\cos(n\pi \cdot (t/\lambda))}{(n)^2 - ((M/2)\cdot(\lambda/s))^2} \right\} \cdot \left\{ \frac{n \cdot \operatorname{cosech}((2\pi n/\lambda)\cdot g)}{(n)^2 - ((m/2)\cdot(\lambda/s))^2} \right\} \times \cos\left\{ n\pi \cdot \frac{(t + 2\delta)}{\lambda} \right\} \tag{4.43e}$$

for $m, M = 1,2,3, \ldots$.

The third infinite series is an even function of n, and since

$$n \cdot \coth\left(\frac{2\pi n}{\lambda} \cdot g \right) \cong |n|, \quad \text{for } n \neq 0 \tag{4.43f}$$

Thus,

$$\mathcal{S}_5(m,M) \cong \sum_{n=-\infty}^{\infty} \left\{ \frac{\cos^2(n\pi \cdot (t/\lambda))}{(n)^2 - ((M/2)\cdot(\lambda/s))^2} \right\} \cdot \left\{ \frac{|n|/2}{(n)^2 - ((m/2)\cdot(\lambda/s))^2} \right\} \tag{4.43g}$$

This infinite series can be readily summed up using Poisson's summation formula:

$$\sum_{n=-\infty}^{n=\infty} f(n) = \sum \text{Residues of} [-\pi \cdot f(z) \cdot \cot(\pi z)], \quad \text{at the poles of } f(z) \tag{4.43h}$$

Comparing Equations 4.18b, 4.32b and 4.46a, we find that

$$\mathcal{S}_5(m,M) = \mathcal{S}_2(m,M) = \mathcal{S}(m,M) \tag{4.43i}$$

Thus,

$$\mathcal{S}_5(m,M) = -\frac{\pi}{2}\cdot\left[\frac{\cos^2(M\pi t/2s)\cdot\cot(M\pi\lambda/2s) - \cos^2((m\pi t/2s))\cdot\cot(m\pi\lambda/2s)}{(M\lambda/2s)^2 - (m\lambda/2s)^2}\right]$$

for $m, M = 1,2,3,\ldots$

$$\tag{4.43j}$$

In view of Equation 4.43, if the infinite series in m and M are terminated after retaining the first $\mathcal{M}$-values, one obtains $\mathcal{M}$ linearly independent algebraic simultaneous equations between equal numbers of unknowns. Numerical solution of these equations determines the unknowns, p_m (i.e., p_1, p_2, p_3, …, $p_{\mathcal{M}}$). Then, on substituting these values of p_M in Equations 4.41a and 4.41b, values for q_o and q_n can be easily found.

4.2.4 Air-Gap Permeance

The air-gap field varies periodically along the peripheral direction with a period of one tooth-pitch λ. In the absence of slotting, the air-gap field does not vary along the peripheral direction. Assuming that the potential at the air-gap surface $z = -g/2$ as $+1/2$, and at the air-gap surface $z = +g/2$ as $-1/2$, the potential distribution in the air gap can be given as

$$\mathcal{V}_o = -\frac{z}{g} \tag{4.44}$$

Thus, net flux over a tooth-pitch λ can be given as

$$\varphi_\lambda = -\mu_o \int_0^\lambda \frac{\partial \mathcal{V}_o}{\partial z}\cdot dy = \mu_o(\lambda/g) \tag{4.44a}$$

Since the potential difference between the two smooth air-gap surfaces is unity, the gap permeance P_λ, over λ is numerically equal to the flux over λ. Thus, in view of Equation 4.48a, we have

$$P_\lambda = \varphi_\lambda = \mu_o(\lambda/g) \tag{4.44b}$$

In Equation 4.36, δ is the distance between tooth-centres of two slotted equipotential surfaces as shown in Figure 4.3. The tooth-opposite-tooth orientation corresponds to δ equal to zero, while for the tooth-opposite-slot orientation the value for δ is one-half the tooth-pitch λ. In the former case, the gap permeance, P_1, is maximum. The value of permeance decreases as δ increases from zero. It reaches to a minimum value P_2, at the tooth-opposite-slot orientation. In view of Equation 4.36, the gap permeance P over λ for the double slotted air-gap surfaces is found as follows:

$$P = -\mu_o \int_0^{\lambda} \frac{\partial V_o}{\partial z} \cdot dy = \mu_o 2q_o(\lambda/g) \tag{4.45}$$

Note that both q_o and P depend on the value of the separation distance δ, between the consecutive stator and rotor tooth-centres. For tooth-opposite-tooth orientation, $\delta = 0$. This corresponds to the maximum permeance, $P|_{\delta=0} = P_1$, while, for the tooth-opposite-slot orientation, $\delta = \lambda/2$. This corresponds to the minimum permeance, $P|_{\delta=\lambda/2} = P_2$.

The reduction in flux (referred to as lost flux) due to the double-slotting, over the period λ, is given by

$$(P_\lambda - P) = \mu_o(\lambda/g) \cdot (1 - 2q_o) \stackrel{def}{=} \bar{\sigma} \cdot (s/g) \tag{4.46a}$$

where $\bar{\sigma}$ the generalised Carter's coefficient for any arbitrary value of δ is defined as

$$\bar{\sigma} = \mu_o \cdot (1 - 2q_o) \cdot \frac{\lambda}{s} \tag{4.46b}$$

And the generalised gap-extension factor ($K_{g\delta}$) is given by

$$K_{g\delta} = \frac{1}{2q_o} \tag{4.46c}$$

Figures 4.4a,b and c illustrate continuous variations of the air-gap permeance with the tooth separation δ; between a rotor-tooth axis and an adjacent stator-tooth axis for rectangular double-slotting. The separation of tooth centres varies from 0 (tooth-opposite-tooth configuration) to half a slot pitch, that is, $\lambda/2$ (tooth-opposite-slot configuration). The curves are plotted on per unit basis. The base quantity chosen is the permeance for the tooth-opposite-tooth configuration. For this configuration the permeance is the maximum, and it is the minimum for the tooth-opposite-slot configuration.

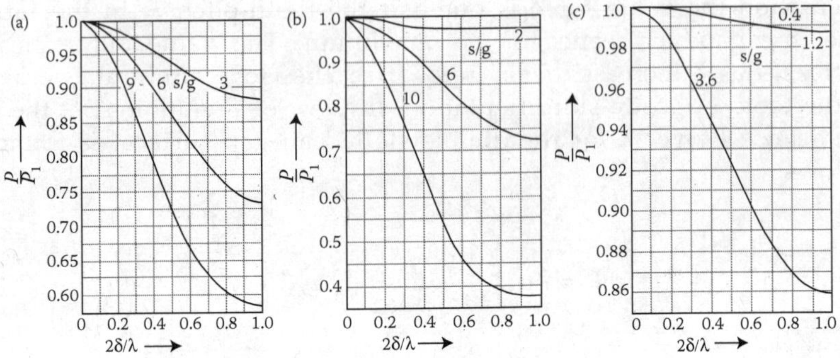

FIGURE 4.4
Variations of air-gap permeance with displacement for t/g = (a) 3, (b) 6 and (c) 1.2.

4.3 Modelling for Aperiodical Field Distributions

The field distribution in identical double-slotting discussed in Section 4.2 identifies periodic variation of air-gap field in the peripheral direction. The potential distributions on the air-gap surfaces, that is, at $z = \pm g/2$, are expressed as Fourier series of space coordinate-y with a period (wavelength) equals two times of the pole-pitch, λ. For wide slots as the slot width $s \to \infty$, the pole-pitch $\lambda \to \infty$. As a result, the Fourier series transforms into the Fourier integrals. A general treatment for such a case is given by Khan.[5,6]

4.3.1 Tooth-Opposite-Tooth Orientation

In the case of tooth-opposite-tooth, the air-gap surface $z = 0$ is a zero potential surface, $\mathcal{V} = 0$. This orientation for wide and deep slots is shown in Figure 4.5. Across this surface, the potential distribution is an odd function of z. Further, the potential distribution is an even function of y. Therefore, the potential distribution in the air gap can be expressed as

$$\mathcal{V}_o = -\frac{2}{\pi}\int_0^\infty F(u)\cdot\cos(u\cdot y)\cdot\frac{\sinh\{u\cdot z\}}{\sinh(u\cdot g/2)}\cdot du - \frac{z}{g} \tag{4.47}$$

over $-\infty \le y < \infty$.

The first term on the right-hand side is the Fourier integral representation that accounts for variation of the potential along the y direction at $z = \pm g/2$. The function $F(u)$ is the Fourier cosine transform of $(\mathcal{V}_o|_{z=-g/2} - 1/2)$, thus

$$F(u) = \int_0^\infty (\mathcal{V}_o|_{z=-g/2} - 1/2)\cdot\cos(u\cdot y)\cdot du \tag{4.48}$$

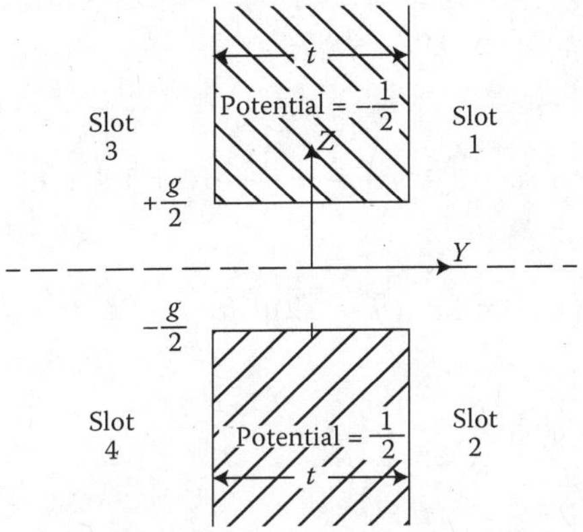

FIGURE 4.5
Tooth-opposite-tooth, wide and deep slots.

The potential distribution in the region for slot 1 can be given as

$$\mathcal{V}_1 = \frac{2}{\pi}\int_0^{\infty} f(w)\cdot \sin\{w\cdot(y-t/2)\}\cdot e^{w(z+g/2)}\cdot dw - \frac{1}{\pi}\cdot \tan^{-1}\left(\frac{z+g/2}{y-t/2}\right) \tag{4.49}$$

over $t/2 \le y < \infty$ and $z \not> -g/2$.

In Equation 4.49 the second term ensures the potential value on the iron surface at $y = t/2$, for $z < -g/2$. The first term permits a general variation in the potential distribution at the slot-opening without disturbing the potential on the iron surface. This term may be considered as a corrective term.

4.3.1.1 Evaluation of Unknown Functions

Equations 4.47 and 4.49 involve two unknown functions, viz. $F(u)$ and $f(w)$. To evaluate these functions, the following boundary conditions are to be used.

$$\begin{aligned} \mathcal{V}_0\big|_{z=-g/2} &= 1/2 \quad \text{over } 0 \le y < t/2 \\ &= \mathcal{V}_1\big|_{z=-g/2} \quad \text{over } t/2 \le y < \infty \end{aligned} \tag{4.50}$$

and

$$\left.\frac{\partial \mathcal{V}_0}{\partial z}\right|_{z=-g/2} = \left.\frac{\partial \mathcal{V}_1}{\partial z}\right|_{z=-g/2} \quad \text{over } t/2 \le y < \infty \tag{4.51}$$

In view of Equations 4.47, 4.49 and 4.50, one gets

$$\frac{2}{\pi}\int_0^\infty F(u)\cdot\cos(u\cdot y)\cdot du + \frac{1}{2} = \frac{1}{2} \quad \text{over } 0 \le y < t/2$$

$$= \frac{2}{\pi}\int_0^\infty f(w)\cdot\sin\{w\cdot(y-t/2)\}\cdot dw \quad \text{over } t/2 \le y < \infty$$

Or

$$\frac{2}{\pi}\int_0^\infty F(u)\cdot\cos(u\cdot y)\cdot du = 0 \quad \text{over } 0 \le y < t/2$$

$$= \frac{2}{\pi}\int_0^\infty f(w)\cdot\sin\{w\cdot(y-t/2)\}\cdot dw - 1/2 \quad \text{over } t/2 \le y < \infty \tag{4.52}$$

Further, on considering Equations 4.48, 4.39 and 4.51, we get

$$-\frac{2}{\pi}\int_0^\infty u\cdot F(u)\cdot\cos(u\cdot y)\cdot\coth(u\cdot g/2)\cdot du - \frac{1}{g}$$

$$= \frac{2}{\pi}\int_0^\infty w\cdot f(w)\cdot\sin\{w\cdot(y-t/2)\}\cdot dw - \frac{1}{\pi}\cdot\left(\frac{1}{y-t/2}\right)$$

over $t/2 \le y < \infty$. (4.53)

Substitution of $Y = y - t/2$ in Equation 4.53 gives

$$-\frac{2}{\pi}\int_0^\infty u\cdot F(u)\cdot\cos\{u\cdot(Y+t/2)\}\cdot\coth(u\cdot g/2)\cdot du - \frac{1}{g}$$

$$= \frac{2}{\pi}\int_0^\infty w\cdot f(w)\cdot\sin(w\cdot Y)\cdot dw - \frac{1}{\pi}\cdot\left(\frac{1}{Y}\right)$$

over $0 \le Y < \infty$. (4.54)

Equations A2.6 and A2.13 of Appendix 2 give the relations between $F(v)$ and $f(w)$ for $0 < v < \infty$. These relations renumbered as Equations 4.55 and 4.56 are

$$F(v) = \int_0^\infty f(w)\cdot\cos(w\cdot t/2)\cdot\frac{1}{2}\left\{\frac{1}{(w+v)}+\frac{1}{(w-v)}\right\}\cdot dw - f(v)\cdot\frac{\pi}{2}\cdot\sin(v\cdot t/2)$$
$$-\int_0^\infty f(w)\cdot\cos(w\cdot t/2)\frac{1}{2}\left\{\frac{1-\cos(w+v)T/2}{(w+v)}+\frac{1-\cos(w-v)T/2}{(w-v)}\right\}\cdot dw$$
$$+\int_0^\infty f(w)\cdot\sin(w\cdot t/2)\cdot\frac{1}{2}\left\{\frac{\sin(w+v)T/2}{(w+v)}+\frac{\sin(w-v)T/2}{(w-v)}\right\}\cdot dw \qquad (4.55)$$

$$\frac{\pi}{2}\cdot v\cdot f(v) = \frac{1}{2}-\frac{1}{\pi}\int_0^\infty u\cdot F(u)\cdot\coth\left(u\cdot\frac{g}{2}\right)\cdot\cos\left(u\cdot\frac{t}{2}\right)\cdot\left\{\frac{1}{(v+u)}-\frac{1}{(v-u)}\right\}\cdot du$$
$$+ v\cdot F(v)\cdot\coth\left(v\cdot\frac{g}{2}\right)\cdot\cos\left(v\cdot\frac{t}{2}\right)-\frac{1}{g}\cdot\frac{1}{v} \qquad (4.56)$$

Numerical solution of Equations 4.55 and 4.56 gives values of $F(v)$ and $f(v)$.

4.3.2 Tooth-Opposite-Slot Orientation

The tooth-opposite-slot orientation for wide and deep slots is shown in Figure 4.6. The tooth is maintained at unity potential. The air gap and the

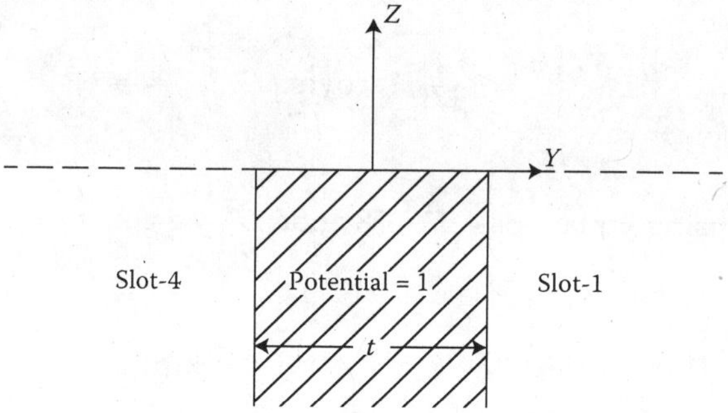

FIGURE 4.6
Tooth-opposite-slot, wide and deep slots.

slot opposite to the tooth are merged together forming an air region. The tooth is located between slots numbered as 1 and 2. The potential tends to zero as z approaches to infinity. In view of the symmetry about the z axis, the scalar magnetic potential $\mathcal{V}_o$, for the air region, $z \geq 0$, can be given as follows:

$$\mathcal{V}_o = \frac{2}{\pi}\int_0^\infty F(u) \cdot \cos(u \cdot y) \cdot e^{-uz} \cdot du \tag{4.57}$$

where F indicates an arbitrary function of u. The potential distribution in the slot 1, for $t/2 \leq y < \infty$ and $0 \geq z \geq -\infty$, can likewise be given in terms of an arbitrary function f, as shown below:

$$\mathcal{V}_1 = \frac{2}{\pi}\int_0^\infty f(w) \cdot \sin\{w \cdot (y - t/2)\} \cdot e^{wz} \cdot dw - \frac{2}{\pi} \cdot \tan^{-1}\left(\frac{z}{y - t/2}\right) \tag{4.58}$$

4.3.2.1 Evaluation of Unknown Functions

The two unknown functions, viz. $F(u)$ and $f(w)$, involved in Equations 4.57 and 4.58 can be evaluated in view of the following boundary conditions.

$$\begin{aligned} \mathcal{V}_o\big|_{z=o} &= 1 \quad \text{over } 0 \leq y < t/2 \\ &= \mathcal{V}_1\big|_{z=o} \quad \text{over } t/2 \leq y < \infty \end{aligned} \tag{4.59}$$

and

$$\left.\frac{\partial \mathcal{V}_o}{\partial z}\right|_{z=0} = \left.\frac{\partial \mathcal{V}_1}{\partial z}\right|_{z=0} \quad \text{over } t/2 \leq y < \infty \tag{4.60}$$

Therefore, using Equations 4.57, 4.58 and 4.59, we get

$$\begin{aligned} \frac{2}{\pi}\int_0^\infty F(u) \cdot \cos(u \cdot y) \cdot du &= 1 \quad \text{over } 0 \leq y < \frac{T}{2} \\ &= \frac{2}{\pi}\int_0^\infty f(w) \cdot \sin\{w \cdot (y - t/2)\} \cdot dw \quad \text{over } \frac{T}{2} \leq y < \infty \end{aligned} \tag{4.61}$$

And from Equations 4.57, 4.58 and 4.60, we get

$$-\frac{2}{\pi}\int_0^\infty u \cdot F(u) \cdot \cos(u \cdot y) \cdot du$$

$$= \frac{2}{\pi}\int_0^\infty wf(w) \cdot \sin\{w \cdot (y - t/2)\} \cdot dw - \frac{2}{\pi} \cdot \frac{1}{y - t/2}$$

$$\text{over } t/2 < y < \infty. \tag{4.62}$$

On substitution of $Y = y - t/2$, Equation 4.62 can be rewritten as

$$-\frac{2}{\pi}\int_0^\infty u \cdot F(u) \cdot \cos\{u \cdot (Y + t/2)\} \cdot du = \frac{2}{\pi}\int_0^\infty wf(w) \cdot \sin(w \cdot Y) \cdot dw - \frac{2}{\pi} \cdot \frac{1}{Y}$$

$$\text{over } 0 < Y < \infty. \tag{4.63}$$

Equations A3.4 and A3.6 of Appendix 3 give the relations between $F(v)$ and $f(v)$ for $0 < v < \infty$. These relations renumbered as Equations 4.64 and 4.65 are

$$F(v) = \frac{\sin(v \cdot T/2)}{v} - f(v) \cdot \sin(v \cdot t/2)$$

$$+ \int_0^\infty f(w) \cdot dw \left[\cos(w \cdot t/2) \cdot \left[\frac{\cos\{(w+v)(T/2)\}}{(w+v)} + \frac{\cos\{(w-v)(T/2)\}}{(w-v)} \right] \right.$$

$$\left. + \sin(w \cdot t/2) \cdot \left[\frac{\sin\{(w+v)(T/2)\}}{(w+v)} + \frac{\sin\{(w-v)(T/2)\}}{(w-v)} \right] \right] \tag{4.64}$$

$$v \cdot F(v) \cdot \sin(v \cdot t/2) - \frac{1}{\pi}\int_0^\infty u \cdot F(u) \cdot \cos(u \cdot t/2) \cdot \left\{ \frac{1}{(v+u)} + \frac{1}{(v-u)} \right\} \cdot du$$

$$= v \cdot f(v) - 1 \tag{4.65}$$

Numerical solution of Equations 4.64 and 4.65 gives values of $F(v)$ and $f(v)$.

4.3.3 Arbitrary Orientation of Two Teeth

In Section 4.4.1, we have considered the potential distribution between two identical coaxial teeth wherein each tooth is located between deep and wide slots. The modelling of potential distributions discussed in this section assumes that the axes of these two teeth are displaced by (2δ) from each other as shown in Figure 4.7. As indicated in this figure, the air space is divided into an air gap and four slot regions referred to as regions 1 to 4. This figure shows deep and wide slots around two identical teeth each of width t, separated by the air gap of length g. Let the tooth above the air gap $(z \geq g/2)$ be at a magnetic potential of -0.5 A and that below the air gap $(z \leq -g/2)$ be at $+0.5$ A. The distribution of scalar magnetic potential $\mathcal{V}$, in each air region, satisfies the Laplace equation.

In region 1, $(z > g/2)$ since the value of the magnetic potential at $y = (\delta + t/2)$ is taken as -0.5 A, we may tentatively express the potential distribution in this region as

$$\mathcal{V}_1' = -\frac{1}{\pi} \cdot \tan^{-1}\left(\frac{z - g/2}{y - \delta - t/2}\right) \tag{4.66a}$$

This expression gives the required potential at $y = (\delta + t/2)$ and also at $z = \infty$. Further, it gives zero potential at $y = \infty$. To this expression, a supplementary solution $\mathcal{V}_1''$ could be added that provides an arbitrary potential distribution at $z = g/2$, and vanishes at $y = (\delta + t/2)$ as well as at $y = \infty$. This supplementary solution must also vanish as $z \to \infty$. Therefore, using Fourier integral representation for an arbitrary function describing the potential distribution at $z = g/2$, we have

$$\mathcal{V}_1'' = -\frac{2}{\pi}\int_0^\infty f_1(w) \cdot \sin(w \cdot y_1) \cdot e^{-w(z-g/2)} \cdot dw \tag{4.66b}$$

Since $\mathcal{V}_1 = \mathcal{V}_1' + \mathcal{V}''$, we get

$$\mathcal{V}_1 = -\frac{1}{\pi} \cdot \tan^{-1}\left(\frac{z - g/2}{y - \delta - t/2}\right) - \frac{2}{\pi}\int_0^\infty f_1(w) \cdot \sin(w \cdot y_1) \cdot e^{-w(z-g/2)} \cdot dw \tag{4.67}$$

for, $z > g/2$ and $(y - \delta - t/2 \geq 0$.

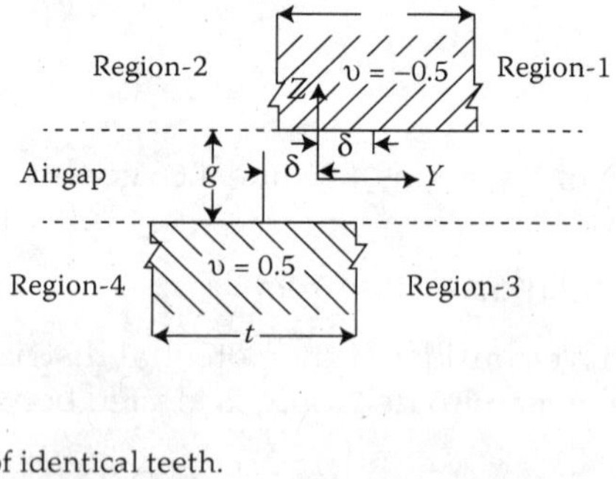

FIGURE 4.7
Air regions near a pair of identical teeth.

where the Fourier coefficient $f_1(w)$ indicates an arbitrary function. Similarly, the potential distribution for regions 2–4 can be written as

$$V_2 = \frac{1}{\pi}\cdot \tan^{-1}\left(\frac{z-g/2}{y-\delta+t/2}\right)+\frac{2}{\pi}\int_0^{\infty} f_2(w)\cdot \sin(w\cdot y_2)\cdot e^{-w(z-g/2)}\cdot dw \tag{4.68}$$

for, $z > g/2$ and $(y-\delta+t/2 \geq 0$.

$$V_3 = -\frac{1}{\pi}\cdot \tan^{-1}\left(\frac{z+g/2}{y+\delta-t/2}\right)+\int_0^{\infty} f_2(w)\cdot \sin(w\cdot y_3)\cdot e^{w(z+g/2)}\cdot dw \tag{4.69}$$

for, $z < -g/2$ and $(y-\delta-t/2 \geq 0$.
and

$$V_4 = \frac{1}{\pi}\cdot \tan^{-1}\left(\frac{z+g/2}{y+\delta+t/2}\right)-\int_0^{\infty} f_1(w)\cdot \sin(w\cdot y_4)\cdot e^{w(z+g/2)} \tag{4.70}$$

for, $z < -g/2$ and $(y+\delta+t/2) \geq 0$.

In Equations 4.67, 4.68, 4.69 and 4.70

$$y_1 = y-\delta-t/2 \tag{4.71a}$$

$$y_2 = y-\delta+t/2 \tag{4.71b}$$

$$y_3 = y+\delta-t/2 \tag{4.71c}$$

and

$$y_4 = y+\delta+t/2 \tag{4.71d}$$

The distribution of scalar magnetic potential in the air-gap region can be expressed in terms of four arbitrary functions, $F_1'(u), F_1''(u), F_2'(u)$ and $F_2''(u)$ as

$$V_0 = \int_0^{\infty}\left[\{F_1'(u)\cdot\cos(u\cdot y)+F_1''(u)\cdot\sin(u\cdot y)\}\cdot\frac{\sinh\{u\cdot(z+g/2)\}}{\sinh(u\cdot g)} - \{F_2'(u)\cdot\cos(u\cdot y)+F_2''(u)\cdot\sin(u\cdot y)\}\cdot\frac{\sinh\{u\cdot(z-g/2)\}}{\sinh(u\cdot g)}\right]\cdot du \tag{4.72}$$

4.3.3.1 Evaluation of Unknown Functions

To find values of potentials $\mathcal{V}_1, \mathcal{V}_2, \mathcal{V}_3, \mathcal{V}_4$ and $\mathcal{V}_o$, the arbitrary functions $f_1(w)$, $f_2(w)$, $F_1'(u), F_1''(u), F_2'(u)$ and $F_2''(u)$ involved in Equations 4.67 through 4.70 and 4.72 are to be evaluated. For evaluation of these functions, first consider the values of $\mathcal{V}_o$ given by Equation 4.72 at $z = g/2$ and $z = -g/2$. These are

$$\mathcal{V}_o\Big|_{z=g/2} = \int_0^\infty \{F_1'(u)\cdot\cos(u\cdot y) + F_1''(u)\cdot\sin(u\cdot y)\}\cdot du \tag{4.73a}$$

$$\mathcal{V}_o\Big|_{z=-g/2} = \int_0^\infty \{F_2'(u)\cdot\cos(u\cdot y) + F_2''(u)\cdot\sin(u\cdot y)\}\cdot du \tag{4.73b}$$

In view of the anti-symmetry:

$$\mathcal{V}_o(y,z)\Big|_{z=-g/2} = -\mathcal{V}_o(-y,z)\Big|_{z=g/2} \tag{4.74}$$

Thus,

$$F_1'(u) = -F_2'(u) \stackrel{def}{=} F'(u) \tag{4.74a}$$

and

$$F_1''(u) = -F_2''(u) \stackrel{def}{=} F''(u) \tag{4.74b}$$

Here, it is to be noted that in view of Equations 4.74a and 4.74b the number of arbitrary functions reduces from six to four viz. $f_1(w)$, $f_2(w)$, $F'(u)$ and $F''(u)$. Thus, Equation 4.72 can be modified to

$$\mathcal{V}_o = \int_0^\infty \left[F'(u)\cdot\cos(u\cdot y)\cdot\left\{\frac{\sinh(u\cdot z)}{\sinh(u\cdot g/2)}\right\} + F''(u)\cdot\sin(u\cdot y) \times\left\{\frac{\cosh(u\cdot z)}{\cosh(u\cdot g/2)}\right\}\right]\cdot du \tag{4.75}$$

Equation 4.73a can also be rewritten as

$$\mathcal{V}_o\big|_{z=g/2} = \int_0^{\infty} \{F'(u)\cdot\cos(u\cdot y) + F''(u)\cdot\sin(u\cdot y)\}\cdot du \tag{4.75a}$$

Continuity of potentials: In view of the continuity of potentials at $z = g/2$, we can write

$$\begin{aligned}\mathcal{V}_o\big|_{z=g/2} &= \mathcal{V}_2\big|_{z=g/2} \quad \text{over } -\infty \le y \le -\delta \\ &= 1/2 \quad \text{over } -\delta \le y \le \delta \\ &= \mathcal{V}_1\big|_{z=g/2} \quad \text{over } \delta \le y \le \infty\end{aligned} \tag{4.76}$$

Therefore, using Equations 4.67, 4.68, 4.75a and 4.76, we get

$$\begin{aligned}&\int_0^{\infty} \{F'(u)\cdot\cos(u\cdot y) + F''(u)\cdot\sin(u\cdot y)\}\cdot du \\ &= \int_0^{\infty} f_2(w)\cdot\sin(w\cdot y_2)\cdot dw \quad \text{over } -\infty \le y \le (\delta - t/2) \\ &= -1/2 \quad \text{over } (\delta - t/2) \le y \le (\delta + t/2) \\ &= -\int_0^{\infty} f_1(w)\cdot\sin(w\cdot y_1)\cdot dw \quad \text{over } (\delta + t/2) \le y \le \infty\end{aligned} \tag{4.76a}$$

Appendix 4, Part A and Part B leads to Equation A4.5 (for all $v > 0$), which is renumbered as Equation 4.77a and is given below:

$$\begin{aligned}\pi\cdot F'(v) = &-\cos\{v\cdot(\delta - t/2)\}\cdot\int_0^{\infty} f_2(w)\cdot\left[\frac{w}{(w^2 - v^2)}\right]\cdot dw \\ &-f_2(v)\cdot\frac{\pi}{2}\cdot\sin\{v\cdot(\delta - t/2)\} - \left[\frac{\sin\{v\cdot(\delta + t/2)\} - \sin\{v\cdot(\delta - t/2)\}}{2v}\right] \\ &-\cos\{v\cdot(\delta + t/2)\}\cdot\int_0^{\infty} f_1(w)\left[\frac{w}{(w^2 - v^2)}\right]\cdot dw + f_1(v)\cdot\frac{\pi}{2} \\ &\times\sin\{v\cdot(\delta + t/2)\}\end{aligned} \tag{4.77a}$$

Appendix 4, Part C and Part D leads to Equation A4.10 (for all $v > 0$), which is renumbered as Equation 4.77b as given below:

$$\pi \cdot F''(v) = f_2(v) \cdot \frac{\pi}{2} \cdot \cos\{v \cdot (\delta - t/2)\} - f_1(v) \cdot \frac{\pi}{2} \cdot \cos\{v \cdot (\delta + t/2)\}$$
$$- \int_0^\infty f_2(w) \cdot \frac{1}{2} \cdot \left[\frac{\sin\{(2w - v) \cdot (\delta - t/2)\}}{(w - v)} - \frac{\sin\{(2w + v) \cdot (\delta - t/2)\}}{(w + v)} \right] \cdot dw$$
$$+ \frac{1}{2} \cdot \left[\frac{\cos\{v \cdot (\delta + t/2)\} - \cos\{v \cdot (\delta - t/2)\}}{v} \right]$$
$$+ \int_0^\infty f_1(w) \cdot \frac{1}{2} \cdot \left[\frac{\sin\{(2w - v) \cdot (\delta + t/2)\}}{(w - v)} - \frac{\sin\{(2w + v) \cdot (\delta + t/2)\}}{(w + v)} \right] \cdot dw \quad (4.77b)$$

Continuity of derivatives of potentials: In view of the continuity of derivatives of potentials at $z = g/2$:

$$\left.\frac{\partial \mathcal{V}_1}{\partial z}\right|_{z=g/2} = \left.\frac{\partial \mathcal{V}_o}{\partial z}\right|_{z=g/2} \quad \text{over } (\delta + t/2) \le y < \infty \quad (4.78a)$$

$$\left.\frac{\partial \mathcal{V}_2}{\partial z}\right|_{z=g/2} = \left.\frac{\partial \mathcal{V}_o}{\partial z}\right|_{z=g/2} \quad \text{over } -\infty < y \le (\delta - t/2) \quad (4.78b)$$

In view of Equation 4.67, the z derivative at $z = g/2$ is

$$\frac{\partial \mathcal{V}_1}{\partial z} = -\frac{1}{\pi} \cdot \frac{1}{1 + ((z - g/2)/(y - \delta - t/2))^2} \cdot \frac{1}{(y - \delta - t/2)}$$
$$+ \int_0^\infty f_1(w) \cdot w \cdot \sin(w \cdot y_1) \cdot e^{-w(z-g/2)} \cdot dw \quad (4.79a)$$

$$\left.\frac{\partial \mathcal{V}_1}{\partial z}\right|_{z=g/2} = -\frac{1}{\pi} \cdot \frac{1}{(y - \delta - t/2)} + \int_0^\infty f_1(w) \cdot w \cdot \sin(w \cdot y_1) \cdot dw \quad (4.79b)$$

In view of Equation 4.68, the z derivative at $z = g/2$ is

$$\frac{\partial \mathcal{V}_2}{\partial z} = \frac{1}{\pi} \cdot \frac{1}{1 + ((z - g/2)/(y - \delta + t/2))^2} \cdot \frac{1}{(y - \delta + t/2)}$$

$$- \int_0^\infty f_2(w) \cdot w \cdot \sin(w \cdot y_2) \cdot e^{-w(z-g/2)} \cdot dw \tag{4.80a}$$

$$\left.\frac{\partial \mathcal{V}_2}{\partial z}\right|_{z=g/2} = \frac{1}{\pi} \cdot \frac{1}{(y - \delta + t/2)} - \int_0^\infty f_2(w) \cdot w \cdot \sin(w \cdot y_2) \cdot dw \tag{4.80b}$$

In view of Equation 4.72, the z derivative at $z = g/2$ is

$$\frac{\partial \mathcal{V}_0}{\partial z} = \int_0^\infty \left[F'(u) \cdot \cos(u \cdot y) \cdot \frac{u \cdot \cosh(u \cdot z)}{\sinh(u \cdot g/2)} + F''(u) \cdot \sin(u \cdot y) \right.$$

$$\left. \times \frac{u \cdot \sinh(u \cdot z)}{\cosh(u \cdot g/2)} \right] \cdot du \tag{4.81a}$$

$$\left.\frac{\partial \mathcal{V}_0}{\partial z}\right|_{z=g/2} = \int_0^\infty [F'(u) \cdot \coth(u \cdot g/2) \cdot \cos(u \cdot y)$$

$$+ F''(u) \cdot \tanh(u \cdot g/2) \cdot \sin(u \cdot y)] \cdot u \cdot du \tag{4.81b}$$

On equating Equations 4.79b, 4.81b and 4.71a, we get

$$-\frac{1}{\pi} \cdot \frac{1}{(y - \delta - t/2)} + \int_0^\infty f_1(w) \cdot w \cdot \sin\{w \cdot (y - \delta - t/2)\} \cdot dw$$

$$= \int_0^\infty [F'(u) \cdot \coth(u \cdot g/2) \cdot \cos(u \cdot y) + F''(u) \cdot \tanh(u \cdot g/2) \cdot \sin(u \cdot y)] \cdot u \cdot du \tag{4.82a}$$

over $(\delta + t/2) < y < \infty$.

Substitution of $Y = (y - \delta - t/2)$, or $y = (Y + \delta + t/2)$ in Equation 4.82a gives

$$-\frac{1}{\pi}\cdot\frac{1}{Y}+\int_0^\infty f_1(w)\cdot w\cdot\sin(w\cdot Y)\cdot dw$$

$$= \int_0^\infty F'(u)\cdot\coth(u\cdot g/2)\cdot[\cos(u\cdot Y)\cdot\cos\{u\cdot(\delta+t/2)\}$$
$$-\sin(u\cdot Y)\cdot\sin\{u\cdot(\delta+t/2)\}]$$
$$+F''(u)\cdot\tanh(u\cdot g/2)\cdot[\sin(u\cdot Y)\cdot\cos\{u\cdot(\delta+t/2)\}$$
$$+\cos(u\cdot Y)\cdot\sin\{u\cdot(\delta+t/2)\}]\cdot u\cdot du \tag{4.82b}$$

over $0 \le Y < \infty$.

Similarly, on equating Equations 4.80b, 4.81b and 4.71b, we get

$$\frac{1}{\pi}\cdot\frac{1}{(y-\delta+t/2)}-\int_0^\infty f_2(w)\cdot w\cdot\sin\{w\cdot(y-\delta+t/2)\}\cdot dw$$

$$= \int_0^\infty [F'(u)\cdot\coth(u\cdot g/2)\cdot\cos(u\cdot y)+F''(u)\cdot\tanh(u\cdot g/2)\cdot\sin(u\cdot y)]\cdot u\cdot du \tag{4.83a}$$

over $-\infty < y \le (\delta - t/2)$.

Substitution of $Y = (y - \delta + t/2)$, or $y = (Y + \delta - t/2)$, in Equation 4.83a gives

$$\frac{1}{\pi}\cdot\frac{1}{Y}-\int_0^\infty f_2(w)\cdot w\cdot\sin(w\cdot Y)\cdot dw$$

$$= \int_0^\infty \Big[F'(u)\cdot\coth(u\cdot g/2)\cdot[\cos(u\cdot Y)\cdot\cos\{u\cdot(\delta-t/2)\}$$
$$-\sin(u\cdot Y)\cdot\sin\{u\cdot(\delta-t/2)\}]$$
$$-F''(u)\cdot\tanh(u\cdot g/2)\cdot\big[\sin(u\cdot Y)\cdot\cos\{u\cdot(\delta-t/2)\}$$
$$+\cos(u\cdot Y)\cdot\sin\{u\cdot(\delta-t/2)\}\big]\Big]\cdot u\cdot du \tag{4.83b}$$

over $0 < Y < \infty$.

Appendix 4, Part E leads to Equations A4.15 and A4.16 (for all $v > 0$). These equations renumbered as Equations 4.84a and 4.84b are given below:

$$
\begin{aligned}
-\frac{1}{2}+\frac{\pi}{2}\cdot v\cdot f_1(v) &= \int_0^{\infty}[F'(u)\cdot\coth(u\cdot g/2)\cdot\cos\{u\cdot(\delta+t/2)\}\\
&\quad + F''(u)\cdot\tanh(u\cdot g/2)\cdot\sin\{u\cdot(\delta+t/2)\}]\cdot\left\{\frac{v\cdot u}{(v^2-u^2)}\right\}\cdot du\\
&\quad - F'(v)\cdot\coth(v\cdot g/2)\cdot\sin\{v\cdot(\delta+t/2)\}\cdot\frac{\pi}{2}\\
&\quad + F''(v)\cdot\tanh(v\cdot g/2)\cdot\cos\{u\cdot(\delta+t/2)\}\cdot\frac{\pi}{2}
\end{aligned}
\tag{4.84a}
$$

$$
\begin{aligned}
\frac{1}{2}-\frac{\pi}{2}\cdot v\cdot f_2(v) &= \int_0^{\infty}[F'(u)\cdot\coth(u\cdot g/2)\cdot\cos\{u\cdot(\delta-t/2)\}\\
&\quad - F''(u)\cdot\tanh(u\cdot g/2)\cdot\sin\{u\cdot(\delta-t/2)\}]\cdot\left\{\frac{v\cdot u}{(v^2-u^2)}\right\}\cdot du\\
&\quad - F'(v)\cdot v\cdot\coth(v\cdot g/2)\cdot\sin\{v\cdot(\delta-t/2)\}\cdot\frac{\pi}{2}\\
&\quad - F''(v)\cdot v\cdot\tanh(v\cdot g/2)\cdot\cos\{v\cdot(\delta-t/2)\}\cdot\frac{\pi}{2}
\end{aligned}
\tag{4.84b}
$$

Numerical solution of Equations 4.77a, 4.77b, 4.84a and 4.84b gives the values of all arbitrary functions describing potential distributions in various regions.

4.4 Fringing Flux for Tooth-Opposite-Tooth Orientation with Small Air Gap

Figure 4.8a illustrates the configuration of tooth-opposite-tooth orientation with small air gap. For machines with very small air gaps, one may assume that in the case of tooth-opposite-tooth orientation, only one edge of the opposing teeth needs to be considered. In view of this figure as $g \ll t$ and also $g \ll s$, both the tooth width t and the slot width s can be treated as semi-infinite. In view of symmetry, we may ignore, say, the bottom half of the configuration. The resulting figure is shown as Figure 4.8b wherein the plane of symmetry is at zero potential and the surface of the tooth located above this plane is at

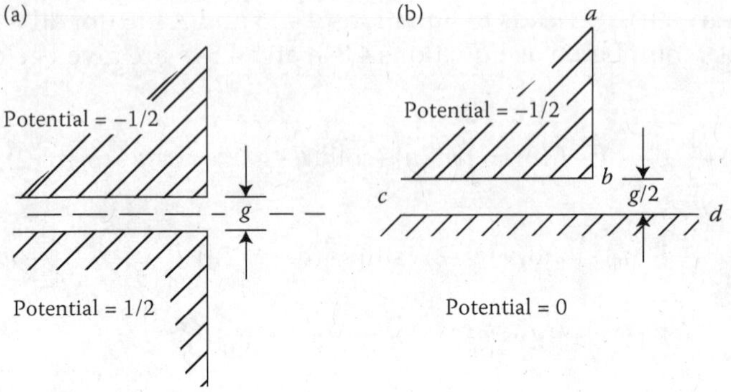

FIGURE 4.8
Equipotential surfaces for (a) symmetrically opposite teeth and (b) one tooth and the plane of symmetry.

minus half a potential. In this figure, the flux lines are to emanate from the plane of symmetry cd and terminate on the equipotential surface abc.

4.4.1 Schwarz–Christoffel Transformation

In Figure 4.8b, there is only one right angle and our objective is to determine the equipotentials and flux lines. Therefore, in this two-dimensional problem involving the solution of the Laplace equation, we can opt for conformal mapping[4] using Schwarz–Christoffel transformation. As per the requirement this transformation is carried out in two steps, that is, from z plane to w plane and from χ plane to w plane.

4.4.1.1 Transformation from z Plane to w Plane

As can be seen from Figure 4.9 the real and imaginary axes of z-plane are represented by x and y axes, and the real and imaginary axes of w-plane by u and v axes, respectively. For transforming z-plane (Figure 4.9a) to w-plane (Figure 4.9b), the real axis in z-plane is to be mapped on the positive real axis in w-plane, while the tooth contour in the z-plane is to be mapped on the negative real axis in w-plane. This calls for placing the real axis in z-plane on the positive part of the real axis in w-plane. Further, straightening the configuration in z-plane and the tooth contour on the z-plane is to be mapped on the negative part of the real axis in w-plane. On z-plane the convenient points $w = 0$, $w = -\infty$ and $w = \infty$ are shown in Figure 4.9a. The resulting points on w-plane for $z = j\omega$, $jg/2$ and ∞ are shown in Figure 4.9b. In pulling out the configuration to a straight line two angles are to be straightened. These angles are shown in Figure 4.9a as α and β.

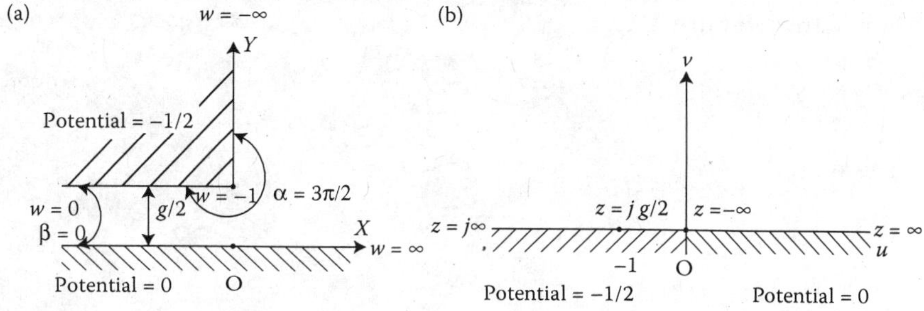

FIGURE 4.9
(a) *z*-plane and (b) *w*-plane.

The Schwarz–Christoffel transformation from *z*-plane to *w*-plane is performed through the relation:

$$\frac{dz}{dw} = A\,(w-a)^{(\alpha-\pi/\pi)} \cdot (w-b)^{(\beta-\pi/\pi)} \tag{4.85}$$

where a and b are the locations of the two internal angles α and β in the *w*-plane. Therefore,

$$a = -1, \quad \alpha = 3\pi/2 \tag{4.86a}$$

and

$$b = 0, \quad \beta = 0 \tag{4.86b}$$

Inserting these values in Equation 4.85, we get

$$dz = A \cdot \frac{(w+1)^{1/2}}{w} \cdot dw \tag{4.87}$$

On integrating, we have

$$z = A \cdot \left[2(w+1)^{1/2} + \log\left\{ \frac{(w+1)^{1/2} - 1}{(w+1)^{1/2} + 1} \right\} \right] + C \tag{4.88}$$

where C indicates the constant of integration. For the origin in the *z*-plane, shown in Figure 4.9a, the value of C is zero. On setting $w = -1$ this equation results:

$$z\big|_{w=-1} = A \cdot \log(-1) = A \cdot j\pi \tag{4.88a}$$

While from Figure 4.9a

$$z\big|_{w=-1} = j\,(g/2) \tag{4.88b}$$

Therefore,

$$A = \frac{g}{2\pi} \tag{4.89}$$

Giving the transformation relation as

$$z = \frac{g}{2\pi}\cdot\left[2(w+1)^{1/2} + \log\left\{\frac{(w+1)^{1/2}-1}{(w+1)^{1/2}+1}\right\}\right] \tag{4.90}$$

In the w-plane, the positive and negative parts of the real (or u) axis are at different equipotential values; that is, zero for positive u and –1/2 for negative u. Therefore, flux lines in this plane are semicircles.

4.4.1.2 Transformation from χ Plane to w Plane

Next, consider the χ $(= \varphi + j\psi)$-plane shown in Figure 4.10. In this plane, the equipotential surfaces are parallel to $\psi = 0$ (*or* 1) plane, while flux lines are parallel to the ψ-axis. The value of potential varies linearly with the distance from $\psi = 0$ plane. This plane is, therefore, called *regular field plane*. The convenient values of w ($-\infty$, –1, 0 and ∞) on this plane are shown in Figure 4.10.

The transformation of χ-plane into w-plane requires straightening only one internal angle, namely, $\alpha = 0$. Therefore,

$$\frac{d\chi}{dw} = A\,(w-a)^{(\alpha-\pi/\pi)} \tag{4.91}$$

where $a = 0$, and $\alpha = 0$.

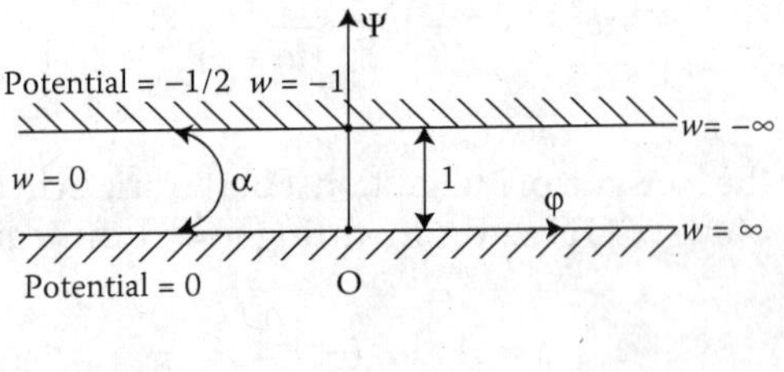

FIGURE 4.10
Regular field plane.

Thus,

$$d\chi = \frac{A}{w} dw \tag{4.92}$$

On integrating, we get

$$\chi = A \log(w) \tag{4.93a}$$

Or

$$w = e^{(\chi/A)} \tag{4.93b}$$

Since, as shown in Figure 4.10 $w = -1$ at $\chi = j$, thus

$$A = 1/\pi \tag{4.94}$$

Therefore,

$$w = e^{\pi \cdot \chi} \tag{4.95}$$

Substituting in Equation 4.92, we get

$$z = \frac{g}{2\pi} \cdot \left[2(e^{\pi \cdot \chi} + 1)^{1/2} + \log \left\{ \frac{(e^{\pi \cdot \chi} + 1)^{1/2} - 1}{(e^{\pi \cdot \chi} + 1)^{1/2} + 1} \right\} \right] \tag{4.96}$$

Since $z = x + jy$ and $\chi = \varphi + j\psi$, plot for a flux line can be obtained by choosing a constant value for φ, and joining points whose coordinates in the z-plane are found from this equation for different values of ψ. The procedure could be repeated for different values of φ, each value results in a different flux line. Similarly, plots for equipotential lines can be obtained by interchanging the roles of φ and ψ.

4.5 Air-Gap Field of a Conductor Deep inside an Open Slot

In the configuration shown in Figure 4.11a, the equipotential and flux lines are to be obtained in the air gap of a smooth rotor machine due to a current-carrying conductor placed deep inside a stator slot. Since the two parts of the stator iron are assigned scalar magnetic potentials of $(I/2)$ and $-(I/2)$, the presence of conductor in the slot can be ignored. By virtue of symmetry

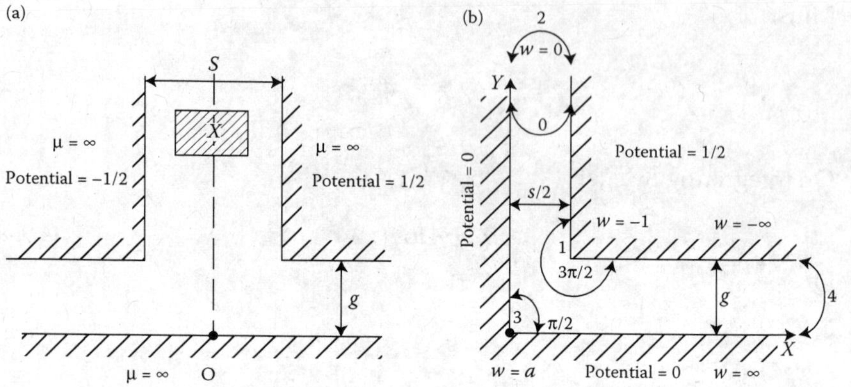

FIGURE 4.11
(a) Showing a current-carrying conductor in a deep slot and (b) right half of the above figure in z-plane with the conductor replaced by equipotential surfaces.

entailed in this figure, it is noted that both the rotor iron and the plane of symmetry must be at zero magnetic potential. Therefore, as shown in Figure 4.11b, only right-half of Figure 4.11a needs to be considered.

4.5.1 Schwarz–Christoffel Transformation

As this problem involves two right angles, it is convenient to use Schwarz–Christoffel transformation. As in Section 4.5, this transformation is carried out in two steps, that is, from z plane to w plane and from χ plane to w plane.

4.5.1.1 Transformation from z-Plane to w-Plane

Figure 4.11b shows the z-plane wherein some values of w, viz. $w = -\infty, -1, 0, a$ and ∞, are indicated. At this state one value of w, viz. $w = a$, remains unspecified as this problem involves one dimension-less pair, that is, $g/(s/2)$. With these values of w and noting various angles in the figure, the equation for Schwarz–Christoffel transformation is given as

$$\frac{dz}{dw} = A \cdot (w - a)^{(\pi/2 - \pi/\pi)} \cdot (w - 0)^{(0 - \pi/\pi)} \cdot (w + 1)^{(3\pi/2 - \pi/\pi)} \tag{4.97}$$

Therefore,

$$dz = A \cdot \sqrt{\frac{w + 1}{w - a}} \cdot \frac{dw}{w} \tag{4.98}$$

On integrating, we get

$$z = A\cdot\left[-\frac{2}{\sqrt{a}}\cdot\tan^{-1}\left\{\sqrt{a}\cdot\sqrt{\frac{w+1}{w-a}}\right\}+\log\left\{\frac{\sqrt{w-a}+\sqrt{w+1}}{\sqrt{w-a}-\sqrt{w+1}}\right\}\right]+C \tag{4.99}$$

In this expression we have three unknowns, *A*, *a* and the constant of integration *C*. To evaluate these consider a contour comprising a semicircle of radius *R* and a straight line indenting the points at $w = -1$, 0, and *a*, in the *w*- plane using small semicircles of radius ρ shown in Figure 4.12.

To perform integration over the semicircle as $R \to \infty$, set $w = Re^{j\theta}$ and let $R \to \infty$. Equate this to the corresponding integration in the *z*-plane,

$$\int_{j0}^{jg} dz = \lim_{R\to\infty}\int_0^{\pi} A\cdot\sqrt{\frac{Re^{j\theta}+1}{Re^{j\theta}-a}}\cdot\frac{jRe^{j\theta}}{Re^{j\theta}}d\theta \tag{4.100}$$

It gives

$$jg = \int_0^{\pi} A\cdot j\,d\theta = A\cdot j\pi \quad \text{or} \quad A = g/\pi \tag{4.101}$$

To perform integration over the semicircle as $\rho \to 0$ around the origin of the *w*-plane, set $w = \rho e^{j\theta}$ and let $\rho \to 0$. Equate this to the corresponding integration in the *z*-plane,

$$\int_{s/2}^{0} dz = \lim_{\rho\to 0}\int_{\pi}^{0} A\cdot\sqrt{\frac{\rho e^{j\theta}+1}{\rho e^{j\theta}-a}}\cdot\frac{j\rho e^{j\theta}}{\rho e^{j\theta}}d\theta \tag{4.102}$$

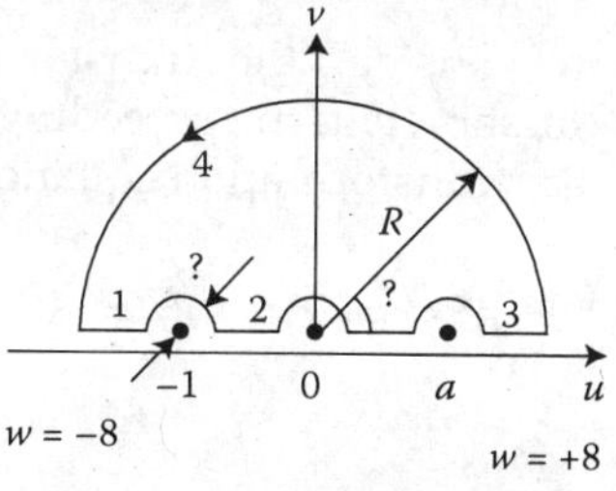

FIGURE 4.12
A contour in *w*-plane.

Thus,

$$s/2 = A \cdot (\pi/\sqrt{a})$$

Or,

$$a = \left(A \cdot \frac{\pi}{s/2}\right)^2 = (2g/s)^2 \tag{4.102a}$$

Lastly, to find C, consider point-1 on the z-plane, which is the same as the point $w = -1$ on the w-plane. In view of Figure 4.11b, we get from Equation 4.99:

$$z = \frac{s}{2} + jg = A \cdot \left[-\frac{2}{\sqrt{a}} \cdot \tan^{-1}\left\{\sqrt{a} \cdot \sqrt{0}\right\} + \log\left\{\frac{\sqrt{w-a} + \sqrt{0}}{\sqrt{w-a} - \sqrt{0}}\right\}\right] + C \tag{4.103}$$

Or,

$$\frac{s}{2} + jg = A \cdot [0] + C$$

Or,

$$C = (s/2) + jg \tag{4.103a}$$

Thus, from Equation 4.99

$$z = \frac{g}{\pi} \cdot \left[-\frac{s}{g} \cdot \tan^{-1}\left\{\frac{2g}{s} \cdot \sqrt{\frac{w+1}{w-(2g/s)^2}}\right\} + \log\left\{\frac{\sqrt{w-(2g/s)^2} + \sqrt{w+1}}{\sqrt{w-(2g/s)^2} - \sqrt{w+1}}\right\}\right] + \left\{\frac{s}{2} + jg\right\} \tag{4.104}$$

4.5.1.2 Transformation from χ-Plane to w-Plane

To complete the solution, consider the transformation from χ-plane to w-plane. For the Figure 4.10, shown in the preceding section (with the potential –1/2 changed to $I/2$), the transformation equation is:

$$\frac{d\chi}{dw} = A' \cdot (w-0)^{(0-\pi/\pi)} = A'/w \tag{4.105}$$

Or,

$$d\chi = A' \cdot \frac{dw}{w} \tag{4.105a}$$

Consider the semicircle of radius R in Figure 4.12. It is seen from this figure that as θ changes from 0 to π, w changes from ∞ to $-\infty$! While from Figure 4.11b, the corresponding change in the potential is from 0 to $I/2$. Therefore, on setting $w = Re^{j\theta}$ and then integrating Equation 4.105a we get

$$\int_{\varphi=0}^{I/2} d\chi = A' \cdot \int_{\theta=0}^{\pi} \frac{jRe^{j\theta}}{Re^{j\theta}} d\theta \tag{4.105b}$$

Equation 4.105b gives

$$\frac{I}{2} = jA' \cdot \pi \quad \text{or} \quad A' = -j\frac{I}{2\pi} \tag{4.105c}$$

The integration of Equation 4.105a results

$$\chi = A' \cdot \log(w) + B' \tag{4.105d}$$

At point-1 on z-plane, which is same as the point $w = -1$ on the w-plane, $\varphi = I/2$.

Thus,

$$\frac{I}{2} = -j\frac{I/2}{\pi}\log(-1) + B' \quad \text{or} \quad B' = 0$$

Thus,

$$\chi = -j\frac{I}{2\pi} \cdot \log(w) \tag{4.106a}$$

Or,

$$w = e^{j(\chi 2\pi/I)} = e^{((-\psi+j\varphi)2\pi/I)} \tag{4.106b}$$

Or,

$$w = e^{-(\psi 2\pi/I)} \cdot \left[\cos\left(\frac{\varphi 2\pi}{I}\right) + j\sin\left(\frac{\varphi 2\pi}{I}\right)\right] \tag{4.107}$$

The final result can be obtained by substituting Equation 4.107 on the RHS of Equation 4.104.

As stated in the preceding section, plot for a flux line can be obtained by choosing a constant value for φ, and joining points whose coordinates in the

z-plane are found from Equations 4.104 and 4.107 for different values of ψ. The procedure could be repeated for different values of φ, each value results in a different flux line. Similarly, plots for equipotential lines can be obtained by interchanging the roles of φ and ψ.

4.6 Magnetic Field near Armature Winding Overhang

Leakage inductance for armature winding of induction machines consists of a number of parts, viz. slot leakage inductance, air-gap leakage inductance, end-winding leakage inductance and so on. In the study of induction machine performance, values of these inductances are required. The end-winding leakage inductance is equal to the flux linking with the end-winding per unit current in the armature winding. Thus, to determine the end-winding leakage inductance, one needs to find the distribution of magnetic field near the axial ends of the machine. In this section, we shall study the distribution of this magnetic field.

End covers are provided at the two axial ends of most rotating electrical machines. On these end covers, bearings supporting the rotor shaft are fixed. These end covers are usually made of cast iron and form parts of the enclosure housing laminated stator and rotor cores with windings. Figure 4.13 shows developed view of air space near an axial end of an induction machine rotating at synchronous speed. If the iron loss in the laminated stator core is neglected, the armature winding will carry magnetising current. This is simulated by a suitable current sheet with surface current density K. The skewing of the stator slots is neglected in this treatment. Therefore, in region 3 this current density has only axial component, while at the boundary between regions 1 and 4 the surface current density simulating overhang part of the armature winding has a peripheral component as well. Distribution of magnetic field due to a known current sheet is found neglecting tangential

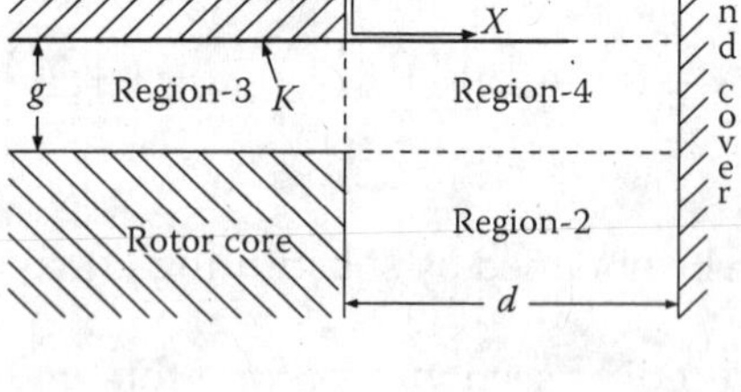

FIGURE 4.13
Axial end regions of an induction machine.

components of magnetic field on various iron surfaces. The magnetic field in each of the four regions shown in Figure 4.13 varies periodically in the peripheral direction with a period equal to two pole-pitches, that is, 2λ. Further, it also varies sinusoidally in time with supply frequency, ω radian per second.

4.6.1 Surface Current Density

Figure 4.13 shows the Cartesian coordinates system with the reference frame, which is stationary relative to the stator. In this system x- indicates axial, y- peripheral and z- radial direction. The surface current density, K, simulating the current carrying armature winding varies as $e^{j((m\pi/\lambda)\cdot y-\omega t)}$ and its divergence is zero. It is given as

$$\mathbf{K} = K_x \mathbf{a}_x \quad \text{over } x \leq 0 \tag{4.108}$$

where

$$K_x = \sum_{m=1}^{\infty} k_m \cdot e^{j((m\pi/\lambda)\cdot y-\omega t)} \tag{4.108a}$$

Further,

$$\mathbf{K} = K_{ox}\boldsymbol{a}_x + K_{oy}\boldsymbol{a}_y \quad \text{over } 0 \leq x \leq d \tag{4.109}$$

where

$$K_{ox} = \sum_{m=1}^{\infty}\sum_{n-odd}^{\infty} \ell_{m\cdot n} \cdot \left(j\frac{m\pi}{\lambda}\right) \cdot \cos\left(\frac{n\pi}{2d}\cdot x\right) \cdot e^{j((m\pi/\lambda)\cdot y-\omega t)} \tag{4.109a}$$

and

$$K_{oy} = \sum_{m=1}^{\infty}\sum_{n-odd}^{\infty} \ell_{m\cdot n} \cdot \left(\frac{n\pi}{2d}\right) \cdot \sin\left(\frac{n\pi}{2d}\cdot x\right) \cdot e^{j((m\pi/\lambda)\cdot y-\omega t)} \tag{4.109b}$$

where, the coefficient for the double Fourier series $\ell_{m.n}$ is known from the armature winding details. The coefficient k_m can be given as

$$k_m = \left(j\frac{m\pi}{\lambda}\right) \cdot \sum_{n-odd}^{\infty} \ell_{m.n} = a_m \tag{4.109c}$$

In each of the following field expressions, the factor $e^{j((m\pi/\lambda)\cdot y-\omega t)}$ and the summation sign $\Sigma_{m=1}^{\infty}$ are suppressed though their presence is recognised.

4.6.2 Magnetic Field Intensity

Figure 4.13 illustrates four air regions all of which are taken to be current-free. Therefore, for each region the curl of magnetic field intensity is zero. Its divergence is also zero, that is,

$$\nabla \times \boldsymbol{H} = 0 \tag{4.110a}$$

$$\nabla \bullet \boldsymbol{H} = 0 \tag{4.110b}$$

Since

$$\nabla \times \nabla \times \boldsymbol{H} \equiv \nabla(\nabla \bullet \boldsymbol{H}) - \nabla^2 \boldsymbol{H} \tag{4.110c}$$

It gives

$$\nabla^2 \boldsymbol{H} = 0 \tag{4.111}$$

The magnetic field intensity in each of the four regions satisfies Equations 4.110a through 4.111. Also, all the components of magnetic field intensity in these regions satisfy the Laplace equation. The tangential components of magnetic field on iron surfaces may thus be neglected. Further, the magnetic field vanishes at far distances from its source, that is, the current-sheet simulating the end-winding located at $z = 0$. In view of the above specified conditions, the field distributions in different regions can thus be obtained.

4.6.2.1 Field in Region 1

The field distribution in region 1 can be written by setting,

$$\alpha_{m,p} = \sqrt{\left(\frac{m\pi}{\lambda}\right)^2 + \left(\frac{p\pi}{d}\right)^2} \tag{4.112}$$

We can write the expressions to represent field components as

$$H_{1x} = \sum_{p=1}^{\infty} P'_{m,p}\left(\frac{p\pi}{d}\right)\cdot \cos\left(\frac{p\pi}{d}\cdot x\right)\cdot e^{-(\alpha_{m,p}\cdot z)} \tag{4.112a}$$

$$H_{1y} = \sum_{p=1}^{\infty} P'_{m,p}\cdot\left(j\frac{m\pi}{\lambda}\right)\cdot \sin\left(\frac{p\pi}{d}\cdot x\right)\cdot e^{-(\alpha_{m,p}\cdot z)} \tag{4.112b}$$

$$H_{1z} = -\sum_{p=1}^{\infty} P'_{m,p} \cdot \alpha_{m,p} \cdot \sin\left(\frac{p\pi}{d} \cdot x\right) \cdot e^{-(\alpha_{m,p} \cdot z)} \tag{4.112c}$$

where $P'_{m,p}$ indicates a set of arbitrary constants.

4.6.2.2 Field in Region 2

The distribution of magnetic field intensity in region 2 can be given as

$$H_{2x} = \sum_{p=1}^{\infty} P''_{m,p} \cdot \left(\frac{p\pi}{d}\right) \cdot \cos\left(\frac{p\pi}{d} \cdot x\right) \cdot e^{\alpha_{m,p} \cdot (z+g)} \tag{4.113a}$$

$$H_{2y} = \sum_{p=1}^{\infty} P''_{m,p} \cdot \left(j\frac{m\pi}{\lambda}\right) \cdot \sin\left(\frac{p\pi}{d} \cdot x\right) \cdot e^{\alpha_{m,p} \cdot (z+g)} \tag{4.113b}$$

$$H_{2z} = \sum_{p=1}^{\infty} P''_{m,p} \cdot \alpha_{m,p} \cdot \sin\left(\frac{p\pi}{d} \cdot x\right) \cdot e^{\alpha_{m,p} \cdot (z+g)} \tag{4.113c}$$

where $P''_{m,p}$ indicates a set of arbitrary constants.

4.6.2.3 Field in Region 3

The magnetic field distribution in region 3 can be expressed by setting

$$\beta_{m,q} = \sqrt{\left(\frac{m\pi}{\lambda}\right)^2 + \left(\frac{q\pi}{g}\right)^2} \tag{4.114}$$

$$H_{3y} = a_m \cdot j \cdot \frac{\sinh(m\pi/\lambda)(z+g)}{\sinh((m\pi/\lambda) \cdot g)} + \sum_{q=1}^{\infty} b_{m,q} \cdot \left(j\frac{m\pi}{\lambda}\right) \cdot \sin\left(\frac{q\pi}{g} \cdot z\right) \cdot e^{\beta_{m,q} \cdot x} \tag{4.114a}$$

$$H_{3x} = \sum_{q=1}^{\infty} b_{m,q} \cdot \beta_{m,q} \cdot \sin\left(\frac{q\pi}{g} \cdot z\right) \cdot e^{\beta_{m,q} \cdot x} \tag{4.114b}$$

$$H_{3z} = a_m \cdot \frac{\cosh(m\pi/\lambda)(z+g)}{\sinh((m\pi/\lambda) \cdot g)} + \sum_{q=1}^{\infty} b_{m,q} \cdot \left(\frac{q\pi}{g}\right) \cdot \cos\left(\frac{q\pi}{g} \cdot z\right) \cdot e^{\beta_{m,q} \cdot x} \tag{4.114c}$$

where, a_m and $b_{m,q}$ indicate two sets of arbitrary constants. The first terms on the right-hand side of Equations 4.114a and 4.114c indicate field components due to the current-sheet simulating the main winding housed in the stator slots. The remaining terms in these field equations vanish deep inside the air gap, far removed from the end-winding, that is, as $x \to -\infty$.

Therefore, as $x \to -\infty$, $H_{3z}\big|_{z=0} = a_m = k_m$.

4.6.2.4 Field in Region 4

In region 4, since H_{4y} and H_{4z} vanish at $x = d$, let the field components, in view of Equations 4.110a, 4.110b and 4.111, be expressed in terms of new sets of arbitrary constants. Thus by setting,

$$\gamma_{m,n} = \sqrt{\left(\frac{m\pi}{\lambda}\right)^2 + \left(\frac{n\pi}{2d}\right)^2} \tag{4.115}$$

$$H_{4y} = \sum_{n-odd}^{\infty}\left(j\frac{m\pi}{\lambda}\right)\cdot\cos\left(\frac{n\pi}{2d}\cdot x\right)\cdot\left[\begin{array}{l} A'_{m,n}\cdot\dfrac{\sinh\{\gamma_{m,n}(z+g)\}}{\sinh(\gamma_{m,n}\cdot g)} \\ -A''_{m,n}\cdot\dfrac{\sinh(\gamma_{m,n}\cdot z)}{\sinh(\gamma_{m,n}\cdot g)}\end{array}\right]$$
$$+\sum_{q=1}^{\infty}A_{m,q}\cdot\left(j\frac{m\pi}{\lambda}\right)\cdot\sin\left(\frac{q\pi}{g}\cdot z\right)\cdot\frac{\sinh\{\beta_{m,q}\cdot(x-d)\}}{\sinh(\beta_{m,q}\cdot d)} \tag{4.115a}$$

$$H_{4z} = \sum_{n-odd}^{\infty}(\gamma_{m,n})\cdot\cos\left(\frac{n\pi}{2d}\cdot x\right)\cdot\left[\begin{array}{l} A'_{m,n}\cdot\dfrac{\cosh\{\gamma_{m,n}(z+g)\}}{\sinh(\gamma_{m,n}\cdot g)} \\ -A''_{m,n}\cdot\dfrac{\cosh(\gamma_{m,n}\cdot z)}{\sinh(\gamma_{m,n}\cdot g)}\end{array}\right]$$
$$+\sum_{q=1}^{\infty}A_{m,q}\cdot\left(\frac{q\pi}{g}\right)\cdot\cos\left(\frac{q\pi}{g}\cdot z\right)\cdot\frac{\sinh\{\beta_{m,q}\cdot(x-d)\}}{\sinh(\beta_{m,q}\cdot d)} \tag{4.115b}$$

$$H_{4x} = -\sum_{n-odd}^{\infty}\left(\frac{n\pi}{2d}\right)\cdot\sin\left(\frac{n\pi}{2d}\cdot x\right)\cdot\left[\begin{array}{l} A'_{m,n}\cdot\dfrac{\sinh\{\gamma_{m,n}(z+g)\}}{\sinh(\gamma_{m,n}\cdot g)} \\ -A''_{m,n}\cdot\dfrac{\sinh(\gamma_{m,n}\cdot z)}{\sinh(\gamma_{m,n}\cdot g)}\end{array}\right]$$
$$+\sum_{q=1}^{\infty}A_{m,q}\cdot\beta_{m,q}\cdot\sin\left(\frac{q\pi}{g}\cdot z\right)\cdot\frac{\cosh\{\beta_{m,q}\cdot(x-d)\}}{\sinh(\beta_{m,q}\cdot d)} \tag{4.115c}$$

where $A_{m,q}$, $A'_{m,n}$ and $A''_{m,n}$ indicate three sets of arbitrary constants.

4.6.3 Boundary Conditions

In view of the assumption that the relative permeability for iron is large (i.e. $\mu_r \gg 1$), two sets of boundary conditions are specified. The first of these is used for the selection of field expressions, whereas the second can be used to evaluate the arbitrary constants.

4.6.3.1 Selection of Field Expressions

For selecting the field expressions, the following boundary conditions are assumed. These boundary conditions are to be identically satisfied by the selected expressions.

$$H_{4y}\Big|_{x=d} = 0 \tag{4.116a}$$

$$H_{4z}\Big|_{x=d} = 0 \tag{4.116b}$$

$$H_{3y}\Big|_{z=-g} = 0 \tag{4.116c}$$

$$H_{3x}\Big|_{z=0} = 0 \tag{4.116d}$$

$$H_{3x}\Big|_{z=-g} = 0 \tag{4.116e}$$

4.6.3.2 Evaluation of Arbitrary Constants

The various arbitrary constants used to describe magnetic fields in different regions can be evaluated by using the following boundary conditions:

$$H_{4x}\Big|_{z=0} = H_{1x}\Big|_{z=0} - K_{oy} \quad \text{over } 0 \le x \le d \tag{4.117a}$$

$$H_{4x}\Big|_{z=-g} = H_{2x}\Big|_{z=-g} \quad \text{over } 0 \le x \le d \tag{4.117b}$$

$$H_{4x}\Big|_{x=0} = H_{3x}\Big|_{x=0} \quad \text{over } -g \le z \le 0 \tag{4.117c}$$

$$H_{4y}\Big|_{z=0} = H_{1y}\Big|_{z=0} + K_{ox} \quad \text{over } 0 \le x \le d \tag{4.117d}$$

$$H_{4y}\Big|_{z=-g} = H_{2y}\Big|_{z=-g} \quad \text{over } 0 \le x \le d \tag{4.117e}$$

$$H_{4y}\Big|_{x=0} = H_{3y}\Big|_{x=0} \quad \text{over} -g \le z \le 0 \tag{4.117f}$$

$$H_{4z}\big|_{z=0} = H_{1z}\big|_{z=0} \quad \text{over } 0 \le x \le d \tag{4.117g}$$

$$H_{4z}\big|_{z=-g} = H_{2z}\big|_{z=-g} \quad \text{over } 0 \le x \le d \tag{4.117h}$$

$$H_{4z}\big|_{x=0} = H_{3z}\big|_{x=0} \quad \text{over} -g \le z \le 0 \tag{4.117i}$$

$$H_{3y}\big|_{z=0} = K_x \quad \text{over } 0 \le x \le d \tag{4.117j}$$

From Equations 4.117j, 4.114a, 4.109c and 4.108a:

$$a_m = k_m = \left(j\frac{m\pi}{\lambda}\right)\cdot\sum_{n-odd}^{\infty}\ell_{m\cdot n} \quad \text{for } m = 1, 2, 3, \ldots. \tag{4.118}$$

Since $\ell_{m.n}$ are known from the armature winding details, a_m can be found from Equation 4.118.

From Equations 4.117a, 4.115c, 4.112a and 4.109b:

$$\sum_{n-odd}^{\infty}(\ell_{m.n} - A'_{m,n})\cdot\left(\frac{n\pi}{2d}\right)\cdot\sin\left(\frac{n\pi}{2d}\cdot x\right) = \sum_{p=1}^{\infty}P'_{m,p}\cdot\left(\frac{p\pi}{d}\right)\cdot\cos\left(\frac{p\pi}{d}\cdot x\right) \tag{4.119a}$$

over $0 \le x \le d$, for $m = 1, 2, 3, \ldots$.

From Equations 4.117b, 4.115c and 4.113a:

$$-\sum_{n-odd}^{\infty}A''_{m,n}\cdot\left(\frac{n\pi}{2d}\right)\cdot\sin\left(\frac{n\pi}{2d}\cdot x\right) = \sum_{p=1}^{\infty}P''_{m,p}\cdot\left(\frac{p\pi}{d}\right)\cdot\cos\left(\frac{p\pi}{d}\cdot x\right) \tag{4.119b}$$

over $0 \le x \le d$, for $m = 1, 2, 3, \ldots$.

From Equations 4.117c, 4.115c and 4.114b:

$$A_{m,q}\cdot\coth(\beta_{m,q}\cdot d) = b_{m,q} \quad \text{for } m, q = 1, 2, 3, \ldots. \tag{4.119c}$$

From Equations 4.117d, 4.115a, 4.112b and 4.109a:

$$\sum_{n-odd}^{\infty}(A'_{m,n} - \ell_{m\cdot n})\cdot\cos\left(\frac{n\pi}{2d}\cdot x\right) = \sum_{p=1}^{\infty}P'_{m,p}\cdot\sin\left(\frac{p\pi}{d}\cdot x\right) \tag{4.119d}$$

over $0 \le x \le d$, for $m = 1, 2, 3, \ldots$.

Note that Equation 4.119a is obtained on differentiating Equation 4.119d.

Now, consider Equations 4.117e, 4.115a and 4.113b:

$$\sum_{n-odd}^{\infty} A''_{m,n} \cdot \cos\left(\frac{n\pi}{2d} \cdot x\right) = \sum_{p=1}^{\infty} P''_{m,p} \cdot \sin\left(\frac{p\pi}{d} \cdot x\right) \tag{4.119e}$$

over $0 \le x \le d$, for $m = 1, 2, 3, \ldots$.

Note that Equation 4.119b is obtained on differentiating Equation 4.119e.

In view of Equations 4.119f, 4.115a and 4.114a:

$$a_m \cdot \frac{\sinh(m\pi/\lambda)(z+g)}{\sinh((m\pi/\lambda)\cdot g)} \cdot \left(\frac{\lambda}{m\pi}\right) + \sum_{q=1}^{\infty} (b_{m,q} + A_{m,q}) \cdot \sin\left(\frac{q\pi}{g} \cdot z\right)$$

$$= \sum_{n-odd}^{\infty} \begin{bmatrix} A'_{m,n} \cdot \dfrac{\sinh\{\gamma_{m,n}(z+g)\}}{\sinh(\gamma_{m,n} \cdot g)} \\ - A''_{m,n} \cdot \dfrac{\sinh(\gamma_{m,n} \cdot z)}{\sinh(\gamma_{m,n} \cdot g)} \end{bmatrix} \tag{4.119f}$$

over $-g \le z \le 0$, for $m = 1, 2, 3, \ldots$.

In view of Equations 4.117g, 4.115b and 4.112c:

$$\sum_{n-odd}^{\infty} (\gamma_{m,n}) \cdot \cos\left(\frac{n\pi}{2d} \cdot x\right) \cdot \begin{bmatrix} A'_{m,n} \cdot \coth(\gamma_{m,n} \cdot g) \\ - A''_{m,n} \cdot \operatorname{cosech}(\gamma_{m,n} \cdot g) \end{bmatrix}$$

$$+ \sum_{q=1}^{\infty} A_{m,q} \cdot \left(\frac{q\pi}{g}\right) \cdot \frac{\sinh\{\beta_{m,q} \cdot (x-d)\}}{\sinh(\beta_{m,q} \cdot d)}$$

$$= -\sum_{p=1}^{\infty} P'_{m,p} \cdot \alpha_{m,p} \cdot \sin\left(\frac{p\pi}{d} \cdot x\right) \tag{4.119g}$$

over $0 \le x \le d$, for $m = 1, 2, 3, \ldots$.

In view of Equations 4.117h, 4.115b and 4.113c, we have

$$\sum_{n-odd}^{\infty} (\gamma_{m,n}) \cdot \cos\left(\frac{n\pi}{2d} \cdot x\right) \cdot \begin{bmatrix} A'_{m,n} \cdot \operatorname{cosech}(\gamma_{m,n} \cdot g) \\ - A''_{m,n} \cdot \coth(\gamma_{m,n} \cdot g) \end{bmatrix} + \sum_{q=1}^{\infty} A_{m,q} \cdot \left(\frac{q\pi}{g}\right) \cdot \cos(q\pi)$$

$$\times \frac{\sinh\{\beta_{m,q} \cdot (x-d)\}}{\sinh(\beta_{m,q} \cdot d)} = \sum_{p=1}^{\infty} P''_{m,p} \cdot \alpha_{m,p} \cdot \sin\left(\frac{p\pi}{d} \cdot x\right) \tag{4.119h}$$

over $0 \le x \le d$, for $m = 1, 2, 3, \ldots$.

In view of Equations 4.117i, 4.115b and 4.114c, we have

$$a_m \cdot \frac{\cosh(m\pi/\lambda)(z+g)}{\sinh((m\pi/\lambda)\cdot g)} + \sum_{q=1}^{\infty}(b_{m,q} + A_{m,q})\cdot\left(\frac{q\pi}{g}\right)\cdot\cos\left(\frac{q\pi}{g}\cdot z\right)$$
$$= \sum_{n-odd}^{\infty}(\gamma_{m,n})\cdot\left[\begin{array}{l} A'_{m,n}\cdot\dfrac{\cosh\{\gamma_{m,n}(z+g)\}}{\sinh(\gamma_{m,n}\cdot g)} \\ -A''_{m,n}\cdot\dfrac{\cosh(\gamma_{m,n}\cdot z)}{\sinh(\gamma_{m,n}\cdot g)}\end{array}\right] \tag{4.119i}$$

over $-g \le z \le 0$, for $m = 1, 2, 3, \ldots$.

Note that Equation 4.119i is obtained on differentiating Equation 4.119f. From the eight sets of Equations, viz. 4.119a through 4.119h, only six are linearly independent. Numerical solutions of these linearly independent algebraic equations determine the unknown arbitrary constants, viz. $A'_{m,n}$, $A''_{m,n}$, $A_{m,q}$, $b_{m,q}$, $P'_{m,p}$ and $P''_{m,p}$.

PROJECT PROBLEMS

1. Evaluate Carter's coefficient. Compare your results with those obtained by Coe, R. T. and H. W. Taylor[2].
2. Evaluate Carter's coefficient. Compare your results with those obtained by Mukherji, K. C. and S. Neville[3].
3. Evaluate the two sets of arbitrary constants, p_m and q_n, involved in the potential distributions for arbitrary orientation.
4. Draw a graph showing the variation of the air-gap permeance with the orientation angle δ over the range $0 \le \delta \le \lambda/4$.
5. For tooth-opposite-tooth orientation find equipotential surfaces in the case of wide and deep slots using Schwarz–Christoffel transformation[4].
6. For tooth-opposite-slot orientation find equipotential surfaces in the case of wide and deep slots using Schwarz–Christoffel transformation[4].
7. Find the tangential component of force between two identical displaced teeth in the case of wide and deep slots.
8. Obtain family of curves showing fringing flux and equipotential lines for tooth-opposite-tooth orientation with small air-gap lengths.
9. For tooth-opposite-tooth orientation with small air-gap lengths, calculate and plot the values of flux density on the $y = 0$ plane.
10. Plot equipotential- and flux-lines in the air gap due to a current-carrying conductor placed deep inside an open slot. Assume (a) that the slot is facing a smooth iron surface, (b) that the air gap is infinitely large.

11. From the eight sets of Equations, viz. 4.119a through 4.119h, identify six linearly independent equations. Draw a program to solve them.

References

1. Carter, F. W., The magnetic field of the dynamoelectric machine, *J. I. E. E.* 64, 1115–1138, 1926.
2. Coe, R. T. and Taylor, H. W., Some problems in electrical machine design involving elliptic functions, *Phil. Mag.*, 6, 100–145, 1928.
3. Mukherji, K. C. and Neville, S., Magnetic permeance of identical double slotting, *Proc. IEE*, 118(9), 1257–1268, 1971.
4. Gibbs, W.J., *Conformal Transformation in Electrical Engineering*, Chapman & Hall Ltd., London, 1958.
5. Khan, A. S. and Mukerji, S. K., Field between two unequal opposite and displaced slots, *IEEE Trans.*, (90IC561-I, EC), 1–7, 1990.
6. Khan, A. S., Distribution of magnetic field between slotted equipotential surfaces, PhD thesis, Department of Electrical Engineering, Aligarh Muslim University, Aligarh, India, 1980.

5

Eddy Currents in Magnetic Cores

5.1 Introduction

The eddy current phenomenon is deeply involved in some electrical machines. The study of this phenomenon gains ground when a conducting medium is subjected to a time-varying field. In electrical machines and devices having magnetic cores, the eddy current phenomenon is quite common. The understanding of the eddy current phenomenon becomes essential in the design of electromechanical clutches, eddy current brakes, solid iron rotor induction machines and self-starting synchronous motors. The penetration of alternating flux into the solid iron also needs proper estimation. The involvement of magnetic cores of different shapes and types in electrical machines demands a thorough study of the eddy current phenomenon. Such a study must encompass induction machines with solid rotors, large plates subjected to single-phase and polyphase excitations. In this chapter, an effort has been made to cover all the above aspects.

5.2 Eddy Current Machines (Solid Rotor Induction Machines)

Eddy currents are induced in conducting regions subjected to time-varying electromagnetic fields. Eddy currents due to transient electromagnetic fields are discussed in Chapter 7. This section is devoted to the induction of eddy currents due to steady-state sinusoidally time-varying electromagnetic fields. For power frequency excitation, the displacement currents are usually neglected. Therefore, the magnetic field intensity, $\boldsymbol{H}$, satisfies the following equations:

$$\nabla^2 \boldsymbol{H} = \eta^2 \boldsymbol{H} \tag{5.1a}$$

$$\nabla \cdot \boldsymbol{H} = \boldsymbol{0} \tag{5.1b}$$

$$\eta^2 = -j\omega_o \cdot \mu\sigma \tag{5.2}$$

where ω_o is the frequency of the sinusoidally time-varying field, and μ is the permeability and σ is the conductivity of the material.

Once Equations 5.1a and b are solved, the eddy current density can be readily found from

$$\boldsymbol{J} = \nabla \times \boldsymbol{H} \tag{5.3}$$

The solution of Equations 5.1a and b for the magnetic field intensity $\boldsymbol{H}$ is discussed through the following boundary-value problems.

5.2.1 Two-Dimensional Model

Figure 5.1 shows a simplified two-dimensional model of a polyphase solid rotor induction machine with its armature winding simulated by a surface current sheet on a smooth highly permeable stator surface at $z = -g$. Let the surface current density in the reference frame fixed on the rotor at $z = 0$ be given as

$$K_x = K_o \cdot e^{j(\ell y - \omega_o \cdot t)} \tag{5.4}$$

where

$$\omega_o = s \cdot \omega \tag{5.4a}$$

$$s = \text{slip} \overset{def}{=} 1 - \frac{\text{rotor speed}}{\text{synchronous speed}} = 1 - \frac{v}{(\omega/\ell)} \tag{5.4b}$$

$$\omega = \text{supply frequency} \tag{5.4c}$$

$$\ell = \frac{\pi}{\text{pole pitch}} = \frac{\pi}{\tau} \tag{5.4d}$$

and $|k_o|$ indicates the amplitude of the surface current density, with currents flowing in the x (or axial) direction. This simplified treatment neglects the

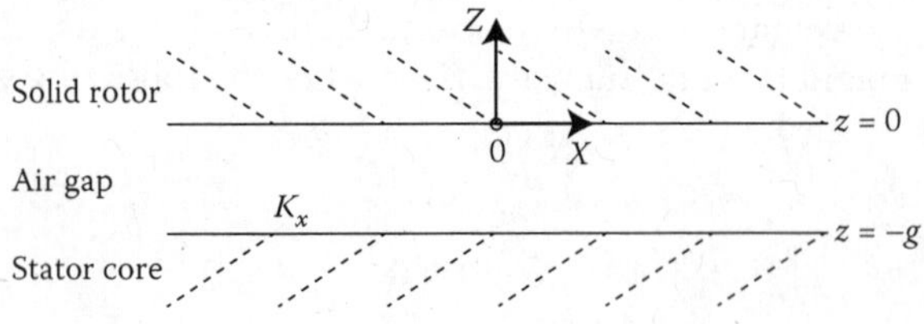

FIGURE 5.1
Components of the solid rotor induction machine.

curvature of air-gap surfaces. The analysis that takes cognizance of curvature is available in the literature.[1]

The infinite half-space, $z > 0$, with conductivity σ and permeability μ represents the solid rotor. The free space, $0 > z > -g$, indicates the air gap of the machine. The highly permeable ($\mu \approx \infty$) stator core fills the infinite half-space $z < -g$. The stator current flowing in the x-direction varies periodically in the y-direction. This results in y- (i.e. perpendicular to the x- and z-axis in Figure 5.1) and z-components of the magnetic field intensity in the air gap as well as in the solid rotor.

5.2.1.1 Field Components

In view of the boundary conditions for the air-gap field

$$H_{1y}\Big|_{z=-g} = -K_x = -K_o \cdot e^{j(\ell y-\omega_o \cdot t)} \tag{5.5a}$$

and

$$H_{1y}\Big|_{z=0} = K_1 \cdot e^{j(\ell y-\omega_o \cdot t)} \tag{5.5b}$$

where K_1 indicates an unknown constant. Since the electromagnetic field is independent of x-coordinate, the solution for Equation 5.1 with zero air-gap conductivity, obtained on neglecting the displacement currents, is

$$H_{1y} = \left[K_o \cdot \frac{\sinh(\ell z)}{\sinh(\ell g)} + K_1 \cdot \frac{\sinh \ell(z+g)}{\sinh(\ell g)}\right] \cdot e^{j(\ell y-\omega_o \cdot t)} \tag{5.6a}$$

$$H_{1z} = -j \cdot \left[K_o \cdot \frac{\cosh(\ell z)}{\sinh(\ell g)} + K_1 \cdot \frac{\cosh \ell(z+g)}{\sinh(\ell g)}\right] \cdot e^{j(\ell y-\omega_o \cdot t)} \tag{5.6b}$$

In view of the boundary conditions for the field in the rotor iron

$$H_{2y}\Big|_{z=0} = H_{1y}\Big|_{z=0} = K_1 \cdot e^{j(\ell y-\omega_o \cdot t)} \tag{5.7a}$$

$$H_{2y}\Big|_{z=\infty} = 0 \tag{5.7b}$$

The solution for Equation 5.1, for the rotor region, obtained on neglecting the displacement currents is

$$H_{2y} = K_1 \cdot e^{-\alpha \cdot z + j(\ell y-\omega_o \cdot t)} \tag{5.8a}$$

$$H_{2z} = j \cdot \frac{\ell}{\alpha} \cdot K_1 \cdot e^{-\alpha \cdot z + j(\ell y - \omega_o \cdot t)} \tag{5.8b}$$

where

$$\alpha = \sqrt{\ell^2 + \eta^2} \tag{5.9}$$

The normal component of magnetic induction, **B**, is continuous at the boundary $z = 0$:

$$B_{2z}\big|_{z=0} = B_{1z}\big|_{z=0} \tag{5.10}$$

Therefore, using Equations 5.8b and 5.6b, we get

$$\mu \cdot j \cdot \frac{\ell}{\alpha} \cdot K_1 \cdot e^{j(\ell y - \omega_o \cdot t)} = -\mu_o \cdot j \cdot \left[\frac{K_o + K_1 \cdot \cosh(\ell g)}{\sinh(\ell g)} \right] \cdot e^{j(\ell y - \omega_o \cdot t)} \tag{5.10a}$$

It gives

$$K_1 = -\frac{(\mu_o/\mu) \cdot K_o}{\left[(\ell/\alpha) \cdot \sinh(\ell g) + (\mu_o/\mu) \cdot \cosh(\ell g) \right]} \tag{5.10b}$$

The distribution of the magnetic field ***H*** in the solid rotor is obtained from Equations 5.8a,b and 5.10b.

5.2.1.2 Eddy Current Density

The eddy current density can be readily obtained from the field intensity ***H*** in view of Equations 5.2, 5.3, 5.4d, 5.8a,b and 5.9 as indicated below.

$$J_{2x} = \frac{\partial H_{2z}}{\partial y} - \frac{\partial H_{2y}}{\partial z} = \frac{\eta^2}{\alpha} \cdot K_1 \cdot e^{-\alpha \cdot z + j(\ell y - \omega_o \cdot t)} = -\frac{j\omega_o \cdot \mu\sigma}{\alpha} \cdot K_1 \cdot e^{-\alpha \cdot z + j(\ell y - \omega_o \cdot t)} \tag{5.11}$$

5.2.1.3 Eddy Current Loss

The eddy current loss, W_E, in the solid rotor is given by

$$W_E = P \cdot \tau \cdot L_R \cdot \frac{1}{2} \int_0^{\infty} \left(E_{2x} \cdot \tilde{j}_{2x} \right) dz \tag{5.12}$$

where P is the number of stator poles, L_R is the rotor length and τ is the pole pitch. Therefore, using Equations 5.11 and 5.12, we get

$$W_E = P \cdot \tau \cdot L_R \cdot \frac{\sigma}{2} \cdot (\omega_o \cdot \mu)^2 \cdot \left[\frac{|K_1|}{|\alpha|}\right]^2 \cdot \int_0^{\infty} e^{-(\alpha+\tilde{\alpha})\cdot z} dz \tag{5.13}$$

Or,

$$W_E = s^2 \cdot \frac{\sigma}{2} \cdot (\omega \cdot \mu)^2 \cdot P \cdot \tau \cdot L_R \cdot \left[\frac{|K_1|}{|\alpha|}\right]^2 \cdot \frac{1}{(\alpha + \tilde{\alpha})} \tag{5.13a}$$

5.2.1.4 Force Density

Since the force density, $\mathcal{F}$, is given by $\boldsymbol{J} \times \boldsymbol{B}$, its peripheral component can be written as

$$\mathcal{F}_y = \mu \cdot (J_z \cdot H_x - J_x \cdot H_Z) \tag{5.13b}$$

In this two-dimensional problem there being only the axial component of eddy current density, the time average of the peripheral force developed in the rotor is given as

$$F_y = -\mu \cdot P \cdot \tau \cdot L_R \cdot \frac{1}{2} \cdot \int_0^{\infty} J_{2x} \cdot \tilde{H}_{2z} \cdot dz \tag{5.14}$$

Therefore, using Equations 5.11 and 5.8b, we get

$$F_y = P \cdot \tau \cdot L_R \cdot \frac{1}{2} \cdot \ell \cdot \omega_o \cdot \mu^2 \cdot \sigma \cdot \left[\frac{|K_1|}{|\alpha|}\right]^2 \cdot \frac{1}{(\alpha + \tilde{\alpha})} \tag{5.14a}$$

5.2.1.5 Mechanical Power Developed

Since $v = (1 - s) \cdot (\omega/\ell)$, the mechanical power developed, P_M, is given by

$$P_M = v \cdot F_y = (1 - s) \cdot s \cdot \frac{\sigma}{2} \cdot P \cdot \tau \cdot L_R \cdot (\omega \cdot \mu)^2 \cdot \left[\frac{|K_1|}{|\alpha|}\right]^2 \cdot \frac{1}{(\alpha + \tilde{\alpha})} \tag{5.15}$$

5.2.1.6 Rotor Power Input

The rotor power input is given as

$$P_R = W_E + P_M = s \cdot \frac{\sigma}{2} \cdot P \cdot \tau \cdot L_R \cdot (\omega \cdot \mu)^2 \cdot \left[\frac{|K_1|}{|\alpha|}\right]^2 \cdot \frac{1}{(\alpha + \tilde{\alpha})} \tag{5.16}$$

Therefore, from Equations 5.13a, 5.15 and 5.16

$$\frac{P_R}{1} = \frac{W_E}{s} = \frac{P_M}{(1-s)} \tag{5.17}$$

This is a well-known relation for any type of induction or hysteresis machine. Involved treatments highlighting certain features of the solid rotor induction machine are presented in Chapter 7. In Chapter 8, a simplified theory for the hysteresis machine is discussed.

5.3 Eddy Currents in Large Plates due to Alternating Excitation Current

Figure 5.2 shows a large conducting plate of thickness *T*. The excitation winding carrying ac current is simulated by current sheets on its two surfaces. Owing to these surface current sheets, time-varying electromagnetic fields will be produced which will induce eddy currents in the plate. In the following subsections, the expressions for the induced eddy currents are obtained for the cases of single-phase and polyphase excitations. It may be seen that in both of these cases, only *z*-component of the magnetic field exists.

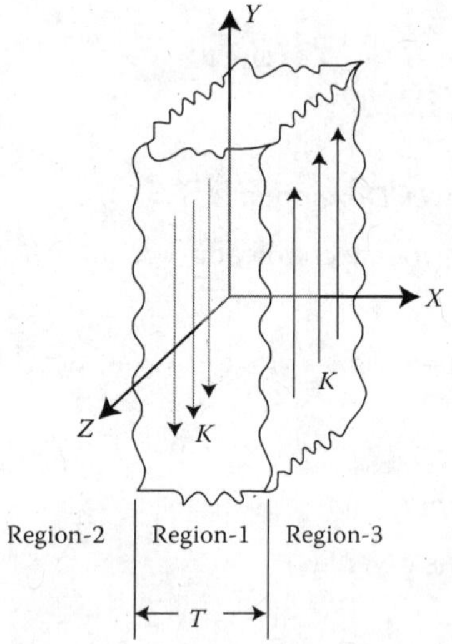

FIGURE 5.2
Large conducting plate. (Courtesy of The Electromagnetics Academy.)

5.3.1 Single-Phase Excitation

The z-component of the magnetic field in phasor form is a function of only the x-coordinate. The eddy current equation for this one-dimensional problem, found from Maxwell's equations, is as follows:

$$\frac{d^2H_z}{dx^2} = j\omega\mu\sigma H_z \stackrel{def}{=} j\frac{2}{d^2}H_z \tag{5.18}$$

Let the known surface current densities be

$$\boldsymbol{K}\big|_{x=T/2} = K_o \cdot e^{-j\omega t}\ \boldsymbol{a}_y \tag{5.19a}$$

and

$$\boldsymbol{K}\big|_{x=-T/2} = -K_o \cdot e^{-j\omega t}\ \boldsymbol{a}_y \tag{5.19b}$$

In all field expressions occurring below, the factor $e^{-j\omega t}$ is suppressed, though its presence is recognised.

At power frequency, ω being small, displacement currents can be ignored. Therefore, fields outside the plate are zero, thus giving

$$\boldsymbol{H}_{1z}\big|_{x=\pm T/2} = K_o \tag{5.20}$$

The distribution of magnetic field in the plate found by solving Equation 5.18, consistent with Equation 5.20, is given as

$$H_{1z} = K_o \cdot \frac{\cos\left(\theta \cdot (x/(T/2))\right)}{\cos(\theta)} \tag{5.21}$$

where

$$\theta = \frac{T}{2} \cdot \frac{(1-j)}{d} \tag{5.21a}$$

The distribution of eddy current density inside the plate is obtained from Maxwell's equation and using Equation 5.21, as indicated below:

$$\boldsymbol{J} = \nabla \times \boldsymbol{H} = \left[K_o \cdot \left(\frac{\theta}{T/2}\right) \cdot \frac{\sin(\theta \cdot (x/(T/2)))}{\cos(\theta)}\right]\boldsymbol{a}_y \tag{5.22}$$

Thus,

$$J_{1y} = K_o \cdot \left(\frac{\theta}{T/2}\right) \cdot \frac{\sin(\theta \cdot (x/(T/2)))}{\cos(\theta)} \tag{5.22a}$$

Clearly, only the y-component of eddy current density exists.

5.3.2 Polyphase Excitation

With polyphase excitation winding, let the known surface current densities be given as

$$\boldsymbol{K}\Big|_{x=\pm T/2} = \pm K_o \cdot e^{j((2\pi/\lambda)\cdot z - \omega t)} \, \boldsymbol{a}_y \tag{5.23}$$

where λ indicates the length in the z-direction of one set of polyphase winding (or winding pitch).

Obviously, in this case, the electromagnetic field inside the plate will be the functions of both, x- and z-coordinates. Further, the electric field will be entirely in the y-direction, that is only y-component of eddy current density exists and it is a function of x as well as z. Now, since

$$\nabla \times \boldsymbol{E} = j\omega \, \boldsymbol{B} \tag{5.24}$$

Thus,

$$\boldsymbol{H} = -j\frac{2}{d^2}\,(\nabla \times \boldsymbol{J}) \tag{5.24a}$$

In view of what is indicated above, it follows that

$$\boldsymbol{H}_1 = -j\frac{2}{d^2}\left(-\frac{\partial J_{1y}}{\partial z}\boldsymbol{a}_x + \frac{\partial J_{1y}}{\partial x}\boldsymbol{a}_z\right) \tag{5.25}$$

This shows that there will be two components for the magnetic field inside the plate. Therefore, even with the displacement currents neglected, there will be electromagnetic fields outside the plate as well.

The relevant boundary conditions are

$$H_{1z}\Big|_{x=T/2} - H_{2z}\Big|_{x=T/2} = K_o \tag{5.26a}$$

and

$$\mu_1 H_{1x}\Big|_{x=T/2} - \mu_o H_{2x}\Big|_{x=T/2} = 0 \tag{5.26b}$$

The eddy current equation for this two-dimensional problem found from Maxwell's equations is as follows:

$$\frac{\partial^2 H_{1z}}{\partial x^2} + \frac{\partial^2 H_{1z}}{\partial z^2} = j\omega\mu\sigma H_{1z} \stackrel{def}{=} j\frac{2}{d^2} H_{1z} \tag{5.27a}$$

and

$$\frac{\partial^2 H_{1x}}{\partial x^2} + \frac{\partial^2 H_{1x}}{\partial z^2} = j\omega\mu\sigma H_{1x} \stackrel{def}{=} j\frac{2}{d^2} H_{1x} \tag{5.27b}$$

In view of the factor $e^{j(((2\pi)/\lambda)\cdot z-\omega t)}$, these equations can be rewritten as

$$\frac{d^2 H_{1z}}{dx^2} = \left[\left(\frac{2\pi}{\lambda}\right)^2 + j\omega\mu\sigma\right] H_{1z} \stackrel{def}{=} \vartheta^2 \cdot H_{1z} \tag{5.28a}$$

and

$$\frac{\partial^2 H_{1x}}{\partial x^2} = \left[\left(\frac{2\pi}{\lambda}\right)^2 + j\omega\mu\sigma\right] H_{1x} \stackrel{def}{=} \vartheta^2 \cdot H_{1x} \tag{5.28b}$$

where

$$\vartheta^2 = \left[\left(\frac{2\pi}{\lambda}\right)^2 + j\omega\mu\sigma\right] \tag{5.29}$$

Since the solution of Equation 5.28a can be given in terms of an arbitrary constant

$$H_{1z} = H_{1o} \cdot \frac{\cosh\left(\varnothing \cdot \cdot \frac{x}{T/2}\right)}{\cosh(\varnothing)} \tag{5.30}$$

where H_{1o} indicates an arbitrary constant and the parameter $\varnothing$ is given as

$$\varnothing = \frac{T}{2} \cdot \vartheta \tag{5.31}$$

over $-T/2 \leq x \leq T/2$.

Now, since divergence of $\boldsymbol{H}_1$ is zero, we have

$$H_{1x} = -H_{1o} \cdot j\frac{2\pi}{\lambda \cdot \vartheta} \cdot \frac{\sinh(\varnothing \cdot \cdot(x/(T/2)))}{\cosh(\varnothing)} \tag{5.32}$$

over $-T/2 \leq x \leq T/2$.

Next, consider the air region $-T/2 \le x \le \infty$. Since the field must vanish as x tends to infinity, the equations for field components are

$$\frac{d^2 H_{2z}}{dx^2} = \left(\frac{2\pi}{\lambda}\right)^2 \cdot H_{2z} \tag{5.33a}$$

and

$$\frac{\partial^2 H_{2x}}{\partial x^2} = \left(\frac{2\pi}{\lambda}\right)^2 \cdot H_{2x} \tag{5.33b}$$

The solution for Equation 5.33a can be given as

$$H_{2z} = H_{2o} \cdot e^{-(2\pi/\lambda)\cdot(x-T/2)} \tag{5.34}$$

where H_{2o} indicates an arbitrary constant. Since divergence of $\boldsymbol{H}_2$ is zero, we have

$$H_{2x} = jH_{2o} \cdot e^{-(2\pi/\lambda)\cdot(x-T/2)} \tag{5.35}$$

For the determination of the two arbitrary constants (H_{1o} and H_{2o}), consider Equations 5.26a and b. These equations give

$$H_{1o} - H_{2o} = K_o \tag{5.36}$$

and

$$\mu_1 \cdot H_{1o} \cdot \frac{2\pi}{\lambda \cdot \vartheta} \cdot \tanh(\varnothing) + \mu_o \cdot H_{2o} = 0 \tag{5.37}$$

On solving these equations, we get

$$H_{1o} = \frac{\mu_o \cdot K_o}{\left[\mu_1 \cdot (2\pi/(\lambda \cdot \vartheta)) \cdot \tanh(\varnothing) + \mu_o\right]} \tag{5.38}$$

$$H_{2o} = \frac{-\mu_1 \cdot (2\pi/(\lambda \cdot \vartheta)) \cdot \tanh(\varnothing)}{\left[\mu_1 \cdot (2\pi/(\lambda \cdot \vartheta)) \cdot \tanh(\varnothing) + \mu_o\right]} \tag{5.39}$$

Having determined the magnetic fields inside and outside the plate, the expression for the eddy current density in the plate can be found from Maxwell's equation:

$$\boldsymbol{J} = \nabla \times \boldsymbol{H} \tag{5.40}$$

and expressions for the magnetic field in the plate, vide Equations 5.30 and 5.32.

$$J_x = \frac{\partial H_{1z}}{\partial y} - \frac{\partial H_{1y}}{\partial z} = 0 \tag{5.41a}$$

$$J_y = \frac{\partial H_{1x}}{\partial z} - \frac{\partial H_{1z}}{\partial x} = H_{1o} \cdot \left[\left(\frac{2\pi}{\lambda}\right)^2 \cdot \frac{1}{\vartheta} - \vartheta\right] \cdot \frac{\sinh(\varnothing \cdot (x/(T/2)))}{\cosh(\varnothing)} \tag{5.41b}$$

$$J_z = \frac{\partial H_{1y}}{\partial x} - \frac{\partial H_{1x}}{\partial y} = 0 \tag{5.41c}$$

5.4 Eddy Currents in Cores with Rectangular Cross-Sections

Figure 5.3 shows a long rectangular solid-conducting core with a uniformly distributed winding carrying alternating current of angular frequency ω. At power frequencies, displacement currents are usually neglected and the excitation winding is simulated by a current sheet on the core surfaces.

Let the density of the surface current be given as

$$\boldsymbol{K}\big|_{x=\pm W/2} = \pm K_o \cdot e^{j\omega t}\boldsymbol{a}_y \quad \text{over } -T/2 \le y \le T/2 \tag{5.42a}$$

$$\boldsymbol{K}\big|_{y=\pm T/2} = \pm K_o \cdot e^{j\omega t}\boldsymbol{a}_x \quad \text{over } -W/2 \le x \le W/2 - W/2 \le x \le W/2 \tag{5.42b}$$

The external field being zero, the magnetic field in the core satisfies the following boundary conditions:

$$H_z\big|_{x=\pm W/2} = K_o \cdot e^{j\omega t} \quad \text{over } -T/2 \le y \le T/2 \tag{5.43a}$$

$$H_z\big|_{y=\pm T/2} = K_o \cdot e^{j\omega t} \quad \text{over } -W/2 \le x \le W/2 \tag{5.43b}$$

Therefore, the solution for the field equations

$$\nabla^2 \boldsymbol{H} = \eta^2 \boldsymbol{H} \tag{5.44a}$$

$$\nabla \bullet \boldsymbol{H} = \boldsymbol{0} \tag{5.44b}$$

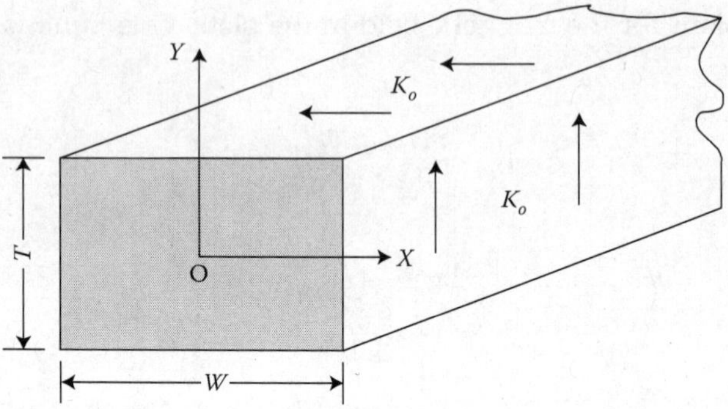

FIGURE 5.3
Solid rectangular core.

is as follows:

$$H_z = \sum_{m-odd}^{\infty} a_m \cdot \cos\left(\frac{m\pi}{T} \cdot y\right) \cdot \frac{\cosh(\alpha_m \cdot x)}{\cosh(\alpha_m \cdot (W/2))} \cdot e^{j\omega t} + \sum_{n-odd}^{\infty} b_n \cdot \cos\left(\frac{n\pi}{W} \cdot x\right) \cdot \frac{\cosh(\beta_n \cdot y)}{\cosh(\beta_n \cdot (T/2))} \cdot e^{j\omega t} \quad (5.45)$$

for $-W/2 < x < W/2$ and $-T/2 < y < T/2$.

In Equation 5.45,

$$\alpha_m = \sqrt{\left(\frac{m\pi}{T}\right)^2 + \eta^2} \quad (5.46a)$$

$$\beta_n = \sqrt{\left(\frac{n\pi}{W}\right)^2 + \eta^2} \quad (5.46b)$$

$$\sum_{m-odd}^{\infty} a_m \cdot \cos\left(\frac{m\pi}{T} \cdot y\right) = K_o \quad \text{over } -T/2 < y < T/2 - T/2 < y < T/2 \quad (5.46c)$$

and

$$\sum_{n-odd}^{\infty} b_n \cdot \cos\left(\frac{n\pi}{W} \cdot x\right) = K_o \quad \text{over } -W/2 < x < W/2 \quad (5.46d)$$

Thus, the Fourier coefficients are

$$a_m = K_o \cdot \frac{4}{m\pi} \cdot \sin\left(\frac{m\pi}{2}\right) \tag{5.47a}$$

$$b_n = K_o \cdot \frac{4}{n\pi} \cdot \sin\left(\frac{n\pi}{2}\right) \tag{5.47b}$$

For cores with square cross-sections, $W = T \overset{def}{=} L$, Equation 5.45 reduces to

$$H_z = \sum_{p-odd}^{\infty} K_o \cdot \frac{4}{p\pi} \cdot \sin\left(\frac{p\pi}{2}\right) \times \left[\cos\left(\frac{p\pi}{L}\cdot x\right)\cdot\frac{\cosh(\gamma_p \cdot y)}{\cosh(\gamma_p \cdot (L/2))} + \cos\left(\frac{p\pi}{L}\cdot y\right)\cdot\frac{\cosh(\gamma_p \cdot x)}{\cosh(\gamma_p \cdot (L/2))}\right]\cdot e^{j\omega t} \tag{5.48}$$

for $-L/2 < x, y < L/2$.

In Equation 5.48,

$$\gamma_p = \sqrt{\left(\frac{p\pi}{L}\right)^2 + \eta^2} \tag{5.48a}$$

Note that Equation 5.48 is symmetrical about x and y, that is the equation remains unaltered if x and y are mutually interchanged. Since $\boldsymbol{J} = \nabla \times \boldsymbol{H}$, from Equations 5.45 and 5.48, expressions for eddy current density can be readily obtained.[1] This example finds application for the determination of the distribution of the magnetic field in cores with triangular cross-sections.

5.5 Eddy Currents in Cores with Triangular Cross-Sections

For the determination of eddy currents in solid-conducting cores of a rectangular cross-section, analytical solutions of eddy current equations are available in the literature.[2] The distribution of the magnetic field in cores with a cross-section in the shape of an arbitrary triangle can best be obtained by numerical methods.[3,4] For cores with right-angled isosceles triangular cross-sections, analytical solutions for the Laplace equation and wave equation are available.[5] This section presents an analytical solution for eddy current equation giving field distribution in cores with right-angled isosceles triangular cross-sections.[6,7]

Consider a long conducting core with a uniformly distributed winding on its surfaces that carries an alternating current with angular frequency ω. Let the winding currents be simulated by a current sheet on core surfaces with surface current density $K_o \cdot e^{j\omega t}$, where $|K_o|$ indicates ampere turns per unit core length. If the displacement currents are neglected, the magnetic field outside the core winding will be zero. The eddy current equation for this two-dimensional problem found from Maxwell's equations is

$$\frac{\partial^2 H_z}{\partial x^2} + \frac{\partial^2 H_z}{\partial y^2} = j\omega\mu\sigma H_z \stackrel{def}{=} j\frac{2}{d^2}H_z \tag{5.49}$$

where μ is the permeability, σ is the conductivity of the core and d is the classical depth of penetration.

Having solved Equation 5.49, the expression for the eddy current density in the core is found from Maxwell's equation, with the displacement currents neglected:

$$\boldsymbol{J} = \nabla \times \boldsymbol{H} \tag{5.50}$$

For the triangular section shown in Figure 5.4, the boundary conditions are

$$H_z\big|_{x=0} = K_o \cdot e^{j\omega t} \quad \text{over } 0 < y < \ell \tag{5.51a}$$

$$H_z\big|_{y=0} = K_o \cdot e^{j\omega t} \quad \text{over } 0 < x < \ell \tag{5.51b}$$

$$H_z\big|_{y=\ell-x} = K_o \cdot e^{j\omega t} \quad \text{over } 0 < x < \ell \tag{5.51c}$$

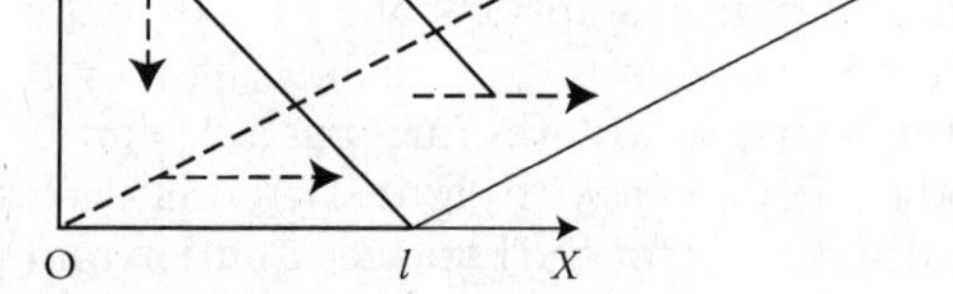

FIGURE 5.4
A core with right angled triangular cross-section.

Let us split the field H_z into two partial field components, and each is a solution of eddy current Equation 5.49:

$$H_z = H_{1z} + H_{2z} \tag{5.52}$$

Consider the square region ABCD in Figure 5.5. Let the boundary conditions for the magnetic field H_{1z} in this region be as indicated below:

$$H_{1z}\big|_{x=\pm\ell/\sqrt{2}} = K_o \cdot e^{j\omega t} \quad \text{over } -\ell/\sqrt{2} < y < \ell/\sqrt{2} \tag{5.53a}$$

$$H_{1z}\big|_{y=\pm\ell/\sqrt{2}} = -K_o \cdot e^{j\omega t} \quad \text{over } -\ell/\sqrt{2} < x < \ell/\sqrt{2} \tag{5.53b}$$

For this set of boundary condition, the distribution of the magnetic field in the square obtained by solving Equation 5.49 is

$$H_{1z} = \sum_{m-odd}^{\infty} K_o \cdot \frac{4}{m\pi} \cdot \left[\cos\left(\frac{m\pi}{\ell\sqrt{2}} \cdot y\right) \cdot \frac{\cosh(\alpha_{1m} \cdot x)}{\cosh(\alpha_{1m} \cdot \ell/\sqrt{2})} - \cos\left(\frac{m\pi}{\ell\sqrt{2}} \cdot x\right) \cdot \frac{\cosh(\alpha_{1m} \cdot y)}{\cosh(\alpha_{1m} \cdot \ell/\sqrt{2})} \right] \cdot e^{j\omega t} \tag{5.54}$$

where

$$\alpha_{1m} = \sqrt{\left(\frac{m\pi}{\ell\sqrt{2}}\right)^2 + \eta^2} \tag{5.54a}$$

and

$$\eta^2 = j\omega \cdot \mu\sigma \tag{5.54b}$$

It may be noted that the field H_{1z}, along the two diagonals $y = \pm x$, is zero.

If the square ABCD in Figure 5.5 is subjected to an angular displacement about its centre in the counter-clock direction by $\pi/4$ (the resulting square ABCD is shown in this figure by dotted lines), the distribution of the magnetic field H_{1z}, in the square ABCD found from Equation 5.54, will be

$$H_{1z} = -\sum_{m-odd}^{\infty} K_o \cdot \frac{4}{m\pi} \cdot \left[\cos\frac{m\pi}{2\ell}(x+y) \cdot \sin\left(\frac{m\pi}{2}\right) \cdot \frac{\cosh \alpha_{1m}(x-y)/\sqrt{2}}{\cosh(\alpha_{1m} \cdot \ell/\sqrt{2})} - \cos\frac{m\pi}{2\ell}(x-y) \cdot \sin\left(\frac{m\pi}{2}\right) \cdot \frac{\cosh \alpha_{1m}(x+\cdot\, y)/\sqrt{2}}{\cosh(\alpha_{1m} \cdot \ell/\sqrt{2})} \right] \cdot e^{j\omega t} \tag{5.55}$$

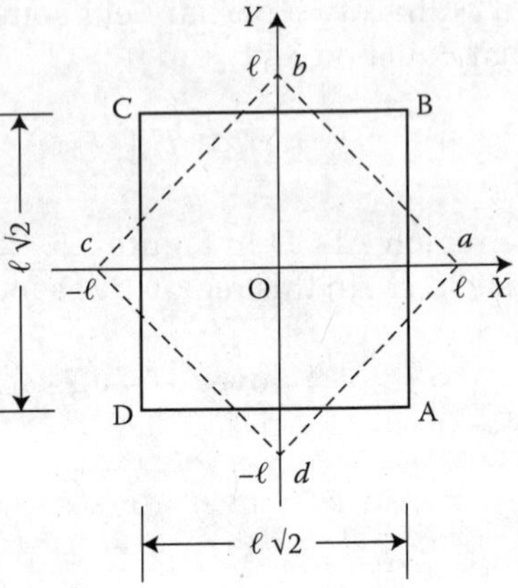

FIGURE 5.5
Square cross-section.

It may be noted that

$$H_{1z}\big|_{x,y=0} = 0 \tag{5.55a}$$

$$H_{1z}\big|_{y=\ell-x} = K_o \cdot e^{j\omega t} \tag{5.55b}$$

Consider the square ABCD shown in Figure 5.5, with the origin shifted to the point D and the length of each side set to ℓ. Further, let the magnetic field H_{2z} in the square core satisfy the following boundary conditions:

$$H_{2z}\big|_{x=0} = K_o \cdot e^{j\omega t} \quad \text{over } 0 < y < l \tag{5.56a}$$

$$H_{2z}\big|_{y=0} = K_o \cdot e^{j\omega t} \quad \text{over } 0 < x < l \tag{5.56b}$$

$$H_{2z}\big|_{x=\ell} = -K_o \cdot e^{j\omega t} \quad \text{over } 0 < y < l \tag{5.56c}$$

$$H_{2z}\big|_{y=\ell} = -K_o \cdot e^{j\omega t} \quad \text{over } 0 < x < l \tag{5.56d}$$

The solution of Equation 5.49 subjected to the above boundary conditions is given as

$$H_{2z} = -\sum_{m-odd}^{\infty} K_o \cdot \frac{4}{m\pi} \cdot \left[\sin\left(\frac{m\pi}{\ell} \cdot y\right) \cdot \frac{\sinh \alpha_{2m} \cdot (x - \ell/2)}{\sinh(\alpha_{2m} \cdot \ell/2)} + \sin\left(\frac{m\pi}{\ell} \cdot x\right) \cdot \frac{\sinh \alpha_{2m} \cdot (y - \ell/2)}{\sinh(\alpha_{2m} \cdot \ell/2)} \right] \cdot e^{j\omega t} \tag{5.57}$$

where

$$\alpha_{2m} = \sqrt{\left(\frac{m\pi}{\ell}\right)^2 + \eta^2} \tag{5.57a}$$

The magnetic field along the diagonal, $y = \ell - x$, found from Equation 5.57 is

$$H_{2z}\big|_{y=\ell-x} = 0 \tag{5.58a}$$

Further, it may be seen that

$$H_{2z}\big|_{x=0,0<y<\ell} = H_{2z}\big|_{y=0,0<x<\ell} = K_o \cdot e^{j\omega t} \tag{5.58b}$$

The distribution of magnetic field in the triangular core obtained in view of Equations 5.52, 5.54 and 5.57 is

$$H_z = -\sum_{m-odd}^{\infty} K_o \cdot \frac{4}{m\pi} \cdot \left[\left\{\sin\left(\frac{m\pi}{\ell}\cdot y\right) \cdot \frac{\sinh\alpha_{2m}\cdot(x-\ell/2)}{\sinh(\alpha_{2m}\cdot\ell/2)} + \sin\left(\frac{m\pi}{\ell}\cdot x\right)\right.\right.$$
$$\times \left.\frac{\sinh\alpha_{2m}\cdot(y-\ell/2)}{\sinh(\alpha_{2m}\cdot\ell/2)}\right\} + \sin\left(\frac{m\pi}{2}\right)\cdot\left\{\cos\frac{m\pi}{2\ell}(x+y)\cdot\frac{\cosh\alpha_{1m}(x-y)/\sqrt{2}}{\cosh(\alpha_{1m}\cdot\ell/\sqrt{2})}\right.$$
$$\left.\left. - \cos\frac{m\pi}{2\ell}(x-y)\cdot\frac{\cosh\alpha_{1m}(x+y)/\sqrt{2}}{\cosh(\alpha_{1m}\cdot\ell/\sqrt{2})}\right\}\right] \cdot e^{j\omega t} \tag{5.59}$$

The expressions for the components of eddy current density in the triangular core can now be readily found by using Equations 5.50 and 5.59.

5.6 Eddy Currents in Cores with Regular Polygonal Cross-Sections

Distributions of magnetic fields in solid cores with rectangular and circular cross-sections due to alternating current excitation have been analytically determined.[2,3] For cores with uncommon cross-sections, field distributions are usually evaluated using numerical methods.[4,8] Analytical solutions are available[5–7] for field distributions in cores with cross-sections in the shape of isosceles right-angled triangles. A quasi-analytical method for the determination of the approximate distribution of magnetic field intensity in cores with regular polygonal cross-sections is presented in this section as an alternative to the existing numerical methods. Although only three types of core sections, namely, cores with triangular, hexagonal and octagonal cross-sections, as shown in Figures 5.6 through 5.8 are considered, the method can be readily extended for other regular polygonal sections.

Consider a long conducting core carrying a surface current sheet with density K simulating a uniformly distributed current-carrying winding wound around the core. The winding current is at power frequency. The magnetic field outside the core will be zero if the displacement currents are neglected. Inside the core, the magnetic field will be axial, that is in the z-direction such that just under the current sheet

$$H_z\big|_{\text{core surface}} = K \tag{5.60}$$

where $|K|$ indicates the root mean square (rms) value of the surface current density on the conductor surface flowing in the anticlockwise direction, and H_z indicates the magnetic field in the axial direction, both in phasor form.

The eddy current equation for the magnetic field is

$$\nabla^2 H_z = \eta^2 H_z \tag{5.61}$$

where

$$\eta^2 = j\omega_o \cdot \mu\sigma \tag{5.61a}$$

$$\omega_o = \text{frequency of the sinusoidally time-varying field} \tag{5.61b}$$

$$\mu = \text{permeability of the core} \tag{5.61c}$$

$$\sigma = \text{conductivity of the core} \tag{5.61d}$$

This is a two-dimensional problem as fields vary along x- and y-directions only. Thus,

$$\frac{\partial H_z}{\partial x^2} + \frac{\partial H_z}{\partial y^2} = \eta^2 H_z \tag{5.62}$$

The solutions of this equation for solid cores with triangular, hexagonal and octagonal cross-sections are discussed in the following three subsections.

5.6.1 Cores with Triangular Cross-Sections

Consider a long solid-conducting core with a triangular cross-section shown in Figure 5.6. Let the length of each side of the triangle be L. A rectangle constructed using the base of this equilateral triangle is shown by dotted lines. Let the torch function be defined by the finite Fourier series:

$$H_z'\big|_{y=L/\sqrt{3}} = \sum_{m-odd}^{(2M-1)} T_m \cdot \cos\left(\frac{m\pi}{L} \cdot x\right) \tag{5.63}$$

where T_m indicates a set of Fourier coefficients.

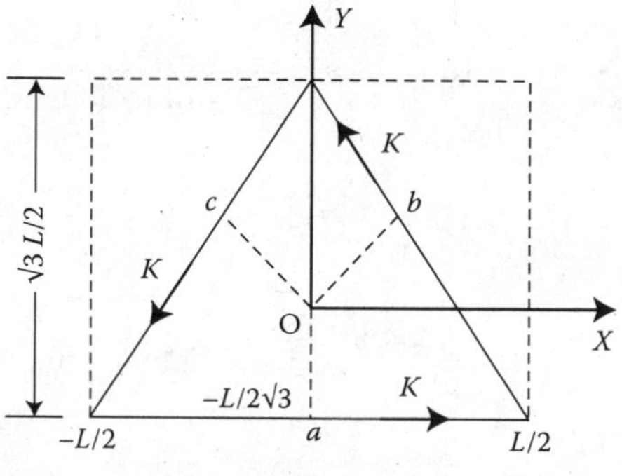

FIGURE 5.6
A core with equilateral triangular cross-section.

On setting

$$H'_z\big|_{x=\pm L/2} = H'_z\big|_{y=-L/(2\sqrt{3})} = 0 \tag{5.63a}$$

The solution of eddy current equation for the rectangular region can be given as

$$H'_z = \sum_{m-odd}^{(2M-1)} T_m \cdot \cos\left(\frac{m\pi}{L} \cdot x'\right) \cdot \frac{\sinh[\alpha_m \cdot \{y' + L/(2\sqrt{3})\}]}{\sinh(\alpha_m \cdot L\sqrt{3}/2)} \tag{5.64}$$

where

$$\alpha_m = \sqrt{\left(\frac{m\pi}{L}\right)^2 + \eta^2} \tag{5.64a}$$

$$x' = x \tag{5.64b}$$

$$y' = y \tag{5.64c}$$

Next, we construct two more similar rectangles, each containing one or the other of the two remaining sides of the equilateral triangle. Let the field distributions in these regions be

$$H''_z = \sum_{m-odd}^{(2M-1)} T_m \cdot \cos\left(\frac{m\pi}{L} \cdot x''\right) \cdot \frac{\sinh[\alpha_m \cdot \{y'' + L/(2\sqrt{3})\}]}{\sinh(\alpha_m \cdot L \cdot \sqrt{3}/2)} \tag{5.65}$$

$$H_z'' = \sum_{m-odd}^{(2M-1)} T_m \cdot \cos\left(\frac{m\pi}{L} \cdot x''\right) \cdot \frac{\sin[\alpha_m \cdot \{y'' + L/(2\sqrt{3})\}]}{\sinh(\alpha_m \cdot L \cdot \sqrt{3}/2)} \tag{5.66}$$

where

$$x'' = y \cdot \frac{\sqrt{3}}{2} - x \cdot \frac{1}{2} \tag{5.65a}$$

$$y'' = -y \cdot \frac{1}{2} - x \cdot \frac{\sqrt{3}}{2} \tag{5.65b}$$

$$x'' = -y \cdot \frac{\sqrt{3}}{2} - x \cdot \frac{1}{2} \tag{5.66a}$$

$$y'' = -y \cdot \frac{1}{2} + x \cdot \frac{\sqrt{3}}{2} \tag{5.66b}$$

It may be noted that the target zone, that is the equilateral triangle, is common to all the three rectangular regions. Further, there is a kind of symmetry in the field distribution inside the equilateral triangle. This condition is met if the resultant field in the triangular region is taken as the sum of the three field expressions given by Equations 5.64, 5.65 and 5.66. Thus,

$$H_z = H_z' + H_z' + H_z' \tag{5.67}$$

In view of Equation 5.67, we get

$$H_z = \sum_{m-odd}^{(2M-1)} T_m \left[\cos\left(\frac{m\pi}{L} \cdot x'\right) \cdot \frac{\sinh[\alpha_m \cdot \{y' + L/(2\sqrt{3})\}]}{\sinh(\alpha_m \cdot L\sqrt{3}/2)} \right.$$
$$+ \cos\left(\frac{m\pi}{L} \cdot x''\right) \cdot \frac{\sinh[\alpha_m \cdot \{y'' + L/(2\sqrt{3})\}]}{\sinh(\alpha_m \cdot L \cdot \sqrt{3}/2)}$$
$$\left. + \cos\left(\frac{m\pi}{L} \cdot x''\right) \cdot \frac{\sinh[\alpha_m \cdot \{y'' + L/(2\sqrt{3})\}]}{\sinh(\alpha_m \cdot L \cdot \sqrt{3}/2)} \right] \tag{5.68}$$

In view of Equations 5.64b,c, 5.65a,b, 5.66a and b, Equation 5.68 reduces to

$$H_z = \sum_{m-odd}^{(2M-1)} T_m \left[\cos\left(\frac{m\pi}{L}\cdot x\right)\cdot\frac{\sinh[\alpha_m\cdot\{y+L/(2\sqrt{3})\}]}{\sinh(\alpha_m\cdot L\sqrt{3}/2)} \right.$$

$$+\cos\left\{\frac{m\pi}{L}\cdot\left(y\cdot\frac{\sqrt{3}}{2}-x\cdot\frac{1}{2}\right)\right\}\cdot\frac{\sinh[\alpha_m\cdot\{(-y\cdot(1/2)-x\cdot(\sqrt{3}/2))+L/(2\sqrt{3})\}]}{\sinh(\alpha_m\cdot L\cdot\sqrt{3}/2)}$$

$$\left.+\cos\left\{\frac{m\pi}{L}\cdot\left(y\cdot\frac{\sqrt{3}}{2}+x\cdot\frac{1}{2}\right)\right\}\cdot\frac{\sinh[\alpha_m\cdot\{(-y\cdot(1/2)+x\cdot(\sqrt{3}/2))+L/(2\sqrt{3})\}]}{\sinh(\alpha_m\cdot L\cdot\sqrt{3}/2)}\right] \tag{5.69}$$

This equation gives the distribution of magnetic field in the triangular core in terms of the unknown coefficient T_m. To determine this coefficient, consider the boundary condition given by Equation 5.51. Thus,

$$H_z\big|_{y=-L/(2\sqrt{3})} = K \quad \text{over} \quad -L/2 < x < L/2 \tag{5.70}$$

Now, using finite Fourier series expansion, we get

$$H_z\big|_{y=-L/(2\sqrt{3})} \cong \sum_{n-odd}^{(2N-1)}\left[K\cdot\frac{4}{n\pi}\cdot\sin\left(\frac{n\pi}{2}\right)\right]\cdot\cos\left(\frac{n\pi}{L}\cdot x\right) \tag{5.71}$$

over $-L/2 < x < L/2$.

The chosen value for N should closely satisfy Equation 5.70.

Thus, from Equations 5.69 and 5.71, we get

$$\sum_{n-odd}^{(2N-1)}\left[K\cdot\frac{4}{\pi}\cdot\frac{1}{n}\cdot\sin\left(\frac{n\pi}{2}\right)\right]\cdot\cos\left(\frac{n\pi}{L}\cdot x\right)$$

$$=\sum_{m-odd}^{(2M-1)} T_m\cdot\left[-\cos\left\{\frac{m\pi}{L}\cdot\left(x\cdot\frac{1}{2}+\frac{L}{4}\right)\right\}\cdot\frac{\sinh\{\alpha_m\cdot(x\cdot(\sqrt{3}/2)-L(\sqrt{3}/4))\}}{\sinh(\alpha_m\cdot L\cdot\sqrt{3}/2)}\right.$$

$$\left.+\cos\left\{\frac{m\pi}{L}\cdot\left(x\cdot\frac{1}{2}-\frac{L}{4}\right)\right\}\cdot\frac{\sinh\{\alpha_m(x\cdot(\sqrt{3}/2)+L(\sqrt{3}/4))\}}{\sinh(\alpha_m\cdot L\cdot\sqrt{3}/2)}\right] \tag{5.72}$$

over $-L/2 < x < L/2$.

In view of the orthogonal property of Fourier series, we have

$$\left[K\cdot\frac{4}{\pi}\cdot\frac{1}{n}\cdot\sin\left(\frac{n\pi}{2}\right)\right]\cdot\frac{L}{2}=\sum_{m-odd}^{(2M-1)}T_m\cdot\int_{-L/2}^{L/2}\left[-\frac{\sinh\{\alpha_m\cdot(x\cdot(\sqrt{3}/2)-L(\sqrt{3}/2))\}}{\sinh(\alpha_m\cdot L\cdot\sqrt{3}/2)}\right.$$

$$\times\cos\left\{\frac{m\pi}{L}\cdot\left(x\cdot\frac{1}{2}+\frac{L}{4}\right)\right\}+\cos\left\{\frac{m\pi}{L}\cdot\left(x\cdot\frac{1}{2}-\frac{L}{4}\right)\right\}\cdot\left.\frac{\sinh\{\alpha_m(x\cdot(\sqrt{3}/2)+L(\sqrt{3}/4))\}}{\sinh(\alpha_m\cdot L\cdot\sqrt{3}/2)}\right]$$

$$\times\cos\left(\frac{n\pi}{L}\cdot x\right)\cdot dx\quad\text{for } n=1,3,5,\ldots,(2N-1).\tag{5.73}$$

On performing the integration, we get

$$\left[K\cdot\frac{2}{\pi}\cdot\frac{1}{n}\right]$$

$$=-\sum_{m-odd}^{(2M-1)}T_m\cdot\left[\frac{(\alpha_m\cdot L(\sqrt{3}/2))\cdot\sin(m\pi/2)+((m\pi/2)-(n\pi/1))\cdot\sinh(\alpha_m\cdot L(\sqrt{3}/2))}{\{((m\pi/2)-(n\pi/1))^2+(\alpha_m\cdot L(\sqrt{3}/2))^2\}\cdot\sinh(\alpha_m\cdot L\cdot\sqrt{3}/2)}\right.$$

$$\left.+\frac{\left(\alpha_m\cdot L(\sqrt{3}/2)\right)\cdot\sin(m\pi/2)+((m\pi/2)+(n\pi/1))\cdot\sinh(\alpha_m\cdot L(\sqrt{3}/2))}{\{((m\pi/2)+(n\pi/1))^2+(\alpha_m\cdot L(\sqrt{3}/2))^2\}\cdot\sinh(\alpha_m\cdot L\cdot\sqrt{3}/2)}\right]\tag{5.74}$$

for $n = 1,3,5,\ldots,(2N-1)$.

Since there are M number of unknowns and N number of equations, we chose $M = N \stackrel{def}{=} P$.

As σ is large, both α_m and $\sinh(\alpha_m\cdot L\cdot\sqrt{3}/2)$ are large quantities; thus, Equation 5.74 reduces to

$$\left[K\cdot\frac{2}{\pi}\cdot\frac{1}{n}\right]\cong-\sum_{m-odd}^{(2P-1)}T_m\cdot\left[\frac{((m\pi/2)-(n\pi/1))}{\left\{((m\pi/2)-(n\pi/1))^2+\left(\alpha_m\cdot L(\sqrt{3}/2)\right)^2\right\}}\right.$$

$$\left.+\frac{((m\pi/2)+(n\pi/1))}{\left\{((m\pi/2)+(n\pi/1))^2+\left(\alpha_m\cdot L(\sqrt{3}/2)\right)^2\right\}}\right]\tag{5.74a}$$

for $n = 1,3,5,\ldots,(2P-1)$.

The numerical solution of this equation gives the values of coefficient T_m, involved in the expression for the torch function and also for the magnetic field, vide Equations 5.63 and 5.69.

5.6.2 Cores with Hexagonal Cross-Sections

Figure 5.7 shows a long solid-conducting core with a hexagonal cross-section having the length of each of its side L. This figure also shows another concentric hexagon with side length ℓ, where

$$\ell = \sqrt{3} \cdot L \tag{5.75}$$

The smaller hexagon, indicating the core section, is provided with a uniformly distributed excitation winding carrying sinusoidal current at power frequency. This current-carrying winding is simulated by uniformly distributed currents flowing in the anticlockwise direction over the core surface with density $|K|$. Consequently, the axial component of the magnetic field H_z is established inside the core and is a negligible field outside. Because of eddy currents, H_z will be nonuniform but symmetrically distributed over the core section. On the core surface, H_z is given by Equation 5.60.

As shown in Figure 5.7, around this core section, the rectangle *abcd* is constructed. The width of this rectangle is ℓ and its height is 3 L. We determine the solution for Equation 5.62 for H_z' using the following boundary conditions:

$$H_z'\big|_{x=\pm\ell/2} = 0 \tag{5.76a}$$

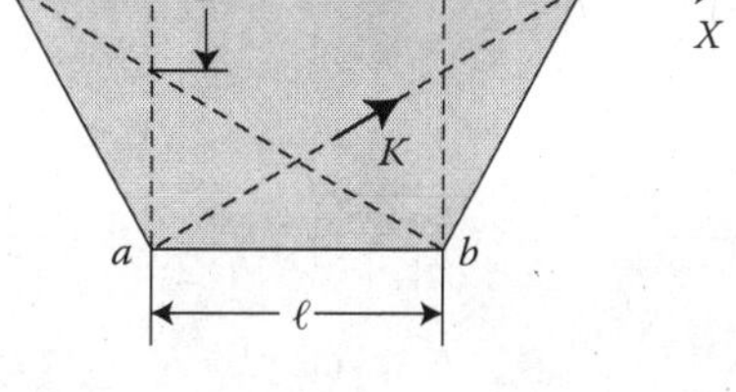

FIGURE 5.7
Core with a hexagonal cross-section.

and

$$H'_z\big|_{y=\pm\sqrt{3}\ell/2} = \sum_{m-odd}^{(2M-1)} T_m \cdot \cos\left(\frac{m\pi}{\ell} \cdot x\right) \tag{5.76b}$$

where T_m indicates a set of Fourier coefficients in the expression for the torch function. Therefore,

$$H'_z = \sum_{m-odd}^{(2M-1)} T_m \cdot \cos\left(\frac{m\pi}{\ell} \cdot x'\right) \cdot \frac{\cosh(\beta_m \cdot y')}{\cosh(\beta_m \cdot L \cdot 3/2)} \tag{5.77}$$

where

$$\beta_m = \sqrt{\left(\frac{m\pi}{\ell}\right)^2 + \eta^2} \tag{5.77a}$$

Next, we construct two more similar rectangles, each containing two opposite sides of the larger hexagon. Let the field distributions in these regions be given as follows:

$$H''_z = \sum_{m-odd}^{(2M-1)} T_m \cdot \cos\left(\frac{m\pi}{L} \cdot x''\right) \cdot \frac{\cosh(\beta_m \cdot y'')}{\cosh(\beta_m \cdot L \cdot 3/2)} \tag{5.78a}$$

$$H'''_z = \sum_{m-odd}^{(2M-1)} T_m \cdot \cos\left(\frac{m\pi}{L} \cdot x'''\right) \cdot \frac{\cosh(\beta_m \cdot y''')}{\cosh(\beta_m \cdot L \cdot 3/2)} \tag{5.78b}$$

where x'', y'', x''' and y''' are given by Equations 5.65a,b, 5.66a and b, respectively. Therefore,

$$H''_z = \sum_{m-odd}^{(2M-1)} T_m \cdot \cos\left\{\frac{m\pi}{L}\left(y \cdot \frac{\sqrt{3}}{2} - x \cdot \frac{1}{2}\right)\right\} \cdot \frac{\cosh\{\beta_m \cdot (-y \cdot (1/2) - x \cdot (\sqrt{3}/2))\}}{\cosh(\beta_m \cdot L \cdot 3/2)} \tag{5.79a}$$

and

$$H'''_z = \sum_{m-odd}^{(2M-1)} T_m \cdot \cos\left\{\frac{m\pi}{L}\left(y \cdot \frac{\sqrt{3}}{2} + x \cdot \frac{1}{2}\right)\right\} \times \frac{\cosh\{\beta_m \cdot (-y \cdot (1/2) + x \cdot (\sqrt{3}/2))\}}{\cosh(\beta_m \cdot L \cdot 3/2)} \tag{5.79b}$$

Since the target zone, that is the smaller hexagon in Figure 5.7, is common to all these rectangular regions, and the magnetic field is symmetrically distributed over the core section, H_z can be given as

$$H_z = H_z' + H_z'' + H_z''' \tag{5.80}$$

Hence, using Equations 5.64b, 5.64c, 5.77, 5.79a, 5.79b and 5.80, we get

$$\begin{aligned} H_z = \sum_{m-odd}^{(2M-1)} T_m \cdot \Bigg[& \cos\left(\frac{m\pi}{\ell}\cdot x\right)\cdot\frac{\cosh(\beta_m\cdot y)}{\cosh(\beta_m\cdot L\cdot 3/2)} \\ & + \cos\left\{\frac{m\pi}{L}\left(y\cdot\frac{\sqrt{3}}{2} - x\cdot\frac{1}{2}\right)\right\}\cdot\frac{\cosh\{\beta_m\cdot(-y.(1/2) - x\cdot(\sqrt{3}/2))\}}{\cosh(\beta_m\cdot L\cdot 3/2)} \\ & + \cos\left\{\frac{m\pi}{L}\left(y\cdot\frac{\sqrt{3}}{2} + x\cdot\frac{1}{2}\right)\right\}\cdot\frac{\cosh\left\{\beta_m\cdot\left(-y\cdot 1/2 + x\cdot\sqrt{3}/2\right)\right\}}{\cosh(\beta_m\cdot L\cdot 3/2)}\Bigg] \end{aligned} \tag{5.81}$$

While, in view of Equation 5.69, we have

$$H_z\big|_{x=\ell/2} = K \quad \text{over} \ -L/2 < y < L/2 \tag{5.82}$$

Expanding the RHS of this equation into finite Fourier series, one gets

$$H_z\big|_{x=\ell/2} \cong \sum_{n-odd}^{(2N-1)}\left[K\cdot\frac{4}{n\pi}\cdot\sin\left(\frac{n\pi}{2}\right)\right]\cdot\cos\left(\frac{n\pi}{L}\cdot y\right) \tag{5.83}$$

over $-L/2 < y < L/2$.

The value of N should be chosen keeping in mind that Equation 5.82 is satisfied within the desired limits of accuracy. Therefore, from Equations 5.81 and 5.83

$$\begin{aligned} & \sum_{n-odd}^{(2N-1)}\left[K\cdot\frac{4}{n\pi}\cdot\sin\left(\frac{n\pi}{2}\right)\right]\cdot\cos\left(\frac{n\pi}{L}\cdot y\right) \\ & = \sum_{m-odd}^{(2M-1)} T_m\cdot\Bigg[\cos\left\{\frac{m\pi}{L}\left(y\cdot\frac{\sqrt{3}}{2} - L\cdot\frac{\sqrt{3}}{4}\right)\right\}\cdot\frac{\cosh\{\beta_m\cdot(-y\cdot(1/2) - L\cdot(3/4))\}}{\cosh(\beta_m\cdot L\cdot 3/2)} \\ & \qquad + \cos\left\{\frac{m\pi}{L}\left(y\cdot\frac{\sqrt{3}}{2} + L\cdot\frac{\sqrt{3}}{4}\right)\right\}\cdot\frac{\cosh\{\beta_m\cdot(-y\cdot(1/2) + L\cdot(3/4))\}}{\cosh(\beta_m\cdot L\cdot 3/2)}\Bigg] \end{aligned} \tag{5.84}$$

over $-L/2 < y < L/2$.

Now, in view of the orthogonal property of Fourier series, we have

$$\left[K\cdot\frac{4}{\pi}\cdot\frac{1}{n}\cdot\sin\left(\frac{n\pi}{2}\right)\right]\cdot\frac{L}{2}$$
$$=\sum_{m-odd}^{(2M-1)}T_m\cdot\int_{-L/2}^{L/2}\left[\cos\left\{\frac{m\pi}{L}\left(y\cdot\frac{\sqrt{3}}{2}-L\cdot\frac{\sqrt{3}}{4}\right)\right\}\cdot\frac{\cosh\{\beta_m\cdot(-y\cdot(1/2)-L\cdot(3/4))\}}{\cosh(\beta_m\cdot L\cdot 3/2)}\right.$$
$$\left.+\cos\left\{\frac{m\pi}{L}\left(y\cdot\frac{\sqrt{3}}{2}+L\cdot\frac{\sqrt{3}}{4}\right)\right\}\cdot\frac{\cosh\{\beta_m\cdot(-y\cdot(1/2)+L\cdot(3/4))\}}{\cosh(\beta_m\cdot L\cdot 3/2)}\right]$$
$$\times\cos\left(\frac{n\pi}{L}\cdot y\right)\cdot dy\quad\text{for } n=1,3,5,\ldots,(2N-1) \tag{5.85}$$

On integrating the RHS of this equation, and some simplifications, one obtains

$$\left[K\cdot\frac{4}{\pi}\cdot\frac{1}{n}\right]=\sum_{m-odd}^{(2M-1)}T_m\cdot\frac{\left\{\begin{array}{l}\cos(m\pi\cdot(\sqrt{3}/2))\cdot\cosh(\beta_m\cdot L(1/2))\\ +\,j\sin(m\pi\cdot(\sqrt{3}/2))\cdot\sinh(\beta_m\cdot L(1/2))+\cosh(\beta_m\cdot L)\end{array}\right\}}{\cosh(\beta_m\cdot L\cdot 3/2)}$$
$$\times\left\{\frac{4\cdot(m\pi\cdot\sqrt{3}+j\beta_m\cdot L)}{4\cdot(n\pi)^2-(m\pi\cdot\sqrt{3}+j\beta_m\cdot L)^2}\right\}$$
$$+\sum_{m-odd}^{(2M-1)}T_m\cdot\frac{\left\{\begin{array}{l}\cos(m\pi\cdot(\sqrt{3}/2))\cdot\cosh\left(\beta_m\cdot L\frac{1}{2}\right)-j\sin(m\pi\cdot(\sqrt{3}/2))\\ \times\sinh(\beta_m\cdot L(1/2))+\cosh(\beta_m\cdot L)\end{array}\right\}}{\cosh(\beta_m\cdot L\cdot 3/2)}$$
$$\times\left\{\frac{4\cdot(m\pi\cdot\sqrt{3}-j\beta_m\cdot L)}{4\cdot(n\pi)^2-(m\pi\cdot\sqrt{3}-j\beta_m\cdot L)^2}\right\}\quad\text{for}\quad n=1,3,5,\ldots(2N-1) \tag{5.86}$$

The series involved in Equation 5.86 is fast convergent due to the hyperbolic function in the denominator. Taking $M = N$, this equation can be solved, numerically or otherwise, giving the Fourier coefficients T_m, occurring in the expression for the torch function and also for the magnetic field, given by Equations 5.76b and 5.81, respectively.

5.6.3 Cores with Octagonal Cross-Sections

Consider a long conducting core with the cross-section in the shape of a regular octagon, and the length of each of its side is L (vide Figure 5.8). On the surface of the core, there is a uniformly distributed winding carrying alternating current. The magnetic field in the octagonal core satisfies the following boundary condition:

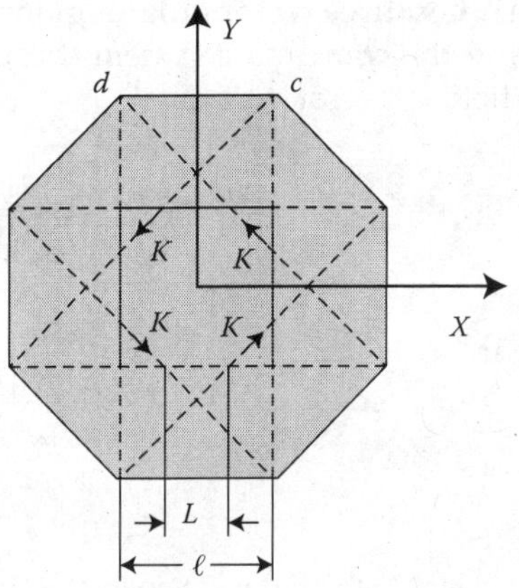

FIGURE 5.8
Core with an octagonal cross-section.

$$H_z\big|_{\text{core surface}} = K_o \tag{5.87}$$

where K_o is the surface current density on the core surface simulating the winding current.

Around this core, imagine a larger octagon symmetrically placed with side length ℓ, as shown in Figure 5.8. Joining the opposite sides of this octagon, construct rectangular regions. One of these four rectangles is labelled as *abcd*. Let the torch function on the side *cd* being an even function, be defined by the following finite Fourier series:

$$H_z\big|_{y=(\ell+L/2)} = H_z\big|_{y=2.914L} = \sum_{m-odd}^{M} T_m \cdot \cos\left(\frac{m\pi}{\ell} \cdot x\right) \tag{5.88}$$

where

$$\ell = L + L\sqrt{2} \tag{5.88a}$$

Since the magnetic field inside the core (or the target region) is an even function of y, the solution of the eddy current equation is obtained as

$$H_z^{(1)} = \sum_{m-odd}^{(2M-1)} T_m \cdot \cos\left(\frac{m\pi}{\ell} \cdot x\right) \cdot \frac{\cosh(\tau_m \cdot y)}{\cosh(\tau_m \cdot 2.914L)} \tag{5.89}$$

where

$$\tau_m = \sqrt{\left(\frac{m\pi}{\ell}\right)^2 + \eta^2} \tag{5.89a}$$

Field distributions in other three rectangular regions can be obtained from Equation 5.89 by rotating the coordinate system through $\pi/4$, $\pi/2$ and $-\pi/4$ radians. The resulting field expressions are

$$H_z^{(2)} = \sum_{m-odd}^{(2M-1)} T_m \cdot \cos\frac{m\pi}{\ell} \cdot \left(\frac{y}{\sqrt{2}} + \frac{x}{\sqrt{2}}\right) \cdot \frac{\cosh \tau_m \cdot ((y/\sqrt{2}) - (x/\sqrt{2}))}{\cosh(\tau_m \cdot 2.914L)} \tag{5.90}$$

$$H_z^{(3)} = \sum_{m-odd}^{(2M-1)} T_m \cdot \cos\left(\frac{m\pi}{\ell} \cdot y\right) \cdot \frac{\cosh(\tau_m \cdot x)}{\cosh(\tau_m \cdot 2.914L)} \tag{5.91}$$

and

$$H_z^{(4)} = \sum_{m-odd}^{(2M-1)} T_m \cdot \cos\frac{m\pi}{\ell} \cdot \left(\frac{y}{\sqrt{2}} - \frac{x}{\sqrt{2}}\right) \cdot \frac{\cosh \tau_m \cdot ((y/\sqrt{2}) + (x/\sqrt{2}))}{\cosh(\tau_m \cdot 2.914L)} \tag{5.92}$$

The distribution of magnetic field in the target region is found as the sum of the above four distributions. Therefore,

$$H_z = H_z^{(1)} + H_z^{(2)} + H_z^{(3)} + H_z^{(4)} \tag{5.93}$$

On the top surface of the target region, that is at $y = \ell/2$, we get from Equations 5.89 through 5.93

$$\begin{aligned} H_z\big|_{y=\ell/2} &= \sum_{m-odd}^{(2M-1)} T_m \cdot \cos\left(\frac{m\pi}{\ell} \cdot x\right) \\ &\times \left[\frac{\cosh(\tau_m \cdot y)}{\cosh(\tau_m \cdot 2.914L)} + \frac{\cosh \tau_m \cdot ((y/\sqrt{2}) - (x/\sqrt{2}))}{\cosh(\tau_m \cdot 2.914L)} \right. \\ &\left. + \frac{\cosh(\tau_m \cdot x)}{\cosh(\tau_m \cdot 2.914L)} + \frac{\cosh \tau_m \cdot ((y/\sqrt{2}) + (x/\sqrt{2}))}{\cosh(\tau_m \cdot 2.914L)}\right]_{y=\ell/2} \end{aligned} \tag{5.94}$$

Since $\ell/2 = 1.207L$, we have

$$\begin{aligned} H_z\big|_{y=\ell/2} &= \sum_{m-odd}^{(2M-1)} T_m \cdot \cos\left(\frac{m\pi}{\ell} \cdot x\right) \\ &\times \left[\frac{\cosh(\tau_m \cdot 1.207L)}{\cosh(\tau_m \cdot 2.914L)} + \frac{\cosh \tau_m \cdot ((1.207L/\sqrt{2}) - (x/\sqrt{2}))}{\cosh(\tau_m \cdot 2.914L)} \right. \\ &\left. + \frac{\cosh(\tau_m \cdot x)}{\cosh(\tau_m \cdot 2.914L)} + \frac{\cosh \tau_m \cdot ((1.207L/\sqrt{2}) + (x/\sqrt{2}))}{\cosh(\tau_m \cdot 2.914L)}\right] \end{aligned} \tag{5.95}$$

In view of Equation 5.88, we also have

$$H_z\Big|_{y=\ell/2} = K_o, \quad \text{over } -\frac{L}{2} < x < \frac{L}{2} \tag{5.96}$$

On expanding the RHS of this equation in half-range finite Fourier series, one gets

$$H_z\Big|_{y=\ell/2} = \sum_{n-odd}^{(2N-1)} \left[K_o \cdot \frac{4}{n\pi} \cdot \sin\left(\frac{n\pi}{2}\right) \right] \cdot \cos\left(\frac{n\pi}{L} \cdot x\right) \tag{5.97}$$

Using the orthogonal property of the Fourier series, Equations 5.95 and 5.96 result in

$$\begin{aligned} \left[K_o \cdot \frac{4}{n\pi} \cdot \sin\left(\frac{n\pi}{2}\right) \right] \cdot \frac{L}{2} &= \sum_{m-odd}^{(2M-1)} T_m \\ &\times \int_{-L/2}^{L/2} \left[\frac{\cosh(\tau_m \cdot 1.207L)}{\cosh(\tau_m \cdot 2.914L)} + \frac{\cosh \tau_m \cdot ((1.207L/\sqrt{2}) - (x/\sqrt{2}))}{\cosh(\tau_m \cdot 2.914L)} \right. \\ &\left. + \frac{\cosh(\tau_m \cdot x)}{\cosh(\tau_m \cdot 2.914L)} + \frac{\cosh \tau_m \cdot ((1.207L/\sqrt{2}) + (x/\sqrt{2}))}{\cosh(\tau_m \cdot 2.914L)} \right] \\ &\times \cos\left(\frac{m\pi}{\ell} \cdot x\right) \cdot \cos\left(\frac{n\pi}{L} \cdot x\right) \cdot dx \quad \text{for } n = 1,3,5,\ldots,(2N-1) \end{aligned} \tag{5.98}$$

On evaluating the integral on the RHS of this equation, the following equation results

$$\left[K_o \cdot \frac{4}{n\pi} \cdot \sin\left(\frac{n\pi}{2}\right) \right] \cdot \frac{L}{2} = \sum_{m-odd}^{(2M-1)} T_m [I_{1m} + I_{2m} + I_{3m} + I_{4m}] \tag{5.99}$$

for $n = 1,3,5,\ldots,(2N-1)$.

In Equation 5.99,

$$I_{1m} = \int_{-L/2}^{L/2} \frac{\cosh(\tau_m \cdot 1.207L)}{\cosh(\tau_m \cdot 2.914L)} \cdot \cos\left(\frac{m\pi}{\ell} \cdot x\right) \cdot \cos\left(\frac{n\pi}{L} \cdot x\right) \cdot dx \tag{5.100a}$$

$$I_{2m} = \int_{-L/2}^{L/2} \frac{\cosh \tau_m \cdot ((1.207L/\sqrt{2}) - (x/\sqrt{2}))}{\cosh(\tau_m \cdot 2.914L)} \cdot \cos\left(\frac{m\pi}{\ell} \cdot x\right) \cdot \cos\left(\frac{n\pi}{L} \cdot x\right) \cdot dx \tag{5.100b}$$

$$I_{3m} = \int_{-L/2}^{L/2} \frac{\cosh(\tau_m \cdot x)}{\cosh(\tau_m \cdot 2.914L)} \cdot \cos\left(\frac{m\pi}{\ell} \cdot x\right) \cdot \cos\left(\frac{n\pi}{L} \cdot x\right) \cdot dx \quad (5.100c)$$

and

$$I_{4m} = \int_{-L/2}^{L/2} \frac{\cosh \tau_m \cdot ((1.207L/\sqrt{2}) + (x/\sqrt{2}))}{\cosh(\tau_m \cdot 2.914L)} \cdot \cos\left(\frac{m\pi}{\ell} \cdot x\right) \cdot \cos\left(\frac{n\pi}{L} \cdot x\right) \cdot dx \quad (5.100d)$$

On integrating

$$I_{1m} = \frac{\cosh(\tau_m \cdot 1.207L)}{\cosh(\tau_m \cdot 2.914L)} \cdot \sin\left(\frac{n\pi}{2}\right) \cdot \cos\left(\frac{m\pi}{2} \cdot \frac{L}{\ell}\right) \cdot \frac{2 \cdot (n\pi/L)}{(n\pi/L)^2 - (m\pi/\ell)^2} \quad (5.101a)$$

$$I_{2m} = -2\left(\frac{n\pi}{L}\right) \cdot \sin\left(\frac{n\pi}{2}\right) \cdot \frac{\cosh(\tau_m \cdot (1.207L/\sqrt{2}))}{\cosh(\tau_m \cdot 2.914L)}$$

$$\times \left[\frac{\cos\left(\frac{m\pi}{2} \cdot \frac{L}{\ell} + j\tau_m \cdot \frac{1}{\sqrt{2}} \cdot \frac{L}{2}\right)}{\left(\frac{m\pi}{\ell} + j\tau_m \cdot \frac{1}{\sqrt{2}}\right)^2 - \left(\frac{n\pi}{L}\right)^2} + \frac{\cos\left(\frac{m\pi}{2} \cdot \frac{L}{\ell} - j\tau_m \cdot \frac{1}{\sqrt{2}} \cdot \frac{L}{2}\right)}{\left(\frac{m\pi}{\ell} - j\tau_m \cdot \frac{1}{\sqrt{2}}\right)^2 - \left(\frac{n\pi}{L}\right)^2} \right] \quad (5.101b)$$

$$I_{3m} = \frac{(n\pi/L) \cdot \sin(n\pi/2)}{\cosh(\tau_m \cdot 2.914L)}$$

$$\times \left[\frac{\cos((m\pi/2) \cdot (L/\ell) + j\tau_m \cdot (L/2))}{((m\pi/\ell) + j\tau_m)^2 - (n\pi/L)^2} + \frac{\cos((m\pi/2) \cdot (L/\ell) - j\tau_m \cdot (L/2))}{((m\pi/\ell) - j\tau_m)^2 - (n\pi/\ell)^2} \right] \quad (5.101c)$$

$$I_{4m} = -2\left(\frac{n\pi}{L}\right) \cdot \sin\left(\frac{n\pi}{2}\right) \cdot \frac{\cosh(\tau_m \cdot (1.207L/\sqrt{2}))}{\cosh(\tau_m \cdot 2.914L)}$$

$$\times \left[\frac{\cos((m\pi/2) \cdot (L/\ell) + j\tau_m \cdot (1/\sqrt{2}) \cdot (L/2))}{((m\pi/\ell) + j\tau_m \cdot (1/\sqrt{2}))^2 - (n\pi/L)^2} \right.$$

$$\left. + \frac{\cos((m\pi/2) \cdot (L/\ell) - j\tau_m \cdot (1/\sqrt{2}) \cdot (L/2))}{\left((m\pi/\ell) - j\tau_m \cdot 1/\sqrt{2}\right)^2 - (n\pi/L)^2} \right] \quad (5.101d)$$

Setting $M = N$, Equation 5.52 can be solved numerically leading to the values of coefficients of the torch function T_m. The distribution of the magnetic field in the core can now be obtained from Equations 5.89 through 5.93.

5.7 Eddy Currents in Circular Cores

Consider a long conducting core with a circular cross-section of radius R, as shown in Figure 5.9. The excitation winding carrying ac currents is simulated by a current sheet with surface current density $\boldsymbol{K}$, placed on the core surface, where

$$\boldsymbol{K} = K_o \cdot e^{j\omega t} \cdot \boldsymbol{a}_\varphi \tag{5.102}$$

At power frequencies, the magnetic field outside the core being negligible, we have

$$H_z\big|_{\rho=R} = K_o \cdot e^{j\omega t} \tag{5.103}$$

Let

$$H_z = F(\rho) \cdot e^{j\omega t} \tag{5.104}$$

Neglecting displacement currents,

$$\nabla \times \boldsymbol{H} \cong \boldsymbol{J} = \sigma \boldsymbol{E} \tag{5.105}$$

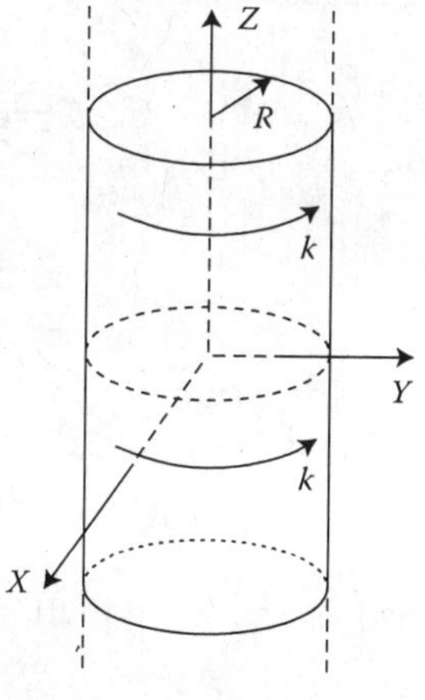

FIGURE 5.9
Cylindrical core with a surface current sheet.

Therefore,

$$\nabla \times \nabla \times \boldsymbol{H} = \sigma \nabla \times \boldsymbol{E} = -\mu\sigma \frac{d\boldsymbol{H}}{dt} = -j\omega\mu\sigma \boldsymbol{H} = -j\omega\mu\sigma F(\rho) \cdot e^{j\omega t} \cdot \boldsymbol{a}_\varphi \tag{5.106}$$

Now, since

$$\nabla \times \boldsymbol{H} = \left[\frac{1}{\rho} \cdot \frac{\partial H_z}{\partial \varphi} - \frac{\partial H_\varphi}{\partial z}\right]\boldsymbol{a}_\rho + \left[\frac{\partial H_\rho}{\partial z} - \frac{\partial H_z}{\partial \rho}\right]\boldsymbol{a}_\varphi + \frac{1}{\rho} \cdot \left[\frac{\partial(\rho H_\varphi)}{\partial \rho} - \frac{\partial H_z}{\partial \varphi}\right]\boldsymbol{a}_z \tag{5.107}$$

and

$$H_\rho = H_\varphi = 0 \tag{5.107a}$$

$$\nabla \times \boldsymbol{H} = -\frac{\partial \boldsymbol{H}_z}{\partial \rho}\boldsymbol{a}_\varphi = \left[-\frac{dF(\rho)}{d\rho} \cdot e^{j\omega t}\right]\boldsymbol{a}_\varphi \tag{5.107b}$$

Thus, $$\nabla \times \nabla \times \boldsymbol{H} = \nabla \times \left[-\frac{dF(\rho)}{d\rho} \cdot e^{j\omega t}\right]\boldsymbol{a}_\varphi = -\frac{1}{\rho} \cdot \left[\frac{d}{d\rho}\left\{\rho \frac{dF(\rho)}{d\rho}\right\}\right] \cdot e^{j\omega t}\boldsymbol{a}_\varphi \tag{5.108}$$

Equating the RHS of Equations 5.106 and 5.108, we get

$$\frac{1}{\rho} \cdot \left[\frac{d}{d\rho}\left\{\rho \frac{dF(\rho)}{d\rho}\right\}\right] = j\omega\mu\sigma F(\rho) \tag{5.108a}$$

or,

$$\frac{d^2F}{d\rho^2} + \frac{1}{\rho} \cdot \frac{dF}{d\rho} + k^2 \cdot F = 0 \tag{5.108b}$$

where $k^2 = -j\omega\mu\sigma$.

Or

$$k = j^{3/2}\sqrt{\omega\mu\sigma} \stackrel{def}{=} j^{3/2} \cdot \ell \tag{5.108c}$$

$$\ell = \sqrt{\omega\mu\sigma} \tag{5.108d}$$

Equation 5.108b is identified as a Bessel equation of zero order. Its complete solution is[9]

$$F = c \cdot J_0(k\rho) + c' \cdot Y_0(k\rho) \tag{5.109}$$

where $J_o(\rho)$ and $Y_o(\rho)$ are Bessel functions of the first and second kind, respectively, while c and c' indicate arbitrary constants. Now, since

$$J_o(\rho)\big|_{\rho=0} = 1 \quad (5.109a)$$

$$Y_o(\rho)\big|_{\rho=0} = -\infty \quad (5.109b)$$

Therefore, the arbitrary constant c' is zero. Thus, Equation 5.109 can be rewritten as

$$F = c \cdot J_o(k\rho) \quad (5.110)$$

Now, in view of Equations 5.103, 5.104 and 5.110, we get

$$H_z = K_o \cdot \frac{J_o(j^{3/2} \cdot \ell \cdot \rho)}{J_o(j^{3/2} \cdot \ell \cdot R)} \cdot e^{j\omega t} \quad (5.111)$$

Bessel function with complex arguments can be resolved into real and imaginary parts using Kelvin functions[9] as

$$J_o(j^{3/2} \cdot \ell \cdot \rho) = \mathrm{ber}(\ell \cdot \rho) + j \cdot \mathrm{bei}(\ell \cdot \rho) \quad (5.112)$$

Thus,

$$H_z = K_o \cdot \frac{\mathrm{ber}(\ell \cdot \rho) + j \cdot \mathrm{bei}(\ell \cdot \rho)}{\mathrm{ber}(\ell \cdot R) + j \cdot \mathrm{bei}(\ell \cdot R)} \cdot e^{j\omega t} \quad (5.113)$$

In view of Equations 5.104 and 5.106b, we get

$$\nabla \times \boldsymbol{H} = -\frac{\partial H_z}{\partial \rho} \boldsymbol{a}_\varphi = \left[-\frac{dF(\rho)}{d\rho} \cdot e^{j\omega t} \right] \boldsymbol{a}_\varphi \quad (5.114)$$

Eddy current density can now be obtained by using the equation

$$\boldsymbol{J} \cong \nabla \times \boldsymbol{H} \quad (5.115)$$

5.8 Distribution of Current Density in Circular Conductors

Skin effects in slot-embedded conductors with trapezoidal,[10] rectangular[11] and circular[12] cross-sections have been obtained. For long isolated conductors with a rectangular cross-section, skin effect has been obtained[13] by assuming constant current density over the conductor surface. These assumptions, however, are not needed for conductors with a circular cross-section located in free space. Consider a long solid conductor with a circular cross-section of radius R, shown in Figure 5.10. This conductor carries an ac current I, at power frequency ω. The current density $\boldsymbol{J}$ in the conductor section is a function of the radial distance, ρ. There is only axial component of this current J_z, satisfying eddy current equation in the cylindrical system of space coordinates as below:

$$\frac{d^2 J_z}{d\rho^2} + \frac{1}{\rho} \cdot \frac{dJ_z}{d\rho} + k^2 \cdot J_z = 0 \tag{5.116}$$

where

$$k = j^{3/2}\sqrt{\omega\mu\sigma} \overset{def}{=} j^{3/2} \cdot \ell \tag{5.116a}$$

Ignoring the factor $e^{j\omega t}$, in the expression of the current density in the circular conductor, the solution of this equation is given as[9]

$$J_z = c \cdot J_o(k\rho) \tag{5.117}$$

where c indicates an arbitrary constant.

Let the total current in the conductor be I, then

$$I = \int_0^R c \cdot J_o(k\rho) \cdot 2\pi\rho d\rho = c \cdot 2\pi \cdot \int_0^R J_o(k\rho) \cdot \rho d\rho = c \cdot \frac{2\pi R}{k} \cdot J_1(kR) \tag{5.118a}$$

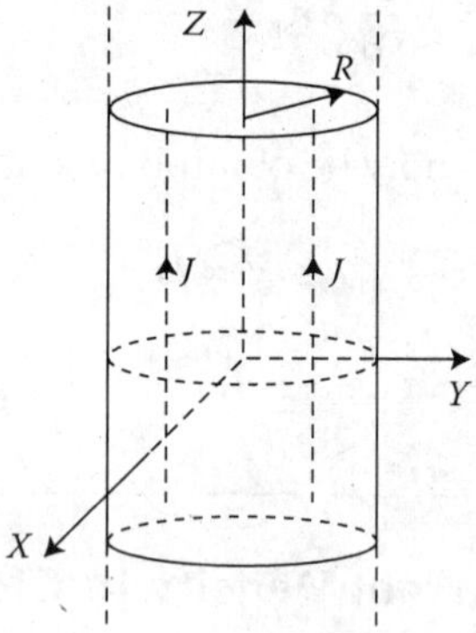

FIGURE 5.10
Cylindrical conductor carrying ac current.

Therefore,

$$c = \frac{I \cdot k}{2\pi R} \cdot \frac{1}{J_1(kR)} \tag{5.118b}$$

Thus, from Equation 5.117, we get

$$J_z = \frac{I \cdot k}{2\pi R} \cdot \frac{J_o(k\rho)}{J_1(kR)} \tag{5.119}$$

since

$$J_o(k\rho)\big|_{\rho=0} = 1 \tag{5.120}$$

In view of Equation 5.119, the current density along the conductor axis is

$$J_z\big|_{\rho=0} = \frac{I \cdot k}{2\pi R} \cdot \frac{1}{J_1(kR)} \tag{5.121}$$

where

$$J_1(kR) = J_1(j^{3/2} \cdot \ell \cdot R) = \text{ber}_1(\ell \cdot R) + j\text{bei}_1(\ell \cdot R) \tag{5.121a}$$

Thus, the expression for the current density can be given in terms of Kelvin functions of the order of zero and one:

$$J_z = I \cdot \frac{j^{3/2} \cdot \ell}{2\pi R} \cdot \frac{\text{ber}(\ell \cdot \rho) + j \cdot \text{bei}(\ell \cdot \rho)}{\text{ber}_1(\ell \cdot R) + j\text{bei}_1(\ell \cdot R)} \tag{5.122}$$

Thus, the magnitude of the current density as a function of the radial distance ρ from the conductor axis is given as

$$|J_z| = \sqrt{J_z \cdot J_z^*} = |I| \cdot \frac{\ell}{2\pi R} \cdot \sqrt{\frac{\text{ber}^2(\ell \cdot \rho) + \text{bei}^2(\ell \cdot \rho)}{\text{ber}_1^2(\ell \cdot R) + \text{bei}_1^2(\ell \cdot R)}} \tag{5.123}$$

5.9 Eddy Currents in Laminated Rectangular Cores

Eddy current loss in an isolated thin-conducting plate is proportional to the square of its thickness.[10] This loss can thus be reduced if laminated cores are used instead of solid iron cores. It has been noticed that the advantage of laminating iron cores is defeated unless a thick insulation coating is given on the two surfaces of each lamination.[3] This is because if laminations are placed close to one another, the interlaminar capacitance predominates, the resulting eddy current loss tends to become linearly proportional to its thickness and not to the thickness squared.

Figure 5.11 shows a rectangular core consisting of n-insulated laminations, each of width W and overall thickness T. Let the insulation thickness on each side of a lamination be $T_1/2$ and its iron thickness be T_2. Further, let the corners of the rectangular core be located at $(-W/2, 0)$, $(W/2, 0)$, $(-W/2, nT)$ and $(W/2, nT)$. In this figure, insulation regions are indicated as Region-$0', 1', 2', 3', \ldots, m', \ldots, n'$. The iron regions are indicated as Region-$1, 2, \ldots, m, \ldots, n$.

The exciting coil is wound around the long rectangular core and carries an alternating current i, where

$$i = Ie^{j\omega t} \tag{5.124}$$

It is simulated by a surface current density K_o:

$$K_o = I \cdot N \tag{5.125}$$

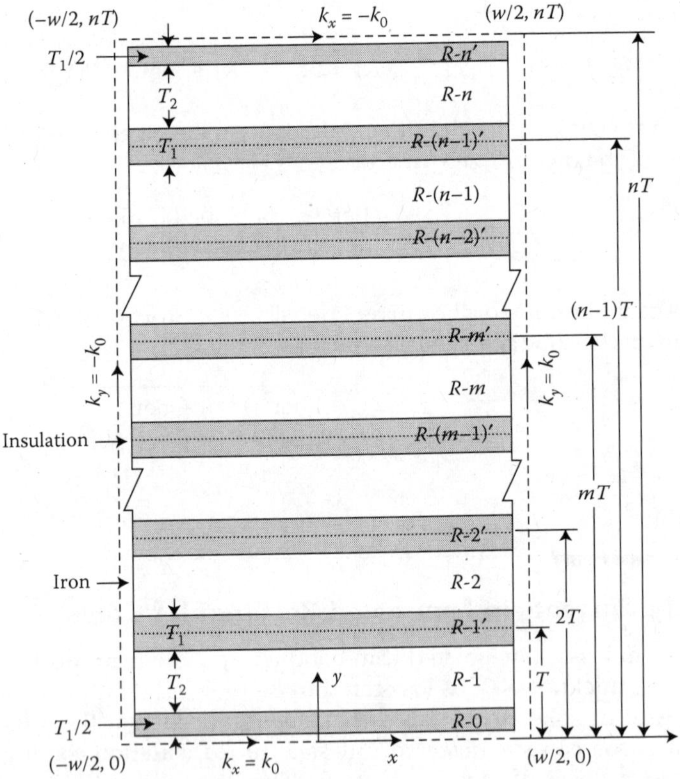

FIGURE 5.11
Cross-sectional view of a laminated rectangular core. (Courtesy of The Electromagnetics Academy.)

where N is the number of turns per unit length of the coil. The current-carrying coil will produce time-varying magnetic field, H_z, in the core and eddy current density with components J_x and J_y, in the conducting regions and displacement currents in the insulation regions of the core. The magnetic field outside the coil is neglected. For the long rectangular core with a uniformly distributed current sheet, the magnetic field is entirely axial and independent of z-coordinate, along the axial direction. It is assumed that the permeability μ, for the iron regions, permittivity ε, for the insulation regions and conductivity (σ, σ′), for both types of regions, are constant. Thus, from Maxwell's equations for harmonic fields, in charge-free regions

$$\frac{\partial^2 H_z}{\partial x^2} + \frac{\partial^2 H_z}{\partial y^2} = -\gamma^2 H_z \tag{5.126}$$

for iron regions , where

$$\gamma = \sqrt{(-j\omega\mu)\cdot(\sigma + j\omega\varepsilon_o)} \tag{5.126a}$$

and

$$\frac{\partial^2 H_z}{\partial x^2} + \frac{\partial^2 H_z}{\partial y^2} = -(\gamma')^2 H_z \tag{5.127}$$

for insulation regions, where

$$\gamma' = \sqrt{(-j\omega\mu_o)\cdot(\sigma' + j\omega\varepsilon)} \tag{5.127a}$$

Solutions of Equations 5.126 and 5.127 can be used to determine the components of eddy current densities in iron and insulation regions, since for iron regions

$$J_x = \delta\cdot\frac{\partial H_z}{\partial y} \tag{5.128a}$$

and

$$J_y = -\delta\cdot\frac{\partial H_z}{\partial x} \tag{5.128b}$$

where

$$\delta = \frac{\sigma}{\sigma + j\omega\varepsilon_o} \tag{5.128c}$$

while for insulation regions

$$J_x = \delta' \cdot \frac{\partial H_z}{\partial y} \tag{5.129a}$$

and

$$J_y = -\delta' \cdot \frac{\partial H_z}{\partial x} \tag{5.129b}$$

where

$$\delta' = \frac{\sigma'}{\sigma' + j\omega\varepsilon} \tag{5.129c}$$

Now consider the region –0′. It extends over $-W/2 \le x \le W/2$, and $0 \le y \le T_1/2$. Since

$$H_z^{0'}\Big|_{x=\pm W/2} = K_o \tag{5.130a}$$

and

$$H_z^{0'}\Big|_{y=0} = K_o \tag{5.130b}$$

Therefore,

$$H_z^{0'} = K_o \cdot \frac{\cos(\gamma' x)}{\cos(\gamma' W/2)} + \sum_{p-odd}^{\infty}\left[a_p^{0'} \cdot \frac{\sinh(\alpha' y)}{\sinh(\alpha'_p T_1/2)} - b_p^{0'} \cdot \frac{\sinh\{\alpha'_p (y - T_1/2)\}}{\sinh(\alpha'_p T_1/2)}\right] \times \cos\left(\frac{p\pi}{W} \cdot x\right) \tag{5.131}$$

where

$$\alpha'_p = \sqrt{\left(\frac{p\pi}{W}\right)^2 - (\gamma')^2} \tag{5.131a}$$

and in view of Equation 5.130b,

$$b_p^{0'} = K_o \cdot \frac{2 \cdot \sin(p\pi/2)}{(p\pi/2)} \cdot \frac{(\gamma' W/\pi)^2}{[(\gamma' W/\pi)^2 - p^2]} \tag{5.131b}$$

for $p = 1,3,5,\ldots$

where $a_p^{0'}$ indicates a set of arbitrary constants.

The components of eddy current density in this region are

$$J_x^{0'} = \sum_{p-odd}^{\infty} (\delta' \cdot \alpha_p') \cdot \left[a_p^{0'} \cdot \frac{\cosh(\alpha_p' y)}{\sinh(\alpha_p' T_1/2)} - b_p^{0'} \cdot \frac{\cosh\{\alpha_p' (y - T_1/2)\}}{\sinh(\alpha_p' T_1/2)} \right] \cdot \cos\left(\frac{p\pi}{W} \cdot x \right) \tag{5.132a}$$

and

$$J_y^{0'} = K_o \cdot (\delta' \cdot \gamma') \cdot \frac{\sin(\gamma' x)}{\cos(\gamma' W/2)} + \sum_{p-odd}^{\infty} \left(\delta' \cdot \frac{P\pi}{W} \right) \cdot \left[a_p^{0'} \cdot \frac{\sinh(\alpha_p' y)}{\sinh(\alpha_p' T_1/2)} - b_p^{0'} \cdot \frac{\sinh\{\alpha_p' (y - T_1/2)\}}{\sinh(\alpha_p' T_1/2)} \right] \cdot \sin\left(\frac{p\pi}{W} \cdot x \right) \tag{5.132b}$$

Next, consider the region $-m'$. This region is extending over $-W/2 \le x \le W/2$, and $(m \cdot T - T_1/2) \le y \le (m \cdot T + T_1/2)$, for $m' = 1', 2', \ldots, n'$.

Since

$$H_z^{m'}\Big|_{x=\pm W/2} = K_o \tag{5.133}$$

Therefore,

$$H_z^{m'} = K_o \cdot \frac{\cos(\gamma' x)}{\cos(\gamma' W/2)} + \sum_{p-odd}^{\infty} \left[a_p^{m'} \cdot \frac{\sinh\{\alpha_p' (y - mT + T_1/2)\}}{\sinh(\alpha_p' T_1)} - b_p^{m'} \cdot \frac{\sinh\{\alpha_p' (y - mT - T_1/2)\}}{\sinh(\alpha_p T_1)} \right] \cdot \cos\left(\frac{p\pi}{W} \cdot x \right) \tag{5.134}$$

for $m' = 1', 2', \ldots (n-1)'$. Where $a_p^{m'}$ and $b_p^{m'}$ indicate two sets of arbitrary constants.

The components of eddy current density in this region are

$$J_x^{m'} = \sum_{p-odd}^{\infty} (\delta' \cdot \alpha_p') \cdot \left[a_p^{m'} \cdot \frac{\cosh\{\alpha_p' (y - mT + T_1/2)\}}{\sinh(\alpha_p' T_1)} - b_p^{m'} \cdot \frac{\cosh\{\alpha_p' (y - mT - T_1/2)\}}{\sinh(\alpha_p' T_1)} \right] \cdot \cos\left(\frac{p\pi}{W} \cdot x \right) \tag{5.135a}$$

for $m' = 1', 2', \ldots, (n-1)'$.

and

$$J_y^{m'} = K_o \cdot (\delta' \cdot \gamma') \cdot \frac{\sin(\gamma' x)}{\cos(\gamma' W/2)} + \sum_{p-odd}^{\infty} \left(\delta' \cdot \frac{P\pi}{W} \right) \cdot \left[a_p^{m'} \cdot \frac{\sinh\{\alpha'_p \, (y - mT + T_1/2)\}}{\sinh(\alpha'_p \, T_1)} \right.$$

$$\left. - \, b_p^{m'} \cdot \frac{\sinh\{\alpha'_p \, (y - mT - T_1/2)\}}{\sinh(\alpha'_p T_1)} \right] \cdot \sin\left(\frac{p\pi}{W} \cdot x \right) \tag{5.135b}$$

for $m' = 1', 2', \ldots, (n-1)'$.

The region $-n'$ extends over $-W/2 \le x \le W/2$, and $(n \cdot T - T_1/2) \le y \le (n \cdot T)$. Since

$$H_z^{n'} \Big|_{x=\pm W/2} = K_o \tag{5.136a}$$

and

$$H_z^{n'} \Big|_{y=n \cdot T} = K_o \tag{5.136b}$$

This gives

$$H_z^{n'} = K_o \cdot \frac{\cos(\gamma' x)}{\cos(\gamma' W/2)} + \sum_{p-odd}^{\infty} \left[a_p^n \cdot \frac{\sinh\{\alpha'_p \, (y - nT + T_1/2)\}}{\sinh(\alpha'_p \, T_1/2)} \right.$$

$$\left. - b_p^{n'} \cdot \frac{\sinh\{\alpha'_p \, (y - nT)\}}{\sinh(\alpha'_p \, T_1/2)} \right] \cdot \cos\left(\frac{p\pi}{W} \cdot x \right) \tag{5.137a}$$

where

$$a_p^{n'} = K_o \cdot \frac{2.\sin((p\pi)/2)}{((p\pi)/2)} \cdot \frac{(\gamma' W/\pi)^2}{[(\gamma' W/\pi)^2 - p^2]} \tag{5.137b}$$

and $b_p^{n'}$ indicates a set of arbitrary constants, for $p = 1, 3, 5, \ldots$

The components of eddy current density in this region are

$$J_x^{n'} = \sum_{p-odd}^{\infty} (\delta' \cdot \alpha'_p) \cdot \left[a_p^{n'} \cdot \frac{\cosh\{\alpha'_p \, (y - nT + T_1/2)\}}{\sinh(\alpha'_p \, T_1/2)} \right.$$

$$\left. - \, b_p^{n'} \cdot \frac{\cosh\{\alpha'_p \, (y - nT)\}}{\sinh(\alpha'_p \, T_1/2)} \right] \cdot \cos\left(\frac{p\pi}{W} \cdot x \right) \tag{5.138a}$$

and

$$J_y^{n'} = K_o \cdot (\delta' \cdot \gamma') \cdot \frac{\sin(\gamma' x)}{\cos(\gamma' W/2)} + \sum_{p-odd}^{\infty} \left(\delta' \cdot \frac{P\pi}{W} \right)$$

$$\times \left[a_p^{n'} \cdot \frac{\sinh\{\alpha_p' (y - nT + T_1/2)\}}{\sinh(\alpha_p' T_1/2)} - b_p^{n'} \cdot \frac{\sinh\{\alpha_p' (y - nT)\}}{\sinh(\alpha_p' T_1/2)} \right] \cdot \sin\left(\frac{p\pi}{W} \cdot x \right) \tag{5.138b}$$

Now, consider the conducting region $-m$, for $m = 1, 2, 3, \ldots, n$. This region extends over $-W/2 \le x \le W/2$, and $(mT - T_1/2 - T_2) \le y \le (mT - T_1/2)$. Since

$$H_z^m \Big|_{x=\pm W/2} = K_o \tag{5.139}$$

thus

$$H_z^m = K_o \cdot \frac{\cos(\gamma x)}{\cos(\gamma W/2)} + \sum_{p-odd}^{\infty} \left[a_p^m \cdot \frac{\sinh\{\alpha_p (y - mT + T_1/2 + T_2)\}}{\sinh(\alpha_p T_2)} - b_p^m \cdot \frac{\sinh\{\alpha_p (y - mT + T_1/2)\}}{\sinh(\alpha_p T_2)} \right] \cdot \cos\left(\frac{p\pi}{W} \cdot x \right) \tag{5.140}$$

for $m = 1, 2, 3, \ldots, n$.

Where

$$\alpha_p = \sqrt{\left(\frac{P\pi}{W} \right)^2 - \gamma^2} \tag{5.140a}$$

while a_p^m and b_p^m indicate two sets of arbitrary constants.

The components of eddy current density in this region are

$$J_x^m = + \sum_{p-odd}^{\infty} \left(\delta \cdot \frac{P\pi}{W} \right) \cdot \left[a_p^m \cdot \frac{\sinh\{\alpha_p (y - mT + T_1/2 + T_2)\}}{\sinh(\alpha_p T_2)} - b_p^m \cdot \frac{\sinh\{\alpha_p (y - mT + T_1/2)\}}{\sinh(\alpha_p T_2)} \right] \cdot \sin\left(\frac{p\pi}{W} \cdot x \right) \tag{5.141a}$$

for $m = 1, 2, 3, \ldots, n$.

and

$$J_y^m = K_o \cdot (\delta \cdot \gamma) \cdot \frac{\sin(\gamma x)}{\cos(\gamma W/2)} + \sum_{p-odd}^{\infty} \left(\delta \cdot \frac{P\pi}{W}\right) \cdot \left[a_p^m \cdot \frac{\sinh\{\alpha_p(y - mT + T_1/2 + T_2)\}}{\sinh(\alpha_p T_2)} - b_p^m \cdot \frac{\sinh\{\alpha_p(y - mT + T_1/2)\}}{\sinh(\alpha_p T_2)} \right] \times \sin\left(\frac{p\pi}{W} \cdot x\right) \quad (5.141b)$$

for $m = 1, 2, 3, \ldots, n$.

Expressions for field distributions are found in terms of arbitrary constants. These arbitrary constants are identified as coefficients of Fourier series describing the distribution of the magnetic field at various boundary surfaces. Those already found are given by Equations 5.131b and 5.137b. In order to evaluate the remaining arbitrary constants, equations relating various arbitrary constants are developed considering the continuity of magnetic fields, H_z and electric field E_x at boundaries between adjacent regions. A detailed treatment leading to a simplified solution is given in Appendix 5.

Approximate treatments neglecting the variation of field distribution from lamination to lamination are found in the literature.[13,14] It has been observed[14] that eddy current loss in *isolated* thin plates is proportional to the square of plate thickness. However, in the case of many insulated thin plates forming a *laminated core,* if the thickness of insulation on these plates is not sufficient, the eddy current distribution in each plate is modified due to *close vicinity of nearby plates.* The resulting eddy current loss in each plate could even be proportional to the plate thickness.[13] This will defeat the very purpose of laminating a core.

Electric charges are deposited on each iron-insulation interface of the laminated core,[14–16] indicating the presence of distributed capacitors in eddy current paths. The laminated cores of phase-shifting transformers and those of many rotating electrical machines are subjected to rotating electromagnetic fields. The interlaminar surface charge distributions in these machines give rise to convection currents on the iron-insulation interfaces.[15] These currents influence the distribution of eddy currents and the resulting eddy current loss.

PROJECT PROBLEMS

1. Show that the eddy current loss in a large isolated plate due to alternating excitation current under certain conditions is proportional to the square of the plate thickness.
2. Obtain the distribution of the magnetic field intensity in a long core with a regular pentagonal cross-section, due to a uniformly distributed winding on its surface carrying alternating current.

3. Obtain the distribution of the eddy current density in a long core with a regular polygonal cross-section with n number of sides, due to a uniformly distributed winding on its surface carrying alternating current (for n-even).
4. Determine the distribution of the eddy current density in a long core with a regular polygonal cross-section with n number of sides, due to a uniformly distributed winding on its surface carrying alternating current (for n-odd).
5. Find the distribution of magnetic field intensity in a long hollow circular conducting core carrying alternating current in its excitation winding. What should be the inner and outer radii of this core if it is to carry the same excitation current as in the excitation winding of a solid core of radius R, so that the inductance is same for the two cores. Plot a graph showing the two radii as functions of frequency.
6. Find the distribution of current density in a long hollow circular conductor carrying alternating current. What should be the inner and outer radii of this conductor if it is to carry the same current as the one flowing in a solid conductor of radius R, so that in both conductors, the ohmic loss per unit length, that is W_L:

$$W_L = \iint_s \left(\frac{1}{2\sigma} \cdot J \bullet J^* \right) ds$$

is the same. Here, the integration is over the cross-sectional area of the conductor. Plot a graph showing the two radii as functions of frequency.
7. For eddy currents in laminated rectangular cores, determine all arbitrary constants without resorting to any additional assumption. Then evaluate eddy current loss in each of n laminations. Use the result to find optimum dimensions to minimise the eddy current loss.
8. Show that the eddy current loss in a laminated core is proportional to its thickness raised to the power p, where $2 > p > 1$. Further, show that as the thickness of the interlaminar insulation increases, the value of p approaches to 2, while as this thickness approaches to zero, the value of p approaches to 1.

References

1. Basu, S., On some problems of electromagnetic field in electrical machines, PhD thesis, Department of Electrical Engineering, University of Roorkee, Roorkee, India, 1969.

2. Mukerji, S. K., George, M., Ramamurthy, M. B. and Asaduzzaman, K., Eddy currents in solid rectangular cores, *Progress in Electromagnetics Research B*, 7, 117–131, 2008.
3. Guru, B. H. and Hiziroglu, H. R., *Electromagnetic Field Theory Fundamentals*, PWS Publishing Company, Boston, MA, pp. 511–535, 1998.
4. Kraus, J. D. and Fleisch, D. A., *Electromagnetics with Applications*, 4th Edn., McGraw-Hill International Editions, New York, pp. 552–557, 1999.
5. Mukerji, S. K., Ramamurthy, M. B. and Goel, S. K., Analytical solutions of electromagnetic fields in prismatic regions, *International Journal of Electric Engineering Education*, 45(1), 17–25, 2008.
6. Mukerji, S. K. and Singh, Y. P., Eddy currents in long solid conducting cores with triangular cross-sections, *National Conference on Recent Advances in Technology and Engineering (RATE 2013)*, Mangalayatan University, Aligarh, India, pp. 1–3, 2013.
7. Mukerji, S. K., Singh, Y. P., George, M. and Pooja, G., Eddy currents in cores with triangular cross-sections, *International Journal of Applied Engineering and Computer Science*, 1(1), 29–32, January 2014.
8. Mukerji, S. K. and Goel, S. K., Electromagnetic fields in isosceles right-angled triangular regions, *Proceedings of the 19th International Conference on Computer Aided Design /Computer Aided Manufacturing, Robotics and Factories of the Future, CARS and FoF 2003*, Kuala Lumpur, Malaysia (SIRIM, Kuala-Lumpur), Vol. 2, pp. 615–619, 2003.
9. Abramowitz, M. and Segun, I. A., *Handbook of Mathematical Functions*, Dover, New York, pp. 253–293, 1965.
10. Buchholz, H., The two dimensional skin-effect in a trapezoidal slot-embedded conductor carrying alternating current, *Arch. Elektrotech. (Germany)*, 49(5), 291–298, 1965, In German.
11. Swann, S. A. and Salmon, J. W., Effective resistance and reactance of a rectangular conductor placed in a semi-closed slot, *Proceedings of the Institution of Electrical Engineers*, UK, Vol. 110, Part C, p. 1656, September 1963.
12. Swann, S. A. and Salmon, J. W., Effective resistance and reactance of a solid cylindrical conductor placed in a semi-closed slot, *Proceedings of the Institution of Electrical Engineers*, UK, Vol. 109, Part C, p. 611, July 1962.
13. Subbarao, V., *Eddy Currents in Linear and Non-Linear Media*, Omega Scientific Publishers, New Delhi, pp. 36–39, 1991.
14. Bewley, L. V., *Two-Dimensional Fields in Electrical Engineering*, Dover, New York, pp. 81–83, 1963.
15. Mukerji, S. K., George, M., Ramamurthy, M. B. and Asaduzzaman, K., Eddy currents in laminated rectangular cores, *Progress in Electromagnetics Research*, 83, 435–445, 2008.
16. Mukerji, S. K., Srivastava, D. S., Singh, Y. P. and Avasthi, D. V., Eddy current phenomena in laminated structures due to travelling electromagnetic fields, *Progress in Electromagnetics Research M*, 18, 159–169, 2011.

6

Laminated-Rotor Polyphase Induction Machines

6.1 Introduction

Polyphase induction machines are generally classified on the basis of construction of their rotors. These rotors may be of laminated or unlaminated forms. The squirrel cage and slip-ring type induction machines belong to the laminated form, while solid rotor induction machines and drag cup type induction machines belong to the unlaminated form.

In the laminated cores used for cage-rotor and wound-rotor induction machines, the slotted region of the rotor is isotropic but inhomogeneous. This region can, however, be represented as an equivalent anisotropic homogeneous region that radially extends from the rotor air-gap surface to the base of rotor slots. For the saturated rotor teeth, the permeability may be chosen to be of a finite constant value to carry out a linear treatment. Also, since the rotor core beyond the slotted region can be presumed to be magnetically unsaturated, the permeability for this region can be taken as infinite. In view of the above, the field analyses of laminated-rotor induction machines need a rigorous mathematical treatment. The field analysis for anisotropic media and its application to the laminated-rotor induction machines also deserves due attention. The effects of skewed rotor slots in laminated-rotor induction machines also need to be properly addressed in terms of field theory.

In a simplified treatment for the eddy current loss in laminated cores, the inhomogeneous laminated core region is normally simulated by an anisotropic homogeneous region. The slotted region of laminated-rotor induction machines can likewise be represented by an equivalent anisotropic homogeneous region. In an exhaustive treatment of squirrel cage induction machines, Mishkin[3] simulated the heterogeneous isotropic slotted regions by homogeneous anisotropic regions. This chapter deals with the field analysis of laminated-rotor induction machines encompassing all the above referred aspects.

6.2 Two-Dimensional Fields in Anisotropic Media

Consider an anisotropic homogeneous medium characterised by conductivity $[\sigma]$, permeability $[\mu]$ and permittivity $[\epsilon]$, such that

$$[\sigma] = (\sigma_x, \sigma_y, \sigma_z) \tag{6.1a}$$

$$[\mu] = (\mu_x, \mu_y, \mu_z) \tag{6.1b}$$

$$[\epsilon] = (\epsilon_x, \epsilon_y, \epsilon_z) \tag{6.1c}$$

while the components of complex conductivity are defined as

$$\bar{\sigma}_x \stackrel{def}{=} \sigma_x + j\omega\epsilon_x \tag{6.2a}$$

$$\bar{\sigma}_y \stackrel{def}{=} \sigma_y + j\omega\epsilon_y \tag{6.2b}$$

$$\bar{\sigma}_z \stackrel{def}{=} \sigma_z + j\omega\epsilon_z \tag{6.2c}$$

Let there be a two-dimensional electromagnetic field that is independent of x-coordinate, varies periodically with y-coordinate as well as with time-t. This variation is given by the factor $e^{j(\omega t - \ell y)}$, where the time period is $2\pi/\omega$ and the wave length is $2\pi/\ell$, that is, two pole-pitches. To determine field variation with z-coordinate, we proceed with the Maxwell equation:

$$\nabla \times \boldsymbol{H} = \boldsymbol{J} + \frac{\partial \boldsymbol{D}}{\partial t} \tag{6.3}$$

Thus,

$$\frac{\partial H_z}{\partial y} - \frac{\partial H_y}{\partial z} = \bar{\sigma}_x E_x \tag{6.3a}$$

$$\frac{\partial H_x}{\partial z} - \frac{\partial H_z}{\partial x} = \bar{\sigma}_y E_y \tag{6.3b}$$

$$\frac{\partial H_y}{\partial x} - \frac{\partial H_x}{\partial y} = \bar{\sigma}_z E_z \tag{6.3c}$$

Since the field is independent of x-coordinate, Equations 6.3b and 6.3c reduce to

$$\frac{\partial H_x}{\partial z} = \bar{\sigma}_y E_y \tag{6.4a}$$

$$\frac{\partial H_x}{\partial y} = -\bar{\sigma}_z E_z \tag{6.4b}$$

Further,

$$\nabla \times \boldsymbol{E} = -\frac{\partial \boldsymbol{B}}{\partial t} \tag{6.5}$$

Thus,

$$\frac{\partial E_z}{\partial y} - \frac{\partial E_y}{\partial z} = -j\omega\mu_x H_x \tag{6.5a}$$

$$\frac{\partial E_x}{\partial z} - \frac{\partial E_z}{\partial x} = -j\omega\mu_y H_y \tag{6.5b}$$

$$\frac{\partial E_y}{\partial x} - \frac{\partial E_x}{\partial y} = -j\omega\mu_z H_z \tag{6.5c}$$

Again, the field being independent of *x*-coordinate, Equations 6.5b and 6.5c reduce to

$$\frac{\partial E_x}{\partial z} = -j\omega\mu_y H_y \tag{6.6a}$$

$$\frac{\partial E_x}{\partial y} = j\omega\mu_z H_z \tag{6.6b}$$

Now, using Equations 6.3a, 6.6a and 6.6b, we get

$$-\frac{1}{\mu_z} \cdot \frac{\partial^2 E_x}{\partial y^2} - \frac{1}{\mu_y} \cdot \frac{\partial^2 E_x}{\partial z^2} = -j\omega\bar{\sigma}_x E_x$$

Thus,

$$\frac{\partial^2 E_x}{\partial z^2} = k_e^2 \cdot E_x \tag{6.7}$$

where

$$k_e^2 = \frac{\mu_y}{\mu_z} \cdot \ell^2 + j\omega\bar{\sigma}_x\mu_y \tag{6.7a}$$

Hence, on solving Equation 6.7, we get

$$E_x = \left[c_1 \cdot e^{k_e \cdot z} + c_2 \cdot e^{-k_e \cdot z}\right] \cdot e^{j(\omega \cdot t - \ell \cdot y)} \tag{6.8}$$

where c_1 and c_2 indicate arbitrary constants.

Also, on solving Equations 6.6a and 6.6b, we get

$$H_y = \frac{jk_e}{\omega\mu_y} \cdot \left[c_1 \cdot e^{k_e \cdot z} - c_2 \cdot e^{-k_e \cdot z}\right] \cdot e^{j(\omega \cdot t - \ell \cdot y)} \tag{6.9}$$

And

$$H_z = -\frac{\ell}{\omega\mu_z} \cdot \left[c_1 \cdot e^{k_e \cdot z} + c_2 \cdot e^{-k_e \cdot z}\right] \cdot e^{j(\omega \cdot t - \ell \cdot y)} \tag{6.10}$$

While using Equations 6.4a, 6.4b and 6.5a, we get

$$-\frac{1}{\bar{\sigma}_z} \cdot \frac{\partial^2 H_x}{\partial y^2} - \frac{1}{\bar{\sigma}_y} \cdot \frac{\partial H_x}{\partial z} = -j\omega\mu_x H_x$$

Thus,

$$\frac{\partial^2 H_x}{\partial z^2} = k_h{}^2 \cdot H_x \tag{6.11}$$

where

$$k_h{}^2 = \frac{\bar{\sigma}_y}{\bar{\sigma}_z} \cdot \ell^2 + j\omega\bar{\sigma}_y\mu_x \tag{6.11a}$$

Hence, on solving Equation 6.11

$$H_x = \left[d_1 \cdot e^{k_h \cdot z} + d_2 \cdot e^{-k_h \cdot z}\right] \cdot e^{j(\omega \cdot t - \ell \cdot y)} \tag{6.12}$$

where d_1 and d_2 indicate arbitrary constants.

Therefore, from Equations 6.3a and 6.3b, we finally get

$$E_y = \frac{k_h}{\bar{\sigma}_y} \cdot \left[d_1 \cdot e^{k_h \cdot z} - d_2 \cdot e^{-k_h \cdot z} \right] \cdot e^{j(\omega \cdot t - \ell \cdot y)} \tag{6.13a}$$

And

$$E_z = \frac{j\ell}{\bar{\sigma}_z} \cdot \left[d_1 \cdot e^{k_h \cdot z} + d_2 \cdot e^{-k_h \cdot z} \right] \cdot e^{j(\omega \cdot t - \ell \cdot y)} \tag{6.13b}$$

6.3 Cage or Wound Rotor Induction Machines

A simplified treatment, based on a homogeneous anisotropic region for the slotted part of the cage or wound rotor, is presented in this section. The developed view for these machines is shown in Figure 6.1. The curvature of air-gap surfaces is neglected and the two-dimensional treatment assumes unskewed slots resulting in zero variation of electromagnetic fields in the axial (or x) direction. Field analysis for machines with skewed rotor slots is presented in Section 6.5. The symbol y indicates the peripheral and z indicates the radial direction. The idealized machine, shown in Figure 6.1, assumes infinitely permeable stator iron and the rotor core beyond its slotted region. The axial length of the machine is considered as infinite with stator and rotor slots running parallel to the axis of the machine. The current-carrying polyphase stator winding is simulated by a suitable current sheet on the smooth stator air-gap surface. The air-gap length is corrected using Carter's coefficient.

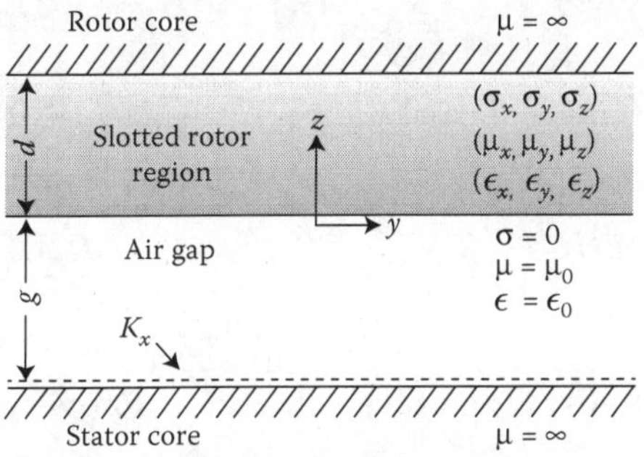

FIGURE 6.1
Sectional view of idealized machine.

Let the stator current sheet in a reference frame moving with the rotor be given as

$$K_x = K_o \cdot e^{j(s\cdot\omega\cdot t-\ell\cdot y)} \tag{6.14}$$

where
$|K_o|$ = peak surface current density of known value
s = slip
ω = angular frequency of the supply

$$\ell \overset{def}{=} \frac{\pi}{\tau} \tag{6.14a}$$

τ = pole pitch.

First, let us consider the air-gap region extended over $0 \geq z \geq -g$, in the radial direction. There being no conduction currents, from Maxwell's equations one gets

$$\nabla \times \boldsymbol{H} = \frac{\partial \boldsymbol{D}}{\partial t} = js \cdot \omega \cdot \epsilon_o \boldsymbol{E} \tag{6.15a}$$

and

$$\nabla \times \boldsymbol{E} = -\frac{\partial \boldsymbol{B}}{\partial t} = -js \cdot \omega \cdot \mu_o \boldsymbol{H} \tag{6.15b}$$

Since

$$\nabla \bullet \boldsymbol{H} = 0 \tag{6.15c}$$

$$\nabla \times \nabla \times \boldsymbol{H} \equiv \nabla(\nabla \bullet \boldsymbol{H}) - \nabla^2 \boldsymbol{H} = -\nabla^2 \boldsymbol{H} \tag{6.16a}$$

Also, using Equations 6.15a and 6.15b

$$\nabla \times \nabla \times \boldsymbol{H} = js \cdot \omega \cdot \epsilon_o \nabla \times \boldsymbol{E} = (s \cdot \omega)^2 \cdot \mu_o \cdot \epsilon_o \boldsymbol{H} = \left(\frac{s \cdot \omega}{c}\right)^2 \boldsymbol{H} \tag{6.16b}$$

where c indicates the velocity of light in free space.

On equating the RHS of Equations 6.16a and 6.16b, we get

$$\nabla^2 \boldsymbol{H} = \frac{\partial^2 \boldsymbol{H}}{\partial y^2} + \frac{\partial^2 \boldsymbol{H}}{\partial z^2} = -\ell^2 \boldsymbol{H} + \frac{\partial^2 \boldsymbol{H}}{\partial z^2} = -\left(\frac{s \cdot \omega}{c}\right)^2 \boldsymbol{H}$$

Therefore,

$$\frac{\partial^2 \boldsymbol{H}}{\partial z^2} = k_o^2 \boldsymbol{H} \tag{6.17}$$

where

$$k_o^2 \stackrel{def}{=} \ell^2 - \left(\frac{s \cdot \omega}{c}\right)^2 \tag{6.17a}$$

and

$$c = \frac{1}{\sqrt{\mu_o \cdot \epsilon_o}} \tag{6.17b}$$

the velocity of light in free space.

The peripheral component of the magnetic field intensity in the air gap, $H_y^{(a)}$, is continuous across the rotor air-gap surface, $z = 0$, and for large value of the stator core permeability, in view of Equation 6.14, it satisfies the following boundary condition on the stator air-gap surface, $z = -g$:

$$H_y^{(a)}\Big|_{z=-g} = -K_x \tag{6.18}$$

Therefore, we can express

$$H_y^{(a)} = \left[a \cdot \frac{\sinh\{k_o(z+g)\}}{\sinh(k_o \cdot g)} + K_o \cdot \frac{\sinh(k_o \cdot z)}{\sinh(k_o \cdot g)} \right] \cdot e^{j(s \cdot w \cdot t - \ell \cdot y)} \tag{6.19}$$

where the symbol '*a*' indicates an arbitrary constant.

Thus,

$$H_y^{(a)}\Big|_{z=0} = a \cdot e^{j(s \cdot \omega \cdot t - \ell \cdot y)} \tag{6.19a}$$

The axial component of the magnetic field intensity in the air gap vanishes at $z = -g$. It may, therefore, be given in view of Equation 6.17 as

$$H_x^{(a)} = a_o \cdot \frac{\sinh\{k_o(z+g)\}}{\sinh(k_o \cdot g)} \cdot e^{j(s \cdot \omega \cdot t - \ell \cdot y)} \tag{6.20}$$

where a_o indicates an arbitrary constant.

Thus,

$$H_x^{(a)}\Big|_{z=0} = a_o \cdot e^{j(s\cdot\omega\cdot t-\ell\cdot y)} \tag{6.20a}$$

Since divergence of the magnetic field intensity in the air gap is zero, the radial component of this field H_z can be given as

$$H_z^{(a)} = j\cdot\frac{\ell}{k_o}\cdot\left[a\cdot\frac{\cosh\{k_o(z+g)\}}{\sinh(k_o\cdot g)} + K_o\cdot\frac{\cosh(k_o\cdot z)}{\sinh(k_o\cdot g)}\right]\cdot e^{j(s\cdot\omega\cdot t-\ell\cdot y)} \tag{6.21}$$

and

$$H_z^{(a)}\Big|_{z=0} = j\cdot\frac{\ell}{k_o}\cdot[a\cdot\coth(k_o\cdot g) + K_o\cdot\operatorname{cosech}(k_o\cdot g)]\cdot e^{j(s\cdot\omega\cdot t-\ell\cdot y)} \tag{6.21a}$$

The components of electric field intensity in the air gap found from Equations 6.15a, 6.17a, 6.19, 6.20 and 6.21 are

$$\begin{aligned} E_x^{(a)} &= -\frac{j}{s\cdot\omega\cdot\epsilon_o}\left[\frac{\partial}{\partial y}H_z^{(a)} - \frac{\partial}{\partial z}H_y^{(a)}\right] \\ &= -j\cdot\frac{\mu_o}{k_o}\cdot s\cdot\omega\left[a\cdot\frac{\cosh\{k_o(z+g)\}}{\sinh(k_o\cdot g)} + K_o\cdot\frac{\cosh(k_o\cdot z)}{\sinh(k_o\cdot g)}\right]\cdot e^{j(s\cdot\omega\cdot t-\ell\cdot y)} \end{aligned} \tag{6.22a}$$

$$\begin{aligned} E_y^{(a)} &= -\frac{j}{s\cdot\omega\cdot\epsilon_o}\left[\frac{\partial}{\partial z}H_x^{(a)} - \frac{\partial}{\partial x}H_z^{(a)}\right] \\ &= -j\frac{k_o}{s\cdot\omega\cdot\epsilon_o}a_o\cdot\frac{\cosh\{k_o(z+g)\}}{\sinh(k_o\cdot g)}\cdot e^{j(s\cdot\omega\cdot t-\ell\cdot y)} \end{aligned} \tag{6.22b}$$

and

$$\begin{aligned} E_z^{(a)} &= -\frac{j}{s\cdot\omega\cdot\epsilon_o}\left[\frac{\partial}{\partial x}H_y^{(a)} - \frac{\partial}{\partial y}H_x^{(a)}\right] \\ &= \frac{\ell}{s\cdot\omega\cdot\epsilon_o}a_o\cdot\frac{\sinh\{k_o(z+g)\}}{\sinh(k_o\cdot g)}\cdot e^{j(s\cdot\omega\cdot t-\ell\cdot y)} \end{aligned} \tag{6.22c}$$

Therefore, on the rotor air-gap surface

$$E_x^{(a)}\Big|_{z=0} = -j\cdot\frac{\mu_o}{k_o}\cdot s\cdot\omega\cdot[a\cdot\coth(k_o\cdot g) + K_o\cdot\operatorname{cosech}(k_o\cdot g)]\cdot e^{j(s\cdot\omega\cdot t-\ell\cdot y)} \tag{6.23a}$$

$$E_y^{(a)}\Big|_{z=0} = -j\frac{k_o}{s\cdot\omega\cdot\epsilon_o}a_o\cdot\coth(k_o\cdot g)\cdot e^{j(s\cdot\omega\cdot t-\ell\cdot y)} \tag{6.23b}$$

$$E_z^{(a)}\Big|_{z=0} = \frac{\ell}{s\cdot\omega\cdot\epsilon_o}a_o\cdot e^{j(s\cdot\omega\cdot t-\ell\cdot y)} \tag{6.23c}$$

Imagine that the rotor core is radially divided into two parts, (i) a 'slot–tooth' belt (over $0 \le z \le d_s$, where d_s indicates the rotor slot depth), and (ii) the highly permeable region beyond this belt (i.e. for $d_s \le z \le \infty$). The latter part can be treated as an infinitely permeable homogeneous medium. The slot–tooth belt with rotor conductors housed in slots will be treated as a homogeneous region with constant anisotropic parameters: $(\mu_x, \mu_y, \mu_z) > 0$, $(\varepsilon_x, \varepsilon_y, \varepsilon_z) > 0$; $\sigma_x = 0$ and $\sigma_y = \sigma_z = 0$. Thus, there will be only axial component of the current density, $J_x^{(r)}$, in this belt. The axial component of the magnetic field intensity in this region, $H_x^{(r)}$, satisfies the following boundary conditions:

$$H_x^{(r)}\Big|_{z=d_s} = 0 \tag{6.24a}$$

$$H_x^{(r)}\Big|_{z=0} = H_x^{(a)}\Big|_{z=0} \tag{6.24b}$$

Therefore, in view of Equations 6.12 and 6.20a, we have

$$H_x^{(r)} = -\,a_o\cdot\frac{\sinh\{k_h(z-d_s)\}}{\sinh(k_h\cdot d_s)}\cdot e^{j(s\cdot\omega\cdot t-\ell\cdot y)} \tag{6.25}$$

where from Equation 6.11a

$$k_h = \sqrt{\frac{\bar{\sigma}_y}{\bar{\sigma}_z}\cdot\ell^2 + js\cdot\omega\cdot\bar{\sigma}_y\cdot\mu_x} \tag{6.25a}$$

$$\bar{\sigma}_y = js\cdot\omega\cdot\epsilon_y \tag{6.25b}$$

and

$$\bar{\sigma}_z = js\cdot\omega\cdot\epsilon_z \tag{6.25c}$$

Therefore,

$$k_h = \sqrt{\frac{\epsilon_y}{\epsilon_z}\cdot\ell^2 - (s\cdot\omega)^2\cdot\epsilon_y\cdot\mu_x} \tag{6.25d}$$

Like the axial component of the magnetic field, its peripheral component also vanishes at $z = d_s$. Thus, in view of Equation 6.9, this component can be defined as

$$H_y^{(r)} = -b \cdot \frac{\sinh\{k_e(z - d_s)\}}{\sinh(k_e \cdot d_s)} \cdot e^{j(s \cdot \omega \cdot t - \ell \cdot y)} \tag{6.26}$$

where in view of Equation 6.7a

$$k_e = \sqrt{\frac{\mu_y}{\mu_z} \cdot \ell^2 + js \cdot \omega \cdot \bar{\sigma}_x \cdot \mu_y} \tag{6.26a}$$

$$\bar{\sigma}_x = \sigma_x + js \cdot \omega \cdot \epsilon_x \tag{6.26b}$$

and b indicates an arbitrary constant.

Therefore,

$$H_y^{(r)}\Big|_{z=0} = b \cdot e^{j(s \cdot \omega \cdot t - \ell \cdot y)} \tag{6.26c}$$

The peripheral component of the magnetic field intensity in this region $H_y^{(r)}$ satisfies the following boundary condition:

$$H_y^{(r)}\Big|_{z=0} = H_y^{(a)}\Big|_{z=0} \tag{6.27}$$

Therefore, in view of Equations 6.19 and 6.26, we have

$$b = a \tag{6.27a}$$

Equation 6.26 can thus be rewritten as

$$H_y^{(r)} = -a \cdot \frac{\sinh\{k_e(z - d_s)\}}{\sinh(k_e \cdot d_s)} \cdot e^{j(s \cdot \omega \cdot t - \ell \cdot y)} \tag{6.28}$$

Now, since divergence of the magnetic induction is zero, we have

$$\mu_x \cdot \frac{\partial}{\partial x} H_x^{(r)} + \mu_y \cdot \frac{\partial}{\partial y} H_y^{(r)} + \mu_z \cdot \frac{\partial}{\partial z} H_z^{(r)} = 0 \tag{6.29}$$

Thus, in view of Equations 6.25 and 6.26,

$$\frac{\partial}{\partial z} H_z^{(r)} = j\ell \cdot \frac{\mu_y}{\mu_z} \cdot H_y^{(r)} = -j\ell \cdot \frac{\mu_y}{\mu_z} \cdot a \cdot \frac{\sinh\{k_e(z - d_s)\}}{\sinh(k_e \cdot d_s)} \cdot e^{j(s \cdot \omega \cdot t - \ell \cdot y)}$$

On integrating both sides with respect to z, and neglecting the constant of integration, we get

$$H_z^{(r)} = -j\frac{\ell}{k_e}\cdot\frac{\mu_y}{\mu_z}\cdot a\cdot\frac{\cosh\{k_e(z-d_s)\}}{\sinh(k_e\cdot d_s)}\cdot e^{j(s\cdot\omega\cdot t-\ell\cdot y)} \tag{6.30}$$

Thus,

$$H_z^{(r)}\Big|_{z=0} = -j\frac{\ell}{k_e}\cdot\frac{\mu_y}{\mu_z}\cdot a\cdot\coth(k_e\cdot d_s)\cdot e^{j(s\cdot\omega\cdot t-\ell\cdot y)} \tag{6.30a}$$

The components of the electric field intensity in the slotted region of the rotor are found from the relation:

$$\nabla\times\boldsymbol{H}^{(r)} = \bar{\sigma}\boldsymbol{E}^{(r)} \tag{6.31}$$

Thus,

$$E_x^{(r)} = \frac{1}{\bar{\sigma}_x}\cdot\left[\frac{\partial}{\partial y}H_z^{(r)} - \frac{\partial}{\partial z}H_y^{(r)}\right] \tag{6.31a}$$

$$E_y^{(r)} = \frac{1}{\bar{\sigma}_y}\cdot\left[\frac{\partial}{\partial z}H_x^{(r)} - \frac{\partial}{\partial x}H_z^{(r)}\right] \tag{6.31b}$$

$$E_z^{(r)} = \frac{1}{\bar{\sigma}_z}\cdot\left[\frac{\partial}{\partial x}H_y^{(r)} - \frac{\partial}{\partial y}H_x^{(r)}\right] \tag{6.31c}$$

Therefore, using Equations 6.25, 6.28 and 6.30, we get

$$E_x^{(r)} = \frac{(k_e^2\cdot\mu_z - \ell^2\cdot\mu_y)}{\bar{\sigma}_x\cdot k_e\cdot\mu_z}\cdot a\cdot\frac{\cosh\{k_e(z-d_s)\}}{\sinh(k_e\cdot d_s)}\cdot e^{j(s\cdot\omega\cdot t-\ell\cdot y)} \tag{6.32a}$$

$$E_y^{(r)} = -\frac{k_h}{\bar{\sigma}_y}\cdot a_o\cdot\frac{\cosh\{k_h(z-d_s)\}}{\sinh(k_h\cdot d_s)}\cdot e^{j(s\cdot\omega\cdot t-\ell\cdot y)} \tag{6.32b}$$

$$E_z^{(r)} = -\frac{jl}{\bar{\sigma}_z}\cdot a_o\cdot\frac{\sinh\{k_h(z-d_s)\}}{\sinh(k_h\cdot d_s)}\cdot e^{j(s\cdot\omega\cdot t-\ell\cdot y)} \tag{6.32c}$$

Thus,

$$E_x^{(r)}\Big|_{z=0} = \frac{(k_e^2 \cdot \mu_z - \ell^2 \cdot \mu_y)}{\bar{\sigma}_x \cdot k_e \cdot \mu_z} \cdot b \cdot \coth(k_e \cdot d_s) \cdot e^{j(s \cdot \omega \cdot t - \ell \cdot y)} \tag{6.33a}$$

$$E_y^{(r)}\Big|_{z=0} = -\frac{k_h}{\bar{\sigma}_y} \cdot a_o \cdot \coth(k_h \cdot d_s) \cdot e^{j(s \cdot \omega \cdot t - \ell \cdot y)} \tag{6.33b}$$

and

$$E_z^{(r)}\Big|_{z=0} = \frac{j\ell}{\bar{\sigma}_z} \cdot a_o \cdot e^{j(s \cdot \omega \cdot t - \ell \cdot y)} \tag{6.33c}$$

Now, considering the boundary condition

$$E_y^{(a)}\Big|_{z=0} = E_y^{(r)}\Big|_{z=0} \tag{6.34}$$

One gets, in view of Equations 6.23b and 6.33b

$$j\frac{k_o}{s \cdot \omega \cdot \epsilon_o} a_o \cdot \coth(k_o \cdot g) = \frac{k_h}{\bar{\sigma}_y} \cdot a_o \cdot \coth(k_h \cdot d_s) \tag{6.34a}$$

Giving,

$$a_o = 0 \tag{6.35}$$

Therefore, from Equations 6.20, 6.22b, 6.22c, 6.25, 6.32b and 6.32c, we find

$$H_x^{(a)} = E_y^{(a)} = E_z^{(a)} = H_x^{(r)} = E_y^{(r)} = E_z^{(r)} = 0 \tag{6.36}$$

Further, since

$$E_x^{(a)}\Big|_{z=0} = E_x^{(r)}\Big|_{z=0} \tag{6.37}$$

Considering Equations 6.23a and 6.31a, we get

$$\begin{aligned} a \cdot [\coth(k_o \cdot g)] + [K_o \cdot \operatorname{cosech}(k_o \cdot g)] \\ = a \cdot \left[j\frac{k_o}{k_e} \cdot \frac{(k_e^2 \cdot \mu_z - \ell^2 \cdot \mu_y)}{s \cdot \omega \cdot \bar{\sigma}_x \cdot \mu_o \cdot \mu_z} \cdot \coth(k_e \cdot h) \right] \end{aligned} \tag{6.38}$$

Now, since

$$\mu_o \cdot H_z^{(a)}\Big|_{z=0} = \mu_z \cdot H_z^{(r)}\Big|_{z=0} \tag{6.39}$$

We get from Equations 6.20a and 6.30a, or from Equations 6.26a and 6.33

$$a \cdot [\coth(k_o \cdot g)] + [K_o \cdot \text{cosech}(k_o \cdot g)] \\ = -a \cdot \left[\frac{k_o}{k_e} \cdot \frac{\mu_y}{\mu_o} \cdot \coth(k_e \cdot d_s)\right] \tag{6.39a}$$

Therefore, on solving for *'a'*, the arbitrary constant found is

$$a = -\frac{K_o \cdot \text{cosech}(k_o \cdot g)}{\left[\coth(k_o \cdot g) + \dfrac{k_o}{k_e} \cdot \dfrac{\mu_y}{\mu_o} \cdot \coth(k_e \cdot d_s)\right]} = b \tag{6.40}$$

Having determined the field distribution in the rotor, rotor copper loss and torque developed can be readily found.

6.3.1 Rotor Parameters

Let us choose the following notations:

d_s = rotor slot depth
w_s = rotor slot width
λ = rotor slot pitch
γ = rotor slot space factor (*copper area/slot area*)
ε = rotor slot insulation permittivity
μ = rotor tooth iron permeability
σ = conductivity for rotor conductors

Now, for the anisotropic homogeneous region, it may be seen that

$$\epsilon_x = \epsilon_o + (\epsilon - \epsilon_o) \cdot (1 - \gamma) \cdot w_s/\lambda \tag{6.41a}$$

$$\epsilon_y = \frac{\epsilon_o \cdot w_s + \epsilon \cdot (1 - \gamma) \cdot (\lambda - w_s)}{\epsilon_o \cdot \epsilon \cdot (1 - \gamma)/\lambda} \tag{6.41b}$$

$$\epsilon_z \cong \epsilon_x \tag{6.41c}$$

$$\sigma_x = \gamma \cdot \frac{\sigma \cdot w_s}{\lambda} \quad (6.41d)$$

$$\mu_x = \mu_o \cdot (w_s/\lambda) + \mu \cdot (1 - w_s/\lambda) \quad (6.41e)$$

$$\mu_y = \frac{\mu \cdot \mu_o \cdot \lambda}{\mu_o \cdot \lambda + (\mu - \mu_o) \cdot w_s} \quad (6.41f)$$

$$\mu_z = \mu - (\mu - \mu_o) \cdot w_s/\lambda \quad (6.41g)$$

where $\gamma < 1$ for wound rotor

$$\cong 1, \text{ for cage rotor} \quad (6.41h)$$

6.4 Induction Machines with Skewed Rotor Slots

When a doubly slotted machine is rotating, the air-gap permeance varies with the change in the slotted rotor position relative to the slotted stator. The rotor, instead of running smoothly, tends to lockup at maximum permeance positions occurring periodically. This causes vibration and noise in the machine. The variation in the air-gap permeance is reduced if the rotor is skewed. The reduced variation of air-gap permeance lowers the level of vibration and noise. Consider a hypothetical induction machine with smooth stator air-gap surface carrying a current sheet simulating the stator winding with balanced polyphase currents. Let this machine be of infinite stator and rotor length. In the case of highly permeable stator core, the tangential components of the magnetic field at the stator air-gap surface do not vary in the axial direction even if the rotor slots are skewed. However, at the rotor air-gap surface, the skewing of the rotor slots results in periodic variation of electromagnetic field in the axial direction with a space period L, where

$$L = 2\tau/\sin(\theta) \quad (6.42)$$

The parameter τ indicates the pole-pitch and θ is the skew angle, that is, the angle between the rotor-axis, X and its slot-axis, x shown in Figure 6.2.

Let a Cartesian system of space coordinate be fixed on the rotor air-gap surface, with $X-$ axial, $Y-$ peripheral and $Z-$ radial direction as shown in Figure 6.2. Further, let the axis of rotor slots indicated as x, subtends an angle

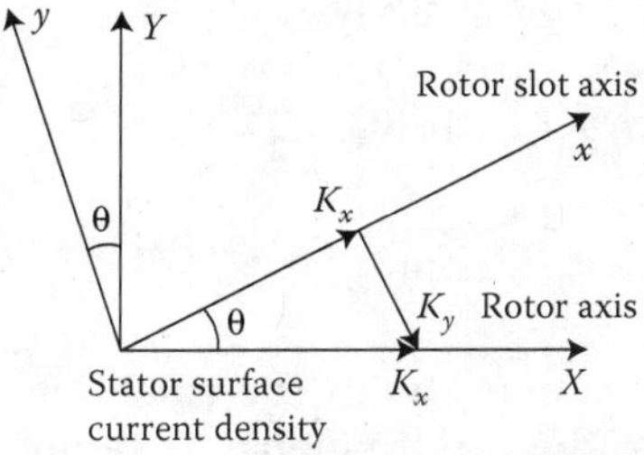

FIGURE 6.2
Axis of skewed rotor slots.

θ with the rotor axis X. With reference to Figure 6.2, the following relations can be readily obtained:

$$X = x \cdot \cos(\theta) - y \cdot \sin(\theta) \tag{6.43a}$$

$$Y = x \cdot \sin(\theta) + y \cdot \cos(\theta) \tag{6.43b}$$

$$Z = z \tag{6.43c}$$

Let the stator surface current density, simulating the current carrying stator winding be given as

$$\boldsymbol{K} = K_X \boldsymbol{a}_X \tag{6.44}$$

where

$$K_X = K_o \cdot e^{j(\omega t - \ell Y)} \tag{6.44a}$$

and $\boldsymbol{a}_X$ indicates the unit vector in the direction of the rotor axis. The vector $\boldsymbol{K}$ can be resolved along two mutually perpendicular directions (vide Figure 6.2), as shown below:

$$\boldsymbol{K} = K_x \boldsymbol{a}_x - K_y \boldsymbol{a}_y \tag{6.45}$$

where

$$K_x = K_X \cdot \cos(\theta) = K_o \cdot \cos(\theta) \cdot e^{j(\omega t - \ell Y)} \tag{6.45a}$$

and

$$K_y = K_X \cdot \sin(\theta) = K_o \cdot \sin(\theta) \cdot e^{j(\omega t - \ell Y)} \tag{6.45b}$$

while $\boldsymbol{a}_x$ and $\boldsymbol{a}_y$ indicate unit vectors in x and y directions, respectively.

In view of Equation 6.43b, Equations 6.45a and 6.45b can be rewritten as

$$K_x = K_o \cdot \cos(\theta) \cdot e^{j(\omega t - \ell \cdot \sin\theta \cdot x - \ell \cdot \cos\theta \cdot y)} \tag{6.46a}$$

$$K_y = K_o \cdot \sin(\theta) \cdot e^{j(\omega t - \ell \cdot \sin\theta \cdot x - \ell \cdot \cos\theta \cdot y)} \tag{6.46b}$$

In a reference frame moving with the rotor, the angular frequency ω must be replaced by $s\omega$, the slip frequency. It can be seen that the velocity of the travelling fields in x and y directions can be respectively given, as follows:

$$\frac{dx}{dt} = \frac{s\omega}{\ell} \cdot \sin(\theta) \tag{6.47a}$$

$$\frac{dy}{dt} = \frac{s\omega}{\ell} \cdot \cos(\theta) \tag{6.47b}$$

6.4.1 Air-Gap Field

Consider the boundary conditions for the air-gap field:

$$H_x^{(a)}\Big|_{z=-g} = -K_y = -K_o \cdot \sin(\theta) \cdot e^{j(s\omega t - \ell \cdot \sin\theta \cdot x - \ell \cdot \cos\theta \cdot y)} \tag{6.48a}$$

$$H_y^{(a)}\Big|_{z=-g} = -K_x = -K_o \cdot \cos(\theta) \cdot e^{j(s\omega t - \ell \cdot \sin\theta \cdot x - \ell \cdot \cos\theta \cdot y)} \tag{6.48b}$$

The exponential factor defining the variation of fields with t, x and y is suppressed in the remaining part of this treatment, though its presence is understood.

Next, consider

$$\nabla \times \nabla \times \boldsymbol{H} = \nabla \times (js\omega \cdot \epsilon_o \boldsymbol{E}) = (s\omega/c)^2 \boldsymbol{H} \tag{6.49a}$$

also

$$\nabla \times \nabla \times \boldsymbol{H} \equiv \nabla(\nabla \bullet \boldsymbol{H}) - \nabla^2 \boldsymbol{H} = -\nabla^2 \boldsymbol{H} \tag{6.49b}$$

where $c = 1/\sqrt{\epsilon_o \cdot \mu_o}$, velocity of light in free space.

Therefore,

$$-\nabla^2 \boldsymbol{H} = (s\omega/c)^2 \boldsymbol{H} \tag{6.50}$$

Tentative solutions for this equation, subjected to the above-mentioned boundary conditions, can be taken as

$$H_x^{(a)} = a_1 \cdot \frac{\sinh\{k \cdot (z+g)\}}{\sinh(k \cdot g)} + K_o \cdot \sin\theta \cdot \frac{\sinh(k \cdot z)}{\sinh(k \cdot g)} \tag{6.50a}$$

$$H_y^{(a)} = a_2 \cdot \frac{\sinh\{k \cdot (z+g)\}}{\sinh(k \cdot g)} + K_o \cdot \cos\theta \cdot \frac{\sinh(k \cdot z)}{\sinh(k \cdot g)} \tag{6.50b}$$

where

$$k^2 = \ell^2 - (s\omega/c)^2 \tag{6.51}$$

Since divergence of the magnetic field intensity in the air gap is zero, we have

$$\begin{aligned} H_z^{(a)} = {} & j(\ell/k) \cdot \sin\theta \cdot \left[a_1 \cdot \frac{\cosh\{k \cdot (z+g)\}}{\sinh(k \cdot g)} + K_o \cdot \sin\theta \cdot \frac{\cosh(k \cdot z)}{\sinh(k \cdot g)} \right] \\ & + j(\ell/k) \cdot \cos\theta \cdot \left[a_2 \cdot \frac{\cosh\{k \cdot (z+g)\}}{\sinh(k \cdot g)} + K_o \cdot \cos\theta \cdot \frac{\cosh(k \cdot z)}{\sinh(k \cdot g)} \right] \end{aligned}$$

or

$$H_z^{(a)} = j(\ell/k) \cdot \left[(\sin\theta \cdot a_1 + \cos\theta \cdot a_2) \cdot \frac{\cosh\{k \cdot (z+g)\}}{\sinh(k \cdot g)} + K_o \cdot \frac{\cosh(k \cdot z)}{\sinh(k \cdot g)} \right] \tag{6.52}$$

Now, since

$$js\omega \cdot \epsilon_o \cdot \boldsymbol{E}^{(a)} = \nabla \times \boldsymbol{H}^{(a)} \tag{6.53}$$

thus

$$js\omega \cdot \epsilon_o \cdot E_x^{(a)} = \frac{\partial}{\partial y} H_z^{(a)} - \frac{\partial}{\partial z} H_y^{(a)} \tag{6.54}$$

Using Equations 6.50b and 6.52, one gets

$$\begin{aligned} js\omega \cdot \epsilon_o \cdot E_x^{(a)} = {} & \left[\frac{\ell^2}{k} \cdot \sin\theta \cdot \cos\theta \cdot a_1 - \frac{1}{k} \cdot (k^2 - \ell^2 \cdot \cos^2\theta) \cdot a_2 \right] \cdot \frac{\cosh\{k \cdot (z+g)\}}{\sinh(k \cdot g)} \\ & + \frac{(\ell^2 - k^2)}{k} \cdot \cos\theta \cdot K_o \cdot \frac{\cosh(k \cdot z)}{\sinh(k \cdot g)} \end{aligned} \tag{6.55a}$$

Thus, in view of Equation 6.51

$$E_x^{(a)} = \frac{-j}{s\omega \cdot \epsilon_o} \cdot \frac{1}{k} \cdot [\ell^2 \cdot \sin\theta \cdot \cos\theta \cdot a_1 + \{(s\omega/c)^2 - \ell^2 \cdot \sin^2\theta\} \cdot a_2] \cdot \frac{\cosh\{k(z+g)\}}{\sinh(k \cdot g)} - js\omega\mu_o \cdot \frac{1}{k} \cdot \cos\theta \cdot K_o \cdot \frac{\cosh(k \cdot z)}{\sinh(k \cdot g)} \tag{6.55b}$$

For the y-component, we have

$$js\omega \cdot \epsilon_o \cdot E_y^{(a)} = \frac{\partial}{\partial z} H_x^{(a)} - \frac{\partial}{\partial x} H_z^{(a)} \tag{6.56}$$

Using Equations 6.50a and 6.52, one gets

$$js\omega \cdot \epsilon_o \cdot E_y^{(a)} = \left[\frac{k^2 - \ell^2 \cdot \sin^2\theta}{k} \cdot a_1 - \frac{\ell^2 \cdot \sin\theta \cdot \cos\theta}{k} \cdot a_2\right] \cdot \frac{\cosh\{k \cdot (z+g)\}}{\sinh(k \cdot g)} - \frac{(k^2 - \ell^2) \cdot \sin\theta}{k} \cdot K_o \cdot \frac{\cosh(k \cdot z)}{\sinh(k \cdot g)} \tag{6.56a}$$

Thus, in view of Equation 6.51

$$E_y^{(a)} = \frac{-j}{s\omega \cdot \epsilon_o} \cdot \frac{1}{k} \cdot [\{\ell^2 \cdot \cos^2\theta - (s\omega/c)^2\} \cdot a_1 - \ell^2 \cdot \sin\theta \cdot \cos\theta \cdot a_2] \cdot \frac{\cosh\{k(z+g)\}}{\sinh(k \cdot g)} + js\omega\mu_o \cdot \frac{1}{k} \cdot \sin\theta \cdot K_o \cdot \frac{\cosh(k \cdot z)}{\sinh(k \cdot g)} \tag{6.56b}$$

And for the Z-component

$$js\omega \cdot \epsilon_o \cdot E_z^{(a)} = \frac{\partial}{\partial x} H_y^{(a)} - \frac{\partial}{\partial y} H_x^{(a)} \tag{6.57}$$

Using Equations 6.50a and 6.50b, one gets

$$js\omega \cdot \epsilon_o \cdot E_z^{(a)} = j\ell \cdot [\cos\theta \cdot a_1 - \sin\theta \cdot a_2] \cdot \frac{\sinh\{k \cdot (z+g)\}}{\sinh(k \cdot g)} \tag{6.57a}$$

Therefore,

$$E_z^{(a)} = \frac{\ell}{s\omega \cdot \epsilon_o} \cdot [\cos\theta \cdot a_1 - \sin\theta \cdot a_2] \cdot \frac{\sinh\{k \cdot (z+g)\}}{\sinh(k \cdot g)} \tag{6.57b}$$

Components of electromagnetic fields on the rotor air-gap surface found from Equations 6.50a, 6.50b, 6.52, 6.55b, 6.56b and 6.57b are

$$H_x^{(a)}\Big|_{z=0} = a_1 \tag{6.58a}$$

$$H_y^{(a)}\Big|_{z=0} = a_2 \tag{6.58b}$$

$$H_z^{(a)}\Big|_{z=0} = j(\ell/k) \cdot [(\sin\theta \cdot a_1 + \cos\theta \cdot a_2) \cdot \coth(k \cdot g) + K_o \cdot \text{cosech}(k \cdot g)] \tag{6.58c}$$

$$E_x^{(a)}\Big|_{z=0} = \frac{-j}{s\omega \cdot \epsilon_o} \cdot \frac{1}{k} \cdot [\ell^2 \cdot \sin\theta \cdot \cos\theta \cdot a_1 - \{\ell^2 \cdot \sin^2\theta - (s\omega/c)^2\} \cdot a_2] \cdot \coth(k \cdot g) - js\omega\mu_o \cdot \frac{1}{k} \cdot \cos\theta \cdot K_o \cdot \text{cosech}\,(k \cdot g) \tag{6.58d}$$

$$E_y^{(a)}\Big|_{z=0} = \frac{-j}{s\omega \cdot \epsilon_o} \cdot \frac{1}{k} \cdot [\{\ell^2 \cdot \cos^2\theta - (s\omega/c)^2\} \cdot a_1 - \ell^2 \cdot \sin\theta \cdot \cos\theta \cdot a_2] \cdot \coth(k \cdot g) + js\omega\mu_o \cdot \frac{1}{k} \cdot \sin\theta \cdot K_o \cdot \text{cosech}\,(k \cdot g) \tag{6.58e}$$

and

$$E_z^{(a)}\Big|_{z=0} = \frac{\ell}{s\omega \cdot \epsilon_o} \cdot [\cos\theta \cdot a_1 - \sin\theta \cdot a_2] \tag{6.58f}$$

6.4.2 Fields in the Anisotropic Rotor Region

Consider the rotor region extending from the rotor air-gap surface to the base of the rotor slot, that is, over $0 \leq z \leq d_s$. Since the rotor teeth are invariably saturated and the rotor core beyond this region being rather unsaturated, we shall assume finite permeability for the slot–tooth region and a large permeability ($\mu \rightarrow \infty$) for the rotor core beyond the slot–tooth region.

The inhomogeneous slot–tooth region is simulated by a suitable homogeneous anisotropic region, as indicated in the preceding section. As a result, the tangential components of the magnetic field vanish at $z = d_s$. Let the components of the magnetic field intensity in this region be expressed as

$$H_x^{(r)} = -b_1 \cdot \frac{\sinh\{k_1 \cdot (z - d_s)\}}{\sinh(k_1 \cdot d_s)} \tag{6.59a}$$

$$H_y^{(r)} = -b_2 \cdot \frac{\sinh\{k_2 \cdot (z - d_s)\}}{\sinh(k_2 \cdot d_s)} \tag{6.59b}$$

where b_1 and b_2 indicate arbitrary constants, while k_1 and k_2 are unknown coefficients. For determination of these coefficients, consider Maxwell's equations:

$$-\frac{\partial \boldsymbol{B}}{\partial t} = \nabla \times \boldsymbol{E} \tag{6.60a}$$

and

$$\boldsymbol{J} + \frac{\partial \boldsymbol{D}}{\partial t} = \nabla \times \boldsymbol{H} \tag{6.60b}$$

or

$$\bar{\sigma}\boldsymbol{E} = \nabla \times \boldsymbol{H} \tag{6.61}$$

where the complex conductivity $\bar{\sigma}$, for harmonic field, is defined as

$$\bar{\sigma} = \sigma + js\omega \cdot \epsilon \tag{6.61a}$$

Now, from Equation 6.60a, one finds

$$-js\omega \cdot \mu_z \cdot H_z = \frac{\partial}{\partial x}E_y - \frac{\partial}{\partial y}E_x = \frac{1}{\bar{\sigma}_y} \cdot \frac{\partial}{\partial x}(\bar{\sigma}_y \cdot E_y) - \frac{1}{\bar{\sigma}_x} \cdot \frac{\partial}{\partial y}(\bar{\sigma}_x \cdot E_x) \tag{6.62}$$

Thus, in view of Equation 6.61

$$-js\omega \cdot \mu_z \cdot H_z = \frac{1}{\bar{\sigma}_y} \cdot \frac{\partial}{\partial x}\left(\frac{\partial}{\partial z}H_x - \frac{\partial}{\partial x}H_z\right) - \frac{1}{\bar{\sigma}_x} \cdot \frac{\partial}{\partial y}\left(\frac{\partial}{\partial y}H_z - \frac{\partial}{\partial z}H_y\right) \tag{6.63}$$

Therefore,

$$-js\omega \cdot \bar{\sigma}_x \cdot \bar{\sigma}_y \cdot \mu_z \cdot H_z = \frac{\partial}{\partial z}\left(\bar{\sigma}_x \cdot \frac{\partial}{\partial x} H_x + \bar{\sigma}_y \cdot \frac{\partial}{\partial y} H_y\right)$$

$$-\left(\bar{\sigma}_x \cdot \frac{\partial^2}{\partial x^2} H_z + \bar{\sigma}_y \cdot \frac{\partial^2}{\partial y^2} H_z\right)$$

or

$$\left[\bar{\sigma}_x \cdot \frac{\partial^2}{\partial x^2} + \bar{\sigma}_y \cdot \frac{\partial^2}{\partial y^2} - js\omega \cdot \bar{\sigma}_x \cdot \bar{\sigma}_y \cdot \mu_z\right] \cdot H_z = \frac{\partial}{\partial z}\left(\bar{\sigma}_x \cdot \frac{\partial}{\partial x} H_x + \bar{\sigma}_y \cdot \frac{\partial}{\partial y} H_y\right) \tag{6.63a}$$

Fields in this region also varies periodically with t, x and y, and this variation is indicated by the factor: $\exp \cdot j(s\omega \cdot t - \ell \sin\theta \cdot x - \ell \cos\theta \cdot y)$. Therefore, from Equations 6.59a, 6.59b and 6.63a, we get

$$H_z^{(r)} = -j\ell \cdot \left[\sin\theta \cdot \bar{\sigma}_x \cdot \frac{k_1}{k_o} \cdot \frac{\cosh\{k_1 \cdot (z - d_s)\}}{\sinh(k_1 \cdot d_s)} \cdot b_1 + \cos\theta \cdot \bar{\sigma}_y \cdot \frac{k_2}{k_o} \cdot \frac{\cosh\{k_2 \cdot (z - d_s)\}}{\sinh(k_2 \cdot d_s)} \cdot b_2\right] \tag{6.64}$$

where

$$k_o = \bar{\sigma}_x \cdot \ell^2 \cdot \sin^2\theta + \bar{\sigma}_y \cdot \ell^2 \cdot \cos^2\theta + js\omega \cdot \bar{\sigma}_x \cdot \bar{\sigma}_y \cdot \mu_z \tag{6.64a}$$

Thus, from Equation 6.64

$$\frac{\partial}{\partial z} H_z^{(r)} = -j\ell \cdot \left[\sin\theta \cdot \bar{\sigma}_x \cdot \frac{k_1^2}{k_o} . \frac{\sinh\{k_1 \cdot (z - d_s)\}}{\sinh(k_1 \cdot d_s)} \cdot b_1 + \cos\theta \cdot \bar{\sigma}_y \cdot \frac{k_2^2}{k_o} \cdot \frac{\sinh\{k_2 \cdot (z - d_s)\}}{\sinh(k_2 \cdot d_s)} \cdot b_2\right] \tag{6.65}$$

Further, since

$$\nabla \bullet \boldsymbol{B} = 0 \tag{6.66}$$

we have

$$\frac{\partial}{\partial z} H_z^{(r)} = -\frac{\mu_x}{\mu_z} \cdot \frac{\partial}{\partial x} H_x^{(r)} - \frac{\mu_y}{\mu_z} \cdot \frac{\partial}{\partial y} H_y^{(r)} \tag{6.66a}$$

Thus, using Equations 6.59a and 6.59b

$$\frac{\partial}{\partial z} H_z^{(r)} = -j\ell \cdot \left[\sin\theta \cdot \frac{\mu_x}{\mu_z} \cdot \frac{\sinh\{k_1 \cdot (z - d_s)\}}{\sinh(k_1 \cdot d_s)} \cdot b_1 \right.$$
$$\left. + \cos\theta \cdot \frac{\mu_y}{\mu_z} \cdot \frac{\sinh\{k_2 \cdot (z - d_s)\}}{\sinh(k_2 \cdot d_s)} \cdot b_2 \right] \tag{6.67}$$

Therefore, on comparing Equations 6.65 and 6.67, we get

$$\bar{\sigma}_x \cdot \frac{k_1^2}{k_o} = \frac{\mu_x}{\mu_z} \tag{6.68a}$$

and

$$\bar{\sigma}_y \cdot \frac{k_2^2}{k_o} = \frac{\mu_y}{\mu_z} \tag{6.68b}$$

This gives

$$k_1 = \sqrt{\frac{\mu_x}{\mu_z} \cdot \frac{k_o}{\bar{\sigma}_x}} \tag{6.69a}$$

and

$$k_2 = \sqrt{\frac{\mu_y}{\mu_z} \cdot \frac{k_o}{\bar{\sigma}_y}} \tag{6.69b}$$

The component of the magnetic field intensity in the slot–tooth region of the rotor is obtained in terms of two arbitrary constants, b_1 and b_2 as given by Equations 6.59a, 6.59b and 6.64. The expressions for the distribution of the components of electric field intensity in this rotor region are readily obtained as below.

Considering Equations 6.59a, 6.59b and 6.61, we have the components of electric field in the slotted rotor region:

$$\bar{\sigma}_x E_x = \frac{\partial}{\partial y} H_z - \frac{\partial}{\partial z} H_y \tag{6.70a}$$

or

$$E_x^{(r)} = -\ell^2 \cdot \sin\theta \cdot \cos\theta \cdot \frac{k_1}{k_o} \cdot \frac{\cosh\{k_1 \cdot (z - d_s)\}}{\sinh(k_1 \cdot d_s)} \cdot b_1$$
$$- \left[\ell^2 \cdot \cos^2\theta \cdot \frac{\bar{\sigma}_y}{\bar{\sigma}_x} \cdot \frac{k_2}{k_o} - \frac{k_2}{\bar{\sigma}_x} \right] \cdot \frac{\cosh\{k_2 \cdot (z - d_s)\}}{\sinh(k_2 \cdot d_s)} \cdot b_2 \tag{6.70b}$$

Thus,

$$E_x^{(r)}\Big|_{z=0} = -\ell^2 \cdot \sin\theta \cdot \cos\theta \cdot \frac{k_1}{k_o} \cdot \coth(k_1 \cdot d_s) \cdot b_1 - \left[\ell^2 \cdot \cos^2\theta \cdot \frac{\bar{\sigma}_y}{\bar{\sigma}_x} \cdot \frac{k_2}{k_o} - \frac{k_2}{\bar{\sigma}_x}\right] \cdot \coth(k_2 \cdot d_s) \cdot b_2 \tag{6.70c}$$

Since

$$J = \nabla \times \boldsymbol{H} \tag{6.71}$$

$$\bar{\sigma}_y E_y = \frac{\partial}{\partial z} H_x - \frac{\partial}{\partial x} H_z \tag{6.71a}$$

or

$$E_y^{(r)} = \left[\ell^2 \cdot \sin^2\theta \cdot \frac{\bar{\sigma}_x}{\bar{\sigma}_y} \cdot \frac{k_1}{k_o} - \frac{k_1}{\bar{\sigma}_y}\right] \cdot \frac{\cosh\{k_1 \cdot (z - d_s)\}}{\sinh(k_1 \cdot d_s)} \cdot b_1 + \ell^2 \cdot \sin\theta \cdot \cos\theta \cdot \frac{k_2}{k_o} \cdot \frac{\cosh\{k_2 \cdot (z - d_s)\}}{\sinh(k_2 \cdot d_s)} \cdot b_2 \tag{6.71b}$$

Thus,

$$E_y^{(r)}\Big|_{z=0} = \left[\ell^2 \cdot \sin^2\theta \cdot \frac{\bar{\sigma}_x}{\bar{\sigma}_y} \cdot \frac{k_1}{k_o} - \frac{k_1}{\bar{\sigma}_y}\right] \cdot \coth(k_1 \cdot d_s) \cdot b_1 + \ell^2 \cdot \sin\theta \cdot \cos\theta \cdot \frac{k_2}{k_o} \cdot \coth(k_2 \cdot d_s) \cdot b_2 \tag{6.71c}$$

and

$$\bar{\sigma}_z E_z = \frac{\partial}{\partial x} H_y - \frac{\partial}{\partial y} H_x \tag{6.72a}$$

or

$$E_z^{(r)} = -\frac{j\ell}{\bar{\sigma}_z} \cdot \left[\cos\theta \cdot \frac{\sinh\{k_1 \cdot (z - d_s)\}}{\sinh(k_1 \cdot d_s)} \cdot b_1 - \sin\theta \cdot \frac{\sinh\{k_2 \cdot (z - d_s)\}}{\sinh(k_2 \cdot d_s)} \cdot b_2\right] \tag{6.72b}$$

Thus,

$$E_z^{(r)}\Big|_{z=0} = \frac{j\ell}{\bar{\sigma}_z} \cdot [\cos\theta \cdot b_1 - \sin\theta \cdot b_2] \tag{6.72c}$$

whereas from Equations 6.59a, 6.59b and 6.64

$$H_x^{(r)}\Big|_{z=0} = b_1 \tag{6.73a}$$

$$H_y^{(r)}\Big|_{z=0} = b_2 \tag{6.73b}$$

and

$$H_z^{(r)}\Big|_{z=0} = -j\ell \cdot \left[\sin\theta \cdot \bar{\sigma}_x \cdot \frac{k_1}{k_o} \cdot \coth(k_1 \cdot d_s) \cdot b_1 + \cos\theta \cdot \bar{\sigma}_y \cdot \frac{k_2}{k_o} \cdot \coth(k_2 \cdot d_s) \cdot b_2\right] \tag{6.73c}$$

Continuity conditions at the rotor air-gap surface results in certain relations between various arbitrary constants. Solution of these equations determines the arbitrary constants associated with the field expressions.

6.4.3 Determination of Arbitrary Constants

Consider the following boundary conditions:

$$H_x^{(a)}\Big|_{z=0} = H_x^{(r)}\Big|_{z=0} \tag{6.74a}$$

$$H_y^{(a)}\Big|_{z=0} = H_y^{(r)}\Big|_{z=0} \tag{6.74b}$$

$$\mu_o \cdot H_z^{(a)}\Big|_{z=0} = \mu \cdot H_z^{(r)}\Big|_{z=0} \tag{6.74c}$$

$$E_x^{(a)}\Big|_{z=0} = E_x^{(r)}\Big|_{z=0} \tag{6.75a}$$

$$E_y^{(a)}\Big|_{z=0} = E_y^{(r)}\Big|_{z=0} \tag{6.75b}$$

$$js\omega \cdot \epsilon_o \cdot E_z^{(a)}\Big|_{z=0} = \bar{\sigma}_z \cdot E_z^{(r)}\Big|_{z=0} \tag{6.75c}$$

From Equations 6.58a, 6.73a and 6.74a, we get

$$a_1 = b_1 \tag{6.76a}$$

From Equations 6.58b, 6.73b and 6.74b, we get

$$a_2 = b_2 \tag{6.76b}$$

From Equations 6.58c, 6.73c and 6.74c, we get

$$(\mu_o/k)\cdot[(\sin\theta\cdot a_1+\cos\theta\cdot a_2)\cdot\coth(k\cdot g)+K_o\cdot\text{cosech}(k\cdot g)]$$
$$=-(\mu_z/k_o)\cdot[\sin\theta\cdot\bar{\sigma}_x\cdot k_1\cdot\coth(k_1\cdot d_s)\cdot b_1+\cos\theta\cdot\bar{\sigma}_y\cdot k_2\cdot\coth(k_2\cdot d_s)\cdot b_2] \tag{6.76c}$$

From Equations 6.58d, 6.70c and 6.75a, we get

$$\frac{j}{s\omega\cdot\epsilon_o\cdot k}\cdot\left[\ell^2\cdot\sin\theta\cdot\cos\theta\cdot a_1+\{(s\omega/c)^2\cdot\cos^2\theta-k^2\cdot\sin^2\theta\}\cdot a_2\right]\cdot\coth(k\cdot g)$$
$$+\frac{j}{s\omega\cdot\epsilon_o\cdot k}\cdot(s\omega/c)^2\cdot\cos\theta\cdot K_o\cdot\text{cosech}(k\cdot g)$$
$$=\ell^2\cdot\sin\theta\cdot\cos\theta\cdot\frac{k_1}{k_o}\cdot\coth(k_1\cdot d_s)\cdot b_1+\left[\ell^2\cdot\cos^2\theta\cdot\frac{\bar{\sigma}_y}{\bar{\sigma}_x}\cdot\frac{k_2}{k_o}-\frac{k_2}{\bar{\sigma}_x}\right]\cdot\coth(k_2\cdot d_s)\cdot b_2 \tag{6.77a}$$

From Equations 6.58e, 6.71c and 6.75b, we get

$$-\frac{j}{s\omega\cdot\epsilon_o\cdot k}\cdot[\{k^2\cdot\cos^2\theta-(s\omega/c)^2\cdot\sin^2\theta\}\cdot a_1-\ell^2\cdot\sin\theta\cdot\cos\theta\cdot a_2]\cdot\coth(k\cdot g)$$
$$+\frac{j}{s\omega\cdot\epsilon_o\cdot k}\cdot(s\omega/c)^2\cdot\sin\theta\cdot K_o\cdot\text{cosech}(k\cdot g)=\left[\ell^2\cdot\sin^2\theta\cdot\frac{\bar{\sigma}_x}{\bar{\sigma}_y}\cdot\frac{k_1}{k_o}\right.$$
$$\left.-\frac{k_1}{\bar{\sigma}_y}\right]\cdot\coth(k_1\cdot d_s)\cdot b_1+\ell^2\cdot\sin\theta\cdot\cos\theta\cdot\frac{k_2}{k_o}\cdot\coth(k_2\cdot d_s)\cdot b_2 \tag{6.77b}$$

From Equations 6.58f, 6.72c and 6.75c, we get

$$[\cos\theta\cdot a_1-\sin\theta\cdot a_2]=[\cos\theta\cdot b_1-\sin\theta\cdot b_2] \tag{6.77c}$$

This equation is identically satisfied in view of Equations 6.76a and 6.76b. An approximate solution for the arbitrary constants can be readily found by noting that the value of $(s\omega \cdot \varepsilon_o)$ is very small at power frequency operations, and so also the value of $(s\omega \cdot \mu_o)$. The $(s\omega/c)^2$ being a product of two small quantities, is still a smaller quantity. Therefore, from Equation 6.51

$$k \cong \ell \tag{6.78}$$

whereas from Equation 6.77a as well as from Equation 6.77b, we get

$$\cos\theta \cdot a_1 \cong \sin\theta \cdot a_2 \overset{def}{=} a_o \tag{6.79}$$

for $0 < \theta < \pi/2$.

From Equations 6.76a, 6.76b and 6.79

$$a_1 = b_1 = a_o/\cos\theta \tag{6.80a}$$

and

$$a_2 = b_2 = a_o/\sin\theta \tag{6.80b}$$

Lastly, from Equation 6.76c, in view of Equations 6.80a and 6.80b

$$a_o \cdot (\tan\theta + \cot\theta) \cdot \frac{\mu_o}{k} \cdot \coth(k \cdot g) + K_o \cdot \frac{\mu_o}{k} \cdot \operatorname{cosech}(k \cdot g)$$
$$+ a_o \cdot \tan\theta \cdot \left[\mu_z \cdot \bar{\sigma}_x \cdot \frac{k_1}{k_o} \cdot \coth(k_1 \cdot d_s)\right] + a_o \cdot \cot\theta \cdot \left[\mu_z \cdot \bar{\sigma}_y \cdot \frac{k_2}{k_o} \cdot \coth(k_2 \cdot d_s)\right] = 0$$

or

$$a_o = -\frac{K_o}{b_o} \cdot \frac{\mu_o}{2k} \cdot \operatorname{cosech}(k \cdot g) \cdot \sin(2\theta) \tag{6.81a}$$

where

$$b_o = \frac{\mu_o}{k} \cdot \coth(k \cdot g) + \sin^2\theta \cdot \left[\mu_z \cdot \bar{\sigma}_x \cdot \frac{k_1}{k_o} \cdot \coth(k_1 \cdot d_s)\right]$$
$$+ \cos^2\theta \cdot \left[\mu_z \cdot \bar{\sigma}_y \cdot \frac{k_2}{k_o} \cdot \coth(k_2 \cdot d_s)\right] \tag{6.81b}$$

Thus, among Equations 6.80a, 6.80b, 6.81a and 6.81b, the values of all arbitrary constants are approximately determined.

For determining the exact values of the four arbitrary constants, viz. a_1, a_2, b_1 and b_2, we consider Equations 6.74c, 6.75a, 6.75b and 6.75c only. Exact solutions of these equations will result in

$$a_1 \neq b_1 \tag{6.82a}$$

and

$$a_2 \neq b_2 \tag{6.82b}$$

These indicate that boundary conditions given by Equations 6.74a and 6.74b are only approximately valid. Clearly, it is evident that surface currents are induced at the rotor air-gap surface. On this surface, the radial component of the rotor eddy current density does not vanish. This results in a deposition of charges on the rotor air-gap surface. Thus, the normal component of electric flux density is discontinuous on this surface, that is,

$$\epsilon_z \cdot E_z^{(r)}\Big|_{z=0} \neq \epsilon_o \cdot E_z^{(a)}\Big|_{z=0} \tag{6.83}$$

Since the electromagnetic field is travelling, the moving surface charge distribution results in the convection surface currents on the rotor air-gap surface. A similar phenomenon has been observed[4] in the study of eddy currents in laminated cores of electrical machines with revolving fields.

PROJECT PROBLEMS

1. Using electromagnetic field theory, obtain an expression for the torque developed in a three-phase induction machine with a slip-ring rotor in terms of the stator supply voltage.
2. Produce a treatment for laminated-rotor induction machines with skewed stator slots.

References

1. Bewley, L. V., *Two-Dimensional Fields in Electrical Engineering*. Dover Publications, Inc., New York, pp. 90–94, 1963.
2. Subbarao, V., *Eddy Currents in Linear and Non-linear Media*, Omega Scientific Publishers, New Delhi, India, pp. 60–63, 1991.

3. Mishkin, E., Theory of the squirrel-cage induction machine derived directly from Maxwell's field equations, *Quarterly Journal of Mechanics and Applied Mathematics*, VII(4), 473–487, 1954.
4. Mukerji, S. K., Srivastava, D. S., Singh, Y. P. and Avasthi, D. V., Eddy current phenomena in laminated structures due to travelling electromagnetic fields, *Progress in Electromagnetics Research M*, 18, 159–169, 2011.

7

Unlaminated Rotor Polyphase Induction Machines

7.1 Introduction

In Chapter 6, field analyses for the laminated rotor induction machines were presented. This chapter presents the analyses of electromagnetic fields in solid-rotor induction machines. The analyses for harmonic fields in solid-rotor induction machines due to tooth-ripples are discussed at length. It encompasses the field distributions in the air gap and solid rotors and the machine performance. The chapter also includes the study of end effects in these machines. This study requires a three-dimensional field analysis for three-phase solid-rotor induction machines.

7.2 Tooth-Ripple Harmonics in Solid-Rotor Induction Machines

In a solid-rotor machine with semi-closed slots, the air-gap field will contain only the winding harmonics if the slot openings are ignored. However, for a machine with open stator slot, the winding harmonics cannot be strictly isolated from the harmonics due to the effect of slot openings. This combined effect, termed as the *tooth-ripple phenomenon*, needs to be studied in particular detail for solid-rotor induction machines, since, unlike wound-rotor or cage-rotor induction machines, the tooth-ripple losses in a solid-rotor induction machine constitute a major part of the stray losses.

7.2.1 Physical Description

Consider a solid-rotor induction machine with three-phase stator winding housed in open rectangular stator slots. The three-phase current flowing in the three-phase winding causes the magnetic field in the air gap as well as in the rotor iron to vary periodically in the peripheral direction. The discrete nature of the winding introduces winding harmonics. These harmonic fields will be present even if the current-carrying conductors are replaced with current strips on smooth stator air-gap surface. Because of the slotted stator air-gap surface, the air-gap permeance varies periodically in the peripheral direction. Owing to this combined effect, the magnetic field, in the air gap as well as in the solid rotor, contains nontriplen odd space harmonics. Each harmonic field rotates at different speeds; some harmonic fields rotate in the direction that the fundamental field rotates in and others in the opposite direction. The former is termed as positive direction and the latter as negative direction. Each harmonic field associated with a specific eddy current density in the rotor iron contributes to the torque developed in the machine. At stand still, torques produced by some of these harmonic fields, called positive, are in the direction that the fundamental field rotates in and the others, called negative, act in the opposite direction. The net torque is the vector sum of all these torques. Consider the machine working as a motor with its rotor rotating at near synchronous speed of the fundamental field. Fields due to 5th, 11th and other negative revolving harmonics cause braking torque. Certain harmonic fields, viz., 7th, 13th and other positive revolving harmonic fields rotate at 1/7th, 1/13th and such fraction of the synchronous speeds of the fundamental field. With reference to these harmonic fields, the machine is rotating at super-synchronous speeds. Therefore, with reference to these harmonic fields, the machine is operating as a synchronous generator; each inducing negative torque. Thus, the effects of each harmonic field are to produce I^2R loss in the rotor iron and to reduce the net torque developed by the machine.

7.2.1.1 Slip/Torque Characteristics

Figure 7.1 shows the slip/torque characteristics[1] of a typical solid-rotor induction machine evaluated with and without considering the harmonic fields. This figure also shows the result of an approximate treatment that accounts for winding harmonics; by simulating current-carrying conductors in stator slots with current filaments on smooth stator air-gap surface. This approximate treatment, therefore, ignores the slotting effect that causes the air-gap permeance to vary periodically in the peripheral direction. In a wound-rotor machine, the design of rotor winding ensures that these unwanted features are reduced.

7.2.1.2 Idealised Configuration

For the analysis presented here, the machine is idealised by assuming infinite permeability for the shaft and the stator iron, constant rotor iron permeability, negligible curvature for the air-gap surfaces and negligible variation of fields in the axial direction. The idealised machine is illustrated in Figure 7.2. The treatment for the magnetic field in stator slots is a modified form of Roth's[1,2]

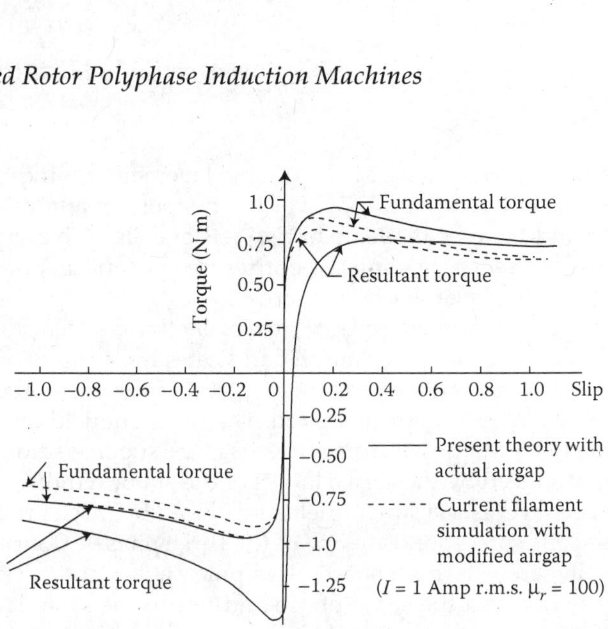

FIGURE 7.1
Slip/torque characteristics of a typical solid-rotor induction machine.

analysis. An alternative treatment based on Rogowski's[3,4] method of analysis is also available.[5]

In a reference frame shown in Figure 7.2, let $z = 0$ be the rotor air-gap surface, and $z = -g$ the stator air-gap surface. In the coordinate system chosen, the axial direction is along the X-axis, the peripheral direction is along the Y-axis and the Z-axis indicates the radial direction. In solid-rotor induction machines, the rotor air-gap surface is smooth, but the stator air-gap surface is slotted. Three-phase winding is housed in the stator slots.

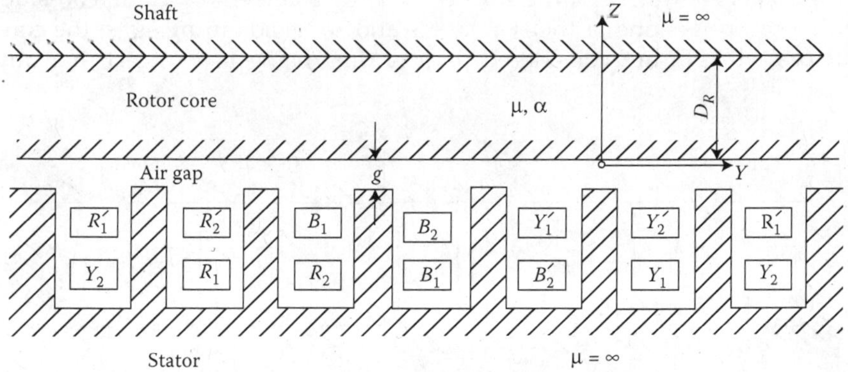

FIGURE 7.2
Idealised machine with three-phase double-layer stator winding (with slots per pole per phase, $\eta = 2$).

Consider a three-phase, integral slot double-layer stator winding with 60° phase spread, as shown in Figure 7.2. For the purpose of analysis, we could suppose this winding to be made up of two sets of coils each carrying three-phase currents, one set being confined entirely to the top layer and the other entirely to the bottom layer. Each layer now consists of full-pitched coils, *the top and bottom layer coils being displaced from each other depending on the extent of the chording of the original coils*. Thus, if ξ indicates the number of slots short chorded, the coils of the two layers are displaced from each other by a distance $(\xi \cdot \lambda)$, where λ indicates a slot pitch. The resultant field distribution in the air gap due to the stator winding currents is the superposition of the field distributions produced by these two layers of the stator winding. Again, the field due to each layer is the sum of fields produced by three groups of coils; each group of coils corresponds to one of the three phases. Consider a series of these full-pitched coils, each spaced two pole pitches apart. This series of coils belongs to one and the same phase and lies in the same layer. Let the number of stator slots per pole per phase be η; then η numbers of similar coil series, placed a slot pitch λ apart, comprise one phase winding of either layer. Thus, the problem is basically reduced to the determination of field due to a coil series similar to that described above.

In Figure 7.3, some stator slots containing red-phase conductors, for a double-layer winding with two slots per pole per phase, are shown. It is assumed that the coils are short pitched by one slot. From these two figures, it is apparent that some of the stator slots contain conductors belonging to the same phase and others may contain conductors of different phases. We shall consider one of the two layers at a time. The air-gap field distribution due to one layer is quite similar to that due to the other layer; except for the reason that (1) the radial position of the two layers is different and (2) there is a phase shift in space between these layers. We shall first consider the bottom layer of the stator winding.

Figure 7.4 shows one of the slots containing a current-carrying coil-side, which comprises one of the two layers and some adjoining slots, the conductors wherein are considered for the time being as not carrying any

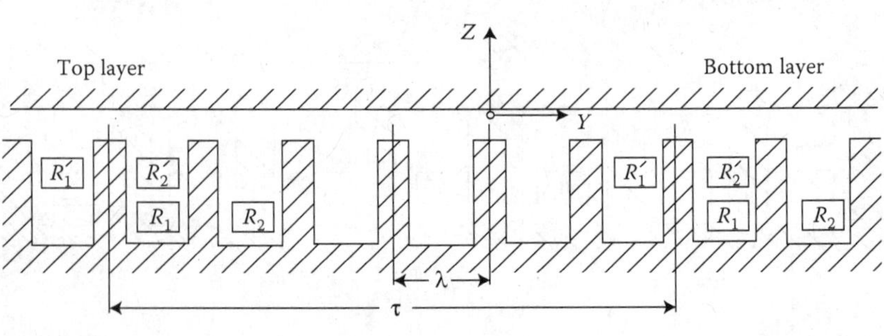

FIGURE 7.3
Top and bottom layer conductors over a pole pitch for phase R.

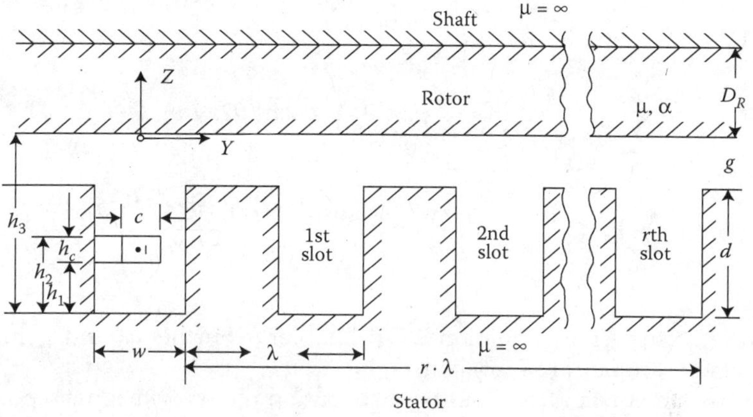

FIGURE 7.4
A current-carrying slot and some adjacent empty slots.

current. This partial figure of the above-mentioned coil series suggests that because of the symmetrical placement of current carrying coil-side with respect to the z-axis chosen, the peripheral component of magnetic field intensity H_y is an even function of y in the air gap as well as in the slot where the conductor is placed. This is represented by the double Fourier series term in the expression for A^o_{sx} (vide Equation 7.1) and for H^o_{sy} (vide Equation 7.3a). It may also be noted that, in general, H_y in current-free slots are neither even nor odd functions of the y coordinate even when the y-axis is chosen to coincide with the respective slot axis. Further, if the origin of the coordinate system is shifted by a distance $\pm 3\eta\tau/2$ (i.e., half the pole pitch), in the peripheral direction, the z-axis will coincide either to a tooth-axis (if η is an odd number) or to a slot axis (if η is an even number). In either case, the peripheral component of magnetic field intensity in the air gap H_{ay} is an odd function of y. Therefore, we need to consider only slots lying over half the pole pitch.

7.2.2 Field Distribution in Stator Slots

In view of the above description, we can now write the field expression in terms of vector magnetic potential and the magnetic field intensity in the region related to the stator slots.

7.2.2.1 Vector Magnetic Potential

With reference to Figure 7.4, the distribution of vector magnetic potential in the current-carrying stator slot can be expressed as a real part of a complex expression, as given below.

$$A_{sx}^{o} = \mathcal{R}e\left[\sum_{m-odd}^{\infty}\sum_{n-odd}^{\infty} A_{m,n}\cdot\cos\left(\frac{m2\pi}{w}\cdot y\right)\cdot\cos\{(n\pi/2d)\cdot(z+g+d)\}\right.$$
$$+\sum_{p-odd}^{\infty} a_p^o\cdot\left(\frac{w}{p2\pi}\right)\cdot\cos\left(\frac{p2\pi}{w}\cdot y\right)\cdot\frac{\cosh\{(p2\pi/w)\cdot(z+g+d)\}}{\cosh((p2\pi/w)\cdot d)}$$
$$\left.+\sum_{q-odd}^{\infty} b_q^o\cdot\left(\frac{w}{q\pi}\right)\cdot\sin\left(\frac{q\pi}{w}\cdot y\right)\cdot\frac{\cosh\{(q\pi/w)\cdot(z+g+d)\}}{\cosh((q\pi/w)\cdot d)}\right]\cdot e^{j\omega t} \quad (7.1)$$

where $A_{m,n}$ is given by Equation 8.7 in Chapter 8. Further, a_p^o and b_q^o indicate two sets of complex arbitrary constants.

For the rth current-free slot, the distribution of vector magnetic potential can now be expressed as

$$A_{sx}^{r} = \mathcal{R}e\left[\sum_{p-odd}^{\infty} a_p^r\cdot\left(\frac{w}{p2\pi}\right)\cdot\cos\left\{\frac{p2\pi}{w}\cdot(y-r\lambda)\right\}\cdot\frac{\cosh\{(p2\pi/w)\cdot(z+g+d)\}}{\cosh((p2\pi/w)\cdot d)}\right.$$
$$\left.+\sum_{q-odd}^{\infty} b_q^r\cdot\left(\frac{w}{q\pi}\right)\cdot\sin\left\{\frac{q\pi}{w}\cdot(y-r\lambda)\right\}\cdot\frac{\cosh\{(q\pi/w)\cdot(z+g+d)\}}{\cosh((q\pi/w)\cdot d)}\right]\cdot e^{j\omega t}$$
$$(7.2)$$

over $(r\lambda - w/2) < y < (r\lambda + w/2)$, for $r = 1,2,\ldots,(3\eta - 1)$, where a_p^r and b_q^r indicate two sets of complex arbitrary constants, for value of each r.

7.2.2.2 Magnetic Field Intensity

The components of the magnetic field intensity in stator slots obtained in view of Equations 7.1 and 7.2 are as follows:

$$H_{sy}^{o} = \frac{1}{\mu_o}\cdot\frac{\partial A_{sx}^{o}}{\partial z}$$
$$= -\mathcal{R}e\sum_{m-odd}^{\infty}\sum_{n-odd}^{\infty}\frac{A_{m,n}}{\mu_o}\cdot\frac{n\pi}{2d}\cdot\cos\left(\frac{m2\pi}{w}\cdot y\right)\cdot\sin\left\{\frac{n\pi}{2d}\cdot(z+g+d)\right\}\cdot e^{j\omega t}$$
$$+\mathcal{R}e\sum_{p-odd}^{\infty}\frac{a_p^o}{\mu_o}\cdot\cos\left(\frac{p2\pi}{w}\cdot y\right)\cdot\frac{\sinh\{(p2\pi/w)\cdot(z+g+d)\}}{\cosh((p2\pi/w)\cdot d)}\cdot e^{j\omega t}$$
$$+\mathcal{R}e\sum_{q-odd}^{\infty}\frac{b_q^o}{\mu_o}\cdot\sin\left(\frac{q\pi}{w}\cdot y\right)\cdot\frac{\sinh\{(q\pi/w)\cdot(z+g+d)\}}{\cosh((q\pi/w)\cdot d)}\cdot e^{j\omega t} \quad (7.3a)$$

over $(-w/2) < y < (+w/2)$,

$$H_{sz}^{o} = -\frac{1}{\mu_o} \cdot \frac{\partial A_{sx}^{o}}{\partial y}$$

$$= -\mathcal{R}e \sum_{m-odd}^{\infty} \sum_{n-odd}^{\infty} \frac{A_{m,n}}{\mu_o} \cdot \frac{n\pi}{2d} \cdot \cos\left(\frac{m2\pi}{w} \cdot y\right)$$

$$\times \sin\left\{\frac{n\pi}{2d} \cdot (z + g + d)\right\} \cdot e^{j\omega t}$$

$$+ \mathcal{R}e \sum_{p-odd}^{\infty} \frac{a_p^{o}}{\mu_o} \cdot \sin\left(\frac{p2\pi}{w} \cdot y\right)$$

$$\times \frac{\cosh\{(p2\pi/w) \cdot (z + g + d)\}}{\cosh((p2\pi/w) \cdot d)} \cdot e^{j\omega t}$$

$$- \mathcal{R}e \sum_{q-odd}^{\infty} \frac{b_q^{o}}{\mu_o} \cdot \cos\left(\frac{q\pi}{w} \cdot y\right)$$

$$\times \frac{\cosh\{(q\pi/w) \cdot (z + g + d)\}}{\cosh((q\pi/w) \cdot d)} \cdot e^{j\omega t} \tag{7.3b}$$

over $(-w/2) < y < (+w/2)$,

$$H_{sy}^{r} = \frac{1}{\mu_o} \cdot \frac{\partial A_{sx}^{r}}{\partial z} = \mathcal{R}e \sum_{p-odd}^{\infty} \frac{a_p^{r}}{\mu_o} \cdot \cos\left\{\frac{p2\pi}{w} \cdot (y - r\lambda)\right\}$$

$$\times \frac{\sinh\{(p2\pi/w) \cdot (z + g + d)\}}{\cosh((p2\pi/w) \cdot d)} \cdot e^{j\omega t}$$

$$+ \mathcal{R}e \sum_{q-odd}^{\infty} \frac{b_q^{r}}{\mu_o} \cdot \sin\left\{\frac{q\pi}{w} \cdot (y - r\lambda)\right\} \cdot \frac{\sinh\{(q\pi/w) \cdot (z + g + d)\}}{\cosh((q\pi/w) \cdot d)} \cdot e^{j\omega t} \tag{7.4a}$$

$$H_{sz}^{r} = -\frac{1}{\mu_o} \cdot \frac{\partial A_{sx}^{r}}{\partial y} = \mathcal{R}e \sum_{p-odd}^{\infty} \frac{a_p^{r}}{\mu_o} \cdot \sin\left\{\frac{p2\pi}{w} \cdot (y - r\lambda)\right\}$$

$$\times \frac{\cosh\{(p2\pi/w) \cdot (z + g + d)\}}{\cosh((p2\pi/w) \cdot d)} \cdot e^{j\omega t}$$

$$- \mathcal{R}e \sum_{q-odd}^{\infty} \frac{b_q^{r}}{\mu_o} \cdot \cos\left\{\frac{q\pi}{w} \cdot (y - r\lambda)\right\} \cdot \frac{\cosh\{(q\pi/w) \cdot (z + g + d)\}}{\cosh((q\pi/w) \cdot d)} \cdot e^{j\omega t} \tag{7.4b}$$

over $(r\lambda - w/2) < y < (r\lambda + w/2)$, for $r = 1,2,\ldots,(3\eta - 1)$.

Therefore, from Equations 7.3a through 7.4b, we get

$$H_{sy}^{o}\Big|_{z=-g} = -\mathcal{R}e\sum_{m-odd}^{\infty}\sum_{n-odd}^{\infty}\frac{A_{m,n}}{\mu_o}\cdot\frac{n\pi}{2d}\cdot\sin\left(\frac{n\pi}{2}\right)\cdot\cos\left(\frac{m2\pi}{w}\cdot y\right)\cdot e^{j\omega t}$$
$$+\mathcal{R}e\sum_{p-odd}^{\infty}\frac{a_p^o}{\mu_o}\cdot\tanh\left(\frac{p2\pi}{w}\cdot d\right)\cdot\cos\left(\frac{p2\pi}{w}\cdot y\right)\cdot e^{j\omega t}$$
$$+\mathcal{R}e\sum_{q-odd}^{\infty}\frac{b_q^o}{\mu_o}\cdot\tanh\left(\frac{q\pi}{w}\cdot d\right)\cdot\sin\left(\frac{q\pi}{w}\cdot y\right)\cdot e^{j\omega t} \tag{7.5a}$$

$$H_{sz}^{o}\Big|_{z=-g} = \mathcal{R}e\sum_{p-odd}^{\infty}\frac{a_p^o}{\mu_o}\cdot\sin\left(\frac{p2\pi}{w}\cdot y\right)\cdot e^{j\omega t}$$
$$-\mathcal{R}e\sum_{q-odd}^{\infty}\frac{b_q^o}{\mu_o}\cdot\cos\left(\frac{q\pi}{w}\cdot y\right)\cdot e^{j\omega t} \tag{7.5b}$$

$$H_{sy}^{r}\Big|_{z=-g} = \mathcal{R}e\sum_{p-odd}^{\infty}\frac{a_p^r}{\mu_o}\cdot\tanh\left(\frac{p2\pi}{w}\cdot d\right)\cdot\cos\left\{\frac{p2\pi}{w}\cdot(y-r\lambda)\right\}\cdot e^{j\omega t}$$
$$-\mathcal{R}e\sum_{q-odd}^{\infty}\frac{b_q^r}{\mu_o}\cdot\tanh\left(\frac{q\pi}{w}\cdot d\right)\cdot\sin\left\{\frac{q\pi}{w}\cdot(y-r\lambda)\right\}\cdot e^{j\omega t} \tag{7.6a}$$

$$H_{sz}^{r}\Big|_{z=-g} = \mathcal{R}e\sum_{p-odd}^{\infty}\frac{a_p^r}{\mu_o}\cdot\sin\left\{\frac{p2\pi}{w}\cdot(y-r\lambda)\right\}\cdot e^{j\omega t}$$
$$+\mathcal{R}e\sum_{q-odd}^{\infty}\frac{b_q^r}{\mu_o}\cdot\cos\left\{\frac{q\pi}{w}\cdot(y-r\lambda)\right\}\cdot e^{j\omega t} \tag{7.6b}$$

over $[(r-1)\lambda + w/2] < y < (r\lambda - w/2)$, for $r = 1,2,\ldots,(3\eta - 1)$.

Further, in view of the high permeability of the stator core, we have, for the top surface of each tooth

$$H_{sy}^{r}\Big|_{z=-g} \cong 0, \tag{7.7}$$

over $(r\lambda - w/2) \le y \le (r\lambda + w/2)$, for $r = 1,2,\ldots,(3\eta)$.

7.2.3 Field Distribution in the Air Gap

With reference to Figure 7.4, it is to be noted that in the air gap, the magnetic field due to the full-pitched coil series is periodic with a period that equals two pole pitches and it contains only odd space harmonics. Further, the peripheral component of the magnetic field intensity is an even function of the variable $-y$. Let this coil series belong to the reference phase $-R$. For the air-gap field $\boldsymbol{H}_a$, due to the entire phase belt, we have

$$\nabla \bullet \boldsymbol{H}_a = 0 \tag{7.8a}$$

and

$$\nabla \times \boldsymbol{H}_a = 0 \tag{7.8b}$$

Therefore, in a reference frame stationary relative to the stator, with its orientation shown in Figure 7.4, the y- and z-components of air-gap field can be expressed as follows:

$$\begin{aligned} H_{ay} = \frac{1}{\mu_o} \cdot \mathcal{R}e \sum_{u-odd}^{\infty} & \left[h_{1u}^R \cdot \frac{\sinh\{(u\pi/\tau)\cdot(z+g)\}}{\sinh((u\pi/\tau)\cdot g)} - h_{1u}^S \cdot \frac{\sinh((u\pi/\tau)\cdot z)}{\sinh((u\pi/\tau)\cdot g)} \right] \\ & \times \cos\left(\frac{u\pi}{\tau}\cdot y\right) \\ & + \left[h_{2u}^R \cdot \frac{\sinh\{(u\pi/\tau)\cdot(z+g)\}}{\sinh((u\pi/\tau)\cdot g)} - h_{2u}^S \cdot \frac{\sinh((u\pi/\tau)\cdot z)}{\sinh((u\pi/\tau)\cdot g)} \right] \cdot \sin\left(\frac{u\pi}{\tau}\cdot y\right) \cdot e^{j\omega t} \end{aligned} \tag{7.9a}$$

and

$$\begin{aligned} H_{az} = \frac{1}{\mu_o} \cdot \mathcal{R}e \sum_{u-odd}^{\infty} & \left[h_{1u}^R \cdot \frac{\cosh\{(u\pi/\tau)(z+g)\}}{\sinh((u\pi/\tau)\cdot g)} - h_{1u}^S \cdot \frac{\cosh((u\pi/\tau)\cdot z)}{\sinh((u\pi/\tau)\cdot g)} \right] \\ & \times \sin\left(\frac{u\pi}{\tau}\cdot y\right) \\ & - \left[h_{2u}^R \cdot \frac{\cosh\{(u\pi/\tau)(z+g)\}}{\sinh((u\pi/\tau)\cdot g)} - h_{2u}^S \cdot \frac{\cosh((u\pi/\tau)\cdot z)}{\sinh((u\pi/\tau)\cdot g)} \right] \cdot \cos\left(\frac{u\pi}{\tau}\cdot y\right) \cdot e^{j\omega t} \end{aligned} \tag{7.9b}$$

where τ is the pole pitch and ω is the supply frequency. Also, h_{1u}^R, h_{2u}^R, h_{1u}^S and h_{2u}^S are the four sets of complex Fourier coefficients defining tangential component of the magnetic field intensity on rotor and stator air-gap surfaces. These are given as

$$H_{ay}\Big|_{z=0} = \frac{1}{\mu_o} \cdot \mathcal{R}e \sum_{u-odd}^{\infty} \left[h_{1u}^R \cdot \cos\left(\frac{u\pi}{\tau}\cdot y\right) + h_{2u}^R \cdot \sin\left(\frac{u\pi}{\tau}\cdot y\right) \right] \cdot e^{j\omega t} \tag{7.10a}$$

$$H_{ay}\Big|_{z=-g} = \frac{1}{\mu_o}\cdot \mathcal{R}e\sum_{u-odd}^{\infty}\left[h_{1u}^{S}\cdot\cos\left(\frac{u\pi}{\tau}\cdot y\right) + h_{2u}^{S}\cdot\sin\left(\frac{u\pi}{\tau}\cdot y\right)\right]\cdot e^{j\omega t} \tag{7.10b}$$

Now, consider the boundary condition

$$H_{ay}\Big|_{z=-g} = H_{sy}^{o}\Big|_{z=-g} \text{ over } (-w/2) < y < (+w/2) \tag{7.11a}$$

$$= H_{sy}^{r}\Big|_{z=-g} \cong 0 \text{ over } \{(r-1)\lambda + w/2\} \le y \le (r\lambda - w/2) \tag{7.11b}$$

for $r = 1,2,\ldots,(3\eta)$.

$$= H_{sy}^{r}\Big|_{z=-g} \text{ over } (r\lambda - w/2) < y < (r\lambda + w/2) \tag{7.11c}$$

for $r = 1,2,\ldots,(3\eta - 1)$.
Or

$$H_{ay}\Big|_{z=-g} = \begin{cases} H_{sy}^{o}\Big|_{z=-g}, & \text{over } (-w/2) < y < (+w/2) \\ \sum_{r=1}^{3\eta} H_{sy}^{r}\Big|_{z=-g} = 0, & \text{over } \{(r-1)\lambda + w/2\} \le y \le (r\lambda - w/2) \\ \sum_{r=1}^{(3\eta-1)} H_{sy}^{r}\Big|_{z=-g}, & \text{over } (r\lambda - w/2) < y < (r\lambda + w/2) \end{cases} \tag{7.11d}$$

Therefore, one gets from Equations 7.10b, 7.5a, 7.6a and 7.11a through 7.11c or 7.11d

$$\begin{aligned} &\frac{1}{\mu_o}\cdot\sum_{u-odd}^{\infty}\left[h_{1u}^{S}\cdot\cos\left(\frac{u\pi}{\tau}\cdot y\right) + h_{2u}^{S}\cdot\sin\left(\frac{u\pi}{\tau}\cdot y\right)\right] \\ &= -\sum_{m-odd}^{\infty}\sum_{n-odd}^{\infty} A_{m,n}\cdot\frac{n\pi}{2d}\cdot\sin\left(\frac{n\pi}{2}\right)\cdot\cos\left(\frac{m2\pi}{w}\cdot y\right) \\ &+ \sum_{p-odd}^{\infty} a_p^{o}\cdot\tanh\left(\frac{p2\pi}{w}\cdot d\right)\cdot\cos\left(\frac{p2\pi}{w}\cdot y\right) \\ &+ \sum_{q-odd}^{\infty} b_q^{o}\cdot\tanh\left(\frac{q\pi}{w}\cdot d\right)\cdot\sin\left(\frac{q\pi}{w}\cdot y\right) \end{aligned} \tag{7.12a}$$

over $(-w/2) < y < (+w/2)$,

$$= 0, \tag{7.12b}$$

over $[(r-1)\lambda + w/2] < y < (r\lambda - w/2)$, for $r = 1,2,\ldots, 3\eta$.

$$= \sum_{p-odd}^{\infty} a_p^r \cdot \tanh\left(\frac{p2\pi}{w} \cdot d\right) \cdot \cos\left\{\frac{p2\pi}{w} \cdot (y - r\lambda)\right\}$$

$$- \sum_{q-odd}^{\infty} b_q^r \cdot \tanh\left(\frac{q\pi}{w} \cdot \mathrm{d}\right) \cdot \sin\left\{\frac{q\pi}{w} \cdot (y - r\lambda)\right\} \tag{7.12c}$$

over $(r\lambda - w/2) \le y \le (r\lambda + w/2)$, for $r = 1,2,\ldots, (3\eta - 1)$.

Equating real and imaginary parts separately, Equations 7.12a through 7.12c can be split into two sets of equations.

For any odd integer U, multiplying both sides of Equations 7.12a through 7.12c by $\cos((U\pi/\tau) \cdot y)$ and then integrating over half the pole pitch (i.e., $\tau/2$)

$$\int_{y=0}^{\tau/2} \left[H_{ay}\Big|_{z=-g} \cdot \cos\left(\frac{U\pi}{\tau} \cdot y\right)\right] \cdot dy = \int_{y=0}^{w/2} \left[H_{sy}^{o}\Big|_{z=-g} \cdot \cos\left(\frac{U\pi}{\tau} \cdot y\right)\right] \cdot dy$$

$$+ \sum_{r=1}^{(3\eta-1)/2} \int_{y=(r\lambda-w/2)}^{(r\lambda+w/2)} \left[H_{sy}^{r}\Big|_{z=-g} \cdot \cos\left(\frac{U\pi}{\tau} \cdot y\right)\right] \cdot dy$$

for odd $-\eta$.

$$\int_{y=0}^{\tau/2} \left[H_{ay}\Big|_{z=-g} \cdot \cos\left(\frac{U\pi}{\tau} \cdot y\right)\right] \cdot dy = \int_{y=0}^{w/2} \left[H_{sy}^{o}\Big|_{z=-g} \cdot \cos\left(\frac{U\pi}{\tau} \cdot y\right)\right] \cdot dy$$

$$+ \sum_{r=1}^{(3\eta-2)/2} \int_{y=(r\lambda-w/2)}^{(r\lambda+w/2)} \left[H_{sy}^{r}\Big|_{z=-g} \cdot \cos\left(\frac{U\pi}{\tau} \cdot y\right)\right] \cdot dy$$

for even $-\eta$.

We get, in view of the identity

$$\int_{0}^{\tau/2} \cos\left(\frac{u\pi}{\tau} \cdot y\right) \cdot \cos\left(\frac{U\pi}{\tau} \cdot y\right) \cdot dy \equiv \begin{cases} \dfrac{\tau}{4} (\text{for } u = U) \\ 0 (\text{for } u \neq U) \end{cases} \tag{7.13}$$

$$\mathrm{LHS} = \frac{\tau}{4} \cdot \frac{1}{\mu_o} \cdot h_{1U}^{S} \tag{7.14}$$

Similarly, for the RHS

$$
\begin{aligned}
\text{RHS} = &-\sum_{m-odd}^{\infty}\sum_{n-odd}^{\infty} A_{m,n}\cdot\frac{n\pi}{2d}\cdot\sin\left(\frac{n\pi}{2}\right)\cdot\int_{0}^{w/2}\cos\left(\frac{m2\pi}{w}\cdot y\right)\cdot\cos\left(\frac{U\pi}{\tau}\cdot y\right)\cdot dy \\
&+\sum_{p-odd}^{\infty} a_p^o\cdot\tanh\left(\frac{p2\pi}{w}\cdot d\right)\cdot\int_{0}^{w/2}\cos\left(\frac{p2\pi}{w}\cdot y\right)\cdot\cos\left(\frac{U\pi}{\tau}\cdot y\right)\cdot dy \\
&+\sum_{q-odd}^{\infty} b_q^o\cdot\tanh\left(\frac{q\pi}{w}\cdot d\right)\cdot\int_{0}^{w/2}\sin\left(\frac{q\pi}{w}\cdot y\right)\cdot\cos\left(\frac{U\pi}{\tau}\cdot y\right)\cdot dy \\
&+\sum_{r=1}^{(3\eta-1)/2}\left[\sum_{p-odd}^{\infty} a_p^r\cdot\tanh\left(\frac{p2\pi}{w}\cdot d\right)\right. \\
&\times\int_{(r\lambda-w/2)}^{(r\lambda+w/2)}\cos\left\{\frac{p2\pi}{w}\cdot(y-r\lambda)\right\}\cdot\cos\left(\frac{U\pi}{\tau}\cdot y\right)\cdot dy \\
&-\sum_{q-odd}^{\infty} b_q^r\cdot\tanh\left(\frac{q\pi}{w}\cdot d\right) \\
&\left.\times\int_{(r\lambda-w/2)}^{(r\lambda+w/2)}\sin\left\{\frac{q\pi}{w}\cdot(y-r\lambda)\right\}\cdot\cos\left(\frac{U\pi}{\tau}\cdot y\right)\cdot dy\right] \quad \text{for odd} -\eta
\end{aligned}
\tag{7.15a}
$$

$$
\begin{aligned}
\text{RHS} = &-\sum_{m-odd}^{\infty}\sum_{n-odd}^{\infty} A_{m,n}\cdot\frac{n\pi}{2d}\cdot\sin\left(\frac{n\pi}{2}\right)\cdot\int_{0}^{w/2}\cos\left(\frac{m2\pi}{w}\cdot y\right)\cdot\cos\left(\frac{U\pi}{\tau}\cdot y\right)\cdot dy \\
&+\sum_{p-odd}^{\infty} a_p^o\cdot\tanh\left(\frac{p2\pi}{w}\cdot d\right)\cdot\int_{0}^{w/2}\cos\left(\frac{p2\pi}{w}\cdot y\right)\cdot\cos\left(\frac{U\pi}{\tau}\cdot y\right)\cdot dy \\
&+\sum_{q-odd}^{\infty} b_q^o\cdot\tanh\left(\frac{q\pi}{w}\cdot d\right)\cdot\int_{0}^{w/2}\sin\left(\frac{q\pi}{w}\cdot y\right)\cdot\cos\left(\frac{U\pi}{\tau}\cdot y\right)\cdot dy \\
&+\sum_{r=1}^{(3\eta-2)/2}\left[\sum_{p-odd}^{\infty} a_p^r\cdot\tanh\left(\frac{p2\pi}{w}\cdot d\right)\right. \\
&\times\int_{(r\lambda-w/2)}^{(r\lambda+w/2)}\cdot\cos\left\{\frac{p2\pi}{w}\cdot(y-r\lambda)\right\}\cdot\cos\left(\frac{U\pi}{\tau}\cdot y\right)\cdot dy
\end{aligned}
$$

$$- \sum_{q-odd}^{\infty} b_q^r \cdot \tanh\left(\frac{q\pi}{w} \cdot d\right)$$

$$\left. \times \int_{(r\lambda - w/2)}^{(r\lambda + w/2)} \sin\left\{\frac{q\pi}{w} \cdot (y - r\lambda)\right\} \cdot \cos\left(\frac{U\pi}{\tau} \cdot y\right) \cdot dy \right] \text{ for even } -\eta \tag{7.15b}$$

Integrals in Equations 7.15a and 7.15b can be readily evaluated, giving for odd values of m, p, q and U:

$$\int_0^{w/2} \sin\left(\frac{q\pi}{w} \cdot y\right) \cdot \cos\left(\frac{U\pi}{\tau} \cdot y\right) \cdot dy \overset{def}{=} I_1^{(q,U)} \tag{7.15c}$$

$$\int_0^{w/2} \cos\left(\frac{m2\pi}{w} \cdot y\right) \cdot \cos\left(\frac{U\pi}{\tau} \cdot y\right) \cdot dy \overset{def}{=} I_2^{(m,U)} \tag{7.15d}$$

$$\int_0^{w/2} \cos\left(\frac{p2\pi}{w} \cdot y\right) \cdot \cos\left(\frac{U\pi}{\tau} \cdot y\right) \cdot dy \overset{def}{=} I_3^{(p,U)} \tag{7.15e}$$

$$\int_{(r\lambda - w/2)}^{(r\lambda + w/2)} \cos\left\{\frac{p2\pi}{w} \cdot (y - r\lambda)\right\} \cdot \cos\left(\frac{U\pi}{\tau} \cdot y\right) \cdot dy \overset{def}{=} I_4^{(p,U)} \tag{7.15f}$$

$$\int_{(r\lambda - w/2)}^{(r\lambda + w/2)} \sin\left\{\frac{q\pi}{w} \cdot (y - r\lambda)\right\} \cdot \cos\left(\frac{U\pi}{\tau} \cdot y\right) \cdot dy \overset{def}{=} I_5^{(q,U)} \tag{7.15g}$$

where

$$I_1^{(q,U)} = \frac{1}{\pi} \cdot \left[\frac{(q/w)}{(q/w)^2 - (U/\tau)^2} - \sin\left(\frac{q\pi}{2}\right) \cdot \sin\left(\frac{U\pi}{\tau} \cdot \frac{w}{2}\right) \cdot \frac{(U/\tau)}{(q/w)^2 - (U/\tau)^2}\right] \tag{7.16a}$$

$$I_2^{(m,U)} = \frac{\pi}{2} \sin\left\{\left(\frac{U}{2} \cdot \frac{w}{\tau}\right)\pi\right\} \cdot \frac{(U/2\tau)}{[(m/w)^2 - (U/2\tau)^2]} \tag{7.16b}$$

$$I_3^{(p,U)} = \frac{\pi}{2}\sin\left\{\left(\frac{U}{2}\cdot\frac{w}{\tau}\right)\pi\right\}\cdot\frac{(U/2\tau)}{[(p/w)^2-(U/2\tau)^2]} \tag{7.16c}$$

$$I_4^{(p,U)} = \frac{1}{2\pi}\cdot\cos\left\{\frac{U}{\tau}\pi r\lambda\right\}\cdot\sin\left\{\frac{U}{\tau}\pi\frac{w}{2}\right\}\cdot\frac{(U/2\tau)}{[(p/w)^2-(U/2\tau)^2]} \tag{7.16d}$$

and

$$I_5^{(q,U)} = \frac{2}{\pi}\sin\left(\frac{q\pi}{2}\right)\cdot\sin\left(\frac{U\pi}{\tau}\cdot r\lambda\right)\cdot\cos\left\{\frac{U\pi}{\tau}\cdot\frac{w}{2}\right\}\cdot\frac{(q/w)}{[(U/\tau)^2-(q/w)^2]} \tag{7.16e}$$

Therefore, in view of Equations 7.15a through 7.15g, we get

$$\begin{aligned}\frac{\tau}{4}\cdot\frac{1}{\mu_o}\cdot h_{1U}^S = &-\sum_{m-odd}^{\infty}\sum_{n-odd}^{\infty}A_{m,n}\cdot\frac{n\pi}{2d}\cdot\sin\left(\frac{n\pi}{2}\right)\cdot I_2^{(m,U)}\\&+\sum_{p-odd}^{\infty}a_p^o\cdot\tanh\left(\frac{p2\pi}{w}\cdot d\right)\cdot I_3^{(p,U)}+\sum_{q-odd}^{\infty}b_q^o\cdot\tanh\left(\frac{q\pi}{w}\cdot d\right)\cdot I_1^{(q,U)}\\&+\sum_{r=1}^{(3\eta-1)/2}\left[\sum_{p-odd}^{\infty}a_p^r\cdot\tanh\left(\frac{p2\pi}{w}\cdot d\right)\cdot I_4^{(p,U)}-\sum_{q-odd}^{\infty}b_q^r\cdot\tanh\left(\frac{q\pi}{w}\cdot d\right)\cdot I_5^{(q,U)}\right]\\&\text{for odd }-\eta\end{aligned} \tag{7.17a}$$

$$\begin{aligned}\frac{\tau}{4}\cdot\frac{1}{\mu_o}\cdot h_{1U}^S = &-\sum_{m-odd}^{\infty}\sum_{n-odd}^{\infty}A_{m,n}\cdot\frac{n\pi}{2d}\cdot\sin\left(\frac{n\pi}{2}\right)\cdot I_2^{(m,U)}\\&+\sum_{p-odd}^{\infty}a_p^o\cdot\tanh\left(\frac{p2\pi}{w}\cdot d\right)\cdot I_3^{(p,U)}+\sum_{q-odd}^{\infty}b_q^o\cdot\tanh\left(\frac{q\pi}{w}\cdot d\right)\cdot I_1^{(q,U)}\\&+\sum_{r=1}^{(3\eta-2)/2}\left[\sum_{p-odd}^{\infty}a_p^r\cdot\tanh\left(\frac{p2\pi}{w}\cdot d\right)\cdot I_4^{(p,U)}-\sum_{q-odd}^{\infty}b_q^r\cdot\tanh\left(\frac{q\pi}{w}\cdot d\right)\cdot I_5^{(q,U)}\right]\\&\text{for even }-\eta\end{aligned} \tag{7.17b}$$

Thus, in view of Equations 7.16a through 7.16e and 7.17a and 7.17b, on replacing U by u, the value for h_{1u}^S for all odd values of u is obtainable in terms of a_p^o, a_p^r and b_q^r. Similarly, for any odd integer U, multiplying the LHS of Equations 7.12a through 7.12c by $\sin((U\pi/\tau)\cdot y)$ and then integrating over half the pole pitch, that is, $\tau/2$, the value for h_{2u}^S for all odd values of u can be found in terms of a_p^o, a_p^r and b_q^r, provided that after the integration, U is replaced by u. Equating real and imaginary parts separately, each resulting equation can be split into two sets of equations.

Next, consider the continuity of the normal component of the magnetic field intensity at slot openings.

$$H_{sz}^{o}\Big|_{z=-g} = H_{az}\Big|_{z=-g} \tag{7.18a}$$

over $(-w/2) < y < (+w/2)$ and

$$H_{sz}^{r}\Big|_{z=-g} = H_{az}\Big|_{z=-g} \tag{7.18b}$$

over $(r\lambda - w/2) \le y \le (r\lambda + w/2)$, for $r = 1,2,\ldots,(3\eta - 1)$.

From Equation 7.9b, we have, ignoring the factor $e^{j\omega t}$

$$H_{az}\Big|_{z=-g} = \frac{1}{\mu_o}\cdot\sum_{u-odd}^{\infty}\left[\left\{h_{1u}^{R}\cdot\text{cosech}\left(\frac{u\pi}{\tau}\cdot g\right) - h_{1u}^{S}\cdot\coth\left(\frac{u\pi}{\tau}\cdot g\right)\right\}\cdot\sin\left(\frac{u\pi}{\tau}\cdot y\right)\right.$$
$$\left. -\left\{h_{2u}^{R}\cdot\text{cosech}\left(\frac{u\pi}{\tau}\cdot g\right) - h_{2u}^{S}\cdot\coth\left(\frac{u\pi}{\tau}\cdot g\right)\right\}\cdot\cos\left(\frac{u\pi}{\tau}\cdot y\right)\right] \tag{7.19}$$

Thus, in view of Equations 7.5b and 7.6b, we have

$$\sum_{p-odd}^{\infty} a_p^{o}\cdot\sin\left(\frac{p2\pi}{w}\cdot y\right) - \sum_{q-odd}^{\infty} b_q^{o}\cdot\cos\left(\frac{q\pi}{w}\cdot y\right)$$
$$= \sum_{u-odd}^{\infty}\left[\left\{h_{1u}^{R}\cdot\text{cosech}\left(\frac{u\pi}{\tau}\cdot g\right) - h_{1u}^{S}\cdot\coth\left(\frac{u\pi}{\tau}\cdot g\right)\right\}\cdot\sin\left(\frac{u\pi}{\tau}\cdot y\right)\right.$$
$$\left. -\left\{h_{2u}^{R}\cdot\text{cosech}\left(\frac{u\pi}{\tau}\cdot g\right) - h_{2u}^{S}\cdot\coth\left(\frac{u\pi}{\tau}\cdot g\right)\right\}\cdot\cos\left(\frac{u\pi}{\tau}\cdot y\right)\right] \tag{7.20a}$$

over $(-w/2) < y < (+w/2)$.

And

$$\sum_{p-odd}^{\infty} a_p^{r}\cdot\sin\left\{\frac{p2\pi}{w}\cdot(y - r\lambda)\right\} + \sum_{q-odd}^{\infty} b_q^{r}\cdot\cos\left\{\frac{q\pi}{w}\cdot(y - r\lambda)\right\}$$
$$= \sum_{u-odd}^{\infty}\left[\left\{h_{1u}^{R}\cdot\text{cosech}\left(\frac{u\pi}{\tau}\cdot g\right) - h_{1u}^{S}\cdot\coth\left(\frac{u\pi}{\tau}\cdot g\right)\right\}\cdot\sin\left(\frac{u\pi}{\tau}\cdot y\right)\right.$$
$$\left. -\left\{h_{2u}^{R}\cdot\text{cosech}\left(\frac{u\pi}{\tau}\cdot g\right) - h_{2u}^{S}\cdot\coth\left(\frac{u\pi}{\tau}\cdot g\right)\right\}\cdot\cos\left(\frac{u\pi}{\tau}\cdot y\right)\right] \tag{7.20b}$$

over $(r\lambda - w/2) < y < (r\lambda + w/2)$, for $r = 1, 2, \ldots,(3\eta - 1)$.

Consider the following identities, for any odd $-P$ and $-Q$:

$$\int_{-w/2}^{w/2} \sin\left(\frac{p2\pi}{w}\cdot y\right)\cdot\sin\left(\frac{P2\pi}{w}\cdot y\right)\cdot dy \equiv \begin{cases} \dfrac{w}{2} & (\text{for } p = P) \\ 0 & (\text{for } p \neq P) \end{cases} \tag{7.21a}$$

$$\int_{-w/2}^{w/2} \sin\left(\frac{P2\pi}{w}\cdot y\right)\cdot\sin\left(\frac{u\pi}{\tau}\cdot y\right)\cdot dy \equiv \sin\left(u\cdot\frac{w}{\tau}\cdot\frac{\pi}{2}\right)\cdot\frac{(2/\pi)\cdot(2P/w)}{\left[(2P/w)^2-(u/\tau)^2\right]} \tag{7.21b}$$

$$\int_{-w/2}^{w/2} \sin\left(\frac{P2\pi}{w}\cdot y\right)\cdot\cos\left(\frac{u\pi}{\tau}\cdot y\right)\cdot dy \equiv 0 \tag{7.21c}$$

$$\int_{-w/2}^{w/2} \sin\left(\frac{p2\pi}{w}\cdot y\right)\cdot\cos\left(\frac{Q\pi}{w}\cdot y\right)\cdot dy \equiv 0 \tag{7.21d}$$

$$\int_{-w/2}^{w/2} \cos\left(\frac{q\pi}{w}\cdot y\right)\cdot\cos\left(\frac{Q\pi}{w}\cdot y\right)\cdot dy \equiv \begin{cases} \dfrac{w}{2} & (\text{for } q = Q) \\ 0 & (\text{for } q \neq Q) \end{cases} \tag{7.21e}$$

$$\int_{-w/2}^{w/2} \sin\left(\frac{u\pi}{\tau}\cdot y\right)\cdot\cos\left(\frac{Q\pi}{w}\cdot y\right)\cdot dy \equiv 0 \tag{7.21f}$$

$$\int_{-w/2}^{w/2} \cos\left(\frac{u\pi}{\tau}\cdot y\right)\cdot\cos\left(\frac{Q\pi}{w}\cdot y\right)\cdot dy \equiv \cos\left(\frac{u\pi}{\tau}\cdot\frac{w}{2}\right)\cdot\sin\left(\frac{Q\pi}{2}\right)\cdot\frac{(2/\pi)\cdot(Q/w)}{[(u/\tau)^2-(Q/w)^2]} \tag{7.21g}$$

$$\int_{(r\lambda-w/2)}^{(r\lambda+w/2)} \sin\left\{\frac{p2\pi}{w}\cdot(y-r\lambda)\right\}\cdot\sin\left\{\frac{P2\pi}{w}\cdot(y-r\lambda)\right\}\cdot dy \equiv \begin{cases} \dfrac{w}{2} & (\text{for } p = P) \\ 0 & (\text{for } p \neq P) \end{cases} \tag{7.21h}$$

$$\int_{(r\lambda-w/2)}^{(r\lambda+w/2)} \sin\left\{\frac{P2\pi}{w}\cdot(y-r\lambda)\right\}\cdot\sin\left(\frac{u\pi}{\tau}\cdot y\right)\cdot dy$$
$$\equiv \frac{\cos((u\pi/\tau)\cdot r\lambda)\sin(u\cdot(w/\tau)\cdot(\pi/2))}{\{(2P/w)^2-(u/\tau)^2\}}\cdot\frac{2}{\pi}\cdot\frac{2P}{w} \tag{7.21i}$$

$$\int_{(r\lambda-w/2)}^{(r\lambda+w/2)} \sin\left\{\frac{P2\pi}{w}\cdot(y-r\lambda)\right\}\cdot\cos\left(\frac{u\pi}{\tau}\cdot y\right)\cdot dy$$

$$\equiv -\cos\left(\frac{u\pi}{\tau}\cdot r\lambda\right)\cdot\cos\left(u\cdot\frac{w}{\tau}\cdot\frac{\pi}{2}\right)\cdot\frac{(2/\pi)\cdot(2P/w)\cdot\cos(P\pi)}{\{(2P/w)^2-(u/\tau)^2\}} \tag{7.21j}$$

$$\int_{(r\lambda-w/2)}^{(r\lambda+w/2)} \sin\left\{\frac{P2\pi}{w}\cdot(y-r\lambda)\right\}\cdot\cos\left\{\frac{q\pi}{w}\cdot(y-r\lambda)\right\}\cdot dy \equiv 0 \tag{7.21k}$$

$$\int_{(r\lambda-w/2)}^{(r\lambda+w/2)} \cos\left\{\frac{q\pi}{w}\cdot(y-r\lambda)\right\}\cdot\cos\left\{\frac{Q\pi}{w}\cdot(y-r\lambda)\right\}\cdot dy \equiv \begin{cases}\dfrac{w}{2} & (\text{for } q=Q),\\ 0 & (\text{for } q\neq Q).\end{cases} \tag{7.21l}$$

$$\int_{(r\lambda-w/2)}^{(r\lambda+w/2)} \cos\left\{\frac{Q\pi}{w}\cdot(y-r\lambda)\right\}\cdot\sin\left(\frac{u\pi}{\tau}\cdot y\right)\cdot dy$$

$$\equiv \sin\left(\frac{u\pi}{\tau}\cdot r\lambda\right)\cdot\cos\left(u\cdot\frac{w}{\tau}\cdot\frac{\pi}{2}\right)\frac{(2/\pi)\cdot(Q/w)\cdot\sin(Q\pi/2)}{[(u/\tau)^2-(Q/w)^2]} \tag{7.21m}$$

$$\int_{(r\lambda-w/2)}^{(r\lambda+w/2)} \cos\left\{\frac{Q\pi}{w}\cdot(y-r\lambda)\right\}\cdot\cos\left(\frac{u\pi}{\tau}\cdot y\right)\cdot dy$$

$$\equiv \cos\left(\frac{u\pi}{\tau}\cdot r\lambda\right)\cdot\cos\left(u\cdot\frac{w}{\tau}\cdot\frac{\pi}{2}\right)\frac{(2/\pi)\cdot(Q/w)\cdot\sin(Q\pi/2)}{[(u/\tau)^2-(Q/w)^2]} \tag{7.21n}$$

Multiply both sides of Equation 7.20a with $\sin((P2\pi/w)\cdot y)$ and then integrate over $(-w/2) < y < (+w/2)$. On replacing P by p, we get, in view of Equations 7.21a through 7.21d

$$a_p^o\cdot\frac{w}{2} = \sum_{u-odd}^{\infty}\left[h_{1u}^R\cdot\operatorname{cosech}\left(\frac{u\pi}{\tau}\cdot g\right)\right.$$

$$\left.-\,h_{1u}^S\cdot\coth\left(\frac{u\pi}{\tau}\cdot g\right)\right]\cdot\sin\left(u\cdot\frac{w}{\tau}\cdot\frac{\pi}{2}\right)\cdot\frac{(2/\pi)\cdot(2p/w)}{[(2p/w)^2-(u/\tau)^2]} \tag{7.22a}$$

for $p = 1, 3, 5, \ldots, \infty$.

Similarly, multiply both sides of Equation 7.20a with $\cos((Q\pi/w)\cdot y)$ and then integrate over $(-w/2) < y < (w/2)$. On replacing Q by q, we get, in view of Equations 7.21d through 7.21g

$$b_q^o \cdot \frac{w}{2} = \sum_{u-odd}^{\infty}\left[\left\{h_{2u}^R \cdot \operatorname{cosech}\left(\frac{u\pi}{\tau}\cdot g\right) - h_{2u}^S \cdot \coth\left(\frac{u\pi}{\tau}\cdot g\right)\right\}\right]\cdot \cos\left(u\cdot\frac{w}{\tau}\cdot\frac{\pi}{2}\right)\cdot \sin\left(\frac{q\pi}{2}\right)\cdot\frac{(2/\pi)\cdot(q/w)}{[(u/\tau)^2-(q/w)^2]} \tag{7.22b}$$

for $q = 1, 3, 5, \ldots, \infty$.

Next, multiply both sides of Equation 7.20b with $\sin\{(P2\pi/w)\cdot(y - r\lambda)\}$ and then integrate over $(r\lambda - w/2) < y < (r\lambda + w/2)$. On replacing P by p one gets, in view of Equations 7.21h through 7.21k

$$\begin{aligned} a_p^r \cdot \frac{w}{2} = \sum_{u-odd}^{\infty} & \left[h_{1u}^R \cdot \operatorname{cosech}\left(\frac{u\pi}{\tau}\cdot g\right) - h_{1u}^S \cdot \coth\left(\frac{u\pi}{\tau}\cdot g\right)\right] \\ & \times \cos\left(\frac{u\pi}{\tau}\cdot r\lambda\right)\sin\left(u\cdot\frac{w}{\tau}\cdot\frac{\pi}{2}\right)\cdot\frac{(2/\pi)\cdot(2p/w)}{\{(2p/w)^2-(u/\tau)^2\}} \\ & + \left[h_{2u}^R \cdot \operatorname{cosech}\left(\frac{u\pi}{\tau}\cdot g\right) - h_{2u}^S \cdot \coth\left(\frac{u\pi}{\tau}\cdot g\right)\right] \\ & \times \cos\left(\frac{u\pi}{\tau}\cdot r\lambda\right)\cdot\cos\left(u\cdot\frac{w}{\tau}\cdot\frac{\pi}{2}\right)\cdot\frac{(2/\pi)\cdot(2p/w)\cdot\cos(p\pi)}{\{(2p/w)^2-(u/\tau)^2\}} \end{aligned} \tag{7.23}$$

for $r = 1, 2, \ldots, (3\eta - 1)$ and for $p = 1, 3, 5, \ldots, \infty$.

Multiply both sides of Equation 7.19b with $\cos\{(Q\pi/w)\cdot(y - r\lambda)\}$ and then integrate over $(r\lambda - w/2) < y < (r\lambda + w/2)$. On replacing Q by q, one gets, in view of Equations 7.20k through 7.20n

$$\begin{aligned} b_q^r \cdot \frac{w}{2} = \sum_{u-odd}^{\infty} & \left[h_{1u}^R \cdot \operatorname{cosech}\left(\frac{u\pi}{\tau}\cdot g\right) - h_{1u}^S \cdot \coth\left(\frac{u\pi}{\tau}\cdot g\right)\right] \\ & \times \sin\left(\frac{u\pi}{\tau}\cdot r\lambda\right)\cdot\cos\left(u\cdot\frac{w}{\tau}\cdot\frac{\pi}{2}\right)\frac{(2/\pi)\cdot(q/w)\cdot\sin(q\pi/2)}{[(u/\tau)^2-(q/w)^2]} \\ & + \left[h_{2u}^R \cdot \operatorname{cosech}\left(\frac{u\pi}{\tau}\cdot g\right) - h_{2u}^S \cdot \coth\left(\frac{u\pi}{\tau}\cdot g\right)\right] \\ & \times \cos\left(\frac{u\pi}{\tau}\cdot r\lambda\right)\cdot\cos\left(u\cdot\frac{w}{\tau}\cdot\frac{\pi}{2}\right)\frac{(2/\pi)\cdot(q/w)\cdot\sin(q\pi/2)}{[(u/\tau)^2-(q/w)^2]} \end{aligned} \tag{7.24}$$

for $r = 1, 2, \ldots, (3\eta - 1)$ and for $q = 1, 3, 5, \ldots, \infty$.

Thus, Equations 7.21 through 7.24 give a_p^o, b_q^o, and b_q^r, in terms of $h_{1u}^R, h_{1u}^S, h_{2u}^R$ and h_{2u}^S.

7.2.4 Field Distribution in the Solid Rotor

Equations 7.9a and 7.9b give the distribution of the components of the magnetic field intensity in the air gap in a reference frame stationary with respect to the stator. Since

$$\cos\left(\frac{u\pi}{\tau}\cdot y\right)\cdot e^{j\omega t} \equiv \frac{1}{2}\cdot\left[e^{-j((u\pi/\tau)\cdot y-\omega t)} + e^{j((u\pi/\tau)\cdot y+\omega t)}\right] \tag{7.25a}$$

$$\sin\left(\frac{u\pi}{\tau}\cdot y\right)\cdot e^{j\omega t} \equiv j\frac{1}{2}\cdot\left[e^{-j((u\pi/\tau)\cdot y-\omega t)} - e^{j((u\pi/\tau)\cdot y+\omega t)}\right] \tag{7.25b}$$

Equations 7.9a and 7.9b can therefore be rewritten as

$$\begin{aligned} H_{ay} = \frac{1}{\mu_o}\cdot \mathcal{R}e\sum_{u-odd}^{\infty}&\left[h_{1u}^R\cdot\frac{\sinh\{(u\pi/\tau)\cdot(z+g)\}}{\sinh((u\pi/\tau)\cdot g)} - h_{1u}^S\cdot\frac{\sinh((u\pi/\tau)\cdot z)}{\sinh((u\pi/\tau)\cdot g)}\right]\\ &\times\frac{1}{2}\cdot\left[e^{-j((u\pi/\tau)\cdot y-\omega\cdot t)} + e^{j((u\pi/\tau)\cdot y+\omega\cdot t)}\right]\\ &+\left[h_{2u}^R\cdot\frac{\sinh\{(u\pi/\tau)\cdot(z+g)\}}{\sinh((u\pi/\tau)\cdot g)} - h_{2u}^S\cdot\frac{\sinh((u\pi/\tau)\cdot z)}{\sinh((u\pi/\tau)\cdot g)}\right]\\ &\times j\frac{1}{2}\cdot\left[e^{-j((u\pi/\tau)\cdot y-\omega\cdot t)} - e^{j((u\pi/\tau)\cdot y+\omega\cdot t)}\right] \end{aligned} \tag{7.26a}$$

$$\begin{aligned} H_{az} = \frac{1}{\mu_o}.\mathcal{R}e\sum_{u-odd}^{\infty}&\left[h_{1u}^R\cdot\frac{\cosh\{(u\pi/\tau)(z+g)\}}{\sinh((u\pi/\tau)\cdot g)} - h_{1u}^S\cdot\frac{\cosh((u\pi/\tau)\cdot z)}{\sinh((u\pi/\tau)\cdot g)}\right]\\ &\times j\frac{1}{2}\cdot\left[e^{-j((u\pi/\tau)\cdot y-\omega\cdot t)} - e^{j((u\pi/\tau)\cdot y+\omega\cdot t)}\right]\\ &-\left[h_{2u}^R\cdot\frac{\cosh\{(u\pi/\tau)(z+g)\}}{\sinh((u\pi/\tau)\cdot g)} - h_{2u}^S\cdot\frac{\cosh((u\pi/\tau)\cdot z)}{\sinh((u\pi/\tau)\cdot g)}\right]\\ &\times\frac{1}{2}\cdot\left[e^{-j((u\pi/\tau)\cdot y-\omega\cdot t)} + e^{j((u\pi/\tau)\cdot y+\omega\cdot t)}\right] \end{aligned} \tag{7.26b}$$

This shows that for each order of the field harmonic $-u$, the magnetic field can be resolved into two revolving fields, one rotating in the positive direction and the other in the negative direction, both with the same rotational speed, that is

$$\frac{dy}{dt} = \pm\frac{\omega\tau}{u\pi} \tag{7.27}$$

Now, if the reference system is rotating with the rotor, the relative velocity of the two revolving fields with respect to the rotor will be

$$v_+ = \frac{\omega\tau}{u\pi} - v \tag{7.28a}$$

$$v_- = -\frac{\omega\tau}{u\pi} - v \tag{7.28b}$$

where v indicates the rotor velocity relative to the stator, while v_+ and v_- indicate velocities of the positive and the negative revolving fields with respect to the rotor. If fractional slip of the rotor, s, is defined as

$$s \stackrel{def}{=} 1 - \frac{\text{speed of the rotor}}{\text{synchronous speed of the fundamental field}} \tag{7.29}$$

or

$$s \stackrel{def}{=} \frac{(\omega\tau/\pi) - v}{(\omega\tau/\pi)} = \frac{(\omega\tau - \pi v)}{\omega\tau} \tag{7.29a}$$

thus,

$$v = (1 - s)\cdot\frac{\omega\tau}{\pi} \tag{7.29b}$$

Therefore, in view of Equations 7.28a and 7.28b

$$v_+ \stackrel{def}{=} \frac{\tau}{u\pi}\cdot[\omega\cdot\{1 - u\cdot(1 - s)\}] \stackrel{def}{=} \frac{\tau}{u\pi}\cdot\omega\cdot s_u^+ \tag{7.30a}$$

$$v_- \stackrel{def}{=} -\frac{\tau}{u\pi}\cdot[\omega\cdot\{1 + u\cdot(1 - s)\}] \stackrel{def}{=} -\frac{\tau}{u\pi}\cdot\omega\cdot s_u^- \tag{7.30b}$$

where

$$s_u^+ \stackrel{def}{=} \{1 - u \cdot (1 - s)\} \tag{7.31a}$$

and

$$s_u^- \stackrel{def}{=} \{1 + u \cdot (1 - s)\} \tag{7.31b}$$

Therefore, the components of the magnetic field intensity in the air gap, *in a reference frame moving with the rotor*, can be given in view of Equations 7.30a through 7.31b as

$$\begin{aligned} H_{ay} = \frac{1}{\mu_o} \cdot \mathcal{R}e \sum_{u-odd}^{\infty} & \left[h_{1u}^R \cdot \frac{\sinh\{(u\pi/\tau)\cdot(z+g)\}}{\sinh((u\pi/\tau)\cdot g)} - h_{1u}^S \cdot \frac{\sinh((u\pi/\tau)\cdot z)}{\sinh((u\pi/\tau)\cdot g)} \right] \\ & \times \frac{1}{2} \cdot \left[e^{-j((u\pi/\tau)\cdot y - \omega\cdot s_u^+ t)} + e^{j((u\pi/\tau)\cdot y + \omega\cdot s_u^- t)} \right] \\ & + \left[h_{2u}^R \cdot \frac{\sinh\{(u\pi/\tau)\cdot(z+g)\}}{\sinh((u\pi/\tau)\cdot g)} - h_{2u}^S \cdot \frac{\sinh((u\pi/\tau)\cdot z)}{\sinh((u\pi/\tau)\cdot g)} \right] \\ & \times j\frac{1}{2} \cdot \left[e^{-j((u\pi/\tau)\cdot y - \omega\cdot s_u^+ t)} - e^{j((u\pi/\tau)\cdot y + \omega\cdot s_u^- t)} \right] \end{aligned} \tag{7.32a}$$

$$\begin{aligned} H_{az} = \frac{1}{\mu_o} \cdot \mathcal{R}e \sum_{u-odd}^{\infty} & \left[h_{1u}^R \cdot \frac{\cosh\{(u\pi/\tau)(z+g)\}}{\sinh\left(\frac{u\pi}{\tau}\cdot g\right)} - h_{1u}^S \cdot \frac{\cosh((u\pi/\tau)\cdot z)}{\sinh((u\pi/\tau)\cdot g)} \right] \\ & \times j\frac{1}{2} \cdot \left[e^{-j((u\pi/\tau)\cdot y - \omega\cdot s_u^+ t)} - e^{j((u\pi/\tau)\cdot y + \omega\cdot s_u^- t)} \right] \\ & - \left[h_{2u}^R \cdot \frac{\cosh\{(u\pi/\tau)(z+g)\}}{\sinh((u\pi/\tau)\cdot g)} - h_{2u}^S \cdot \frac{\cosh((u\pi/\tau)\cdot z)}{\sinh((u\pi/\tau)\cdot g)} \right] \\ & \times \frac{1}{2} \cdot \left[e^{-j((u\pi/\tau)\cdot y - \omega\cdot s_u^+ t)} + e^{j((u\pi/\tau)\cdot y + \omega\cdot s_u^- t)} \right] \end{aligned} \tag{7.32b}$$

These field distributions on the rotor air-gap surface, in a reference frame moving with the rotor, can therefore be written as

$$\begin{aligned} H_{ay}\Big|_{z=0} = \frac{1}{\mu_o} \cdot \mathcal{R}e \sum_{u-odd}^{\infty} \frac{1}{2} \cdot & \left[h_{1u}^R \cdot \left\{ e^{-j((u\pi/\tau)\cdot y - \omega\cdot s_u^+ t)} + e^{j((u\pi/\tau)\cdot y + \omega\cdot s_u^- t)} \right\} \right. \\ & \left. + j h_{2u}^R \cdot \left\{ e^{-j((u\pi/\tau)\cdot y - \omega\cdot s_u^+ t)} - e^{j((u\pi/\tau)\cdot y + \omega\cdot s_u^- t)} \right\} \right] \end{aligned} \tag{7.33a}$$

$$H_{az}\Big|_{z=0} = \frac{1}{\mu_o}\cdot \mathcal{R}e\sum_{u-odd}^{\infty}\frac{1}{2}\cdot\Bigg[\left\{h_{1u}^{R}\cdot\coth\left(\frac{u\pi}{\tau}\cdot g\right) - h_{1u}^{S}\cdot\text{cosech}\left(\frac{u\pi}{\tau}\cdot g\right)\right\}$$
$$\times j\cdot\left\{e^{-j((u\pi/\tau)\cdot y-\omega\cdot s_u^{+}t)} - e^{j((u\pi/\tau)\cdot y+\omega\cdot s_u^{-}t)}\right\}\Bigg]$$
$$-\Bigg[\left\{h_{2u}^{R}\cdot\coth\left(\frac{u\pi}{\tau}\cdot g\right) - h_{2u}^{S}\cdot\text{cosech}\left(\frac{u\pi}{\tau}\cdot g\right)\right\}$$
$$\times\left\{e^{-j((u\pi/\tau)\cdot y-\omega\cdot s_u^{+}t)} + e^{j((u\pi/\tau)\cdot y+\omega\cdot s_u^{-}t)}\right\}\Bigg] \tag{7.33b}$$

where arbitrary constants, $h_{1u}^R, h_{2u}^R, h_{1u}^S$ and h_{2u}^S, are, in general, complex quantities.

Now, consider the rotor region. At the power frequency operation, the displacement currents are invariably ignored. As a result, electromagnetic fields obey eddy current equations. Let

$$H_{ry}\Big|_{z=0} = \mathcal{R}e\sum_{u-odd}^{\infty}\frac{1}{2}\cdot\left\{g_u^{+}\cdot e^{-j((u\pi/\tau)\cdot y-\omega\cdot s_u^{+}t)} + g_u^{-}\cdot e^{j((u\pi/\tau)\cdot y+\omega\cdot s_u^{-}t)}\right\} \tag{7.34}$$

where g_u^+ and g_u^- indicate two sets of complex arbitrary constants.

In view of the boundary condition

$$H_{ry}\Big|_{z=0} = H_{ay}\Big|_{z=0} \tag{7.35}$$

By using Equations 7.34 and 7.33a, we get

$$\mathcal{R}e\sum_{u-odd}^{\infty}\frac{1}{2}\cdot\left\{g_u^{+}\cdot e^{-j((u\pi/\tau)\cdot y-\omega\cdot s_u^{+}t)} + g_u^{-}\cdot e^{j((u\pi/\tau)\cdot y+\omega\cdot s_u^{-}t)}\right\}$$
$$= \frac{1}{\mu_o}\cdot\mathcal{R}e\sum_{u-odd}^{\infty}\frac{1}{2}\cdot\Big[h_{1u}^{R}\cdot\left\{e^{-j((u\pi/\tau)\cdot y-\omega\cdot s_u^{+}t)} + e^{j((u\pi/\tau)\cdot y+\omega\cdot s_u^{-}t)}\right\}$$
$$+ jh_{2u}^{R}\cdot\left\{e^{-j((u\pi/\tau).y-\omega.s_u^{+}t)} - e^{j((u\pi/\tau).y+\omega.s_u^{-}t)}\right\}\Big] \tag{7.35a}$$

Now, on equating the coefficients of $e^{-j((u\pi/\tau)\cdot y-\omega.s_u^{+}t)}$ and $e^{j((u\pi/\tau)\cdot y+\omega\cdot s_u^{-}t)}$ in the above equation, one gets

$$g_u^{+} = \frac{1}{\mu_o}\cdot(h_{1u}^{R} + jh_{2u}^{R}) \tag{7.36}$$

$$g_u^- = \frac{1}{\mu_o} \cdot (h_{1u}^R - jh_{2u}^R) \tag{7.37}$$

for $u = 1, 3, 5, 7, \ldots, \infty$.

Further, let us assume that this field component vanishes at the surface of the highly permeable shaft, that is, at $z = D_R$; the peripheral component of magnetic field intensity in the rotor region can, therefore, be given as

$$H_{ry} = -\mathcal{R}e \sum_{u-odd}^{\infty} \frac{1}{2} \cdot \left\{ g_u^+ \cdot \frac{\sinh\{k_u^+ \cdot (z - D_R)\}}{\sinh(k_u^+ \cdot D_R)} \cdot e^{-j((u\pi/\tau)\cdot y - \omega \cdot s_u^+ t)} + g_u^- \cdot \frac{\sinh\{k_u^- \cdot (z - D_R)\}}{\sinh(k_u^- \cdot D_R)} \cdot e^{j((u\pi/\tau)\cdot y + \omega \cdot s_u^- t)} \right\} \tag{7.38}$$

Now, since H_{ry} satisfies the eddy current equation

$$\nabla^2 H_{ry} = \mu\sigma \cdot \frac{\partial H_{ry}}{\partial t} \tag{7.39}$$

Thus,

$$k_u^{\pm} = \sqrt{\left(\frac{u\pi}{\tau}\right)^2 + j\omega \cdot s_u^{\pm} \cdot \mu\sigma} \tag{7.39a}$$

Also, since

$$\nabla \bullet \boldsymbol{H}_r = 0 \tag{7.40}$$

the z-component of the magnetic field intensity H_{rz} in the rotor iron can be expressed as

$$H_{rz} = -\mathcal{R}e \left[\sum_{u-odd}^{\infty} \frac{j}{2} \cdot \frac{u\pi}{\tau} \cdot \left\{ g_u^+ \cdot \frac{\cosh\{k_u^+ \cdot (z - D_R)\}}{k_u^+ \cdot \sinh(k_u^+ \cdot D_R)} \cdot e^{-j((u\pi/\tau)\cdot y - \omega \cdot s_u^+ t)} - g_u^- \cdot \frac{\cosh\{k_u^- \cdot (z - D_R)\}}{k_u^- \cdot \sinh(k_u^- \cdot D_R)} \cdot e^{j((u\pi/\tau)\cdot y + \omega \cdot s_u^- t)} \right\} \right] \tag{7.41}$$

Therefore,

$$H_{rz}\Big|_{z=0} = -\mathcal{R}e\left[\sum_{u-odd}^{\infty}\frac{j}{2}\cdot\frac{u\pi}{\tau}\cdot\left\{g_u^+\cdot\frac{\coth(k_u^+\cdot D_R)}{k_u^+}\cdot e^{-j((u\pi/\tau)\cdot y-\omega\cdot s_u^+ t)}\right.\right.$$
$$\left.\left.-g_u^-\cdot\frac{\coth(k_u^-\cdot D_R)}{k_u^-}\cdot e^{j((u\pi/\tau)\cdot y+\omega\cdot s_u^- t)}\right\}\right] \tag{7.41a}$$

Next, consider the boundary condition

$$\mu\cdot H_{rz}\Big|_{z=0} = \mu_o\cdot H_{az}\Big|_{z=0} \tag{7.42}$$

In view of Equations 7.33b and 7.41a, we get the following relation between the six unknowns, g_u^+, g_u^-, h_{1u}^R, h_{1u}^S, h_{2u}^R and h_{2u}^S:

$$-\mathcal{R}e\left[\mu\cdot\sum_{u-odd}^{\infty}\frac{j}{2}\cdot\frac{u\pi}{\tau}\cdot\left\{g_u^+\cdot\frac{\coth(k_u^+\cdot D_R)}{k_u^+}\cdot e^{-j\left(\frac{u\pi}{\tau}\cdot y-\omega\cdot s_u^+ t\right)}\right.\right.$$
$$\left.\left.-\,g_u^-\cdot\frac{\coth(k_u^-\cdot D_R)}{k_u^-}\cdot e^{j\left(\frac{u\pi}{\tau}\cdot y+\omega\cdot s_u^- t\right)}\right\}\right]$$
$$=\mathcal{R}e\sum_{u-odd}^{\infty}\frac{1}{2}\cdot\left[\left[\left\{h_{1u}^R\cdot\coth\left(\frac{u\pi}{\tau}\cdot g\right)-h_{1u}^S\cdot\mathrm{cosech}\left(\frac{u\pi}{\tau}\cdot g\right)\right\}\right.\right.$$
$$\left.\times j\cdot\left\{e^{-j\left(\frac{u\pi}{\tau}\cdot y-\omega\cdot s_u^+ t\right)}-e^{j\left(\frac{u\pi}{\tau}\cdot y+\omega\cdot s_u^- t\right)}\right\}\right]$$
$$-\left[\left\{h_{2u}^R\cdot\coth\left(\frac{u\pi}{\tau}\cdot g\right)-h_{2u}^S\cdot\mathrm{cosech}\left(\frac{u\pi}{\tau}\cdot g\right)\right\}\right.$$
$$\left.\left.\times\left\{e^{-j\left(\frac{u\pi}{\tau}\cdot y-\omega\cdot s_u^+ t\right)}+e^{j\left(\frac{u\pi}{\tau}\cdot y+\omega\cdot s_u^- t\right)}\right\}\right]\right] \tag{7.43}$$

On equating the coefficients of $e^{-j((u\pi/\tau)\cdot y-\omega\cdot s_u^+ t)}$ and $e^{j((u\pi/\tau)\cdot y+\omega\cdot s_u^- t)}$ in Equation 7.43, one gets

$$-\mu\cdot\frac{u\pi}{\tau}\cdot g_u^+\cdot\frac{\coth(k_u^+\cdot D_R)}{k_u^+}=\left\{h_{1u}^R\cdot\coth\left(\frac{u\pi}{\tau}\cdot g\right)-h_{1u}^S\cdot\mathrm{cosech}\left(\frac{u\pi}{\tau}\cdot g\right)\right\}$$
$$+\,j\left\{h_{2u}^R\cdot\coth\left(\frac{u\pi}{\tau}\cdot g\right)-h_{2u}^S\cdot\mathrm{cosech}\left(\frac{u\pi}{\tau}\cdot g\right)\right\} \tag{7.44a}$$

and

$$\mu \cdot j \cdot \frac{u\pi}{\tau} \cdot g_u^- \cdot \frac{\coth(k_u^- \cdot D_R)}{k_u^-} = \left\{ h_{1u}^R \cdot \coth\left(\frac{u\pi}{\tau} \cdot g\right) - h_{1u}^S \cdot \text{cosech}\left(\frac{u\pi}{\tau} \cdot g\right) \right\}$$
$$- j\left\{ h_{2u}^R \cdot \coth\left(\frac{u\pi}{\tau} \cdot g\right) - h_{2u}^S \cdot \text{cosech}\left(\frac{u\pi}{\tau} \cdot g\right) \right\} \qquad (7.44b)$$

for $u = 1, 3, 5, 7, \ldots, \infty$.

In each of these two equations, real and imaginary parts on the two sides must be equated separately; thus each resulting in two linearly independent equations.

Between six sets of unknowns, viz., g_u^+, g_u^-, h_{1u}^R, h_{2u}^R, h_{1u}^S and h_{2u}^S, we have four sets of equations, namely, Equations 7.36, 7.37, 7.44a and 7.44b. As discussed above, two more sets of equations can be generated using boundary conditions given by Equations 7.11a through 7.11c. Equations between various arbitrary constants developed above can be solved numerically. This completes the determination of field distributions in different regions due to a series of coils belonging to the same phase and same layer. Field distribution in the solid rotor due to all the coils of the same phase and same layer can be readily obtained if the RHS of Equations 7.38 and 7.41 are multiplied by the number of slots per pole per phase, η, and the *distribution factor* k_{du}, for the harmonic field of order u. It is given as[6]

$$k_{du} = \frac{\sin(\eta \cdot (u\pi\lambda/2\tau))}{[\eta \cdot \sin(u\pi\lambda/2\tau)]} = \frac{\sin(u \cdot \pi/6)}{[\eta \cdot \sin((1/\eta) \cdot (u \cdot \pi/6))]} \qquad (7.45)$$

Therefore, the field distribution in the solid rotor due to the entire red-phase winding of one layer can be given in view of Equations 7.38 and 7.41 as

$$H_{ry}^{(1)} = -\mathcal{R}e\left[\sum_{u-odd}^{\infty} \frac{1}{2} \cdot \eta \cdot k_{du} \cdot \left\{ g_u^{+(1)} \cdot \frac{\sinh\{k_u^+ \cdot (z - D_R)\}}{\sinh(k_u^+ \cdot D_R)} \cdot e^{-j((u\pi/\tau)\cdot y - \omega \cdot s_u^+ t)} \right.\right.$$
$$\left.\left. + g_u^{-(1)} \cdot \frac{\sinh\{k_u^- \cdot (z - D_R)\}}{\sinh(k_u^- \cdot D_R)} \cdot e^{j((u\pi/\tau)\cdot y + \omega \cdot s_u^- t)} \right\} \right] \qquad (7.46a)$$

and

$$H_{rz}^{(1)} = -\mathcal{R}e\left[\sum_{u-odd}^{\infty} \frac{j}{2} \cdot \frac{u\pi}{\tau} \cdot \eta \cdot k_{du} \cdot \left\{ g_u^{+(1)} \cdot \frac{\cosh\{k_u^+ \cdot (z - D_R)\}}{k_u^+ \cdot \sinh(k_u^+ \cdot D_R)} \cdot e^{-j((u\pi/\tau)\cdot y - \omega \cdot s_u^+ t)} \right.\right.$$
$$\left.\left. - g_u^{-(1)} \cdot \frac{\cosh\{k_u^- \cdot (z - D_R)\}}{k_u^- \cdot \sinh(k_u^- \cdot D_R)} \cdot e^{j((u\pi/\tau)\cdot y + \omega \cdot s_u^- t)} \right\} \right] \qquad (7.46b)$$

For the second layer of the red-phase winding, Equations 7.46a and 7.46b need to be rewritten with a different set of values for g_u^+ and for g_u^-, say $g_u^{+(2)}$ and $g_u^{-(2)}$, calculated by using revised values for h_1 and h_2; consistent with the second layer (vide Figures 7.3 and 7.4). Also, it is needed to shift the space phase in the peripheral direction by '$\xi \cdot \lambda$', the original coils being short chorded by ξ slot pitches. Further, if the z-axis is shifted in the y-direction by $((1/2)\xi \cdot \lambda)$, then y in Equations 7.46a and 7.46b needs to be replaced by $(y - (1/2)\xi \cdot \lambda)$. Thus, the revised expressions are

$$H_{ry}^{(1)} = -\mathcal{R}e \sum_{u-odd}^{\infty} \frac{1}{2} \cdot \eta \cdot k_{du} \cdot \left[g_u^{+(1)} \cdot \frac{\sinh\{k_u^+ \cdot (z - D_R)\}}{\sinh(k_u^+ \cdot D_R)} \cdot e^{-j\{((u\pi/\tau)\cdot(y-(1/2)\xi\cdot\lambda)-\omega\cdot s_u^+ t\}} \right.$$
$$\left. + g_u^{-(1)} \cdot \frac{\sinh\{k_u^- \cdot (z - D_R)\}}{\sinh(k_u^- \cdot D_R)} \cdot e^{j\{(u\pi/\tau)\cdot(y-1/2)\xi\cdot\lambda)+\omega\cdot s_u^- t\}} \right] \quad (7.47a)$$

and

$$H_{rz}^{(1)} = -\mathcal{R}e \sum_{u-odd}^{\infty} \frac{j}{2} \cdot \frac{u\pi}{\tau} \cdot \eta \cdot k_{du} \cdot \left[g_u^{+(1)} \cdot \frac{\cosh\{k_u^+ \cdot (z - D_R)\}}{k_u^+ \cdot \sinh(k_u^+ \cdot D_R)} \cdot e^{-j\{(u\pi/\tau)\cdot(y-(1/2)\xi\cdot\lambda)-\omega\cdot s_u^+ t\}} \right.$$
$$\left. - g_u^{-(1)} \cdot \frac{\cosh\{k_u^- \cdot (z - D_R)\}}{k_u^- \cdot \sinh(k_u^- \cdot D_R)} \cdot e^{j\{(u\pi/\tau)\cdot(y-(1/2)\xi\cdot\lambda)+\omega\cdot s_u^- t\}} \right] \quad (7.47b)$$

The corresponding expressions for field components due to the second layer may be given as

$$H_{ry}^{(2)} = -\mathcal{R}e \sum_{u-odd}^{\infty} \frac{1}{2} \cdot \eta \cdot k_{du} \cdot \left[g_u^{+(2)} \cdot \frac{\sinh\{k_u^+ \cdot (z - D_R)\}}{\sinh(k_u^+ \cdot D_R)} \cdot e^{-j\{(u\pi/\tau)\cdot(y+(1/2)\xi\cdot\lambda)-\omega\cdot s_u^+ t\}} \right.$$
$$\left. + g_u^{-(2)} \cdot \frac{\sinh\{k_u^- \cdot (z - D_R)\}}{\sinh(k_u^- \cdot D_R)} \cdot e^{j\{(u\pi/\tau)\cdot(y+(1/2)\xi\cdot\lambda)+\omega\cdot s_u^- t\}} \right] \quad (7.48a)$$

and

$$H_{rz}^{(2)} = -\mathcal{R}e \sum_{u-odd}^{\infty} \frac{j}{2} \cdot \frac{u\pi}{\tau} \cdot \eta \cdot k_{du} \cdot \left[g_u^{+(2)} \cdot \frac{\cosh\{k_u^+ \cdot (z - D_R)\}}{k_u^+ \cdot \sinh(k_u^+ \cdot D_R)} \cdot e^{-j\{(u\pi/\tau)\cdot(y+(1/2)\xi\cdot\lambda)-\omega\cdot s_u^+ t\}} \right.$$
$$\left. - g_u^{-(2)} \cdot \frac{\cosh\{k_u^- \cdot (z - D_R)\}}{k_u^- \cdot \sinh(k_u^- \cdot D_R)} \cdot e^{j\{(u\pi/\tau)\cdot(y+(1/2)\xi\cdot\lambda)+\omega\cdot s_u^- t\}} \right] \quad (7.48b)$$

Therefore, the components of the magnetic field in the solid rotor due to the two layers of red-phase winding can be expressed as

$$H_{ry} = H_{ry}^{(1)} + H_{ry}^{(2)} \tag{7.49a}$$

and

$$H_{rz} = H_{rz}^{(1)} + H_{rz}^{(2)} \tag{7.49b}$$

In view of above equations, it transpires that four sets of arbitrary constants, viz., $g_u^{+(1)}, g_u^{-(1)}, g_u^{+(2)}$ and $g_u^{-(2)}$, need to be numerically evaluated. Considering Figure 7.4, it can be noted that $g_u^{\pm(1)}$ and $g_u^{\pm(2)}$ differ from each other because the values for both h_1 and h_2 are different for the two layers. If we chose values for these parameters as average values of the two layers, and with these values numerically evaluate the arbitrary constants, $g_u^{\pm}$, then

$$g_u^{\pm(1)} \cong g_u^{\pm(2)} \overset{def}{=} g_u^{\pm} \tag{7.50}$$

This approximate treatment will considerably reduce the volume of numerical evaluation. The resulting simplified forms for Equations 7.49a and 7.49b are

$$H_{ry} = -\mathcal{R}e\sum_{u-odd}^{\infty} \eta \cdot k_{du} \cdot k_{pu} \cdot \left[g_u^{+} \cdot \frac{\sinh\{k_u^{+} \cdot (z - D_R)\}}{\sinh(k_u^{+} \cdot D_R)} \cdot e^{-j\{(u\pi/\tau)\cdot y - \omega \cdot s_u^{+} t\}} \right.$$
$$\left. + g_u^{-} \cdot \frac{\sinh\{k_u^{-} \cdot (z - D_R)\}}{\sinh(k_u^{-} \cdot D_R)} \cdot e^{j\{(u\pi/\tau)\cdot y + \omega \cdot s_u^{-} t\}} \right] \tag{7.51a}$$

and

$$H_{rz} = -\mathcal{R}e\sum_{u-odd}^{\infty} j \cdot \frac{u\pi}{\tau} \cdot \eta \cdot k_{du} \cdot k_{pu} \cdot \left[g_u^{+} \cdot \frac{\cosh\{k_u^{+} \cdot (z - D_R)\}}{k_u^{+} \cdot \sinh(k_u^{+} \cdot D_R)} \cdot e^{-j\{(u\pi/\tau)\cdot y - \omega \cdot s_u^{+} t\}} \right.$$
$$\left. - g_u^{-} \cdot \frac{\cosh\{k_u^{-} \cdot (z - D_R)\}}{k_u^{-} \cdot \sinh(k_u^{-} \cdot D_R)} \cdot e^{j\{(u\pi/\tau)\cdot y + \omega \cdot s_u^{-} t\}} \right] \tag{7.51b}$$

where the *pitch factor*, k_{pu}, for the harmonic field of order u, is[6]

$$k_{pu} = \cos\left(\frac{u\pi\xi\lambda}{2\tau}\right) \tag{7.51c}$$

The components of the magnetic field due to the other two phases will be identical to those given above, except that there will be phase shifts in space phase (i.e., in the y-direction) as well as in time phase, each by ±120° electrical (or $\pm(2\pi/3)$ electrical radian). Consequently, in a reference frame *fixed on the stator*, the triplen harmonic magnetic fields in the air gap as well as in the rotor iron vanish. Harmonic fields of order $u = 1, 7, 13, \ldots$ rotate in the same direction, called 'positive' direction, while the harmonics of order $u = 5, 11, 17, \ldots$ rotate in the so-called 'negative' direction. Expressions of magnetic field due to the three-phase stator winding currents in a reference frame fixed on the rotor are

$$H_{ry} = -\mathcal{R}e \sum_{u=1,7,13,\ldots}^{\infty} 3 \cdot \eta \cdot k_{du} \cdot k_{pu} \cdot g_u^+ \cdot \frac{\sinh\{k_u^+ \cdot (z - D_R)\}}{\sinh(k_u^+ \cdot D_R)} \cdot e^{-j\{(u\pi/\tau)\cdot y - \omega \cdot s_u^+ t\}}$$
$$-\mathcal{R}e \sum_{u=5,11,17,\ldots}^{\infty} 3 \cdot \eta \cdot k_{du} \cdot k_{pu} \cdot g_u^- \cdot \frac{\sinh\{k_u^- \cdot (z - D_R)\}}{\sinh(k_u^- \cdot D_R)} \cdot e^{j\{(u\pi/\tau)\cdot y + \omega \cdot s_u^- t\}} \tag{7.52a}$$

and

$$H_{rz} = -\mathcal{R}e \sum_{u=1,7,13,\ldots}^{\infty} j \cdot \frac{u\pi}{\tau} \cdot 3 \cdot \eta \cdot k_{du} \cdot k_{pu} \cdot g_u^+ \cdot \frac{\cosh\{k_u^+ \cdot (z - D_R)\}}{k_u^+ \cdot \sinh(k_u^+ \cdot D_R)} \cdot e^{-j\{(u\pi/\tau)\cdot y - \omega \cdot s_u^+ t\}}$$
$$+\mathcal{R}e \sum_{u=5,11,17,\ldots}^{\infty} j \cdot \frac{u\pi}{\tau} \cdot 3 \cdot \eta \cdot k_{du} \cdot k_{pu} \cdot g_u^- \cdot \frac{\cosh\{k_u^- \cdot (z - D_R)\}}{k_u^- \cdot \sinh(k_u^- \cdot D_R)} \cdot e^{j\{(u\pi/\tau)\cdot y + \omega \cdot s_u^- t\}} \tag{7.52b}$$

The expression for the eddy current density in the solid rotor can be readily found if the displacement currents are neglected. Since

$$\boldsymbol{J} = \nabla \times \boldsymbol{H} \tag{7.53}$$

Thus

$$J_{rx} = \frac{\partial H_{rz}}{\partial y} - \frac{\partial H_{ry}}{\partial z} \tag{7.53a}$$

It gives

$$J_{rx} = -\mathcal{R}e \sum_{u=1,7,13,\ldots}^{\infty} \left(\frac{u\pi}{\tau}\right)^2 \cdot 3 \cdot \eta \cdot k_{du} \cdot k_{pu} \cdot g_u^+ \cdot \frac{\cosh\{k_u^+ \cdot (z - D_R)\}}{k_u^+ \cdot \sinh(k_u^+ \cdot D_R)} \cdot e^{-j\{(u\pi/\tau)\cdot y - \omega\cdot s_u^+ t\}}$$
$$- \mathcal{R}e \sum_{u=5,11,17,\ldots}^{\infty} \left(\frac{u\pi}{\tau}\right)^2 \cdot 3 \cdot \eta \cdot k_{du} \cdot k_{pu} \cdot g_u^- \cdot \frac{\cosh\{k_u^- \cdot (z - D_R)\}}{k_u^- \cdot \sinh(k_u^- \cdot D_R)} \cdot e^{j\{(u\pi/\tau)\cdot y + \omega\cdot s_u^- t\}}$$
$$+ \mathcal{R}e \sum_{u=1,7,13,\ldots}^{\infty} (k_u^+)^2\, 3 \cdot \eta \cdot k_{du} \cdot k_{pu} \cdot g_u^+ \cdot \frac{\cosh\{k_u^+ \cdot (z - D_R)\}}{k_u^+ \cdot \sinh(k_u^+ \cdot D_R)} \cdot e^{-j\{(u\pi/\tau)\cdot y - \omega\cdot s_u^+ t)\}}$$
$$+ \mathcal{R}e \sum_{u=5,11,17,\ldots}^{\infty} (k_u^-)^2\, 3 \cdot \eta \cdot k_{du} \cdot k_{pu} \cdot g_u^- \cdot \frac{\cosh\{k_u^- \cdot (z - D_R)\}}{k_u^- \cdot \sinh(k_u^- \cdot D_R)} \cdot e^{j\{(u\pi/\tau)\cdot y + \omega\cdot s_u^- t\}} \tag{7.53b}$$

This equation, in view of Equation 7.39a, can be rewritten as

$$J_{rx} = \mathcal{R}e \sum_{u=1,7,13,\ldots}^{\infty} j\omega \cdot s_u^+ \cdot \mu\sigma \cdot 3 \cdot \eta \cdot k_{du} \cdot k_{pu} \cdot g_u^+ \cdot \frac{\cosh\{k_u^+ \cdot (z - D_R)\}}{k_u^+ \cdot \sinh(k_u^+ \cdot D_R)}$$
$$\times e^{-j\{(u\pi/\tau)\cdot y - \omega\cdot s_u^+ t\}}$$
$$+ \mathcal{R}e \sum_{u=5,11,17,\ldots}^{\infty} j\omega \cdot s_u^- \cdot \mu\sigma \cdot 3 \cdot \eta \cdot k_{du} \cdot k_{pu} \cdot g_u^- \cdot \frac{\cosh\{k_u^- \cdot (z - D_R)\}}{k_u^- \cdot \sinh(k_u^- \cdot D_R)}$$
$$\times e^{j\{(u\pi/\tau)\cdot y + \omega\cdot s_u^- t\}} \tag{7.54}$$

From which

$$E_{rx} = \mathcal{R}e \sum_{u=1,7,13,\ldots}^{\infty} j\omega \cdot s_u^+ \cdot \mu \cdot 3 \cdot \eta \cdot k_{du} \cdot k_{pu} \cdot g_u^+ \cdot \frac{\cosh\{k_u^+ \cdot (z - D_R)\}}{k_u^+ \cdot \sinh(k_u^+ \cdot D_R)}$$
$$\times e^{-j\{(u\pi/\tau)\cdot y - \omega\cdot s_u^+ t\}}$$
$$+ \mathcal{R}e \sum_{u=5,11,17,\ldots}^{\infty} j\omega \cdot s_u^- \cdot \mu \cdot 3 \cdot \eta \cdot k_{du} \cdot k_{pu} \cdot g_u^- \cdot \frac{\cosh\{k_u^- \cdot (z - D_R)\}}{k_u^- \cdot \sinh(k_u^- \cdot D_R)}$$
$$\times e^{j\{(u\pi/\tau)\cdot y + \omega\cdot s_u^- t\}} \tag{7.55}$$

7.2.5 Machine Performances

In view of the above analysis, the performance parameters of the machine can be obtained. These parameters are as given below.

7.2.5.1 Eddy Current Loss

The eddy current loss, W_E, in the solid rotor is given as

$$W_E = P \cdot L_R \cdot \tau \cdot \frac{1}{2}\int_0^{D_R} (E_{rx} \cdot \tilde{j}_{rx})dz \tag{7.56}$$

where
- P = number of poles
- L_R = rotor axial length
- D_R = rotor radial depth
- τ = pole pitch

Therefore, from Equations 7.54 and 7.55, we get

$$\begin{aligned} W_E = P \cdot L_R \cdot \tau \cdot \frac{1}{2} \cdot \sigma \cdot (\omega \cdot \mu \cdot 3 \cdot \eta)^2 \Bigg[& \sum_{u=1,7,13,\ldots}^{\infty} (s_u^+ \cdot k_{du} \cdot k_{pu})^2 \cdot \left|g_u^+\right|^2 \\ & \times \int_0^{D_R} \left\{ \frac{\cosh\{k_u^+ \cdot (z - D_R)\}}{k_u^+ \sinh(k_u^+ \cdot D_R)} \cdot \frac{\cosh\{\tilde{k}_u^+ \cdot (z - D_R)\}}{\tilde{k}_u^+ \cdot \sinh(\tilde{k}_u^+ \cdot D_R)} \right\} \cdot dz \\ & + \sum_{u=5,11,17,\ldots}^{\infty} (s_u^- \cdot k_{du} \cdot k_{pu})^2 \cdot \left|g_u^-\right|^2 \\ & \times \int_0^{D_R} \left\{ \frac{\cosh\{k_u^- \cdot (z - D_R)\}}{k_u^- \sinh(k_u^- \cdot D_R)} \cdot \frac{\cosh\{\tilde{k}_u^- \cdot (z - D_R)\}}{\tilde{k}_u^- \cdot \sinh(\tilde{k}_u^- \cdot D_R)} \right\} \cdot dz \Bigg] \end{aligned} \tag{7.56a}$$

or

$$\begin{aligned} W_E = P \cdot L_R \cdot \frac{\tau}{2} \cdot \sigma \cdot (\omega \cdot \mu \cdot 3 \cdot \eta)^2 \Bigg[& \sum_{u=1,7,13,\ldots}^{\infty} (s_u^+ \cdot k_{du} \cdot k_{pu})^2 \cdot \left|g_u^+\right|^2 \\ & \times \frac{\coth(k_u^+ \cdot D_R)/k_u^+ - \coth(\tilde{k}_u^+ \cdot D_R)/\tilde{k}_u^+}{[(\tilde{k}_u^+)^2 - (k_u^+)^2]} \\ & + \sum_{u=5,11,17,\ldots}^{\infty} (s_u^- \cdot k_{du} \cdot k_{pu})^2 \cdot \left|g_u^-\right|^2 \\ & \times \frac{\coth(k_u^- \cdot D_R)/k_u^- - \coth(\tilde{k}_u^- \cdot D_R)/\tilde{k}_u^-}{[(\tilde{k}_u^-)^2 - (k_u^-)^2]} \Bigg] \end{aligned} \tag{7.57}$$

7.2.5.2 Force Density

As the force density is given by $\boldsymbol{J} \times \boldsymbol{B}$, the time average of the peripheral force is given by

$$F_y = -\mu \cdot P \cdot L_R \cdot \tau \cdot \frac{1}{2} \cdot \int_0^{D_R} J_{rx} \cdot \tilde{H}_{rz} \cdot dz \tag{7.58}$$

Therefore, using Equations 7.54 and 7.52b, we get

$$\begin{aligned} F_y = P \cdot L_R \cdot \frac{\pi}{2}(\omega \cdot \sigma) \cdot (\mu \cdot 3 \cdot \eta)^2 \cdot \Bigg[& \sum_{u=1,7,13,\ldots}^{\infty} u \cdot s_u^+ \cdot (k_{du} \cdot k_{pu})^2 \cdot \left|g_u^+\right|^2 \\ & \times \int_0^{D_R} \left\{ \frac{\cosh\{k_u^+ \cdot (z - D_R)\}}{k_u^+ \cdot \sinh(k_u^+ \cdot D_R)} \cdot \frac{\cosh\{\tilde{k}_u^+ \cdot (z - D_R)\}}{\tilde{k}_u^+ \cdot \sinh(\tilde{k}_u^+ \cdot D_R)} \right\} dz \\ & - \sum_{u=5,11,17,\ldots}^{\infty} u \cdot s_u^- \cdot (k_{du} \cdot k_{pu})^2 \cdot \left|g_u^-\right|^2 \\ & \times \int_0^{D_R} \left\{ \frac{\cosh\{k_u^- \cdot (z - D_R)\}}{k_u^- \cdot \sinh(k_u^- \cdot D_R)} \cdot \frac{\cosh\{\tilde{k}_u^- \cdot (z - D_R)\}}{\tilde{k}_u^- \cdot \sinh(\tilde{k}_u^- \cdot D_R)} \right\} dz \Bigg] \end{aligned} \tag{7.58a}$$

or

$$\begin{aligned} F_y = P \cdot L_R \cdot \frac{\pi}{2}(\omega \cdot \sigma) \cdot (\mu \cdot 3 \cdot \eta)^2 \cdot \Bigg[& \sum_{u=1,7,13,\ldots}^{\infty} u \cdot s_u^+ \cdot (k_{du} \cdot k_{pu})^2 \cdot \left|g_u^+\right|^2 \\ & \times \frac{\coth(k_u^+ D_R)/k_u^+ - \coth(\tilde{k}_u^+ \cdot D_R)/\tilde{k}_u^+}{[(\tilde{k}_u^+)^2 - (k_u^+)^2]} \\ & - \sum_{u=5,11,17,\ldots}^{\infty} u \cdot s_u^- \cdot (k_{du} \cdot k_{pu})^2 \cdot \left|g_u^-\right|^2 \\ & \times \frac{\coth(k_u^- \cdot D_R)/k_u^- - \coth(\tilde{k}_u^- \cdot D_R)/\tilde{k}_u^-}{[(\tilde{k}_u^-)^2 - (k_u^-)^2]} \Bigg] \end{aligned} \tag{7.58b}$$

7.2.5.3 Mechanical Power Developed

The mechanical power developed, P_M, is

$$P_M = v \cdot F_y \tag{7.59}$$

Since

$$v = \frac{\omega\tau}{u\pi}\cdot(1 - s_u^+) = -\frac{\omega\tau}{u\pi}\cdot(1 - s_u^-) \tag{7.59a}$$

we get, in view of Equation 7.58b

$$\begin{aligned}P_M &= P\cdot L_R\cdot\frac{\tau}{2}\cdot\sigma\cdot(\omega\cdot\mu\cdot 3\cdot\eta)^2\cdot\left[\sum_{u=1,7,13,\ldots}^{\infty}(1 - s_u^+)\cdot s_u^+\cdot(k_{du}\cdot k_{pu})^2\cdot\left|g_u^+\right|^2\right.\\&\times\frac{\coth(k_u^+ D_R)/k_u^+ - \coth(\tilde{k}_u^+\cdot D_R)/\tilde{k}_u^+}{[(\tilde{k}_u^+)^2 - (k_u^+)^2]}\\&+\sum_{u=5,11,17,\ldots}^{\infty}(1 - s_u^-)\cdot s_u^-\cdot(k_{du}\cdot k_{pu})^2\cdot\left|g_u^-\right|^2\\&\left.\times\frac{\coth(k_u^-\cdot D_R)/k_u^- - \coth(\tilde{k}_u^-\cdot D_R)/\tilde{k}_u^-}{[(\tilde{k}_u^-)^2 - (k_u^-)^2]}\right]\end{aligned} \tag{7.60}$$

7.2.5.4 Rotor Input Power

From Equations 7.57 and 7.60, the power input to the rotor, P_R, can be obtained as

$$\begin{aligned}P_R &= W_E + P_M = P\cdot L_R\cdot\frac{\tau}{2}\cdot\sigma\cdot(\omega\cdot\mu\cdot 3\cdot\eta)^2\cdot\left[\sum_{u=1,7,13,\ldots}^{\infty}s_u^+\cdot(k_{du}\cdot k_{pu})^2\cdot\left|g_u^+\right|^2\right.\\&\times\frac{\coth(k_u^+ D_R)/k_u^+ - \coth(\tilde{k}_u^+\cdot D_R)/\tilde{k}_u^+}{[(\tilde{k}_u^+)^2 - (k_u^+)^2]}\\&+\sum_{u=5,11,17,\ldots}^{\infty}s_u^-\cdot(k_{du}\cdot k_{pu})^2\cdot\left|g_u^-\right|^2\\&\left.\cdot\frac{\coth(k_u^-\cdot D_R)/k_u^- - \coth(\tilde{k}_u^-\cdot D_R)/\tilde{k}_u^-}{[(\tilde{k}_u^-)^2 - (k_u^-)^2]}\right]\end{aligned} \tag{7.61}$$

Thus, in view of Equations 7.57, 7.60 and 7.61, for each order of harmonic, u, we find

$$\frac{P_R^u}{1} = \frac{W_E^u}{S_u} = \frac{P_M^u}{(1 - S_u)} \tag{7.62}$$

This is a well-known relation for all induction and hysteresis machines.

7.2.5.5 Torque

The torque developed, T, can be obtained by using the relation

$$T = F_y \cdot \frac{P \cdot \tau}{2\pi} \tag{7.63}$$

7.3 Three-Dimensional Fields in Solid-Rotor Induction Machines

In a wound- or cage-rotor induction machine, the induced rotor currents flow along discrete conductors laid out in the axial direction over the active rotor length and the return paths are provided by end-windings or by end-rings, as the case may be. In a solid-rotor induction machine, however, the induced rotor currents are distributed throughout the volume of the rotor iron, and, in general, all the three components of the current density vector exist, in the rotor. In common with the phenomena familiar in wound- or cage-rotor induction machines, it is the axial current that contributes substantially to the electromagnetic torque developed in the solid-rotor induction machine. This axial rotor current necessarily diminishes in magnitude in the solid rotor as the two longitudinal ends of the rotor are approached from the central parts. The performance characteristics of the solid-rotor induction machine may, therefore, be expected to differ from those of the common induction machine and this difference is ascribed to what are commonly called *end-effects*. The treatment presented here is based on the analyses given by Roth.[1]

7.3.1 Idealised Model

A model machine, based on a set of simplifying assumptions, is shown in Figure 7.5a. As the curvature of air-gap surfaces is ignored, no special functions are needed for the field expressions. The stator and rotor cores, in their developed forms, are represented by two infinitely long regions with rectangular cross-sections. The current-carrying three-phase stator winding is simulated by a current sheet, as given in Appendix 6. This current sheet is placed on the smooth stator surface, as indicated in Figure 7.5b. The separation between the stator and rotor cores is the effective air gap of the machine. The stator and rotor cores are enclosed on one side by the

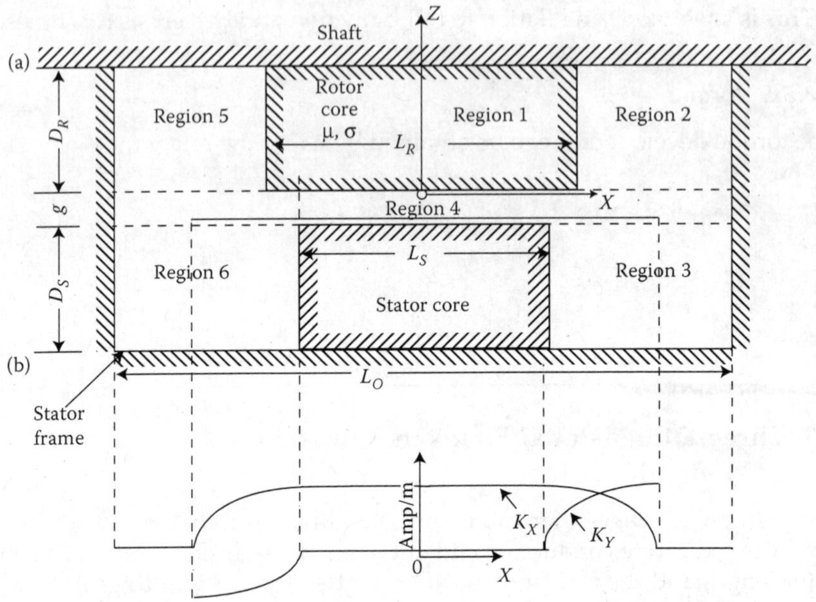

FIGURE 7.5
(a) Sectional view of the idealised machine. (b) Current sheet simulating stator winding.

developed surface of the shaft and on the remaining three sides by the inner surface of the stator frame, which includes end covers. It is assumed that the stator frame, stator core and the rotor shaft are highly permeable so that the tangential components of magnetic field intensity on these surfaces are negligible.

The stator current sheet varies periodically in the peripheral direction. A sinusoidal variation is assumed; thus, effects of winding harmonics are neglected. The magnetic saturation is neglected and the rotor iron permeability is taken as a positive real number. In the machine thus idealised, rotor core is the only conducting region. In Figure 7.5a, the rotor core is shown as region 1. The remaining five regions in this figure are air regions. Because of the symmetry between these regions, field distributions in regions 5 and 6 need not be considered. The primary source for the magnetic field in all regions is the known stator current sheet.

In view of Figure 7.5a, X is taken parallel to the axial, Y to the peripheral and Z to the radial direction in the rectangular Cartesian system of space coordinates. The smooth rotor surface at the air gap is taken as the surface $z = 0$; $z = -g$ and $z = D_R$ represent the stator air-gap surface and the shaft surface, respectively. The middle of the axial length of the machine is taken as the surface $x = 0$ and $x = \pm L_R/2$ and $x = \pm L_S/2$ represent the rotor and the stator end surfaces, respectively. The two end covers are presumed to be located at $x = \pm L_O/2$.

7.3.2 Field Distributions

The stator current sheet simulating the stator winding, with balanced three-phase currents, is the primary source for the magnetic field in all regions, that is, from region 1 to region 6. Therefore, the distribution of magnetic field in each region is characterised by the exponential factor: $\exp \cdot j(\omega t - \ell y)$, where $\ell = \pi/\tau$, τ being the pole pitch. Field expressions in this section are written in complex form without the exponential factor. Complete expressions for all field quantities can be obtained by inserting the exponential factor and then selecting the real part.

The rotor region is the only conducting region. All the remaining regions are air regions. The eddy current density is governed by the following field equations:

$$\nabla \bullet \boldsymbol{J}_1 = 0 \tag{7.64}$$

and

$$\nabla^2 \boldsymbol{J}_1 = \frac{j}{d^2} \boldsymbol{J}_1 \tag{7.65}$$

where

$$d^2 = \frac{1}{s \cdot \omega \cdot \mu \cdot \sigma} \tag{7.65a}$$

The magnetic field intensity $\boldsymbol{H}_1$ in the solid rotor in terms of eddy current density $\boldsymbol{J}_1$ can be written as

$$\boldsymbol{H}_1 = jd^2 \nabla \times \boldsymbol{J}_1 \tag{7.66}$$

The axial component of the rotor eddy current density vanishes at the rotor end surfaces and is an even function of x. Thus, in a reference frame fixed to the rotor, the axial component of the rotor eddy current density at the rotor air-gap surface can be given by the following half-range Fourier series:

$$J_{x1}\Big|_{z=0} = \sum_{p-odd}^{\infty} a_p \cdot \cos\left(\frac{p\pi}{L_R} \cdot x\right), \tag{7.67}$$

where a_p indicates a set of arbitrary constants.

Since the tangential components of the magnetic field is assumed to vanish on the highly permeable shaft surface, that is

$$H_{x1}\big|_{z=D_R} \cong 0 \tag{7.68a}$$

and

$$H_{y1}\big|_{z=D_R} \cong 0 \tag{7.68b}$$

Thus

$$J_{z1}\big|_{z=D_R} = \left(\frac{\partial H_{y1}}{\partial x} - \frac{\partial H_{x1}}{\partial y}\right)\Bigg|_{z=D_R} = 0 \tag{7.69a}$$

Further, since

$$J_{z1}\big|_{z=0} = 0 \tag{7.69b}$$

Therefore, in view of Equations 7.65, 7.69a and 7.69b

$$J_{z1} = \sum_{q=1}^{\infty} b_q \cdot \sin\left(\frac{q\pi}{D_R} \cdot z\right) \cdot \frac{\sinh(\beta_q \cdot x)}{\sinh(\beta_q \cdot L_R/2)} \tag{7.70}$$

where

$$\beta_q = \sqrt{\left(\frac{q\pi}{D_R}\right)^2 + \ell^2 + \frac{j}{d^2}} \tag{7.70a}$$

and b_q indicates a set of arbitrary constants.

In view of Equations 7.66, 7.68b and 7.70, the distribution of the axial component of eddy current density in the solid rotor can be given as

$$J_{x1} = \sum_{p-odd}^{\infty} a_p \cdot \cos\left(\frac{p\pi}{L_R} \cdot x\right) \cdot \frac{\cosh\{\alpha_p \cdot (z - D_R)\}}{\cosh(\alpha_p \cdot D_R)} \tag{7.71}$$

where

$$\alpha_p = \sqrt{\left(\frac{p\pi}{L_R}\right)^2 + \ell^2 + \frac{j}{d^2}} \tag{7.71a}$$

From Equations 7.64, 7.70 and 7.71, we get

$$J_{y1} = \sum_{p-odd}^{\infty} a_p \cdot \left(\frac{p\pi}{L_R} \cdot \frac{j}{\ell}\right) \cdot \sin\left(\frac{p\pi}{L_R} \cdot x\right) \cdot \frac{\cosh\{\alpha_p \cdot (z - D_R)\}}{\cosh(\alpha_p \cdot D_R)}$$
$$- \sum_{q=1}^{\infty} b_q \cdot \left(\frac{q\pi}{D_R} \cdot \frac{j}{\ell}\right) \cdot \cos\left(\frac{q\pi}{D_R} \cdot z\right) \cdot \frac{\sinh(\beta_q \cdot x)}{\sinh(\beta_q \cdot L_R/2)} \tag{7.72}$$

In view of Equations 7.66, 7.70, 7.71 and 7.72, the components of the magnetic field intensity in the rotor iron can be given as

$$H_{x1} = \sum_{p-odd}^{\infty} a_p \cdot \left(\frac{p\pi}{L_R} \cdot \frac{d^2}{\ell} \cdot \alpha_p\right) \cdot \sin\left(\frac{p\pi}{L_R} \cdot x\right) \cdot \frac{\sinh\{\alpha_p \cdot (z - D_R)\}}{\cosh(\alpha_p \cdot D_R)}$$
$$+ \sum_{q=1}^{\infty} b_q \cdot \left[\left(\frac{q\pi}{D_R}\right)^2 + \ell^2\right] \cdot \frac{d^2}{\ell} \cdot \sin\left(\frac{q\pi}{D_R} \cdot z\right) \cdot \frac{\sinh(\beta_q \cdot x)}{\sinh(\beta_q \cdot L_R/2)} \tag{7.73}$$

$$H_{y1} = \sum_{p-odd}^{\infty} a_p \cdot (jd^2 \cdot \alpha_p) \cdot \cos\left(\frac{p\pi}{L_R} \cdot x\right) \cdot \frac{\sinh\{\alpha_p \cdot (z - D_R)\}}{\cosh(\alpha_p \cdot D_R)}$$
$$- \sum_{q=1}^{\infty} b_q \cdot (jd^2 \cdot \beta_q) \cdot \sin\left(\frac{q\pi}{D_R} \cdot z\right) \cdot \frac{\cosh(\beta_q \cdot x)}{\sinh(\beta_q \cdot L_R/2)} \tag{7.74}$$

$$H_{z1} = -\sum_{p-odd}^{\infty} a_p \cdot \left[\left(\frac{p\pi}{L_R}\right)^2 + \ell^2\right] \cdot \frac{d^2}{\ell} \cdot \cos\left(\frac{p\pi}{L_R} \cdot x\right) \cdot \frac{\cosh\{\alpha_p \cdot (z - D_R)\}}{\cosh(\alpha_p \cdot D_R)}$$
$$+ \sum_{q=1}^{\infty} b_q \cdot \left(\frac{q\pi}{D_R} \cdot \frac{d^2}{\ell} \cdot \beta_q\right) \cdot \cos\left(\frac{q\pi}{D_R} \cdot z\right) \cdot \frac{\cosh(\beta_q \cdot x)}{\sinh(\beta_q \cdot L_R/2)} \tag{7.75}$$

Thus, the field in this region is obtained in terms of two sets of arbitrary constants, viz., a_p and b_q. These arbitrary constants were introduced to define the distribution of eddy current density in the solid rotor (vide Equations 7.70 and 7.71).

In Figure 7.5a, the air space adjacent to the rotor surface is labelled as region 2. The magnetic field intensity in this region satisfies the following equations:

$$\nabla^2 \boldsymbol{H}_2 = 0 \tag{7.76a}$$

and

$$\nabla \bullet \boldsymbol{H}_2 = 0 \tag{7.76b}$$

For this current-free region, let us define scalar magnetic potential $\mathcal{V}$ as

$$\boldsymbol{H}_2 \overset{def}{=} -\nabla \mathcal{V}_2 \tag{7.77a}$$

Thus

$$\nabla^2 \mathcal{V}_2 = 0 \tag{7.77b}$$

The tangential components of this field, that is, $\boldsymbol{H}_2$, on the rotor end surface are continuous and are zero on the end cover surface as well as on the shaft surface. Thus

$$H_{y2}\big|_{x=L_R/2} = H_{y1}\big|_{x=L_R/2} \text{ over } 0 < z < D_R \tag{7.78a}$$

$$H_{z2}\big|_{x=L_R/2} = H_{z1}\big|_{x=L_R/2} \text{ over } 0 < z < D_R \tag{7.78b}$$

$$H_{y2}\big|_{x=L_O/2} = H_{z2}\big|_{x=L_O/2} = 0 \text{ over } 0 < z < D_R \tag{7.78c}$$

$$H_{x2}\big|_{z=D_R} = H_{y2}\big|_{z=D_R} = 0 \text{ over } L_R/2 < z < L_O/2 \tag{7.78d}$$

The shaft and the stator-frame surfaces can be assumed to be at zero magnetic potential. Therefore

$$\mathcal{V}_2\big|_{z=L_R/2,\ L_O/2} = 0 \text{ over } 0 < z < D_R \tag{7.79a}$$

and

$$\mathcal{V}_2\big|_{z=D_R} = 0 \text{ over } L_R/2 < z < L_O/2 \tag{7.79b}$$

Therefore, from Equations 7.74 and 7.78a

$$H_{y2}\big|_{x=L_R/2} = -\frac{\partial \mathcal{V}_2}{\partial y}\bigg|_{x=L_R/2} = -\sum_{q=1}^{\infty} b_q \cdot (jd^2 \cdot \beta_q) \cdot \sin\left(\frac{q\pi}{D_R} \cdot z\right) \cdot \coth(\beta_q \cdot L_R/2) \tag{7.80a}$$

and from Equations 7.75 and 7.78b

$$H_{z2}\Big|_{x=L_R/2} = -\frac{\partial \mathcal{V}_2}{\partial z}\Big|_{x=L_R/2} = \sum_{q=1}^{\infty} b_q \cdot \left(\frac{q\pi}{D_R} \cdot \frac{d^2}{\ell} \cdot \beta_q\right) \cdot \cos\left(\frac{q\pi}{D_R} \cdot z\right) \cdot \coth(\beta_q \cdot L_R/2) \tag{7.80b}$$

In view of Equations 7.79a, 7.79b, 7.80a and 7.80b, one gets

$$\begin{aligned}\mathcal{V}_2 = &\sum_{q=1}^{\infty} b_q \cdot \left(\frac{d^2}{\ell} \cdot \beta_q\right) \cdot \sin\left(\frac{q\pi}{D_R} \cdot z\right) \cdot \coth(\beta_q \cdot L_R/2) \cdot \frac{\sinh\{\gamma_q \cdot (x - L_O/2)\}}{\sinh\{\gamma_q \cdot (L_O/2 - L_R/2)\}} \\ &+ \sum_{m=1}^{\infty} c_m \cdot \sin\left\{\frac{m2\pi}{(L_O - L_R)} \cdot (x - L_R/2)\right\} \cdot \frac{\sinh\{\delta_m \cdot (z - D_R)\}}{\sinh(\delta_m \cdot D_R)}\end{aligned} \tag{7.81}$$

where

$$\gamma_q = \sqrt{\left(\frac{q\pi}{D_R}\right)^2 + \ell^2} \tag{7.81a}$$

and

$$\delta_m = \sqrt{\left(\frac{m2\pi}{L_O - L_R}\right)^2 + \ell^2} \tag{7.81b}$$

while c_m indicates a set of arbitrary constants.

Therefore, in view of Equations 7.77a and 7.81

$$\begin{aligned}H_{x2} = &-\sum_{q=1}^{\infty} b_q \cdot \left(\frac{d^2}{\ell} \cdot \beta_q \cdot \gamma_q\right) \cdot \sin\left(\frac{q\pi}{D_R} \cdot z\right) \cdot \coth(\beta_q \cdot L_R/2) \\ &\times \frac{\cosh\{\gamma_q \cdot (x - L_O/2)\}}{\sinh\{\gamma_q \cdot (L_O/2 - L_R/2)\}} \\ &- \sum_{m=1}^{\infty} c_m \cdot \frac{m2\pi}{(L_O - L_R)} \cdot \cos\left\{\frac{m2\pi}{(L_O - L_R)} \cdot (x - L_R/2)\right\} \cdot \frac{\sinh\{\delta_m \cdot (z - D_R)\}}{\sinh(\delta_m \cdot D_R)}\end{aligned} \tag{7.82a}$$

$$H_{y2} = \sum_{q=1}^{\infty} b_q \cdot \left(j\ell \cdot \frac{d^2}{\ell} \cdot \beta_q \right) \cdot \sin\left(\frac{q\pi}{D_R} \cdot z \right) \cdot \coth(\beta_q \cdot L_R/2)$$
$$\times \frac{\sinh\{\gamma_q \cdot (x - L_O/2)\}}{\sinh\{\gamma_q \cdot (L_O/2 - L_R/2)\}}$$
$$+ \sum_{m=1}^{\infty} c_m \cdot (j\ell) \cdot \sin\left\{ \frac{m2\pi}{(L_O - L_R)} \cdot (x - L_R/2) \right\} \cdot \frac{\sinh\{\delta_m \cdot (z - D_R)\}}{\sinh(\delta_m \cdot D_R)} \tag{7.82b}$$

and

$$H_{z2} = -\sum_{q=1}^{\infty} b_q \cdot \left(\frac{q\pi}{D_R} \cdot \frac{d^2}{\ell} \cdot \beta_q \right) \cdot \cos\left(\frac{q\pi}{D_R} \cdot z \right) \cdot \coth(\beta_q \cdot L_R/2)$$
$$\times \frac{\sinh\{\gamma_q \cdot (x - L_O/2)\}}{\sinh\{\gamma_q \cdot (L_O/2 - L_R/2)\}}$$
$$- \sum_{m=1}^{\infty} c_m \cdot \delta_m \cdot \sin\left\{ \frac{m2\pi}{(L_O - L_R)} \cdot (x - L_R/2) \right\} \cdot \frac{\cosh\{\delta_m \cdot (z - D_R)\}}{\sinh(\delta_m \cdot D_R)} \tag{7.82c}$$

Further, we have

$$\mu_o \cdot H_{x2}\big|_{x=L_R/2} = \mu \cdot H_{x1}\big|_{x=L_R/2} \text{ over } 0 < z < D_R \tag{7.83}$$

or

$$H_{x2}\big|_{x=L_R/2} = \mu_r \cdot H_{x1}\big|_{x=L_R/2} \text{ over } 0 < z < D_R \tag{7.83a}$$

Therefore, in view of Equations 7.73 and 7.82a

$$\mu_r \cdot \left[\sum_{p-odd}^{\infty} a_p \cdot \left(\frac{p\pi}{L_R} \cdot \frac{d^2}{\ell} \cdot \alpha_p \right) \cdot \sin\left(\frac{p\pi}{2} \right) \cdot \frac{\sinh\{\alpha_p \cdot (z - D_R)\}}{\cosh(\alpha_p \cdot D_R)} \right.$$
$$\left. + \sum_{q=1}^{\infty} b_q \cdot \gamma_q^2 \cdot \frac{d^2}{\ell} \cdot \sin\left(\frac{q\pi}{D_R} \cdot z \right) \right]$$
$$= -\sum_{q=1}^{\infty} b_q \cdot \left(\frac{d^2}{\ell} \cdot \beta_q \cdot \gamma_q \right) \cdot \sin\left(\frac{q\pi}{D_R} \cdot z \right) \cdot \coth(\beta_q \cdot L_R/2) \cdot \coth\{\gamma_q \cdot (L_O/2 - L_R/2)\}$$
$$- \sum_{m=1}^{\infty} c_m \cdot \frac{m2\pi}{(L_O - L_R)} \cdot \frac{\sinh\{\delta_m \cdot (z - D_R)\}}{\sinh(\delta_m \cdot D_R)} \tag{7.83b}$$

Multiplying both sides of this equation by $\sin((Q\pi/D_R)\cdot z)$ and then on integrating over $0 < z < D_R$, we get, on replacing Q by q

$$b_q \cdot \left[(\mu_r \cdot \gamma_q + \beta_q) \cdot \gamma_q \cdot \frac{d^2}{\ell} \cdot \frac{D_R}{2} \right]$$
$$= \sum_{p-odd}^{\infty} a_p \cdot \left(\mu_r \cdot \frac{p\pi}{L_R} \cdot \frac{d^2}{\ell} \cdot \alpha_p \right) \cdot \sin\left(\frac{p\pi}{2}\right) \cdot \tanh\{\alpha_p \cdot D_R\} \cdot \frac{(D_R/\pi)\cdot q}{[q^2 + (\alpha_p \cdot D_R/\pi)^2]}$$
$$+ \sum_{m=1}^{\infty} c_m \cdot \frac{m2\pi}{(L_O - L_R)} \cdot \frac{(D_R/\pi)\cdot q}{[q^2 + (\delta_m \cdot D_R/\pi)^2]} \tag{7.84}$$

For $q = 1, 2, 3, \ldots$.

In Figure 7.5a, the air space adjacent to the stator end surface is labelled as region 3. The magnetic field intensity in this region satisfies the following equations:

$$\nabla^2 \boldsymbol{H}_3 = 0 \tag{7.85a}$$

and

$$\nabla \bullet \boldsymbol{H}_3 = 0 \tag{7.85b}$$

For this current-free region, we define scalar magnetic potential, $\mathcal{V}$ as

$$\boldsymbol{H}_3 \overset{def}{=} -\nabla \mathcal{V}_3 \tag{7.86a}$$

Thus

$$\nabla^2 \mathcal{V}_3 = 0 \tag{7.86b}$$

The tangential components of this field on the stator end surface and also on the inner surface of the stator frame are zero. Thus

$$H_{y3}\Big|_{x=L_S/2, L_O/2} = 0 \text{ over } -(g + D_S) < z < -g \tag{7.87a}$$

$$H_{z3}\Big|_{x=L_S/2, L_O/2} = 0 \text{ over } -(g + D_S) < z < -g \tag{7.87b}$$

$$H_{y3}\Big|_{z=-(g+D_S)} = 0 \text{ over } L_S/2 < x < L_O/2 \tag{7.87c}$$

$$H_{x3}\big|_{z=-(g+D_S)} = 0 \text{ over } L_S/2 < z < L_O/2 \tag{7.87d}$$

Let the potential be zero on these surfaces. The solution for Equation 7.86b can therefore be given as

$$\mathcal{V}_3 = \sum_{n=1}^{\infty} d_n \cdot \sin\left\{\frac{n2\pi}{(L_O - L_S)} \cdot (x - L_S/2)\right\} \cdot \frac{\sinh\{\zeta_n \cdot (z + g + D_S)\}}{\sinh(\zeta_n \cdot D_S)} \tag{7.88}$$

where

$$\zeta_n = \sqrt{\left(\frac{n2\pi}{L_O - L_S}\right)^2 + \ell^2} \tag{7.88a}$$

while d_n indicates a set of arbitrary constants.

Thus, the components of magnetic field intensity in this region, obtained from Equations 7.86a and 7.88, are

$$H_{x3} = -\sum_{n=1}^{\infty} d_n \cdot \left(\frac{n2\pi}{L_O - L_S}\right) \cdot \cos\left\{\frac{n2\pi}{(L_O - L_S)} \cdot (x - L_S/2)\right\} \cdot \frac{\sinh\{\zeta_n \cdot (z + g + D_S)\}}{\sinh(\zeta_n \cdot D_S)} \tag{7.89a}$$

$$H_{y3} = \sum_{n=1}^{\infty} d_n \cdot j\ell \cdot \sin\left\{\frac{n2\pi}{(L_O - L_S)} \cdot (x - L_S/2)\right\} \cdot \frac{\sinh\{\zeta_n \cdot (z + g + D_S)\}}{\sinh(\zeta_n \cdot D_S)} \tag{7.89b}$$

$$H_{z3} = -\sum_{n=1}^{\infty} d_n \cdot \zeta_n \cdot \sin\left\{\frac{n2\pi}{(L_O - L_S)} \cdot (x - L_S/2)\right\} \cdot \frac{\cosh\{\zeta_n \cdot (z + g + D_S)\}}{\sinh(\zeta_n \cdot D_S)} \tag{7.89c}$$

Lastly, consider the air-gap region, labelled as region 4 in Figure 7.5a. For this current-free region, we define scalar magnetic potential $\mathcal{V}$ as

$$\boldsymbol{H}_4 \stackrel{def}{=} -\nabla\mathcal{V}_4 \tag{7.90a}$$

Thus,

$$\nabla^2\mathcal{V}_4 = 0 \tag{7.90b}$$

In this region, the scalar magnetic potential distribution is an even function of the x coordinate. This potential on the end-cover surfaces, that is, at $x = -L_O/2$ and $x = L_O/2$, can be taken as zero. Therefore, air-gap potential can be expressed as

$$\mathcal{V}_4 = \sum_{r-odd}^{\infty} \cos\left(\frac{r\pi}{L_O} \cdot x\right) \cdot \left[a_r' \frac{\sinh\{\eta_r \cdot (z+g)\}}{\sinh(\eta_r \cdot g)} - a_r'' \frac{\sinh(\eta_r \cdot z)}{\sinh(\eta_r \cdot g)} \right] \tag{7.91}$$

where

$$\eta_r = \sqrt{\left(\frac{r\pi}{L_O}\right)^2 + \ell^2} \tag{7.91a}$$

while a_r' and a_r'' indicate two sets of arbitrary constants.

Components of the magnetic field intensity in this region obtained from Equations 7.90a and 7.90b are

$$H_{x4} = \sum_{r-odd}^{\infty} \left(\frac{r\pi}{L_O}\right) \cdot \sin\left(\frac{r\pi}{L_O} \cdot x\right) \cdot \left[a_r' \frac{\sinh\{\eta_r \cdot (z+g)\}}{\sinh(\eta_r \cdot g)} - a_r'' \frac{\sinh(\eta_r \cdot z)}{\sinh(\eta_r \cdot g)} \right] \tag{7.92a}$$

$$H_{y4} = \sum_{r-odd}^{\infty} (j\ell) \cdot \cos\left(\frac{r\pi}{L_O} \cdot x\right) \cdot \left[a_r' \frac{\sinh\{\eta_r \cdot (z+g)\}}{\sinh(\eta_r \cdot g)} - a_r'' \frac{\sinh(\eta_r \cdot z)}{\sinh(\eta_r \cdot g)} \right] \tag{7.92b}$$

and

$$H_{z4} = -\sum_{r-odd}^{\infty} \eta_r \cdot \cos\left(\frac{r\pi}{L_O} \cdot x\right) \cdot \left[a_r' \frac{\cosh\{\eta_r \cdot (z+g)\}}{\sinh(\eta_r \cdot g)} - a_r'' \frac{\cosh(\eta_r \cdot z)}{\sinh(\eta_r \cdot g)} \right] \tag{7.92c}$$

The boundary conditions at the rotor air-gap surface, that is, at $z = 0$ are

$$H_{y4}\big|_{z=0} = H_{y1}\big|_{z=0} \text{ over } 0 < x < L_R/2 \tag{7.93a}$$

$$= H_{y2}\big|_{z=0} \text{ over } L_R/2 < z < L_O/2 \tag{7.93b}$$

and

$$H_{z4}\big|_{z=0} = \mu_r \cdot H_{z1}\big|_{z=0} \text{ over } 0 < x < L_R/2 \tag{7.93c}$$

$$H_{z4}\big|_{z=0} = H_{z2}\big|_{z=0} \text{ over } L_R/2 < x < L_O/2 \tag{7.93d}$$

The boundary conditions at the stator air-gap surface, that is, at $z = -g$ are

$$H_{y4}\big|_{z=-g} + K_x = 0 \text{ over } 0 < x < L_S/2 \tag{7.94a}$$

$$= H_{y3}\big|_{z=-g} \text{ over } L_S/2 < z < L_O/2 \tag{7.94b}$$

and

$$H_{z4}\big|_{z=-g} = H_{z3}\big|_{z=-g} \text{ over } L_S/2 < x < L_O/2 \tag{7.94c}$$

where K_x indicates the axial component of the stator current sheet simulating the stator winding carrying known armature currents. Let this surface current distribution be given by the following Fourier series expansion:

$$K_x = \sum_{r-odd}^{\infty} K_r \cdot \cos\left(\frac{r\pi}{L_O}\cdot x\right) \text{ over } 0 < x < L_O/2 \tag{7.94d}$$

In view of Equations 7.74, 7.82b, 7.92b, 7.93a and 7.94b, we get

$$\sum_{r-odd}^{\infty} [a'_r \cdot (j\ell)] \cdot \cos\left(\frac{r\pi}{L_O}\cdot x\right) = -\sum_{p-odd}^{\infty} [a_p \cdot (jd^2 \cdot \alpha_p) \times \tanh(\alpha_p \cdot D_R)] \cdot \cos\left(\frac{p\pi}{L_R}\cdot x\right) \text{ over } 0 < x < L_R/2 \tag{7.95a}$$

$$= -\sum_{m=1}^{\infty} [c_m \cdot (j\ell)] \cdot \sin\left\{\frac{m2\pi}{(L_O - L_R)}\cdot (x - L_R/2)\right\} \text{ over } L_R/2 < z < L_O/2 \tag{7.95b}$$

Multiply both sides of these equations with $\cos((R\pi/L_O)\cdot x)$, where R is an odd integer, and then integrate over $0 < x < L_O/2$, to get

$$\begin{aligned}\sum_{r-odd}^{\infty} [a'_r \cdot (j\ell)] \cdot \int_0^{L_O/2} \cos\left(\frac{R\pi}{L_O}\cdot x\right)\cdot\cos\left(\frac{r\pi}{L_O}\cdot x\right)\cdot dx \\ = -\sum_{p-odd}^{\infty} [a_p \cdot (jd^2\cdot\alpha_p)\cdot\tanh(\alpha_p\cdot D_R)]\cdot\int_0^{L_R/2}\cos\left(\frac{R\pi}{L_O}\cdot x\right)\cdot\cos\left(\frac{p\pi}{L_R}\cdot x\right)\cdot dx \\ -\sum_{m=1}^{\infty}[c_m\cdot(j\ell)]\cdot\int_{L_R/2}^{L_O/2}\cos\left(\frac{R\pi}{L_O}\cdot x\right)\cdot\sin\left\{\frac{m2\pi}{(L_O - L_R)}\cdot(x - L_R/2)\right\}\cdot dx\end{aligned} \tag{7.96a}$$

The LHS will be zero for all values of $r \neq R$; thus, after completing the integrations and on setting $R = r$, one obtains

$$a'_r \cdot \left\{\ell \cdot \frac{L_O}{4}\right\} = -\sum_{p-odd}^{\infty} a_p \cdot \left\{d^2 \cdot \alpha_p \cdot \tanh(\alpha_p \cdot D_R) \cdot \frac{1}{\pi} \cdot \cos\left(\frac{r\pi}{2}\frac{L_R}{L_O}\right) \times \frac{(p/L_R)\cdot \sin(p\pi/2)}{[(p/L_R)^2 - (r/L_O)^2]}\right\} - \sum_{m=1}^{\infty} c_m \cdot \left\{\ell \cdot \frac{1}{\pi} \cdot \cos\left(\frac{r\pi}{2} \cdot \frac{L_R}{L_O}\right) \cdot \frac{(2m/L_O - L_R)}{[2m/(L_O - L_R)^2 - (r/L_O)^2]}\right\} \tag{7.96b}$$

for $r = 1, 3, 5, \ldots$.

From Equations 7.75 and 7.92c

$$H_{z1}\big|_{z=0} = -\sum_{p-odd}^{\infty} a_p \cdot \left[\left(\frac{p\pi}{L_R}\right)^2 + \ell^2\right] \cdot \frac{d^2}{\ell} \cdot \cos\left(\frac{p\pi}{L_R} \cdot x\right) + \sum_{q=1}^{\infty} b_q \cdot \left(\frac{q\pi}{D_R} \cdot \frac{d^2}{\ell} \cdot \beta_q\right) \cdot \frac{\cosh(\beta_q \cdot x)}{\sinh(\beta_q \cdot L_R/2)} \tag{7.97a}$$

and

$$H_{z4}\big|_{z=0} = -\sum_{r-odd}^{\infty} \eta_r \cdot \cos\left(\frac{r\pi}{L_O} \cdot x\right) \cdot [a'_r \coth(\eta_r \cdot g) - a''_r \operatorname{cosech}(\eta_r \cdot g)] \tag{7.97b}$$

Therefore, in view of the boundary condition given by Equation 7.93c

$$\sum_{p-odd}^{\infty} a_p \cdot \mu_r \left[\left(\frac{p\pi}{L_R}\right)^2 + \ell^2\right] \cdot \frac{d^2}{\ell} \cdot \cos\left(\frac{p\pi}{L_R} \cdot x\right) - \sum_{q=1}^{\infty} b_q \cdot \mu_r \left(\frac{q\pi}{D_R} \cdot \frac{d^2}{\ell} \cdot \beta_q\right) \times \frac{\cosh(\beta_q \cdot x)}{\sinh(\beta_q \cdot L_R/2)} = \sum_{r-odd}^{\infty} \eta_r \cdot \cos\left(\frac{r\pi}{L_O} \cdot x\right) \cdot [a'_r \cdot \coth(\eta_r \cdot g) - a''_r \cdot \operatorname{cosech}(\eta_r \cdot g)] \tag{7.98}$$

over $0 \leq x < L_R/2$.

Multiply both sides of this equation with $\cos(P\pi/L_R \cdot x)$, where P is an odd integer, and then integrate over $0 \le x < L_R/2$, to get

$$\sum_{p-odd}^{\infty} a_p \cdot \mu_r \cdot \left[\left(\frac{p\pi}{L_R}\right)^2 + \ell^2\right] \cdot \frac{d^2}{\ell} \cdot \int_0^{L_R/2} \cos\left(\frac{P\pi}{L_R} \cdot x\right) \cdot \cos\left(\frac{p\pi}{L_R} \cdot x\right) \cdot dx$$

$$- \sum_{q=1}^{\infty} b_q \cdot \mu_r \cdot \left(\frac{q\pi}{D_R} \cdot \frac{d^2}{\ell} \cdot \beta_q\right) \cdot \int_0^{L_R/2} \cos\left(\frac{P\pi}{L_R} \cdot x\right) \cdot \frac{\cosh(\beta_q \cdot x)}{\sinh(\beta_q \cdot L_R/2)} \cdot dx$$

$$= \sum_{r-odd}^{\infty} \eta_r \cdot \left[a'_r \coth(\eta_r \cdot g) - a''_r \operatorname{cosech}(\eta_r \cdot g)\right] \cdot \int_0^{L_R/2} \cos\left(\frac{P\pi}{L_R} \cdot x\right) \cdot \cos\left(\frac{r\pi}{L_O} \cdot x\right) \cdot dx \tag{7.99a}$$

The first term on the LHS will be zero for all values of $p \ne P$. Thus, after completing the integrations and on setting $P = p$, one obtains

$$a_p \cdot \mu_r \cdot \left\{\left[\left(\frac{p\pi}{L_R}\right)^2 + \ell^2\right] \cdot \left(\frac{d^2}{\ell} \cdot \frac{L_R}{4}\right)\right\}$$

$$- \sum_{q=1}^{\infty} b_q \cdot \mu_r \cdot \left(\frac{q\pi}{D_R} \cdot \frac{d^2}{\ell} \cdot \beta_q\right) \cdot \coth(\beta_q \cdot L_R/2) \cdot \frac{(p\pi/L_R) \cdot \sin(p\pi/2)}{[\beta_q^2 + (p\pi/L_R)^2]}$$

$$= \sum_{r-odd}^{\infty} \eta_r \cdot [a'_r \coth(\eta_r \cdot g) - a''_r \operatorname{cosech}(\eta_r \cdot g)]$$

$$\times \frac{1}{\pi} \cdot \sin\left(\frac{p\pi}{2}\right) \cdot \cos\left(\frac{r\pi}{2} \cdot \frac{L_R}{L_O}\right) \cdot \frac{(p/L_R)}{[(p/L_R)^2 - (r/L_O)^2]} \tag{7.99b}$$

for $p = 1, 3, 5, \ldots$.

In view of Equation 7.82c, we get

$$H_{z2}\big|_{z=0} = -\sum_{q=1}^{\infty} b_q \cdot \left(\frac{q\pi}{D_R} \cdot \frac{d^2}{\ell} \cdot \beta_q\right) \cdot \coth(\beta_q \cdot L_R/2) \cdot \frac{\sinh\{\gamma_q \cdot (x - L_O/2)\}}{\sinh\{\gamma_q \cdot (L_O/2 - L_R/2)\}}$$

$$- \sum_{m=1}^{\infty} c_m \cdot \delta_m \cdot \sin\left\{\frac{m2\pi}{(L_O - L_R)} \cdot (x - L_R/2)\right\} \cdot \coth(\delta_m \cdot D_R) \tag{7.100}$$

Therefore, using Equations 7.93d, 7.97b and 7.100

$$\sum_{m=1}^{\infty} c_m \cdot \delta_m \cdot \sin\left\{\frac{m2\pi}{(L_O - L_R)} \cdot (x - L_R/2)\right\} \cdot \coth(\delta_m \cdot D_R)$$
$$+ \sum_{q=1}^{\infty} b_q \cdot \left(\frac{q\pi}{D_R} \cdot \frac{d^2}{\ell} \cdot \beta_q\right) \cdot \coth(\beta_q \cdot L_R/2) \cdot \frac{\sinh\{\gamma_q \cdot (x - L_O/2)\}}{\sinh\{\gamma_q \cdot (L_O/2 - L_R/2)\}}$$
$$= \sum_{r-odd}^{\infty} \eta_r \cdot \cos\left(\frac{r\pi}{L_O} \cdot x\right) \cdot [a'_r \coth(\eta_r \cdot g) - a''_r \operatorname{cosech}(\eta_r \cdot g)] \qquad (7.101)$$

over $L_R/2 < x < L_O/2$.

Multiply both sides of this equation with $\sin\{M2\pi/(L_O - L_R) \cdot (x - L_R/2)\}$, where M is an integer, and then integrate over $L_R/2 < x < L_O/2$ to get

$$\sum_{m=1}^{\infty} c_m \cdot \delta_m \cdot \coth(\delta_m \cdot D_R) \cdot \int_{L_R/2}^{L_O/2} \left[\sin\left\{\frac{M2\pi}{(L_O - L_R)} \cdot (x - L_R/2)\right\}\right.$$
$$\left. \times \sin\left\{\frac{m2\pi}{(L_O - L_R)} \cdot (x - L_R/2)\right\}\right] \cdot dx + \sum_{q=1}^{\infty} b_q \cdot \left[\frac{(q\pi/D_R) \cdot (\beta_q \cdot d^2/\ell)}{\tanh(\beta_q \cdot L_R/2)}\right.$$
$$\left. \times \int_{L_R/2}^{L_O/2} \sin\left\{\frac{M2\pi}{(L_O - L_R)} \cdot (x - L_R/2)\right\} \cdot \frac{\sinh\{\gamma_q \cdot (x - L_O/2)\}}{\sinh\{\gamma_q \cdot (L_O/2 - L_R/2)\}} \cdot dx\right]$$
$$= \sum_{r-odd}^{\infty} \eta_r \cdot \left[\frac{a'_r \cdot \cosh(\eta_r \cdot g) - a''_r}{\sinh(\eta_r \cdot g)}\right] \int_{L_R/2}^{L_O/2} \sin\left\{\frac{M2\pi}{(L_O - L_R)} \cdot (x - L_R/2)\right\}$$
$$\times \cos\left(\frac{r\pi}{L_O} \cdot x\right) dx \qquad (7.101a)$$

The first term on the LHS will be zero for all values of $m \neq M$. Thus, after completing the integrations and on setting $M = m$, one obtains

$$c_m \cdot \left\{\delta_m \cdot \coth(\delta_m \cdot D_R) \cdot \frac{(L_O - L_R)}{4}\right\}$$
$$- \sum_{q=1}^{\infty} b_q \cdot \left\{\frac{((q\pi/D_R) \cdot \beta_q \cdot d^2/\ell)}{\tanh(\beta_q \cdot L_R/2)} \cdot \frac{\{(M2\pi/(L_O - L_R)\}}{[\gamma_q^2 + \{M2\pi/(L_O - L_R)\}^2]}\right\}$$
$$= \sum_{r-odd}^{\infty} \eta_r \cdot \left[\frac{a'_r \cdot \cosh(\eta_r \cdot g) - a''_r}{\sinh(\eta_r \cdot g)}\right]$$
$$\times \frac{[\sin((r\pi/2) \cdot (L_R/L_O)) - \sin((r/2) \cdot \pi) \cdot \cos(M\pi)](r/L_O)}{[(2M/(L_O - L_R))^2 - (r/L_O)^2] \cdot \pi} \qquad (7.101b)$$

for $m = 1, 2, 3, \ldots$.

From Equations 7.92b and 7.89b, we get

$$H_{y4}\Big|_{z=-g} = \sum_{r-odd}^{\infty} a_r'' \cdot (j\ell) \cdot \cos\left(\frac{r\pi}{L_O} \cdot x\right) \tag{7.102a}$$

$$H_{y3}\Big|_{z=-g} = \sum_{n=1}^{\infty} d_n \cdot j\ell \cdot \sin\left\{\frac{n2\pi}{(L_O - L_S)} \cdot (x - L_S/2)\right\} \tag{7.102b}$$

Therefore, in view of Equations 7.94a, 7.94b and 7.94d

$$\sum_{r-odd}^{\infty} a_r'' \cdot \cos\left(\frac{r\pi}{L_O} \cdot x\right) = \sum_{r-odd}^{\infty} K_r \cdot \frac{j}{\ell} \cdot \cos\left(\frac{r\pi}{L_O} \cdot x\right) \tag{7.103a}$$

over $0 < x < L_R/2$.

$$\sum_{r-odd}^{\infty} a_r'' \cdot \cos\left(\frac{r\pi}{L_O} \cdot x\right) = \sum_{r-odd}^{\infty} K_r \cdot \frac{j}{\ell} \cdot \cos\left(\frac{r\pi}{L_O} \cdot x\right) + \sum_{n=1}^{\infty} d_n \cdot \sin\left\{\frac{n2\pi}{(L_O - L_S)} \cdot (x - L_S/2)\right\} \tag{7.103b}$$

over $L_S/2 < z < L_O/2$.

The Fourier coefficient a_r'' is obtained as

$$a_r'' = K_r \cdot \frac{j}{\ell} + \sum_{n=1}^{\infty} d_n \cdot \frac{4}{L_O} \cdot \int_{L_S/2}^{L_O/2} \sin\left\{\frac{n2\pi}{(L_O - L_S)} \cdot (x - L_S/2)\right\} \cdot \cos\left(\frac{r\pi}{L_O} \cdot x\right) \cdot dx \tag{7.104a}$$

Thus

$$a_r'' = K_r \cdot \frac{j}{\ell} + \sum_{n=1}^{\infty} d_n \cdot \frac{[\sin((r\pi/2) \cdot (L_S/L_O)) - \sin((r/2) \cdot \pi) \cdot \cos(n\pi)](r/L_O)}{[(2n/(L_O - L_S))^2 - (r/L_O)^2] \cdot \pi} \tag{7.104b}$$

for $r = 1, 3, 5, \ldots$.

Lastly, considering Equations 7.92c and 7.89c, we get

$$H_{z4}\big|_{z=-g} = -\sum_{r-odd}^{\infty} \eta_r \cdot \cos\left(\frac{r\pi}{L_O} \cdot x\right) \cdot [a_r' \cdot \text{cosech}(\eta_r \cdot g) - a_r'' \cdot \cot(\eta_r \cdot g)] \tag{7.105a}$$

$$H_{z3}\big|_{z=-g} = -\sum_{n=1}^{\infty} d_n \cdot \zeta_n \cdot \sin\left\{\frac{n2\pi}{(L_O - L_S)} \cdot (x - L_S/2)\right\} \cdot \coth(\zeta_n \cdot D_S) \tag{7.105b}$$

Therefore, in view of Equation 7.94c, we have

$$\begin{aligned} &\sum_{n=1}^{\infty} d_n \cdot \zeta_n \cdot \sin\left\{\frac{n2\pi}{(L_O - L_S)} \cdot \left(x - \frac{L_S}{2}\right)\right\} \cdot \coth(\zeta_n \cdot D_S) \\ &= \sum_{r-odd}^{\infty} \eta_r \cdot \cos\left(\frac{r\pi}{L_O} \cdot x\right) \cdot \left[\frac{a_r'}{\sinh(\eta_r \cdot g)} - \frac{a_r''}{\tanh(\eta_r \cdot g)}\right] \end{aligned} \tag{7.106}$$

over $L_S/2 < x < L_O/2$.

Multiply both sides of this equation with $\sin\{(N2\pi/(L_O - L_S) \cdot (x - L_S/2)\}$, where N is an integer; and then integrate over $L_S/2 < x < L_O/2$, to get

$$\begin{aligned} &\sum_{n=1}^{\infty} \frac{d_n \cdot \zeta_n}{\tanh(\zeta_n \cdot D_S)} \cdot \int_{L_S/2}^{L_O/2} \sin\left\{\frac{N2\pi}{(L_O - L_S)} \cdot \left(x - \frac{L_S}{2}\right)\right\} \sin\left\{\frac{n2\pi}{(L_O - L_S)} \cdot \left(x - \frac{L_S}{2}\right)\right\} \cdot dx \\ &= \sum_{r-odd}^{\infty} \eta_r [a_r' \cdot \text{cosech}(\eta_r \cdot g) - a_r'' \cdot \cot(\eta_r \cdot g)] \\ &\times \int_{L_S/2}^{L_O/2} \sin\left\{\frac{N2\pi}{(L_O - L_S)} \cdot (x - L_S/2)\right\} \cdot \cos\left(\frac{r\pi}{L_O} \cdot x\right) \cdot dx \end{aligned} \tag{7.106a}$$

Thus, after completing the integrations and on setting $N = n$, one gets

$$\begin{aligned} &d_n \cdot \left\{\frac{\zeta_n \cdot (L_O - L_S)/4}{\tanh(\zeta_n \cdot D_S)}\right\} = \sum_{r-odd}^{\infty} \eta_r [a_r' \cdot \text{cosech}(\eta_r \cdot g) - a_r'' \cdot \cot(\eta_r \cdot g)] \\ &\times \frac{[\sin(r\pi/2) \cdot (L_S/L_O) - \sin(r/2 \cdot \pi) \cdot \cos(N\pi)](r/L_O)}{[(2N/(L_O - L_S))^2 - (r/L_O)^2] \cdot \pi} \end{aligned} \tag{7.106b}$$

for $n = 1, 2, 3, \ldots$

Numerical solution of Equations 7.84, 7.96b, 7.99b, 7.101b, 7.103b and 7.106b determines the values of all arbitrary constants.

7.3.3 Effects of Finite Machine Length

Effects of finite machine length on the distributions of magnetic field intensity and eddy current density at different depths in the solid rotor of a typical machine with stator and rotor axial length, each equalling 4 inches, are demonstrated in Figures 7.6 through 7.11. These figures correspond to a fixed value of stator current, viz., 1 A.

7.3.4 Effect of Different Rotor and Stator Lengths

Three-dimensional analysis permits to compare the field distributions with those obtained from two-dimensional treatments. Further, one can study the variations in field distributions if the rotor length is different from the stator length. These features are illustrated in Figures 7.12 and 7.13.

7.3.5 Performance Parameters

The parameters that spell the performance of a machine can be evaluated as indicated below.

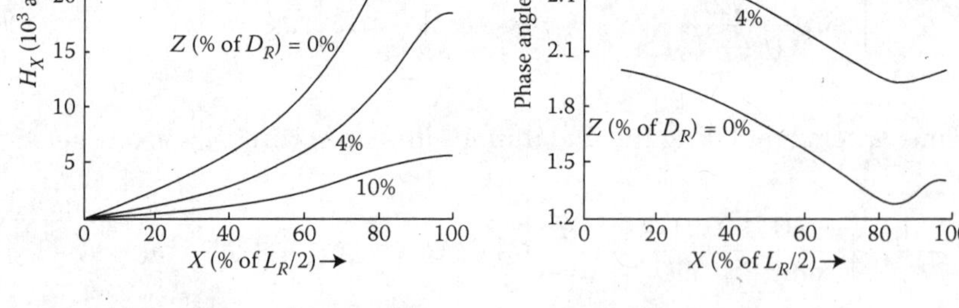

FIGURE 7.6
(a) Magnitude of the axial component of rotor magnetic field. (b) Phase angle of the axial component of rotor magnetic field.

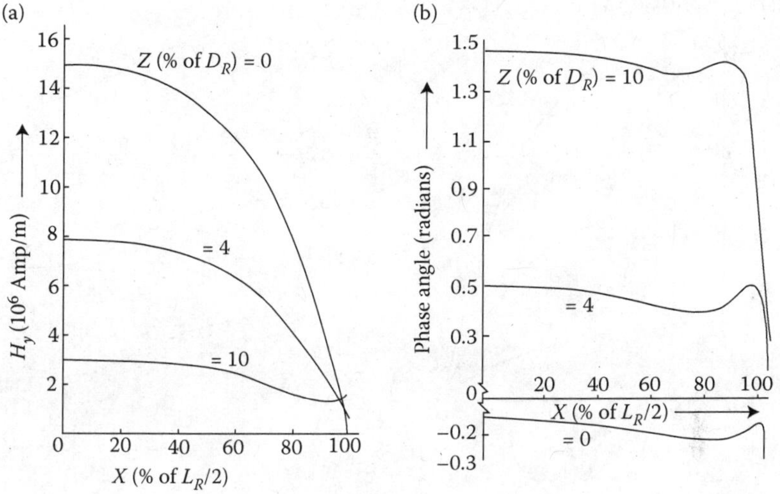

FIGURE 7.7
(a) Magnitude of the peripheral component of rotor magnetic field. (b) Phase angle of the peripheral component of rotor magnetic field.

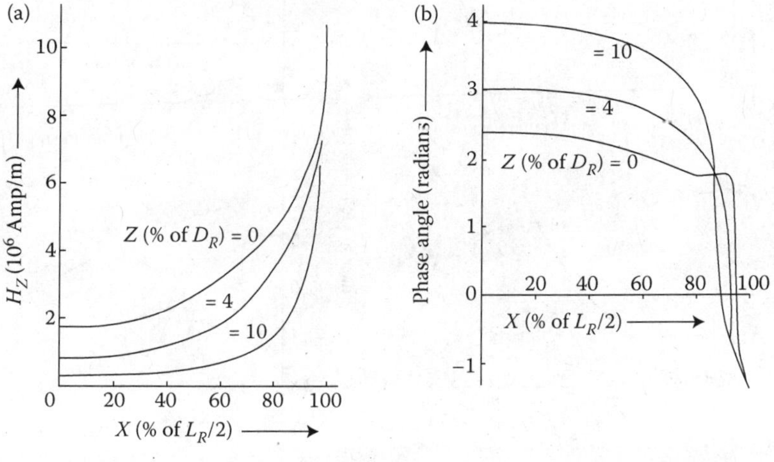

FIGURE 7.8
(a) Magnitude of the radial component of rotor magnetic field. (b) Phase angle of the radial component of rotor magnetic field.

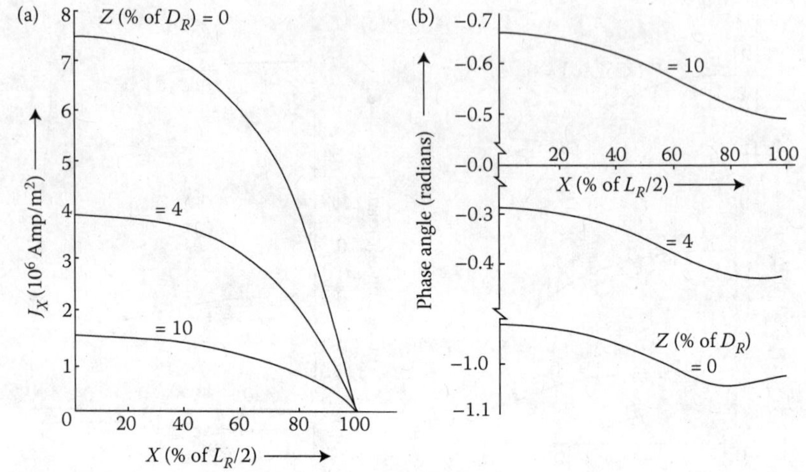

FIGURE 7.9
(a) Magnitude of the axial component of eddy current density. (b) Phase angle of the axial component of eddy current density.

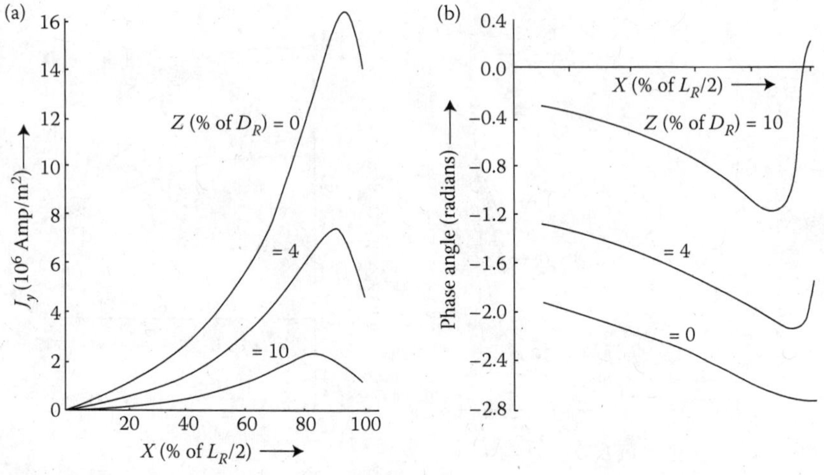

FIGURE 7.10
(a) Magnitude of the peripheral component of eddy current density. (b) Phase angle of the peripheral component of eddy current density.

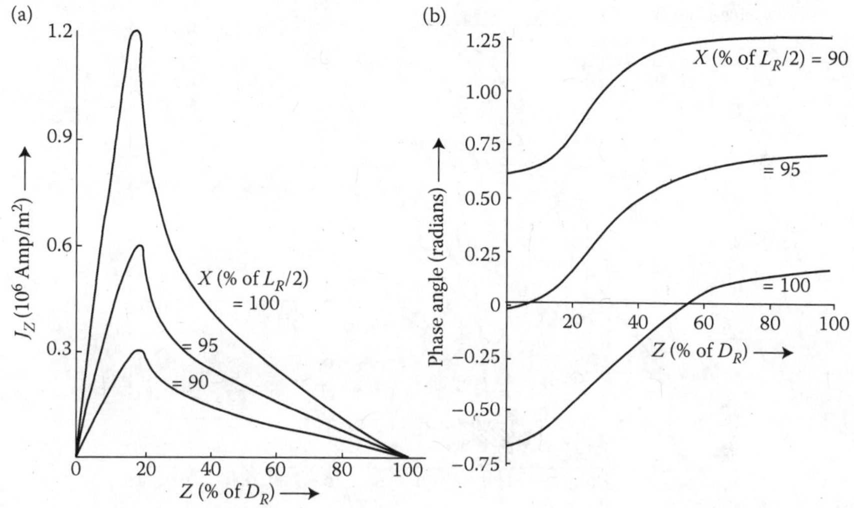

FIGURE 7.11
(a) Magnitude of the radial component of eddy current density. (b) Phase angle of the radial component of eddy current density.

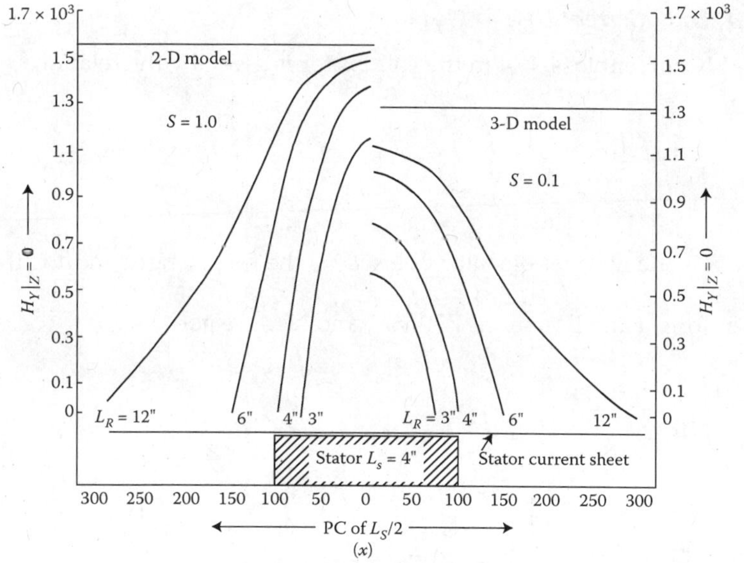

FIGURE 7.12
Peripheral component of magnetic field on rotor surface with ($\mu_r = 500$, $\sigma = 5.5036 \times 10^6$, stator current = 1 A).

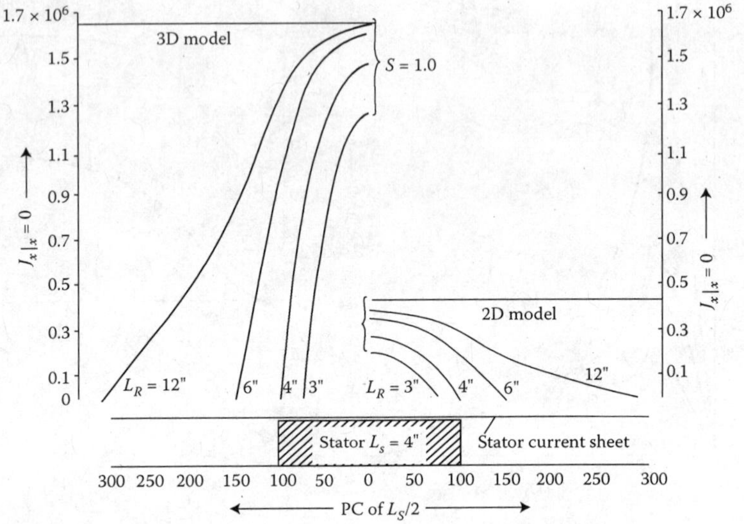

FIGURE 7.13
Axial component of eddy current density on rotor surface with $\mu_r = 500$, $\sigma = 5.5036 \times 10^6$, stator current = 1 A.

7.3.5.1 Eddy Current Loss

The eddy current loss, W_E, in the solid rotor is given by the relation

$$W_E = P \cdot \tau \cdot \frac{1}{2\sigma} \int_{-L_R}^{L_R/2} \int_{0}^{D_R} (J_{1x} \cdot \tilde{J}_{1x} + J_{1y} \cdot \tilde{J}_{1y} + J_{1z} \cdot \tilde{J}_{1z}) \cdot dz \cdot dx \tag{7.107}$$

where P is the number of stator poles, L_R is the rotor length and τ is the pole pitch.

Therefore, using Equations 7.70, 7.71 and 7.72, we get

$$\int_{-L_R/2}^{L_R/2} \int_{0}^{D_R} J_{1x} \cdot \tilde{J}_{1x} \cdot dz \cdot dx$$

$$= \sum_{p-odd}^{\infty} \sum_{P-odd}^{\infty} a_p \tilde{a}_P \cdot \left[\left\{ \int_{-L_R/2}^{L_R/2} \cos\left(\frac{p\pi}{L_R} \cdot x\right) \cdot \cos\left(\frac{P\pi}{L_R} \cdot x\right) \cdot dx \right\} \right.$$

$$\left. \times \left\{ \int_{0}^{D_R} \frac{\cosh\{\alpha_p \cdot (z - D_R)\}}{\cosh(\alpha_p \cdot D_R)} \cdot \frac{\cosh\{\tilde{\alpha}_P \cdot (z - D_R)\}}{\cosh(\tilde{\alpha}_P \cdot D_R)} \cdot dz \right\} \right] \tag{7.108a}$$

$$
\begin{aligned}
&\int_{-L_R/2}^{L_R/2}\int_{0}^{D_R} J_{1y}\cdot\tilde{j}_{1y}\cdot dz\cdot dx \\
&=\sum_{p-odd}^{\infty}\sum_{P-odd}^{\infty} a_p\tilde{a}_P\cdot\left(\frac{p\pi}{L_R}\cdot\frac{j}{\ell}\right)\cdot\left(\frac{P\pi}{L_R}\cdot\frac{j}{\ell}\right) \\
&\quad\times\left[\left\{\int_{-L_R/2}^{L_R/2}\sin\left(\frac{p\pi}{L_R}\cdot x\right)\cdot\sin\left(\frac{P\pi}{L_R}\cdot x\right)\cdot dx\right\}\right. \\
&\quad\left.\times\left\{\int_{0}^{D_R}\frac{\cosh\{\alpha_p\cdot(z-D_R)\}}{\cosh(\alpha_p\cdot D_R)}\cdot\frac{\cosh\{\tilde{\alpha}_P\cdot(z-D_R)\}}{\cosh(\tilde{\alpha}_P\cdot D_R)}\cdot dz\right\}\right] \\
&-\sum_{q=1}^{\infty}\sum_{Q=1}^{\infty} b_q\tilde{b}_Q\cdot\left(\frac{q\pi}{D_R}\cdot\frac{j}{\ell}\right)\cdot\left(\frac{Q\pi}{D_R}\cdot\frac{j}{\ell}\right) \\
&\quad\times\left[\left\{\int_{-L_R/2}^{L_R/2}\frac{\sinh(\beta_q\cdot x)}{\sinh(\beta_q\cdot L_R/2)}\cdot\frac{\sinh(\tilde{\beta}_Q\cdot x)}{\sinh(\tilde{\beta}_Q\cdot L_R/2)}\cdot dx\right\}\right. \\
&\quad\left.\times\left\{\int_{0}^{D_R}\cos\left(\frac{q\pi}{D_R}\cdot z\right)\cdot\cos\left(\frac{Q\pi}{D_R}\cdot z\right)\cdot dz\right\}\right] \qquad (7.108b) \\
&+\sum_{p-odd}^{\infty}\sum_{Q=1}^{\infty} a_p\cdot\tilde{b}_Q\cdot\left(\frac{p\pi}{L_R}\cdot\frac{j}{\ell}\right)\cdot\left(\frac{Q\pi}{D_R}\cdot\frac{j}{\ell}\right) \\
&\quad\times\left[\left\{\int_{-L_R/2}^{L_R/2}\sin\left(\frac{p\pi}{L_R}\cdot x\right)\cdot\frac{\sinh(\tilde{\beta}_Q\cdot x)}{\sinh(\tilde{\beta}_Q\cdot L_R/2)}\cdot dx\right\}\right. \\
&\quad\left.\times\left\{\int_{0}^{D_R}\cos\left(\frac{Q\pi}{D_R}\cdot z\right)\cdot\frac{\cosh\{\alpha_p\cdot(z-D_R)\}}{\cosh(\alpha_p\cdot D_R)}\cdot dz\right\}\right] \\
&+\sum_{q=1}^{\infty}\sum_{P-odd}^{\infty} b_q\cdot\tilde{a}_P\cdot\left(\frac{q\pi}{D_R}\cdot\frac{j}{\ell}\right)\cdot\left(\frac{P\pi}{L_R}\cdot\frac{j}{\ell}\right) \\
&\quad\times\left[\left\{\int_{-L_R/2}^{L_R/2}\sin\left(\frac{P\pi}{L_R}\cdot x\right)\cdot\frac{\sinh(\beta_q\cdot x)}{\sinh(\beta_q\cdot L_R/2)}\cdot dx\right\}\right. \\
&\quad\left.\times\left\{\int_{0}^{D_R}\cos\left(\frac{q\pi}{D_R}\cdot z\right)\cdot\frac{\cosh\{\tilde{\alpha}_P\cdot(z-D_R)\}}{\cosh(\tilde{\alpha}_P\cdot D_R)}\cdot dz\right\}\right]
\end{aligned}
$$

$$\int_{-L_R/2}^{L_R/2}\int_{0}^{D_R} J_{1z}\cdot\tilde{j}_{1z}\cdot dz\cdot dx$$

$$= \sum_{q=1}^{\infty}\sum_{Q=1}^{\infty} b_q\tilde{b}_Q\cdot\left[\left\{\int_{-L_R/2}^{L_R/2}\frac{\sinh(\beta_q\cdot x)}{\sinh(\beta_q\cdot L_R/2)}\cdot\frac{\sinh(\tilde{\beta}_Q\cdot x)}{\sinh(\tilde{\beta}_Q\cdot L_R/2)}\cdot dx\right\}\right.$$

$$\left.\times\left\{\int_{0}^{D_R}\sin\left(\frac{q\pi}{D_R}\cdot z\right)\cdot\sin\left(\frac{Q\pi}{D_R}\cdot z\right)\cdot dz\right\}\right] \tag{7.108c}$$

Now, consider the following integrals:

$$\int_{-L_R/2}^{L_R/2}\cos\left(\frac{p\pi}{L_R}\cdot x\right)\cdot\cos\left(\frac{P\pi}{L_R}\cdot x\right)\cdot dx = \frac{L_R}{2}\quad\text{for}\quad p = P;$$

$$= 0,\ \text{for}\ p \neq P \tag{7.109a}$$

$$\int_{0}^{D_R}\frac{\cosh\{\alpha_p\cdot(z-D_R)\}}{\cosh(\alpha_p\cdot D_R)}\cdot\frac{\cosh\{\tilde{\alpha}_P\cdot(z-D_R)\}}{\cosh(\tilde{\alpha}_P\cdot D_R)}$$

$$\times dz = \left[\frac{\alpha_p\cdot\tanh(\alpha_p\cdot D_R)}{(\alpha_p^2-\tilde{\alpha}_P^2)}+\frac{\tilde{\alpha}_P\cdot\tanh(\tilde{\alpha}_P\cdot D_R)}{(\tilde{\alpha}_P^2-\alpha_p^2)}\right] \tag{7.109b}$$

$$\int_{-L_R/2}^{L_R/2}\sin\left(\frac{p\pi}{L_R}\cdot x\right)\cdot\sin\left(\frac{P\pi}{L_R}\cdot x\right)\cdot dx = \frac{L_R}{2},\ \text{for}\quad p = P;$$

$$= 0,\ \text{for}\ p \neq P \tag{7.109c}$$

$$\int_{-L_R/2}^{L_R/2}\frac{\sinh(\beta_q\cdot x)}{\sinh(\beta_q\cdot L_R/2)}\cdot\frac{\sinh(\tilde{\beta}_Q\cdot x)}{\sinh(\tilde{\beta}_Q\cdot L_R/2)}$$

$$\times dx = 2\cdot\left[\frac{\beta_q\cdot\coth(\tilde{\beta}_Q(L_R/2))}{(\beta_q^2-\tilde{\beta}_Q^2)}+\frac{\tilde{\beta}_Q\cdot\coth(\beta_q(L_R/2))}{(\tilde{\beta}_Q^2-\beta_q^2)}\right] \tag{7.109d}$$

$$\int_{0}^{D_R}\cos\left(\frac{q\pi}{D_R}\cdot z\right)\cdot\cos\left(\frac{Q\pi}{D_R}\cdot z\right)\cdot dz = \frac{D_R}{2}\quad\text{for}\quad q = Q;$$

$$= 0,\ \text{for}\ q \neq Q \tag{7.109e}$$

$$\int_{-L_R/2}^{L_R/2} \sin\left(\frac{p\pi}{L_R}\cdot x\right)\cdot\frac{\sinh(\tilde{\beta}_Q\cdot x)}{\sinh(\tilde{\beta}_Q\cdot L_R/2)}\cdot dx = 2\cdot\left[\frac{\tilde{\beta}_Q\cdot\sin(p\pi/2)\coth(\tilde{\beta}_Q\cdot L_R/2)}{(p\pi/L_R)^2+\tilde{\beta}_Q^2}\right] \tag{7.109f}$$

$$\int_{0}^{D_R} \cos\left(\frac{Q\pi}{D_R}\cdot z\right)\cdot\frac{\cosh\{\alpha_p\cdot(z-D_R)\}}{\cosh(\alpha_p\cdot D_R)}\cdot dz = \left[\frac{\alpha_p\cdot\tanh(\alpha_p D_R)}{(p\pi/L_R)^2+\alpha_p^2}\right] \tag{7.109g}$$

$$\int_{0}^{D_R} \sin\left(\frac{q\pi}{D_R}\cdot z\right)\cdot\sin\left(\frac{Q\pi}{D_R}\cdot z\right)\cdot dz = \frac{D_R}{2} \quad \text{for } q = Q;$$
$$= 0, \quad \text{for } q \neq Q \tag{7.109h}$$

$$\int_{0}^{D_R} \sin\left(\frac{q\pi}{D_R}\cdot z\right)\cdot\frac{\sinh\{\tilde{\alpha}_p\cdot(z-D_R)\}}{\cosh(\tilde{\alpha}_p\cdot D_R)}\cdot dz = \frac{(q\pi/D_R)\cdot\tanh(\tilde{\alpha}_p\cdot D_R)}{[(q\pi/D_R)^2+\tilde{\alpha}_p^2]} \tag{7.109i}$$

$$\int_{-L_R/2}^{L_R/2} \cos\left(\frac{p\pi}{L_R}\cdot x\right)\cdot\frac{\cosh(\tilde{\beta}_q\cdot x)}{\sinh(\tilde{\beta}_q\cdot L_R/2)}$$
$$\times\, dx = 2\cdot\left[\frac{(p\pi/L_R)\cdot\sin(p\pi/2)\cdot\coth(\tilde{\beta}_Q\cdot L_R/2)}{(p\pi/L_R)^2+\tilde{\beta}_q^2}\right] \tag{7.109j}$$

From Equations 7.108a, 7.109a and 7.109b and on setting $p = P$, we get

$$\int_{-L_R/2}^{L_R/2}\int_{0}^{D_R} J_{1x}\cdot\tilde{j}_{1x}\cdot dz\cdot dx$$
$$= \sum_{p-odd}^{\infty} a_p\tilde{a}_p\cdot\frac{L_R}{2}\cdot\left[\frac{\alpha_p\cdot\tanh(\alpha_p\cdot D_R)}{(\alpha_p^2-\tilde{\alpha}_p^2)}+\frac{\tilde{\alpha}_p\cdot\tanh(\tilde{\alpha}_p\cdot D_R)}{(\tilde{\alpha}_p^2-\alpha_p^2)}\right] \tag{7.110a}$$

From Equations 7.108b, 7.109b and 7.109g and on setting $P = p$ and $Q = q$, we get

$$\int_{-L_R/2}^{L_R/2}\int_{0}^{D_R} J_{1y}\cdot\tilde{j}_{1y}\cdot dz\cdot dx$$
$$= \sum_{p-odd}^{\infty} a_p\tilde{a}_P\cdot\left(\frac{p\pi}{L_R}\cdot\frac{1}{\ell}\right)^2\cdot\frac{L_R}{2}\cdot\left[\frac{\alpha_p\cdot\tanh(\alpha_p\cdot D_R)}{(\alpha_p^2-\tilde{\alpha}_p^2)}+\frac{\tilde{\alpha}_p\cdot\tanh(\tilde{\alpha}_p\cdot D_R)}{(\tilde{\alpha}_p^2-\alpha_p^2)}\right]$$

$$+\sum_{q=1}^{\infty} b_q\tilde{b}_q\cdot\left(\frac{q\pi}{D_R}\cdot\frac{1}{\ell}\right)\cdot\left(\frac{q\pi}{D_R}\cdot\frac{1}{\ell}\right)$$

$$\times D_R\cdot\left[\frac{\beta_q\cdot\coth(\tilde{\beta}_q(L_R/2))}{(\beta_q^2-\tilde{\beta}_q^2)}+\frac{\tilde{\beta}_q\cdot\coth(\beta_q(L_R/2))}{(\tilde{\beta}_q-\beta_q^2)}\right]$$

$$-\sum_{p-odd}^{\infty}\sum_{q=1}^{\infty} a_p\cdot\tilde{b}_q\cdot((p\pi/L_R)\cdot(1/\ell))\cdot((q\pi/D_R)\cdot(1/\ell))\cdot 2\cdot\left[\frac{\alpha_p\cdot\tanh(\alpha_p D_R)}{(p\pi/L_R)^2+\alpha_p^2}\right]$$

$$\times\left[\frac{\tilde{\beta}_q\cdot\sin((p\pi/2)\coth(\tilde{\beta}_q\cdot(L_R/2))}{(p\pi/L_R)^2+\tilde{\beta}_q^2}\right]$$

$$-\sum_{q=1}^{\infty}\sum_{p-odd}^{\infty} b_q\cdot\tilde{a}_p\cdot\left(\frac{q\pi}{D_R}\cdot\frac{1}{\ell}\right)\cdot\left(\frac{p\pi}{L_R}\cdot\frac{1}{\ell}\right)\cdot 2\cdot\left[\frac{\tilde{\alpha}_p\cdot\tanh(\tilde{\alpha}_p D_R)}{(p\pi/L_R)^2+\tilde{\alpha}_p^2}\right]$$

$$\times\left[\frac{\beta_q\cdot\sin(p\pi/2)\coth(\beta_q\cdot(L_R/2))}{(p\pi/L_R)^2+\beta_q^2}\right] \tag{7.110b}$$

whereas from Equations 7.108c, 7.109d and 7.109h and on setting $Q = q$, we get

$$\int_{-L_R/2}^{L_R/2}\int_0^{D_R} J_{1z}\cdot\tilde{j}_{1z}\cdot dz\cdot dx$$

$$=\sum_{q=1}^{\infty} b_q\tilde{b}_q\cdot D_R\cdot\left[\frac{\beta_q\cdot\coth(\tilde{\beta}_q(L_R/2))}{(\beta_q^2-\tilde{\beta}_q^2)}+\frac{\tilde{\beta}_q\cdot\coth(\beta_q(L_R/2))}{(\tilde{\beta}_q^2-\beta_q^2)}\right] \tag{7.110c}$$

Therefore, eddy current loss, W_E, in the solid rotor is given by the equation

$$W_E = P\cdot\tau\cdot\frac{1}{2\sigma}\cdot\sum_{p-odd}^{\infty} a_p\tilde{a}_p\cdot\left\{1+\left(\frac{p\pi}{L_R}\cdot\frac{1}{\ell}\right)^2\right\}\frac{L_R}{2}$$

$$\times\left[\frac{\alpha_p\cdot\tanh(\alpha_p\cdot D_R)}{(\alpha_p^2-\tilde{\alpha}_p^2)}+\frac{\tilde{\alpha}_p\cdot\tanh(\tilde{\alpha}_p\cdot D_R)}{(\tilde{\alpha}_p^2-\alpha_p^2)}\right]$$

$$+\sum_{q=1}^{\infty} b_q\tilde{b}_q\cdot\left\{1+\left(\frac{q\pi}{D_R}\cdot\frac{1}{\ell}\right)^2\right\}$$

$$\times D_R\cdot\left[\frac{\beta_q\cdot\coth(\tilde{\beta}_q(L_R/2))}{(\beta_q^2-\tilde{\beta}_q^2)}+\frac{\tilde{\beta}_q\cdot\coth(\beta_q(L_R/2))}{(\tilde{\beta}_q-\beta_q^2)}\right]$$

$$
-\sum_{p-odd}^{\infty}\sum_{q=1}^{\infty} a_p \cdot \tilde{b}_q \cdot \left(\frac{p\pi}{L_R}\cdot\frac{1}{\ell}\right)\cdot\left(\frac{q\pi}{D_R}\cdot\frac{1}{\ell}\right)\cdot 2\cdot\left[\frac{\alpha_p \cdot \tanh(\alpha_p D_R)}{(p\pi/L_R)^2+\alpha_p^2}\right]
\times\left[\frac{\tilde{\beta}_q \cdot \sin(p\pi/2)\coth(\tilde{\beta}_q\cdot(L_R/2))}{(p\pi/L_R)^2+\tilde{\beta}_q^2}\right]
$$

$$
-\sum_{q=1}^{\infty}\sum_{p-odd}^{\infty} b_q \cdot \tilde{a}_p \cdot \left(\frac{q\pi}{D_R}\cdot\frac{1}{\ell}\right)\cdot\left(\frac{p\pi}{L_R}\cdot\frac{1}{\ell}\right)\cdot 2\cdot\left[\frac{\tilde{\alpha}_p \cdot \tanh(\tilde{\alpha}_p D_R)}{(p\pi/L_R)^2+\tilde{\alpha}_p^2}\right]
\times\left[\frac{\beta_q \cdot \sin(p\pi/2)\coth(\beta_q\cdot(L_R/2))}{(p\pi/L_R)^2+\beta_q^2}\right] \tag{7.111}
$$

7.3.5.2 Force Density

The force density, $\mathcal{F}$, in general, is given by $\boldsymbol{J}\times\boldsymbol{B}$. Thus, its peripheral component is

$$
\mathcal{F}_y = \mu\cdot(J_z\cdot H_x - J_x\cdot H_z) \tag{7.112}
$$

The time average of the peripheral force developed in the rotor is given as

$$
F_y = \mu\cdot P\cdot\tau\cdot\frac{1}{2}\cdot\mathcal{R}e\left[\int_0^{D_R}\int_{-L_R/2}^{L_R/2}(J_{z1}\cdot\tilde{H}_{x1} - J_{x1}\cdot\tilde{H}_{z1})\cdot dx\cdot dz\right] \tag{7.113}
$$

Therefore, using Equations 7.70, 7.71, 7.73 and 7.75, we get

$$
F_y = \mu\cdot P\cdot\tau\cdot\frac{1}{2}\cdot\mathcal{R}e\int_0^{D_R}\int_{-L_R/2}^{L_R/2}\sum_{Q=1}^{\infty} b_Q\cdot\sin\left(\frac{Q\pi}{D_R}\cdot z\right)\cdot\frac{\sinh(\beta_Q\cdot x)}{\sinh(\beta_Q\cdot L_R/2)}
$$

$$
\times\left[\sum_{p-odd}^{\infty}\tilde{a}_p\cdot\left(\frac{p\pi}{L_R}\cdot\frac{d^2}{\ell}\cdot\tilde{\alpha}_p\right)\cdot\sin\left(\frac{p\pi}{L_R}\cdot x\right)\cdot\frac{\sinh\{\tilde{\alpha}_p\cdot(z-D_R)\}}{\cosh(\tilde{\alpha}_p\cdot D_R)}\right.
$$

$$
\left.+\sum_{q=1}^{\infty}\tilde{b}_q\cdot\left\{\left(\frac{q\pi}{D_R}\right)^2+\ell^2\right\}\cdot\frac{d^2}{\ell}\cdot\sin\left(\frac{q\pi}{D_R}\cdot z\right)\cdot\frac{\sinh(\tilde{\beta}_q\cdot x)}{\sinh(\tilde{\beta}_q\cdot L_R/2)}\right]
$$

$$
+\sum_{P-odd}^{\infty} a_P\cdot\cos\left(\frac{P\pi}{L_R}\cdot x\right)\cdot\frac{\cosh\{\alpha_P\cdot(z-D_R)\}}{\cosh(\alpha_P\cdot D_R)}
$$

$$
\times\left[\sum_{p-odd}^{\infty}\tilde{a}_p\cdot\left\{\left(\frac{p\pi}{L_R}\right)^2+\ell^2\right\}\cdot\frac{d^2}{\ell}\cdot\cos\left(\frac{p\pi}{L_R}\cdot x\right)\cdot\frac{\cosh\{\tilde{\alpha}_p\cdot(z-D_R)\}}{\cosh(\tilde{\alpha}_p\cdot D_R)}\right.
$$

$$
\left.-\sum_{q=1}^{\infty}\tilde{b}_q\cdot\left(\frac{q\pi}{D_R}\cdot\frac{d^2}{\ell}\cdot\tilde{\beta}_q\right)\cdot\cos\left(\frac{q\pi}{D_R}\cdot z\right)\cdot\frac{\cosh(\tilde{\beta}_q\cdot x)}{\sinh(\tilde{\beta}_q\cdot L_R/2)}\right]\cdot dx\cdot dz \tag{7.114}
$$

On integrating and on replacing P by p, and Q by q, we get

$$F_y = \mu \cdot P \cdot \tau \cdot \frac{1}{2} \cdot \mathcal{R}e \sum_{p-odd}^{\infty} a_p \cdot \tilde{a}_p \cdot \left\{ \left(\frac{p\pi}{L_R} \right)^2 + \ell^2 \right\} \cdot \frac{d^2}{\ell} \cdot \frac{L_R}{2} \cdot \left\{ \frac{\alpha_p \cdot \tanh(\alpha_p \cdot D_R)}{(\alpha_p^2 - \tilde{\alpha}_P^2)} \right.$$
$$\left. + \frac{\tilde{\alpha}_P \cdot \tanh(\tilde{\alpha}_P \cdot D_R)}{(\tilde{\alpha}_P^2 - \alpha_p^2)} \right\} + \sum_{q=1}^{\infty} b_q \cdot \tilde{b}_q \cdot \left\{ \left(\frac{q\pi}{D_R} \right)^2 + \ell^2 \right\} \cdot \frac{d^2}{\ell}$$
$$\times D_R \cdot \left\{ \frac{\beta_q \cdot \coth(\tilde{\beta}_q \cdot L_R/2)}{(\beta_q^2 - \tilde{\beta}_q^2)} \right.$$
$$\left. + \frac{\tilde{\beta}_q \cdot \coth(\beta_q \cdot L_R/2)}{(\tilde{\beta}_q^2 - \beta_q^2)} \right\} + \sum_{p-odd}^{\infty} \sum_{q=1}^{\infty} \left[\tilde{a}_p \cdot b_q \cdot 2 \cdot \left\{ \frac{\sin(p\pi/2)\coth(\beta_q \cdot (L_R/2))}{(p\pi/L_R)^2 + \beta_q{}^2} \right\} \right.$$
$$\times \left(\frac{p\pi}{L_R} \cdot \frac{d^2}{\ell} \cdot \tilde{\alpha}_p \cdot \beta_q \right) - a_p \cdot \tilde{b}_q \cdot \left(\frac{q\pi}{D_R} \cdot \frac{d^2}{\ell} \cdot \tilde{\beta}_q \right)$$
$$\left. \times 2 \cdot \left\{ \frac{(p\pi/L_R) \cdot \sin(p\pi/2) \cdot \coth(\tilde{\beta}_q \cdot (L_R/2))}{(p\pi/L_R)^2 + \tilde{\beta}_q{}^2} \right\} \right] \tag{7.115}$$

7.3.5.3 Mechanical Power Developed

Thus, mechanical power developed, P_M, is

$$P_M = v \cdot F_y = (1 - s) \cdot \frac{\omega}{\ell} \cdot F_y \tag{7.116}$$

Therefore,

$$\frac{P_M}{(1 - s)} = \frac{\omega}{\ell} \cdot F_y = \frac{W_E}{s} \tag{7.117a}$$

or

$$F_y = \frac{\ell}{\omega \cdot s} \cdot W_E \tag{7.117b}$$

or

$$F_y = P \cdot \tau \cdot L_R \cdot \frac{1}{2} \cdot \ell \cdot \omega_o \cdot \mu^2 \cdot \sigma \cdot \left[\frac{|K_1|}{|\alpha|} \right]^2 \cdot \frac{1}{(\alpha + \tilde{\alpha})} \tag{7.118}$$

Now, since $v = (1 - s) \cdot (\omega/\ell)$, the mechanical power developed, P_M, is given by

$$P_M = v \cdot F_y = (1 - s) \cdot s \cdot \frac{\sigma}{2} \cdot P \cdot \tau \cdot L_R \cdot (\omega \cdot \mu)^2 \cdot \left[\frac{|K_1|}{|\alpha|}\right]^2 \cdot \frac{1}{(\alpha + \tilde{\alpha})} \tag{7.119a}$$

7.3.5.4 Rotor Input Power

The rotor power input, P_R, is given as

$$P_R = W_E + P_M = s \cdot \frac{\sigma}{2} \cdot P \cdot \tau \cdot L_R \cdot (\omega \cdot \mu)^2 \cdot \left[\frac{|K_1|}{|\alpha|}\right]^2 \cdot \frac{1}{(\alpha + \tilde{\alpha})} \tag{7.119b}$$

7.3.5.5 Slip-Power Relation

In view of Equations 7.116, 7.119b and 7.120, we get

$$\frac{P_R}{1} = \frac{W_E}{s} = \frac{P_M}{(1 - s)} \tag{7.120}$$

This equation relates the slip to the eddy current loss, mechanical power developed and the rotor input power.

PROJECT PROBLEMS

1. Discuss the distribution of electromagnetic fields in a hollow-rotor induction machine with the stator placed (a) outside the rotor and (b) inside the rotor.
2. Present a treatment for the theory of drag-cup-type induction machine with the excitation winding placed (a) outside the drag-cup rotor and (b) inside the drag-cup rotor.

References

1. Edouard, R., Etude analytique du champ propre d'une encoche, *Rev. Gin. de l'El.*, XXII, 417–424, 1927.
2. Bewley, L. V., *Two-Dimensional Fields in Electrical Engineering*, Dover, New York, pp. 81–83, 1963.
3. Bewley, L. V., *Two-Dimensional Fields in Electrical Engineering*, Dover, New York, pp. 73–81, 1963.

4. Rogowski, W., Ueber das Streufeld und denStreuinduktionskoeffizienten eines Transformators mit Schiebenwicklung und geteilten Endspulen, *V. D. I. Heft*, 71, 1–36, 1909.
5. Mukerji, S. K., Linear electromagnetic field analysis of the solid rotor induction machine, Ph.D. thesis, Department of Electrical Engineering, Indian Institute of Technology, Bombay, India, June 1967.
6. Kesavamurthy, N. and Bedford, R. E., *Fields and Circuits in Electrical Machines*, Thacker Spike & Co. (1933) Private Ltd., Calcutta, pp. 223–228, 1966.

8

Case Studies

8.1 Introduction

In this chapter some assorted examples covering slot leakage inductance of rotating electrical machines, transformer leakage inductance and theory of hysteresis machines are discussed. Modelling of fields for a potentially new type of single-phase induction motor with composite poles is also presented. This chapter also describes the electromagnetic transients in solid conducting plates due to current and voltage impact excitations.

8.2 Slot Leakage Inductance for Conductors in Open Slots

Edouard Roth in his work[1,2] expressed the vector magnetic potential due to current-carrying conductor housed in an open rectangular slot, in terms of a double Fourier series. The configuration used by Roth is illustrated in Figure 8.1.

8.2.1 Physical Configuration

Figure 8.1 shows a slot with width w and depth d. It contains a current-carrying conductor of thickness c inside this slot. The bottom and top surfaces of this conductor are located at $y = h_1$ and $y = h_2$, respectively. In view of this configuration, the evaluation of slot leakage inductance requires the following steps.

8.2.2 Current Density Distribution

The current density distribution in the slot is given by the following double Fourier series:

$$J_z = \sum_{m-odd}^{\infty} \sum_{n-odd}^{\infty} J_{m,n} \cdot \cos\left(\frac{m2\pi}{w} \cdot x\right) \cdot \cos\left(\frac{n\pi}{2d} \cdot y\right) \tag{8.1}$$

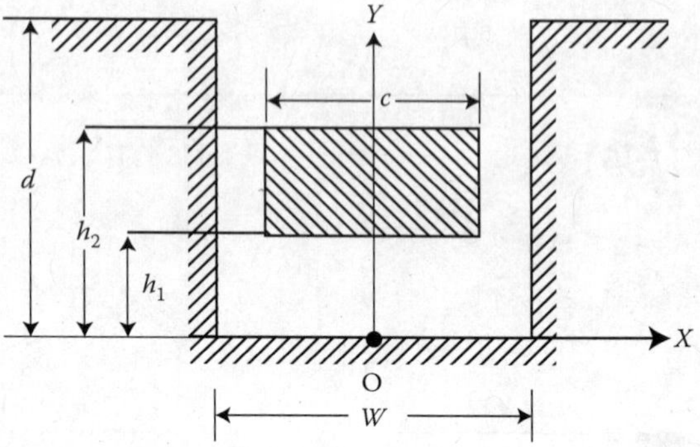

FIGURE 8.1
A current-carrying conductor in an open slot.

The Fourier coefficient $J_{m,n}$, involved in Equation 8.1, can be obtained by multiplying both sides of this equation with $\cos((p2\pi/w) \cdots x) \cdots \cos((q\pi/2d) \cdots y)$, and then integrating the resulting expression over $-w/2 < x < w/2$ and $0 < y < d$. Thus, on setting $p = m$ and $q = n$, we finally get

$$LHS = \int_{h_1}^{h_2}\int_{-C/2}^{C/2}\left[\frac{I}{c\cdot(h_2-h_1)}\right]\cdot\cos\left(\frac{p2\pi}{w}\cdot x\right)\cdot\cos\left(\frac{q\pi}{2d}\cdot y\right)dx\cdot dy$$

$$=\left[\frac{2I}{c\cdot(h_2-h_1)}\right]\cdot\frac{\sin((m\pi/w)\cdot c)}{m\pi/w}\cdot\frac{\{\sin((n\pi/2d)\cdot h_2)-\sin((n\pi/2d)\cdot h_2)\}}{n\pi/d} \tag{8.2}$$

where I indicates the current flowing in the conductor. Note that the current density $[I/(c \cdots (h_2 - h_1))]$ is zero in the slot region beyond the conductor cross-section. Similarly,

$$RHS = \sum_{m-odd}^{\infty}\sum_{n-odd}^{\infty}J_{m,n}\cdot\int_{0}^{d}\int_{-w/2}^{w/2}\cos\left(\frac{m\pi\cdot x}{w/2}\right)\cdot\cos\left(\frac{n\pi\cdot y}{2d}\right)\cdot\cos\left(\frac{p\pi\cdot x}{w/2}\right)\cdot\cos\left(\frac{q\pi\cdot y}{2d}\right)$$

$$=\frac{d\cdot w}{4}\cdot J_{m,n} \tag{8.3}$$

Thus, we finally get the current density distribution by the following expression:

$$J_{m,n}=\left[\frac{8I}{c\cdot(h_2-h_1)}\right]\cdot\frac{\sin((m\pi/w)\cdot c)}{m\pi}\cdot\frac{\{\sin((n\pi/2d)\cdot h_2)-\sin((n\pi/2d)\cdot h_2)\}}{n\pi} \tag{8.4}$$

8.2.3 Vector Magnetic Potential

Once the current density distribution is known, the vector magnetic potential in the slot region can be obtained from the relation

$$\nabla^2 A_z = -\mu_o J_z \tag{8.5}$$

The distribution of vector magnetic potential (A_z) in the slot consists of two parts. These include the particular integral (A_{z1}) and the complementary function (A_{z2}). The expression for A_{z1} is an even function of x. Thus, in view of Equation 8.1, A_{z1} can be expressed as

$$A_{z1} = \sum_{m-odd}^{\infty} \sum_{n-odd}^{\infty} A_{m,n} \cdot \cos\left(\frac{m2\pi}{w} \cdot x\right) \cdot \cos\left(\frac{n\pi}{2d} \cdot y\right) \tag{8.6}$$

The coefficient $A_{m,n}$ can be determined by substituting this expression in Equation 8.5. Thus, in view of Equation 8.1, we get

$$A_{m,n} = J_{m,n} \cdot \frac{\mu_o}{[(m2\pi/w)^2 + (n\pi/2d)^2]} \tag{8.7}$$

The complementary function A_{z2} that describes the potential distribution in the open rectangular slot can be determined in view of the flux density distribution.

8.2.4 Flux Density

In view of the relation

$$\boldsymbol{B} = \nabla \times \boldsymbol{A} \tag{8.8}$$

The components of the flux density in the slot region can be found as

$$B_x = \frac{\partial A_z}{\partial y} \tag{8.9a}$$

and

$$B_y = -\frac{\partial A_z}{\partial x} \tag{8.9b}$$

In view of Equations 8.7 and 8.9, we get

$$B_x = -\sum_{m-odd}^{\infty}\sum_{n-odd}^{\infty} A_{m,n} \cdot \frac{n\pi}{2d} \cdot \cos\left(\frac{m2\pi}{w} \cdot x\right) \cdot \sin\left(\frac{n\pi}{2d} \cdot y\right) \tag{8.10a}$$

$$B_y = \sum_{m-odd}^{\infty}\sum_{n-odd}^{\infty} A_{m,n} \cdot \frac{m2\pi}{w} \cdot \sin\left(\frac{m2\pi}{w} \cdot x\right) \cdot \cos\left(\frac{n\pi}{2d} \cdot y\right) \tag{8.10b}$$

It may be noted that these field expressions ensure that the tangential component of magnetic field on the air–iron interface is zero, that is,

$$B_x\Big|_{y=0} = 0 \tag{8.11a}$$

$$B_y\Big|_{x=\pm w/2} = 0 \tag{8.11b}$$

Roth[1,2] neglected the complementary function A_{z2} to describe the potential distribution in the open rectangular slot. He assumed that the flux lines go straight across the slot opening. Thus,

$$B_y\Big|_{y=d} = 0 \tag{8.12}$$

The complementary function A_{z2} depends on external sources. The expression for A_{z2} may contain, in general, both even and odd functions of x and the sum of these components must satisfy the boundary conditions given by Equations 8.11a and 8.11b. Since the complementary function obeys the Laplace equation, a general solution for A_{z2} may be given as

$$\begin{aligned} A_{z2} = &\sum_{p-odd}^{\infty} b_p \cdot \left(\frac{w}{p2\pi}\right) \cdot \cos\left(\frac{p2\pi}{w}x\right) \cdot \frac{\cosh((p2\pi/w)\cdot y)}{\cosh((p2\pi/w)\cdot d)} \\ &- \sum_{q-odd}^{\infty} c_q \cdot \left(\frac{w}{q\pi}\right) \cdot \sin\left(\frac{q\pi}{w}x\right) \cdot \frac{\cosh((q\pi/w)\cdot y)}{\cosh((q\pi/w)\cdot d)} \end{aligned} \tag{8.13}$$

where b_p and c_q indicate two sets of arbitrary constants.

8.2.5 Inductance

For determining an approximate value of the slot leakage inductance, the complementary function could be neglected. Consider the flux linking with a differential area of region occupied by the conductor

$$d\varphi = A_z \cdot \frac{1}{c \cdot (h_2 - h_1)} \cdot dx \cdot dy \tag{8.14}$$

The total flux linkage, therefore, is

$$\varphi = \frac{1}{c \cdot (h_2 - h_1)} \int_{0}^{h_2} \int_{-w/2}^{w/2} A_z \cdot dx \cdot dy \tag{8.15a}$$

Thus,

$$\varphi = \frac{1}{c \cdot (h_2 - h_1)} \int_{h_1}^{h_2} \int_{-c/2}^{c/2} \sum_{m-odd}^{\infty} \sum_{n-odd}^{\infty} A_{m,n} \cdot \cos\left(\frac{m2\pi}{w} \cdot x\right) \cdot \cos\left(\frac{n\pi}{2d} \cdot y\right) \cdot dx \cdot dy \tag{8.15b}$$

or

$$\varphi = \frac{1}{c \cdot (h_2 - h_1)} \sum_{m-odd}^{\infty} \sum_{n-odd}^{\infty} A_{m,n} \int_{h_1}^{h_2} \int_{-c/2}^{c/2} \cos\left(\frac{m2\pi}{w} \cdot x\right) \cdot \cos\left(\frac{n\pi}{2d} \cdot y\right) \cdot dx \cdot dy \tag{8.15c}$$

Therefore,

$$\varphi = \frac{2d \cdot w}{c \cdot (h_2 - h_1)} \sum_{m-odd}^{\infty} \sum_{n-odd}^{\infty} \left[A_{m,n} \cdot \frac{\sin((m\pi/w) \cdot c)}{m\pi} \times \left\{ \frac{\sin((n\pi/2d) \cdot h_2) - \sin((n\pi/2d) \cdot h_1)}{n\pi} \right\} \right] \tag{8.16}$$

and the leakage inductance is

$$L_\ell = \varphi / I \tag{8.17}$$

8.3 Leakage Inductance of Transformers

Rogowski has developed an expression for leakage inductance of transformer with sandwich windings.[2,3] His treatment is based on the analysis of vector magnetic potential due to the primary and secondary winding currents.

8.3.1 Physical Configuration

Figure 8.2 shows a single-phase transformer with 'sandwich winding', wherein high voltage (HV) and low voltage (LV) windings are divided into HV and LV coils. As illustrated in the figure, these coils are alternately placed on the transformer core. A coil group consists of a HV coil of width a_1. It is sandwiched between two LV coils, which are separated by two ducts, each of width g. As shown in Figure 8.2, only half of the two LV coils, each half of width $(a_2/2)$, is included in a coil group of axial length ℓ. Further, it is assumed that both HV and LV windings have the same radial thickness (i.e. c). In view of this figure, the space outside the core is divided into three regions. Region 1 is the insulation region between coil conductors and the iron core. Region 2 contains HV and LV coils alternately placed along the axial direction. Region 3 is the air space beyond coils.

8.3.2 Current Density Distribution

In region 2 the alternate placement of HV and LV coils along the axial direction results in a periodic distribution of current density with the length of a period ℓ. Neglecting interturn insulation in coils, the distribution of current density in the axial direction is shown in Figure 8.2. If no-load current is ignored, ampere-turn in both HV and LV winding will be the same. However, currents will flow in opposite directions. The current density distribution in region 2 is also shown in Figure 8.2. The Fourier series expansion for this distribution is

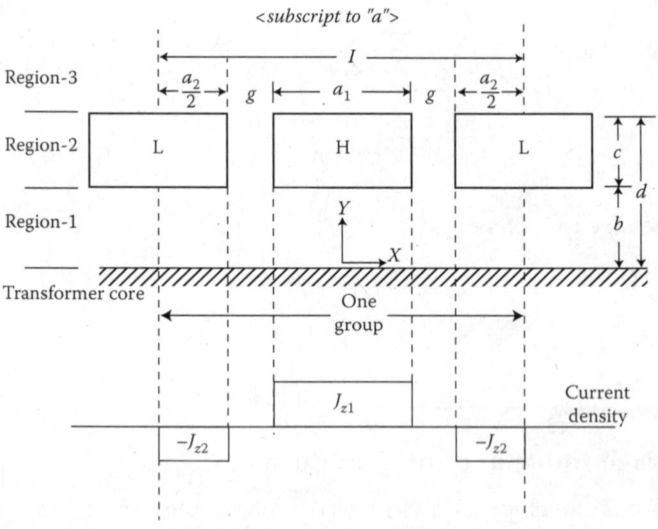

FIGURE 8.2
Sandwich winding of a single-phase shell-type transformer.

$$J_z = \sum_{m=1}^{\infty} J_m \cdot \cos\left(\frac{m2\pi}{\ell} \cdot x\right) \tag{8.18a}$$

where the Fourier coefficient is

$$J_m = \frac{2}{m\pi} \cdot \left[J_{z1} \cdot \sin\left(\frac{m\pi a_1}{\ell}\right) - J_{z2} \cdot \cos(m\pi) \cdot \sin\left(\frac{m\pi a_2}{\ell}\right) \right] \tag{8.18b}$$

8.3.3 Vector Magnetic Potential

In view of the relation

$$\nabla^2 \mathbf{A} = -\mu \mathbf{J}_o \tag{8.19}$$

We get the equation for the *z*-component of the vector potential as

$$\nabla^2 \mathbf{A}_z = -\mu_o \cdot \mathbf{J}_z \tag{8.20a}$$

or

$$\frac{\partial^2 A_z}{\partial x^2} + \frac{\partial^2 A_z}{\partial y^2} = -\mu_o \cdot \sum_{m=1}^{\infty} J_m \cdot \cos\left(\frac{m2\pi}{\ell} \cdot x\right), \quad \text{in region-2} \tag{8.20b}$$

$$= 0, \quad \text{in regions-1 and 3} \tag{8.20c}$$

A tentative solution for Equation 8.20b can be written as follows

$$A_{2z} = \mu_o \cdot \sum_{m=1}^{\infty} A_{2m}(y) \cdot \cos\left(\frac{m2\pi}{\ell} \cdot x\right) \tag{8.21a}$$

On substituting this expression in Equation 8.20b and equating coefficients of like harmonics, we get

$$\frac{d^2 A_{2m}(y)}{dy^2} - \left(\frac{m2\pi}{\ell}\right)^2 \cdot A_{2m}(y) = -J_m \tag{8.21b}$$

Therefore,

$$A_{2m}(y) = C'_{2m} \cdot e^{((m2\pi)/\ell)\cdot y} + C''_{2m} \cdot e^{-((m2\pi)/\ell)\cdot y} + \left(\frac{\ell}{m2\pi}\right)^2 \cdot J_m \tag{8.21c}$$

where C'_{2m} and C''_{2m} indicate two sets of arbitrary constants. Now, in view of Equations 8.21a and 8.21c, we have the distribution of vector potential in region 2

$$A_{2z} = \mu_o \cdot \sum_{m=1}^{\infty} \left[C'_{2m} \cdot e^{((m2\pi)/\ell)\cdot y} + C''_{2m} \cdot e^{-((m2\pi)/\ell)\cdot y} + \left(\frac{\ell}{m2\pi} \right)^2 \cdot J_m \right] \cdot \cos\left(\frac{m2\pi}{\ell} \cdot x \right) \tag{8.22}$$

The transformer core being highly permeable, the tangential component of the magnetic field intensity in region 1 will be zero at $y = 0$. Thus, let

$$A_{1z} = \mu_o \cdot \sum_{m=1}^{\infty} C_{1m} \cdot \cosh\left(\frac{m2\pi}{\ell} \cdot y \right) \cdot \cos\left(\frac{m2\pi}{\ell} \cdot x \right) \tag{8.23}$$

where C_{1m} indicates a set of arbitrary constants.

In region 3, since the field must vanish at $y = \infty$, the expression for the distribution of the magnetic vector potential A_{3z} can be given as

$$A_{3z} = \mu_o \cdot \sum_{m=1}^{\infty} C_{3m} \cdot e^{-((m2\pi)/\ell).y} \cdot \cos\left(\frac{m2\pi}{\ell} \cdot x \right) \tag{8.24}$$

where C_{3m} indicates a set of arbitrary constants.

The distribution of the vector magnetic potentials is expressed in terms of four sets of arbitrary constants $C'_{2m}, C''_{2m}, C_{1m}$ and C_{3m}. Determination of these constants is described in Section 8.3.5. Figure 8.3 shows the constant vector potential surfaces.

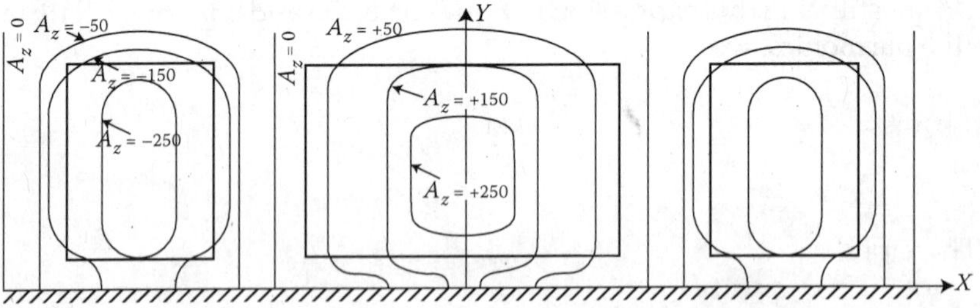

FIGURE 8.3
Constant vector potential surfaces.

8.3.4 Magnetic Flux Density

The components of magnetic flux density, **B**, for the three regions can be given as follows:

$$B_{1x} = \frac{\partial A_{1z}}{\partial y} = \mu_o \cdot \sum_{m=1}^{\infty} \frac{m2\pi}{\ell} \cdot C_{1m} \cdot \sinh\left(\frac{m2\pi}{\ell} \cdot y\right) \cdot \cos\left(\frac{m2\pi}{\ell} \cdot x\right) \tag{8.25}$$

$$B_{1y} = -\frac{\partial A_{1z}}{\partial x} = \mu_o \cdot \sum_{m=1}^{\infty} \frac{m2\pi}{\ell} \cdot C_{1m} \cdot \cosh\left(\frac{m2\pi}{\ell} \cdot y\right) \cdot \sin\left(\frac{m2\pi}{\ell} \cdot x\right) \tag{8.26}$$

$$B_{2x} = \frac{\partial A_{2z}}{\partial y} = \mu_o \cdot \sum_{m=1}^{\infty} \frac{m2\pi}{\ell} \cdot \left[C'_{2m} \cdot e^{((m2\pi)/\ell)\cdot y} - C''_{2m} \cdot e^{-((m2\pi)/\ell)\cdot y}\right] \cdot \cos\left(\frac{m2\pi}{\ell} \cdot x\right) \tag{8.27}$$

$$B_{2y} = -\frac{\partial A_{2z}}{\partial x} = \mu_o \cdot \sum_{m=1}^{\infty} \frac{m2\pi}{\ell} \cdot \left[C'_{2m} \cdot e^{((m2\pi)/\ell)\cdot y} + C''_{2m} \cdot e^{-((m2\pi)/\ell)\cdot y} + \left(\frac{\ell}{m2\pi}\right)^2 \cdot J_m\right] \times \sin\left(\frac{m2\pi}{\ell} \cdot x\right) \tag{8.28}$$

$$B_{3x} = \frac{\partial A_{3z}}{\partial y} = -\mu_o \cdot \sum_{m=1}^{\infty} \frac{m2\pi}{\ell} \cdot C_{3m} \cdot e^{-((m2\pi)/\ell)\cdot y} \cdot \cos\left(\frac{m2\pi}{\ell} \cdot x\right) \tag{8.29}$$

$$B_{3y} = -\frac{\partial A_{3z}}{\partial x} = \mu_o \cdot \sum_{m=1}^{\infty} \frac{m2\pi}{\ell} \cdot C_{3m} \cdot e^{-(m2\pi)/\ell)\cdot y} \cdot \sin\left(\frac{m2\pi}{\ell} \cdot x\right) \tag{8.30}$$

Since all the three regions are nonmagnetic regions, the vector **B** (and its components) must be continuous across adjacent regions. Therefore,

$$B_{1x}\big|_{y=b} = B_{2x}\big|_{y=b} \tag{8.31}$$

$$B_{1y}\big|_{y=b} = B_{2y}\big|_{y=b} \tag{8.32}$$

$$B_{2x}\big|_{y=d} = B_{3x}\big|_{y=d} \tag{8.33}$$

$$B_{2y}\big|_{y=d} = B_{3y}\big|_{y=d} \tag{8.34}$$

8.3.5 Arbitrary Constants

In view of the boundary conditions given by Equations 8.31 through 8.34, we get

$$C_{1m} \cdot \sinh\left(\frac{m2\pi}{\ell} \cdot b\right) = \left[C'_{2m} \cdot e^{((m2\pi)/\ell)\cdot b} - C''_{2m} \cdot e^{-((m2\pi)/\ell)\cdot b}\right] \tag{8.35}$$

$$C_{1m} \cdot \cosh\left(\frac{m2\pi}{\ell} \cdot b\right) = \left[C'_{2m} \cdot e^{((m2\pi)/\ell)\cdot b} + C''_{2m} \cdot e^{-((m2\pi)/\ell)\cdot b} + \left(\frac{\ell}{m2\pi}\right)^2 \cdot J_m\right] \tag{8.36}$$

$$\left[C'_{2m} \cdot e^{((m2\pi)/\ell)\cdot d} - C''_{2m} \cdot e^{-((m2\pi)/\ell)\cdot d}\right] = -C_{3m} \cdot e^{-((m2\pi)/\ell)\cdot d} \tag{8.37}$$

$$\left[C'_{2m} \cdot e^{((m2\pi)/\ell)\cdot d} + C''_{2m} \cdot e^{-((m2\pi)/\ell)\cdot d} + \left(\frac{\ell}{m2\pi}\right)^2 \cdot J_m\right] = C_{3m} \cdot e^{-((m2\pi)/\ell)\cdot d} \tag{8.38}$$

On adding Equations 8.37 and 8.38

$$C'_{2m} = -\frac{1}{2} \cdot \left(\frac{\ell}{m2\pi}\right)^2 \cdot J_m \cdot e^{-((m2\pi)/\ell)\cdot d} \tag{8.39}$$

Also, the addition of Equations 8.35 and 8.36 gives

$$C_{1m} \cdot e^{((m2\pi)/\ell)\cdot b} = 2 \cdot C'_{2m} \cdot e^{((m2\pi)/\ell)\cdot b} + \left(\frac{\ell}{m2\pi}\right)^2 \cdot J_m = -\left(\frac{\ell}{m2\pi}\right)^2 \cdot e^{-((m2\pi)/\ell)\cdot c} \cdot J_m + \left(\frac{\ell}{m2\pi}\right)^2 \cdot J_m$$

Thus,

$$C_{1m} = -\left(\frac{\ell}{m2\pi}\right)^2 \cdot e^{-((m2\pi)/\ell)\cdot d} \cdot J_m + \left(\frac{\ell}{m2\pi}\right)^2 \cdot e^{-((m2\pi)/\ell)\cdot b} \cdot J_m$$

or

$$C_{1m} = \left(\frac{\ell}{m2\pi}\right)^2 \cdot J_m \cdot [1 - e^{-((m2\pi)/\ell)\cdot c}] \cdot e^{-((m2\pi)/\ell)\cdot b} \tag{8.40a}$$

From Equation 8.35, we get

$$C''_{2m} \cdot e^{-((m2\pi)/\ell)\cdot b} = C'_{2m} \cdot e^{((m2\pi)/\ell)\cdot b} - C_{1m} \cdot \sinh\left(\frac{m2\pi}{\ell} \cdot b\right)$$

or

$$C''_{2m} = C'_{2m} \cdot e^{2\cdot((m2\pi)/\ell)\cdot b} - C_{1m} \cdot [e^{2\cdot((m2\pi)/\ell)\cdot b} - 1] \cdot \frac{1}{2} \tag{8.40b}$$

Substitution of Equations 8.39 and 8.40a, in Equation 8.40b, gives

$$C''_{2m} = -\left(\frac{\ell}{m2\pi}\right)^2 \cdot \frac{J_m}{2} \cdot e^{((m2\pi)/\ell)\cdot(b-c)} - \left(\frac{\ell}{m2\pi}\right)^2 \cdot J_m \cdot [1 - e^{-((m2\pi)/\ell)\cdot c}] \times \sinh\left(\frac{m2\pi}{\ell} \cdot b\right) \tag{8.41}$$

Equation 8.37 can be simplified to

$$C_{3m} = C''_{2m} - C'_{2m} \cdot e^{2\cdot((m2\pi)/\ell)\cdot d}$$

Thus, using Equations 8.39 and 8.41, we get

$$C_{3m} = -\left(\frac{\ell}{m2\pi}\right)^2 \cdot J_m \cdot \left[(1 - e^{-((m2\pi)/\ell)\cdot c}) \cdot \sinh\left(\frac{m2\pi}{\ell} \cdot b\right) - e^{((m2\pi)/\ell)\cdot b} \cdot \sinh\left(\frac{m2\pi}{\ell} \cdot c\right)\right]$$

or

$$C_{3m} = \left(\frac{\ell}{m2\pi}\right)^2 \cdot J_m \cdot \left[\sinh\left\{\frac{m2\pi}{\ell} \cdot (b + c)\right\} - \sinh\left(\frac{m2\pi}{\ell} \cdot b\right)\right] \tag{8.42}$$

Thus, all the four arbitrary constants are obtained and are given by Equations 8.39 through 8.42.

8.3.6 Leakage Inductance

Consider the flux linking with the turns in a differential area of region 2:

$$\text{For HV:} \quad d\varphi_1 = \varphi \cdot dn_1 = A_{2z} \cdot \frac{N_1}{c \cdot a_1} \cdot dx \cdot dy$$

$$\text{For LV:}\quad d\varphi_2 = \varphi \cdot dn_2 = A_{2z} \cdot \frac{N_2}{c \cdot a_2} \cdot dx \cdot dy$$

Since no-load current is ignored, the ampere-turns of LV and HV windings are equal. The current densities are

$$J_{1z} = \frac{I_1 \cdot N_1}{c \cdot a_1} = \frac{I_2 \cdot N_2}{c \cdot a_1} \tag{8.43}$$

and

$$J_{2z} = -\frac{I_1 \cdot N_1}{c \cdot a_2} = -\frac{I_2 \cdot N_2}{c \cdot a_2} \tag{8.44}$$

The total flux linkages of region 2, referred to LV side, for one coil group therefore is

$$\varphi = \frac{N_2}{c \cdot a_1} \int_b^d \int_{-a_1/2}^{a_1/2} A_{2z} \cdot dx \cdot dy - \frac{N_2}{c \cdot a_2} \int_b^d \int_{g+a_1/2}^{g+a_1/2+a_2} A_{2z} \cdot dx \cdot dy \tag{8.45}$$

Now, in view of Equation 8.22 since

$$\int_b^d A_{2z} \cdot dy = \mu_o \cdot \sum_{m=1}^{\infty} \left(\frac{\ell}{m2\pi}\right)\Big[C'_{2m} \cdot \left(e^{((m2\pi)/\ell)\cdot d} - e^{((m2\pi)/\ell)\cdot b}\right)$$
$$- C''_{2m} \cdot \left(e^{-((m2\pi)/\ell)\cdot d} - e^{-((m2\pi)/\ell)\cdot b}\right) + \left(\frac{\ell}{m2\pi}\right) \cdot J_m \cdot c\Big] \cdot \cos\left(\frac{m2\pi}{\ell} \cdot x\right) \tag{8.46a}$$

$$\int_b^d \int_{-a_1/2}^{a_1/2} A_{2z} \cdot dx \cdot dy = \mu_o \cdot \sum_{m=1}^{\infty} F_{1m} \cdot 2 \cdot \sin\left(\frac{m\pi}{\ell} \cdot a_1\right) \tag{8.46b}$$

and

$$\int_b^d \int_{g+a_1/2}^{g+a_1/2+a_2} A_{2z} \cdot dx \cdot dy = \mu_o \cdot \sum_{m=1}^{\infty} F_{1m} \cdot 2 \cdot \cos\left\{\frac{m2\pi}{\ell} \cdot \left(g + \frac{a_1 + a_2}{2}\right)\right\} \cdot \sin\left(\frac{m2\pi}{\ell} \cdot \frac{a_2}{2}\right) \tag{8.46c}$$

Therefore,

$$\varphi = \sum_{m=1}^{\infty} F_{1m} \cdot F_{2m} \tag{8.47a}$$

where

$$F_{1m} = \left(\frac{\ell}{m2\pi}\right)^2 \cdot \left[C'_{2m} \cdot \left(e^{((m2\pi)/\ell)\cdot d} - e^{((m2\pi)/\ell)\cdot b}\right) - C''_{2m} \cdot \left(e^{-((m2\pi)/\ell)\cdot d} - e^{-((m2\pi)/\ell)\cdot b}\right) + \left(\frac{\ell}{m2\pi}\right) \cdot J_m \cdot c \right] \tag{8.47b}$$

$$F_{2m} = \mu_o \cdot \left(\frac{N_2}{c}\right)^2 \left[\frac{2}{(a_1)^2} \cdot \sin\left(\frac{m\pi}{\ell} \cdot a_1\right) - \frac{2}{(a_2)^2} \cdot \cos\left\{\frac{m2\pi}{\ell}\left(g + \frac{a_1 + a_2}{2}\right)\right\} \times \sin\left(\frac{m\pi}{\ell} \cdot a_2\right) \right] \tag{8.47c}$$

while the expressions for J_m, C'_{2m} and C''_{2m} are given by Equations 8.18b, 8.39 and 8.41, respectively.

The function F_{1m} decreases at a fast rate due to the factor m^{-4}. While, the variation in F_{2m} is rather small. Therefore, one may assume

$$\varphi = \sum_{m=1}^{\infty} F_{1m} \cdot F_{2m} \cong F_{21} \cdot \sum_{m=1}^{\infty} F_{1m} \tag{8.48}$$

The infinite series can be summed up.[2] Therefore, the leakage inductance referred to LV side for one coil group is

$$\ell_L = \varphi / I_2 \tag{8.49}$$

8.4 Field Theory of Hysteresis Machines

Magnetising force, ***H***, in a medium produces magnetic induction, ***B***. The relation between ***B*** and ***H*** could be linear or nonlinear, depending on the magnetic property of the medium. If the magnetising force is time varying, so will the magnetic induction. In a medium if the variations of $\boldsymbol{H}$ and $\boldsymbol{B}$ are not simultaneous, it is said to be a hysteretic medium. Although hysteresis and nonlinearity are two distinct features, these are invariably found together in various degrees.[4]

8.4.1 Simplifying Assumptions

The simplifying assumptions, commonly made in such treatments,[5–7] are listed below:

1. End effects are neglected. This results in a two-dimensional problem with no variation of fields in the axial direction.
2. Curvature of air-gap surfaces is neglected. Thus, no special functions are needed to express the field distributions.
3. Conductivity of the hysteresis ring, σ_1, is constant.
4. At every point in the ring, the phasor $\boldsymbol{B}$ is proportional to the phasor $\boldsymbol{H}$. The constant of proportionality is a complex number.[8] Thus, each hysteresis loop is of elliptical shape with same slope of the axis and produces the same lag angle.
5. For the rotor base, the conductivity, σ_2, and the permeability, μ_2, are all constant.
6. Highly permeable stator iron ($\mu \approx \infty$), thus fields in the stator core need not be considered.
7. Smooth stator air-gap surface. Stator slot opening is neglected.
8. A current sheet sinusoidally distributed in the peripheral direction simulates armature winding with balanced three-phase currents. This current sheet is located on the stator air-gap surface and the surface currents are in axial direction. This neglects all space harmonics in the field expressions.
9. The machine is connected to a balanced three-phase ac voltage supply.
10. Only steady-state operation is considered and displacement currents are neglected.

8.4.2 Field Distributions

Figure 8.4 shows an idealised machine obtained in view of the assumptions enumerated above. In this figure, X represents axial, Y peripheral and Z radial directions. Let the stator current sheet be given as the surface current density in the axial direction:

$$K_x = K_o \cdot e^{j(\omega t - \ell y)} \tag{8.50}$$

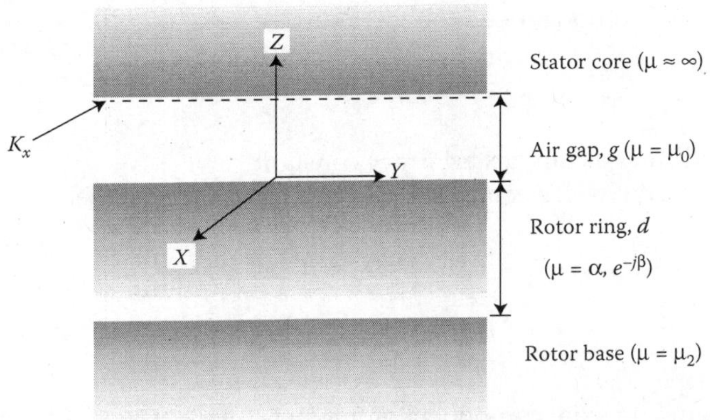

FIGURE 8.4
Idealised machine.

where the amplitude of current density is $|K_0|$, ω is the supply frequency, $\ell = \pi/\lambda$ and λ is the pole pitch. As shown in Figure 8.4, the permeability (μ) is assumed to have the following values:

$$\begin{aligned} \mu &\approx \infty \quad \text{for the stator } (z > g) \\ &= \mu_o \quad \text{for the air-gap } (g > z > 0) \\ &= \alpha \cdot e^{-j\beta} \quad \text{for the rotor ring } (0 > z > -d) \\ &= \mu_2 \quad \text{for the rotor base } (-d > z > -\infty) \end{aligned}$$

where α and β are real positive constants. Let the rotor be moving with a velocity u_y in the peripheral direction. Therefore, the slip is given by

$$s = \frac{\omega/\ell - u_y}{\omega/\ell} = 1 - \frac{\ell}{\omega} \cdot u_y \tag{8.51}$$

The stator current sheet, in a reference frame moving with the rotor, can be given as

$$K_x = K_o \cdot e^{j(s\omega t - \ell y)} \tag{8.52}$$

Thus, the peripheral component of the air-gap field on the stator surface can be given as

$$H_{oy}\Big|_{z=g} = K_o \cdot e^{j(s\omega t - \ell y)} \tag{8.53a}$$

Let this field component on the rotor surface be

$$H_{oy}\Big|_{z=0} = a \cdot e^{j(s\omega t-\ell y)} \tag{8.53b}$$

where a indicates a complex arbitrary constant.

Now, from Maxwell's equations, we have for the air-gap field

$$\nabla^2 \boldsymbol{H}_o = 0 \tag{8.54a}$$

$$\nabla \bullet \boldsymbol{H}_o = 0 \tag{8.54b}$$

There being no axial component of magnetic field, we get, from Equations 8.53a through 8.54b

$$H_{oy} = \left[K_o \cdot \frac{\sinh(\ell z)}{\sinh(\ell g)} - a \cdot \frac{\sinh\{\ell(z-g)\}}{\sinh(\ell g)} \right] \cdot e^{j(s\omega t-\ell y)} \tag{8.55a}$$

$$H_{oz} = j \cdot \left[K_o . \frac{\cosh(\ell z)}{\sinh(\ell g)} - a \cdot \frac{\cosh\{\ell(z-g)\}}{\sinh(\ell g)} \right] . e^{j(s\omega t-\ell y)} \tag{8.55b}$$

In view of Maxwell's equations, the magnetic field in the rotor region can be shown to satisfy

$$\nabla^2 \boldsymbol{H}_1 = \sigma_1 \cdot \alpha \cdot e^{-j\beta} . \frac{\partial \boldsymbol{H}_1}{\partial t} \tag{8.56a}$$

$$\nabla \bullet \boldsymbol{H}_1 = 0 \quad \text{for} \quad 0 > z > -d \tag{8.56b}$$

$$\nabla^2 \boldsymbol{H}_2 = \sigma_2 \cdot \mu_2 \cdot \frac{\partial \boldsymbol{H}_2}{\partial t} \tag{8.56c}$$

$$\nabla \bullet \boldsymbol{H}_2 = 0 \quad \text{for} \quad -d > z > -\infty \tag{8.56d}$$

Since magnetic field vanishes as z tends to minus infinity, we get on solving Equations 8.56c and 8.56d

$$H_{2y} = b \cdot e^{\gamma_2(z+d)+j(s\omega t-\ell y)} \tag{8.57a}$$

$$H_{2z} = j \cdot \frac{\ell}{\gamma_2} \cdot b \cdot e^{\gamma_2(z+d)+j(s\omega t-\ell y)} \tag{8.57b}$$

and

$$\gamma_2 = \sqrt{\ell^2 + js\omega\sigma_2\mu_2} \tag{8.57c}$$

where b indicates an arbitrary constant. Further, since H_{1y} is continuous at $z = 0$ and $-d$, we get on solving Equations 8.56a and 8.56b

$$H_{1y} = \left[a \cdot \frac{\sinh\{\gamma_1(z+d)\}}{\sinh(\gamma_1 d)} - b \cdot \frac{\sinh(\gamma_1 z)}{\sinh(\gamma_1 d)}\right] \cdot e^{j(s\omega t - \ell y)} \tag{8.58a}$$

$$H_{1z} = j \cdot \frac{\ell}{\gamma_1} \cdot \left[a \cdot \frac{\cosh\{\gamma_1(z+d)\}}{\sinh(\gamma_1 d)} - b \cdot \frac{\cosh(\gamma_1 z)}{\sinh(\gamma_1 d)}\right] \cdot e^{j(s\omega t - \ell y)} \tag{8.58b}$$

where

$$\gamma_1 = \sqrt{\ell^2 + js\omega\sigma_1\alpha \cdot e^{-j\beta}} \tag{8.58c}$$

To find the arbitrary constants, a and b, the following boundary conditions can be used:

$$(\mu_o \cdot H_{oz})\Big|_{z=0} = (\alpha \cdot e^{-j\beta} \cdot H_{1z})\Big|_{z=0} \tag{8.59a}$$

$$(\alpha \cdot e^{-j\beta} \cdot H_{1z})\Big|_{z=-d} = (\mu_2 \cdot H_{2z})\Big|_{z=-d} \tag{8.59b}$$

Therefore, in view of Equations 8.55b, 8.57b and 8.58b, we get two equations. Solution of these equations results in

$$a = \frac{K_o \cdot C_1}{C_1 \cdot \cosh(\ell g) + C_2 \cdot \sinh(\ell g)} \tag{8.60a}$$

and

$$b = \frac{K_o}{C_1 \cdot \cosh(\ell g) + C_2 \cdot \sinh(\ell g)} \tag{8.60b}$$

where

$$C_1 = \left[\frac{\gamma_1}{\gamma_2} \cdot \frac{\mu_2}{\alpha} \cdot e^{j\beta}\right] \cdot \sinh(\gamma_1 d) + \cosh(\gamma_1 d) \tag{8.61a}$$

$$C_2 = \left[\frac{\ell}{\gamma_1} \cdot \frac{\alpha}{\mu_o} \cdot e^{-j\beta}\right] \cdot \sinh(\gamma_1 d) + \left[\frac{\ell}{\gamma_2} \cdot \frac{\mu_2}{\mu_o}\right] \cdot \cosh(\gamma_1 d) \tag{8.61b}$$

8.4.3 Induction Machine Action

The idealised machine shown in Figure 8.4 is again the basis of the study of induction machine. As with the hysteresis machine, the displacement currents are again neglected and the eddy current densities in the rotor are found from Maxwell's equation

$$\boldsymbol{J} = \nabla \times \boldsymbol{H} \tag{8.62}$$

8.4.3.1 Eddy Current Density

Maxwell's equation of Equation 8.62 gives the x component of eddy current density as

$$J_x = \frac{\partial H_z}{\partial y} - \frac{\partial H_y}{\partial z} \tag{8.63a}$$

From Equation 8.63a, we can write the expressions for eddy current densities in two regions as

$$J_{1x} = -js\omega \cdot \frac{\sigma_1}{\gamma_1} \cdot \alpha \cdot e^{-j\beta} \cdot \left[a \cdot \frac{\cosh\{\gamma_1(z+d)\}}{\sinh(\gamma_1 d)} - b \cdot \frac{\cosh(\gamma_1 z)}{\sinh(\gamma_1 d)}\right] \cdot e^{j(s\omega t - \ell y)} \tag{8.63b}$$

over $0 > z > -d$
and

$$J_{2x} = -js\omega \cdot \frac{\sigma_2}{\gamma_2} \cdot \mu_2 \cdot b \cdot e^{\gamma_2(z+d) + j(s\omega t - \ell y)} \tag{8.63c}$$

over $-d > z > -\infty$

8.4.3.2 Eddy Current Loss

The eddy current loss per unit rotor volume can be written as

$$\mathcal{P}_{EL} = \frac{1}{2} \cdot \frac{1}{\sigma} \cdot J_x \cdot J_x^* \tag{8.64a}$$

Thus, in view of Equations 8.63b and 8.63c, the losses in the two regions are

$$\mathcal{P}_{1EL} = \frac{1}{2} \cdot \sigma_1 \cdot \left(\frac{s\omega\alpha}{|\gamma_1|} \right)^2 \cdot \left| a \cdot \frac{\cosh\{\gamma_1(z+d)\}}{\sinh(\gamma_1 d)} - b \cdot \frac{\cosh(\gamma_1 z)}{\sinh(\gamma_1 d)} \right|^2 \tag{8.64b}$$

over $0 > z > -d$
and

$$\mathcal{P}_{2EL} = \frac{|b|^2}{2} \cdot \sigma_2 \cdot \left(\frac{s\omega\alpha}{|\gamma_2|} \right)^2 \cdot e^{(\gamma_2 + \gamma_2^*)\cdot(z+d)} \tag{8.64c}$$

over $-d > z > -\infty$

8.4.3.3 Force

The force per unit volume on eddy currents is given as

$$\mathcal{F}_E = \mathcal{R}e\left[\frac{1}{2} \cdot \boldsymbol{J} \times \boldsymbol{B}^* \right] \tag{8.65a}$$

The y component of this force is

$$\mathcal{F}_{yE} = \mathcal{R}e\left[-\frac{1}{2} J_x B_z^* \right] \tag{8.65b}$$

Thus, in view of Equations 8.63b and 8.63c, the forces on eddy currents in two regions are

$$\mathcal{F}_{1yE} = \frac{1}{2} \cdot \sigma_1 s\omega \cdot \left(\frac{\alpha}{|\gamma_1|} \right)^2 \cdot \ell \cdot \left| a \cdot \frac{\cosh\{\gamma_1(z+d)\}}{\sinh(\gamma_1 d)} - b \cdot \frac{\cosh(\gamma_1 z)}{\sinh(\gamma_1 d)} \right|^2 \tag{8.66a}$$

over $0 > z > -d$.
And

$$\mathcal{F}_{2yE} = \frac{|b|^2}{2} \cdot \sigma_2 \cdot s\omega \cdot \left(\frac{\mu_2}{|\gamma_2|} \right)^2 \cdot \ell \cdot e^{(\gamma_2 + \gamma_2^*)\cdot(z+d)} \tag{8.66b}$$

over $-d > z > -\infty$.

8.4.3.4 Mechanical Power

The mechanical power developed per unit volume is

$$\mathcal{P}_{EM} = u_y \cdot \mathcal{F}_{yE} = (1-s) \cdot \frac{\omega}{\ell} \cdot \mathcal{F}_{yE} \tag{8.67a}$$

Thus, the mechanical power developed for the two regions in the rotor are

$$\mathcal{P}_{1EM} = \frac{1}{2} \cdot \sigma_1 \cdot (1-s) \cdot s \cdot \left(\frac{\omega\alpha}{|\gamma_1|}\right)^2 \cdot \left| a \cdot \frac{\cosh\{\gamma_1(z+d)\}}{\sinh(\gamma_1 d)} - b \cdot \frac{\cosh(\gamma_1 z)}{\sinh(\gamma_1 d)} \right|^2 \tag{8.67b}$$

over $0 > z > -d$.
And

$$\mathcal{P}_{2EM} = \frac{|b|^2}{2} \cdot \sigma_2 \cdot (1-s) \cdot s \cdot \left(\omega \cdot \frac{\cdot \mu_2}{|\gamma_2|}\right)^2 \cdot e^{(\gamma_2+\gamma_2^*)\cdot(z+d)} \tag{8.67c}$$

over $-d > z > -\infty$.

8.4.3.5 Slip-Power Relation

From Equations 8.64b, 8.64c, 8.67b and 8.67c, we get

$$\frac{\mathcal{P}_{1EM}}{\mathcal{P}_{1EL}} = \frac{\mathcal{P}_{2EM}}{\mathcal{P}_{2EL}} = \frac{1-s}{s} \tag{8.68}$$

Now, if v_1 and v_2 indicate volume of the hysteresis ring and that of the base on which the ring is mounted, respectively, then from Equation 8.68

$$\int_{v_1} \mathcal{P}_{1EM} dv + \int_{v_2} \mathcal{P}_{2EM} dv = \frac{1-s}{s} \cdot \left[\int_{v_1} \mathcal{P}_{1EL} dv + \int_{v_2} \mathcal{P}_{2EL} dv\right]$$

Therefore,

$$\frac{P_{EM}}{P_{EL}} = \frac{1-s}{s} \tag{8.69}$$

where P_{EM} indicates the total mechanical power developed due to induction machine action, and P_{EL} indicates total eddy current loss in the rotor of the machine.

The treatment so far has been concentrated on the operation of an induction machine. The influence of hysteresis is reflected through the expression for γ_1 in Equation 8.58c, where the term β indicates the lag angle due to hysteresis.

8.4.4 Hysteresis Machine Action

To investigate the occurrence of electromechanical energy conversion in the rotor[9], we can proceed with the Poynting theorem. The complex Poynting vector, $\mathbb{P}$, at any point in the hysteresis ring is given as

$$\mathbb{P} = \frac{1}{2} \cdot (\boldsymbol{E}_1 - \boldsymbol{u} \times \boldsymbol{B}_1) \times \boldsymbol{H}_1^* \tag{8.70}$$

In view of the vector identity

$$\nabla \bullet (\mathbf{A} \times \mathbf{B}) \equiv \mathbf{B} \bullet (\nabla \times \mathbf{A}) - \mathbf{A} \bullet (\nabla \times \mathbf{B}) \tag{8.71a}$$

we get

$$-\nabla \bullet \mathbb{P} \equiv -\frac{1}{2}\boldsymbol{H}_1^* \bullet [\nabla \times (\boldsymbol{E}_1 - \boldsymbol{u} \times \boldsymbol{B}_1)] + \frac{1}{2}(\boldsymbol{E}_1 - \boldsymbol{u} \times \boldsymbol{B}_1) \bullet [\nabla \times \boldsymbol{H}_1^*] \tag{8.71b}$$

Now, from Maxwell's equations for moving medium, if displacement currents are neglected

$$\nabla \times (\boldsymbol{E}_1 - \boldsymbol{u} \times \boldsymbol{B}_1) = -\frac{\partial \boldsymbol{B}_1}{\partial t} \tag{8.72a}$$

and

$$\nabla \times \boldsymbol{H}_1^* = \boldsymbol{J}_1^* \tag{8.72b}$$

Therefore, in a reference frame fixed on the stator

$$-\nabla \bullet \mathbb{P} = \frac{1}{2}\boldsymbol{H}_1^* \bullet \frac{\partial \boldsymbol{B}_1}{\partial t} + \frac{1}{2}(\boldsymbol{E}_1 - \boldsymbol{u} \times \boldsymbol{B}_1) \bullet \boldsymbol{J}_1^*$$

or

$$-\nabla \bullet \mathbb{P} = j\omega\frac{1}{2}\boldsymbol{H}_1^* \bullet \boldsymbol{B}_1 + \frac{1}{2} \bullet \frac{1}{\sigma_1}\boldsymbol{J}_1 \bullet \boldsymbol{J}_1^* + \frac{1}{2}\boldsymbol{u} \bullet (\boldsymbol{J}_1^* \times \boldsymbol{B}_1) \tag{8.73}$$

Since, in view of Equation 8.59a

$$\boldsymbol{B}_1 = \alpha \cdot e^{-j\beta}\boldsymbol{H}_1 \tag{8.74}$$

One may get from Equation 8.73

$$-\nabla \bullet \mathbb{P} = j\omega\frac{1}{2}\alpha \cdot e^{-j\beta}(H_{1y} \cdot H_{1y}^* + H_{1z} \cdot H_{1z}^*) + \frac{1}{2} \cdot \frac{1}{\sigma_1}J_{1x} \cdot J_{1x}^* - u_y \cdot \frac{1}{2}\alpha \cdot e^{-j\beta}J_{1x}^* \cdot H_{1z} \tag{8.75}$$

Equating real parts, we find in view of Equations 8.64b and 8.67b

$$\mathcal{R}e[-\nabla \bullet \mathbb{P}] \stackrel{def}{=} \mathcal{P}_{HL} + \mathcal{P}_{EL} + \mathcal{P}_{EM} + \mathcal{P}_{HM} \tag{8.76a}$$

8.4.4.1 Power Components

Equation 8.76a contains the following four components:

$$\mathcal{P}_{HL} = (s \cdot \omega) \cdot \frac{1}{2} \cdot \alpha \cdot \sin(\beta) \cdot (H_{1y} \cdot H_{1y}^* + H_{1z} \cdot H_{1z}^*) \tag{8.76b}$$

$$\mathcal{P}_{EL} = \frac{1}{2} \cdot \frac{1}{\sigma_1}J_{1x} \cdot J_{1x}^* \tag{8.76c}$$

$$\mathcal{P}_{EM} = u_y \cdot \frac{1}{2}\alpha \cdot \mathcal{R}e[-e^{-j\beta}J_{1x}^* \cdot H_{1z}] \tag{8.76d}$$

$$\mathcal{P}_{HM} = u_y \cdot \frac{1}{2}\ell \cdot \alpha \cdot \sin(\beta) \cdot (H_{1y} \cdot H_{1y}^* + H_{1z} \cdot H_{1z}^*) \tag{8.76e}$$

These four terms bear the following meaning:

1. The first term ($\mathcal{P}_{HL}$) given by Equation 8.76b represents the power density proportional to the slip frequency. For hysteresis-free media, this term is zero. Therefore, it can be considered as hysteresis loss per unit volume of the rotor ring.
2. The second term ($\mathcal{P}_{EL}$) given by Equation 8.76c represents the eddy current loss per unit rotor ring volume. This term vanishes for zero conductivity resulting in the absence of eddy currents.
3. The third term ($\mathcal{P}_{EM}$) is due to eddy currents in the rotor ring. As it is proportional to the rotor speed, it indicates the mechanical power developed due to induction machine action.

4. The fourth term ($\mathcal{P}_{HM}$) is also proportional to the rotor speed and thus indicates the mechanical power developed due to hysteresis machine action. This term vanishes for zero value of the hysteretic angle, β.

8.4.4.2 Slip-Power Relation

From Equations 8.76b and 8.76e, we get

$$\frac{\mathcal{P}_{HM}}{\mathcal{P}_{HL}} = \frac{1-s}{s} \tag{8.77}$$

The total hysteretic power is given as

$$\mathcal{P}_H = \mathcal{P}_{HL} + \mathcal{P}_{HM} = \frac{1}{2} \cdot \omega \cdot \alpha \cdot \sin(\beta) \cdot (H_{1y} \cdot H_{1y}^* + H_{1z} \cdot H_{1z}^*) \tag{8.78}$$

For a hysteresis machine with zero conductivity of the rotor, this term in view of Equations 8.57c, 8.58a, 8.58b, 8.58c, 8.60a and 8.61b becomes slip-independent for a given stator current, whereas the remaining two terms on the right-hand side (RHS) of Equation 8.76a disappear if eddy currents in the rotor ring are absent. Thus, for an ideal hysteresis machine with zero eddy currents, we have

$$\frac{P_{HM}}{P_{HL}} = \frac{1-s}{s} \tag{8.79}$$

where P_{HM} indicates the total mechanical power developed due to hysteresis machine action, and P_{HL} indicates total hysteresis loss in the rotor of the machine. Rotor power input, P_R, being the sum of power loss, P_L, and mechanical power developed, P_M, we have

$$\frac{P_R}{1} = \frac{P_L}{s} = \frac{P_M}{1-s} \tag{8.80}$$

It may be noted that Equations 8.69 and 8.79 indicate that induction machines and hysteresis machines belong to the same class of machines, both satisfying Equation 8.80.

8.4.5 Impact of Different Parameters

The presence of rotor hysteresis modifies the slip-torque characteristic of an induction machine. The presence of rotor conductivity likewise modifies the asynchronous performance of a hysteresis machine. In the case of induction machines, the effects of rotor conductivity predominates, while in the case of a hysteresis machine the predominating factor is the rotor hysteresis. However, Equation 8.80 gives the basic relation for this class of machines.[10,11]

Imagine an ideal hysteresis machine with zero rotor conductivity and zero mechanical losses due to friction and windage. For this machine, in view of Equations 8.57c and 8.58c, both γ_1 and γ_2 are slip independent. Thus, from Equations 8.61a and 8.61b, C_1 and C_2 are also slip independent. In view of Equation 8.60a, this renders the arbitrary constant a as proportional to K_o for all values of slip. Therefore, in view of Equation 8.55b, H_{oz} is also proportional to K_o for all values of slip. It results in a slip-independent flux to current ratio. If this machine is given a constant voltage supply, it draws a constant current from the source. In view of Equations 8.58a, 8.58b and 8.78, the rotor receives a constant hysteresis power from the stator across the air gap. Thus, in view of Equation 8.76e, the rotor develops slip-independent force and torque. If the load torque is less than the motor torque, the rotor will accelerate from standstill to the synchronous speed but not beyond, for the hysteresis force is a variable quantity at synchronous speed.

In a hysteretic medium, the time variation of $\boldsymbol{H}$ is not simultaneous with $\boldsymbol{B}$. The variation of $\boldsymbol{B}$ is delayed, so it lags behind $\boldsymbol{H}$ by a constant time-phase angle β. In the case of a hysteresis machine,[12] both $\boldsymbol{H}$ and $\boldsymbol{B}$ are travelling waves, moving in the peripheral direction at slip speed, $s \cdot u_y$, relative to the rotor. $\boldsymbol{B}$-wave, however, lags behind the $\boldsymbol{H}$-wave by a space-phase angle δ. In electrical radian, δ is equal to β. As the rotor speed equals synchronous speed, $\boldsymbol{H}$-wave stops moving. If the $\boldsymbol{B}$-wave also stops simultaneously, the hysteresis torque remains constant and the rotor continues to accelerate attaining super synchronous speed. Two waves are now travelling in the negative peripheral direction with $\boldsymbol{B}$-wave leading the $\boldsymbol{H}$-wave by the same space-phase angle δ. At super synchronous speed with negative value for the space-phase angle δ (time-phase angle β), both the hysteresis power $\mathcal{P}_H$ as well as the mechanical power developed $\mathcal{P}_{HM}$ are negative. However, there being no mechanical power input to the rotor, the super synchronous speed is unattainable. Therefore, at synchronous speed though $\boldsymbol{H}$-wave stops, $\boldsymbol{B}$-wave continues to move as it is pulled by the $\boldsymbol{H}$-wave into a position that corresponds to a space-phase angle δ so as to produce a torque that balances the load torque on the motor. The corresponding time-phase angle δ is now less than β. For an ideal machine at no-load, the value of δ is zero and the two waves coincide. Thus, for the operation at synchronous speed, the constant angle β in the above equations must be replaced by the variable angle δ, $-\beta \leq \delta \leq \beta$. The speed–torque curve for an ideal hysteresis machine is shown as *curve 1*, as in Figure 8.5. The curve is discontinuous at the synchronous speed, n_o. The machine develops a constant torque T_o at subsynchronous speeds; and $-T_o$ at super synchronous speeds.

If a mechanical load is placed on the ideal machine operating at synchronous speed with no load, the rotor momentarily slows down. The $\boldsymbol{B}$-wave,

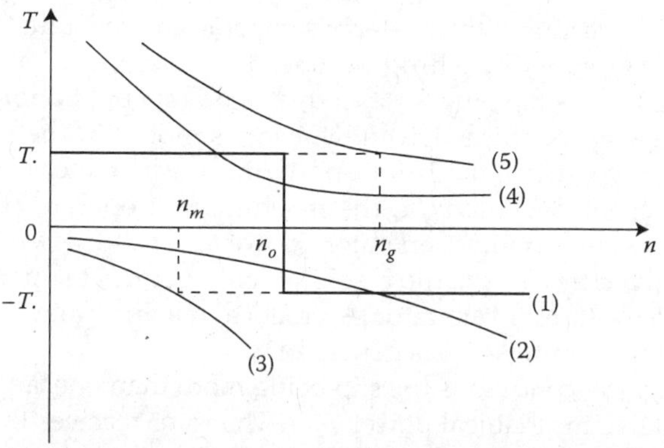

FIGURE 8.5
Asynchronous operation of an ideal hysteresis machine.

representing induced rotor poles, falls back by a space-phase angle δ. The value of δ is positive and is just sufficient to develop a torque to counterbalance the load torque. The machine is operating as a permanent magnet synchronous motor. With the increase in load, the value of δ increases until it becomes equal to β, the hysteretic angle for the rotor material. Any further increase in load causes the machine to run at subsynchronous speeds, with δ equal to β. This will reduce the mechanical power developed since a part of the constant hysteresis power input will be dissipated in the hysteresis loss. At subsynchronous speeds, stable operation of hysteresis motor is possible only for speed-dependent load torque.

Figure 8.5 illustrates the asynchronous operation of an ideal hysteresis machine driving a compressor. If the load torque at synchronous speed is less than the torque developed by the machine (*curve* 2), the machine will operate at synchronous speed, n_o, as a synchronous motor. However, if the load torque at this speed is more than the machine torque (*curve* 3), the machine will operate as a hysteresis motor at subsynchronous speed n_m. A practical machine may, however, operate as induction-cum-hysteresis motor.

Consider the ideal machine rotating at synchronous speed with zero value for δ at no-load. If a small torque is applied to the rotor in the direction of rotation, it will momentarily accelerate. As a result, the ***B***-wave will lead the ***H***-wave, causing a negative value for δ, though the machine may continue to run at synchronous speed. This results in negative value for the hysteresis power input $\mathcal{P}_H$, and the mechanical power developed, $\mathcal{P}_{HM}$. The machine will operate as a permanent magnet synchronous generator. The phase angle δ will decrease with increase in the mechanical power input, until δ becomes equal to −β. Any further increase in mechanical power input will cause the machine to rotate at super synchronous speeds with negative values of slip, *s*. The resulting hysteresis loss will be equal to the increase in mechanical power input. The phase angle δ being fixed to its minimum value, −β, the electrical power output of the hysteresis generator is constant for all negative values of slip. At super synchronous speeds stable operation of the hysteresis generator is possible only for speed-dependent applied torque. Consider an ideal hysteresis machine driven by a series motor, in the same direction as that of the revolving field. If the speed/torque characteristic of the series motor is represented by *curve 4*, the machine will operate at synchronous speed, n_o, as a synchronous generator. However, if the series motor characteristic is as given by *curve 5*, the machine will operate at super synchronous speed, n_g, as a hysteresis generator. A practical machine may, however, operate as induction-cum-hysteresis generator.

During plugging conditions the slip being more than one, an ideal machine develops negative mechanical power $\mathcal{P}_{HM}$. The rotor receives this power from the rotating system. As seen from Equation 8.77, both the constant hysteresis power $\mathcal{P}_H$ received from the stator as well as this mechanical power $\mathcal{P}_{HM}$ are dissipated in the rotor as hysteresis loss. The braking torque of an ideal hysteresis machine is constant, while that of an induction machine varies with the slip.

Unlike ordinary synchronous machines, poles are induced in the rotor of a hysteresis machine. These poles are not rigidly fixed even at the synchronous speed. Therefore, a hysteresis machine operating at synchronous speed is almost free from hunting. It needs no synchronisation or starting device. The limits of the phase angle δ for its stable operation as a synchronous machine are fixed by the hysteretic angle β of the rotor ring material ($-\beta \leq \delta \leq \beta$). The rotor air-gap surface being smooth, hysteresis machines are comparatively free from noise and vibration. Lastly, it can be mentioned that hysteresis machines, with or without rotor conductivity, can also be operated as self-excited hysteresis generators. The induction machines without rotor hysteresis cannot be operated as self-excited induction generators.

8.5 Single-Phase Induction Motors with Composite Poles

Shaded pole-type single-phase induction motors with one set of concentric stator windings and squirrel cage rotor are widely used for small power applications. With the availability of cheap capacitors, capacitor-type induction motors with two sets of distributed stator windings and cage rotor are often considered as a viable alternate to the shaded pole type single-phase induction motors. The concept of shaded pole type motor is extended leading to a capacitor-free cheap single-phase induction motor. Consider a single-phase shaded pole type induction motor. If the cage rotor is replaced by a solid (or hollow) mild steel rotor with same axial length and diameter, the machine will still work as a motor with higher starting torque and reduced efficiency. This will, however, render the machine cheaper. Further, let the shaded part of the laminated pole be replaced by solid iron, as shown in

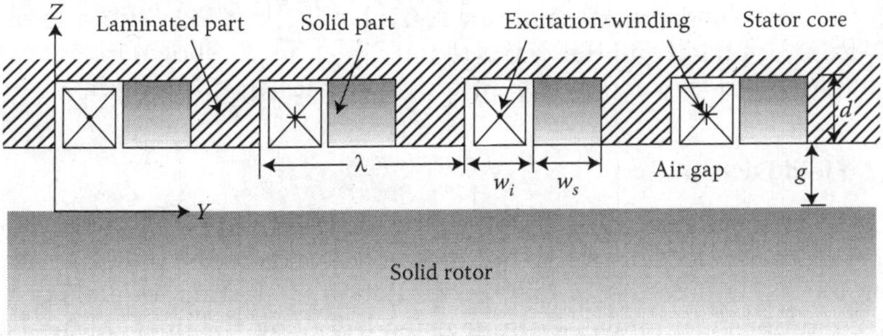

FIGURE 8.6
Idealised machine.

Figure 8.6. The role of the circulating current in the shading ring will be taken up by eddy currents induced in the solid part of the stator pole. As the laminated iron is costlier than solid iron, the stator thus modified will be cheaper. A model for the distributions of electromagnetic field for this proposed machine is presented here.

8.5.1 Simplifying Assumptions

This treatment is based on the following simplifying assumptions:

1. End effects and the curvature of air-gap surfaces are ignored.
2. Resistivity and permeability for the laminated stator iron are very large.
3. Conductivity and permeability for the solid part of the stator iron, σ_S and μ_S, are constants.
4. Conductivity and permeability for the solid rotor iron, σ_R and μ_R, are constants.
5. Coil sides of the excitation winding housed between adjacent poles are of rectangular cross-section with uniformly distributed alternating current of angular frequency ω.
6. The rotor is at stand still ($s = 0$).

8.5.2 Idealised Machine Structure

An idealised machine obtained in view of the above assumptions is shown in Figure 8.6. It shows the solid and laminated parts of stator poles, coil sides of excitation windings with currents flowing in opposite directions; laminated stator core, air gap, solid rotor and Cartesian system of space coordinates with the rotor air-gap surface on the X–Y plane. The X-axis indicates axial, the Y-axis peripheral and the Z-axis radial direction. As shown in Figure 8.6, the air gap extends over $0 \le z \le g$, where g indicates air-gap length.

8.5.3 Field Distribution

The treatment presented here is of an introductory nature. The steady-state analysis is restricted for unity slip that corresponds to the blocked rotor at starting. This will be useful for the calculation of starting torque. For running conditions, the analysis can be extended along the lines discussed in Section 7.2, that is, tooth-ripple harmonics in solid rotor induction machines, presented in Chapter 7.

The electromagnetic field varies periodically in the peripheral, that is, y-direction, with a wave length of 2λ, where λ indicates the pole pitch. Along the rotor and stator air-gap surfaces, vide Figure 8.6, let the distribution of the peripheral component of magnetic field intensity be expressed, respectively, as

$$H_y^a\Big|_{z=0} = \sum_{m=1}^{\infty}\left\{a_m \cdot \cos\left(\frac{m\pi}{\lambda}y\right) + b_m \cdot \sin\left(\frac{m\pi}{\lambda}y\right)\right\} \tag{8.81}$$

$$H_y^a\Big|_{z=g} = \sum_{m=1}^{\infty}\left\{c_m \cdot \cos\left(\frac{m\pi}{\lambda}y\right) + d_m \cdot \sin\left(\frac{m\pi}{\lambda}y\right)\right\} \tag{8.82}$$

over $-\lambda \le y \le \lambda$.

Where a_m, b_m, c_m and d_m are unknown Fourier coefficients, while $z = 0$ and $z = g$ are, respectively, the rotor and the stator air-gap surfaces. In the above and subsequent field expressions, the time-dependent factor $e^{j\omega t}$ has been suppressed.

Since air-gap field satisfies the Laplace equation, for the air-gap region the distribution of H_y^a can, therefore, be given as

$$\begin{aligned} H_y^a = -\sum_{m=1}^{\infty}\Bigg[&\left\{a_m \cdot \cos\left(\frac{m\pi}{\lambda}y\right) + b_m \cdot \sin\left(\frac{m\pi}{\lambda}y\right)\right\} \cdot \frac{\sinh(m\pi/\lambda)(z-g)}{\sinh(m\pi/\lambda \cdot g)} \\ &-\left\{c_m \cdot \cos\left(\frac{m\pi}{\lambda}y\right) + d_m \cdot \sin\left(\frac{m\pi}{\lambda}y\right)\right\} \cdot \frac{\sinh((m\pi/\lambda)\cdot z)}{\sinh((m\pi/\lambda)\cdot g)}\Bigg] \end{aligned} \tag{8.83}$$

Now, since

$$\nabla \cdot \boldsymbol{H}^a = 0 \tag{8.84}$$

and

$$H_x^a = 0 \tag{8.85}$$

we get

$$H_z^a = -\sum_{m=1}^{\infty}\left[\left\{a_m \cdot \sin\left(\frac{m\pi}{\lambda}y\right) - b_m \cdot \cos\left(\frac{m\pi}{\lambda}y\right)\right\} \cdot \frac{\cosh(m\pi/\lambda)(z-g)}{\sinh((m\pi/\lambda)\cdot g)}\right.$$
$$\left. - \left\{c_m \cdot \sin\left(\frac{m\pi}{\lambda}y\right) - d_m \cdot \cos\left(\frac{m\pi}{\lambda}y\right)\right\} \cdot \frac{\cosh((m\pi/\lambda)\cdot z)}{\sinh((m\pi/\lambda)\cdot g)}\right] \tag{8.86}$$

Therefore,

$$H_z^a\Big|_{z=0} = -\sum_{m=1}^{\infty}\left[\left\{a_m \cdot \sin\left(\frac{m\pi}{\lambda}y\right) - b_m \cdot \cos\left(\frac{m\pi}{\lambda}y\right)\right\} \cdot \coth\left(\frac{m\pi}{\lambda}\cdot g\right)\right.$$
$$\left. - \left\{c_m \cdot \sin\left(\frac{m\pi}{\lambda}y\right) - d_m \cdot \cos\left(\frac{m\pi}{\lambda}y\right)\right\} \cdot \operatorname{cosech}\left(\frac{m\pi}{\lambda}\cdot g\right)\right] \tag{8.87}$$

and

$$H_z^a\Big|_{z=g} = -\sum_{m=1}^{\infty}\left[\left\{a_m \cdot \sin\left(\frac{m\pi}{\lambda}y\right) - b_m \cdot \cos\left(\frac{m\pi}{\lambda}y\right)\right\} \cdot \operatorname{cosech}\left(\frac{m\pi}{\lambda}\cdot g\right)\right.$$
$$\left. - \left\{c_m \cdot \sin\left(\frac{m\pi}{\lambda}y\right) - d_m \cdot \cos\left(\frac{m\pi}{\lambda}y\right)\right\} \cdot \coth\left(\frac{m\pi}{\lambda}\cdot g\right)\right] \tag{8.88a}$$

The magnetic field in the solid rotor ($z \le 0$) satisfies eddy current equation and vanishes at $z = -\infty$. Its tangential component is continuous at the rotor air-gap surface, that is,

$$H_y^r\Big|_{z=0} = H_y^a\Big|_{z=0} \tag{8.88b}$$

Therefore, the expression for the peripheral component of the magnetic field intensity in the rotor region H_y^r can be given as

$$H_y^r = \sum_{m=1}^{\infty}\left\{a_m \cdot \cos\left(\frac{m\pi}{\lambda}y\right) + b_m \cdot \sin\left(\frac{m\pi}{\lambda}y\right)\right\} \cdot e^{\alpha_m \cdot z} \tag{8.89}$$

The eddy current equation for the magnetic field at unity slip is

$$\nabla^2 H_y^r = \eta_r^2 H_y^r \tag{8.90}$$

where

$$\eta_r^2 = j\omega \cdot \mu_R \cdot \sigma_R \tag{8.91a}$$

and

$$\alpha_m = \sqrt{\left(\frac{m\pi}{\lambda}\right)^2 + \eta_r^2} \tag{8.91b}$$

Now, since

$$\nabla \bullet \boldsymbol{H}^r = 0 \tag{8.92}$$

and

$$H_x^r = 0 \tag{8.93}$$

we get

$$H_z^r = \sum_{m=1}^{\infty} \frac{(m2\pi/\lambda)}{\alpha_m} \cdot \left\{ a_m \cdot \sin\left(\frac{m\pi}{\lambda} y\right) - b_m \cdot \cos\left(\frac{m\pi}{\lambda} y\right) \right\} \cdot e^{\alpha_m \cdot z} \tag{8.94}$$

Therefore,

$$H_z^r\Big|_{z=0} = \sum_{m=1}^{\infty} \frac{(m2\pi/\lambda)}{\alpha_m} \cdot \left\{ a_m \cdot \sin\left(\frac{m\pi}{\lambda} y\right) - b_m \cdot \cos\left(\frac{m\pi}{\lambda} y\right) \right\} \tag{8.95}$$

Considering the boundary condition

$$H_z^a\Big|_{z=0} = \frac{\mu_R}{\mu_o} \cdot H_z^r\Big|_{z=0} \tag{8.96}$$

we get, from Equations 8.87 and 8.94,

$$\begin{aligned} -\sum_{m=1}^{\infty} \Bigg[& \left\{ a_m \cdot \sin\left(\frac{m\pi}{\lambda} y\right) - b_m \cdot \cos\left(\frac{m\pi}{\lambda} y\right) \right\} \cdot \coth\left(\frac{m\pi}{\lambda} \cdot g\right) \\ & - \left\{ c_m \cdot \sin\left(\frac{m\pi}{\lambda} y\right) - d_m \cdot \cos\left(\frac{m\pi}{\lambda} y\right) \right\} \cdot \operatorname{cosech}\left(\frac{m\pi}{\lambda} \cdot g\right) \Bigg] \\ = \sum_{m=1}^{\infty} & \frac{(m\pi/\lambda)}{\alpha_m} \cdot \frac{\mu_R}{\mu_o} \cdot \left\{ a_m \cdot \sin\left(\frac{m\pi}{\lambda} y\right) - b_m \cdot \cos\left(\frac{m\pi}{\lambda} y\right) \right\} \end{aligned} \tag{8.97a}$$

On equating coefficients of $\sin((m\pi/\lambda)y)$ on both sides, we get

$$c_m = a_m \cdot \left\{ \cosh\left(\frac{m\pi}{\lambda}\cdot g\right) + \frac{(m\pi/\lambda)}{\alpha_m}\cdot\frac{\mu_R}{\mu_o}\cdot\sinh\left(\frac{m\pi}{\lambda}\cdot g\right)\right\} \overset{def}{=} a_m \cdot f_m \tag{8.97b}$$

and on equating coefficients of $\cos((m\pi/\lambda)y)$ on both sides, we get

$$d_m = b_m \cdot \left\{ \cosh\left(\frac{m\pi}{\lambda}\cdot g\right) + \frac{(m\pi/\lambda)}{\alpha_m}\cdot\frac{\mu_R}{\mu_o}\cdot\sinh\left(\frac{m\pi}{\lambda}\cdot g\right)\right\} \overset{def}{=} b_m \cdot f_m \tag{8.97c}$$

where

$$f_m \overset{def}{=} \left\{ \cosh\left(\frac{m\pi}{\lambda}\cdot g\right) + \frac{(m\pi/\lambda)}{\alpha_m}\cdot\frac{\mu_R}{\mu_o}\cdot\sinh\left(\frac{m\pi}{\lambda}\cdot g\right)\right\} \tag{8.97d}$$

for $m = 1, 2, 3, \ldots, \infty$.

Therefore, Equations 8.82 and 8.88a can be rewritten as

$$H_y^a\Big|_{z=g} = \sum_{m=1}^{\infty}\left\{ a_m \cdot \cos\left(\frac{m\pi}{\lambda}y\right) + b_m \cdot \sin\left(\frac{m\pi}{\lambda}y\right)\right\}\cdot f_m \tag{8.98a}$$

and

$$H_z^a\Big|_{z=g} = \sum_{m=1}^{\infty}\left\{ a_m \cdot \sin\left(\frac{m\pi}{\lambda}y\right) - b_m \cdot \cos\left(\frac{m\pi}{\lambda}y\right)\right\} g_m \tag{8.98b}$$

Where,

$$g_m \overset{def}{=} \left\{ f_m \cdot \coth\left(\frac{m\pi}{\lambda}\cdot g\right) - \operatorname{cosech}\left(\frac{m\pi}{\lambda}\cdot g\right)\right\} \tag{8.98c}$$

The interpolar region is occupied by current-carrying conductors with total ampere-turn ι. Let the peripheral width and radial depth of this region be w_i and d, respectively, vide Figure 8.6. Neglecting insulation space, the density of excitation current can be given as

$$J_x = \frac{\iota}{w_i \cdot d} \tag{8.99}$$

The vector magnetic potential defined for this region satisfies the following equation:

$$\nabla^2 A_x^i = \frac{\partial^2 A_x^i}{\partial y^2} + \frac{\partial^2 A_x^i}{\partial z^2} = -\mu_o \cdot J_x \tag{8.100a}$$

while

$$B_y^i = \mu_o \cdot H_y^i = \frac{\partial A_x^i}{\partial z} \tag{8.100b}$$

and

$$B_z^i = \mu_o \cdot H_z^i = -\frac{\partial A_x^i}{\partial y} \tag{8.100c}$$

The components of magnetic field intensity in this region satisfy the following boundary conditions:

$$H_y^i\Big|_{z=g+d} = 0 \quad \text{over} \quad 0 \le y \le w_i \tag{8.101a}$$

$$H_z^i\Big|_{y=0} = 0 \quad \text{over} \quad g \le z \le (g+d) \tag{8.101b}$$

Let the particular integral for Equation 8.100a be given as

$$A_x' = -y^2 \cdot \frac{1}{2}\mu_o \cdot J_x \tag{8.102a}$$

Thus,

$$H_y' = 0 \tag{8.102b}$$

$$H_z' = y \cdot J_x \tag{8.102c}$$

The complementary function for Equation 8.100a can be given as

$$A_x'' = -\mu_o \cdot \sum_{p=1}^{\infty} c_p' \cdot \cos\left(\frac{p\pi}{w_i} \cdot y\right) \cdot \frac{\cosh\{(p\pi/w_i) \cdot (z-d-g)\}}{\sinh(p\pi/w_i \cdot d)}$$

$$-\mu_o \cdot \sum_{q=1}^{\infty} d_q' \cdot \cos\left\{\frac{q\pi}{d} \cdot (z-d-g)\right\} \cdot \frac{\cosh((q\pi/d) \cdot y)}{\sinh((q\pi/d) \cdot w_i)} \tag{8.103a}$$

Thus,

$$H_y'' = -\sum_{p=1}^{\infty} c_p' \cdot \frac{p\pi}{w_i} \cdot \cos\left(\frac{p\pi}{w_i} \cdot y\right) \cdot \frac{\sinh\{(p\pi/w_i) \cdot (z-d-g)\}}{\sinh((p\pi/w_i) \cdot d)}$$

$$+\sum_{q=1}^{\infty} d_q' \cdot \frac{q\pi}{d} \cdot \sin\left\{\frac{q\pi}{d} \cdot (z-d-g)\right\} \cdot \frac{\cosh((q\pi/d) \cdot y)}{\sinh((q\pi/d) \cdot w_i)} \tag{8.103b}$$

and

$$H_z'' = -\sum_{p=1}^{\infty} c_p' \cdot \frac{p\pi}{w_i} \cdot \sin\left(\frac{p\pi}{w_i} \cdot y\right) \cdot \frac{\cosh\{(p\pi/w_i) \cdot (z-d-g)\}}{\sinh((p\pi/w_i) \cdot d)}$$

$$+\sum_{q=1}^{\infty} d_q' \cdot \frac{q\pi}{d} \cdot \cos\left\{\frac{q\pi}{d} \cdot (z-d-g)\right\} \cdot \frac{\sinh\left((q\pi/d) \cdot y\right)}{\sinh\left((q\pi/d) \cdot w_i\right)} \tag{8.103c}$$

where c_p' and d_q' indicate two sets of arbitrary constants.

Therefore, components of magnetic field intensity in this region are

$$H_y^i = H_y' + H_y'' = -\sum_{p=1}^{\infty} c_p' \cdot \frac{p\pi}{w_i} \cdot \cos\left(\frac{p\pi}{w_i} \cdot y\right) \cdot \frac{\sinh\{(p\pi/w_i) \cdot (z-d-g)\}}{\sinh((p\pi/w_i) \cdot d)}$$

$$+\sum_{q=1}^{\infty} d_q' \cdot \frac{q\pi}{d} \cdot \sin\left\{\frac{q\pi}{d} \cdot (z-d-g)\right\} \cdot \frac{\cosh((q\pi/d) \cdot y)}{\sinh((q\pi/d) \cdot w_i)} \tag{8.104a}$$

$$H_z^i = H_z' + H_z'' = y \cdot J_x - \sum_{p=1}^{\infty} c_p' \cdot \frac{p\pi}{w_i} \cdot \sin\left(\frac{p\pi}{w_i} \cdot y\right) \cdot \frac{\cosh\{(p\pi/w_i) \cdot (z-d-g)\}}{\sinh((p\pi/w_i) \cdot d)}$$

$$+\sum_{q=1}^{\infty} d_q' \cdot \frac{q\pi}{d} \cdot \cos\left\{\frac{q\pi}{d} \cdot (z-d-g)\right\} \cdot \frac{\sinh((q\pi/d) \cdot y)}{\sinh((q\pi/d) \cdot w_i)} \tag{8.104b}$$

Field expressions on the remaining two boundaries are

$$H_y^i\Big|_{z=g} = \sum_{p=1}^{\infty} c_p' \cdot \frac{p\pi}{w_i} \cdot \cos\left(\frac{p\pi}{w_i} \cdot y\right) \tag{8.105a}$$

$$H_z^i\Big|_{y=w_i} = w_i \cdot J_x + \sum_{q=1}^{\infty} d_q' \cdot \frac{q\pi}{d} \cdot \cos\frac{q\pi}{d} \cdot (z-d-g) \tag{8.105b}$$

$$\begin{aligned} H_y^i\Big|_{y=w_i} &= -\sum_{p=1}^{\infty} c_p' \cdot \frac{p\pi}{w_i} \cdot \cos(p\pi) \cdot \frac{\sinh\{(p\pi/w_i)\cdot(z-d-g)\}}{\sinh((p\pi/w_i)\cdot d)} \\ &\quad + \sum_{q=1}^{\infty} d_q' \cdot \frac{q\pi}{d} \cdot \sin\left\{\frac{q\pi}{d} \cdot (z-d-g)\right\} \cdot \coth\left(\frac{q\pi}{d} \cdot w_i\right) \end{aligned} \tag{8.105c}$$

$$\begin{aligned} H_z^i\Big|_{z=g} &= y \cdot J_x - \sum_{p=1}^{\infty} c_p' \cdot \frac{p\pi}{w_i} \cdot \sin\left(\frac{p\pi}{w_i} \cdot y\right) \cdot \coth\left(\frac{p\pi}{w_i} \cdot d\right) \\ &\quad + \sum_{q=1}^{\infty} d_q' \cdot \frac{q\pi}{d} \cdot \cos(q\pi) \cdot \frac{\sinh\big((q\pi/d)\cdot y\big)}{\sinh\big((q\pi/d)\cdot w_i\big)} \end{aligned} \tag{8.105d}$$

Consider the region occupied by the solid part of the pole, that is, for $w_i \le y \le (w_i + w_s)$ and $g \le z \le (g + d)$. Let the peripheral width and radial depth of this region be w_s and d, respectively, vide Figure 8.6.

In view of high value for the permeability of laminated iron, with reference to Figure 8.6, the following boundary conditions are assumed for field components in this region:

$$H_y^s\Big|_{z=g+d} = 0 \quad \text{over} \quad w_i \le y \le (w_i + w_s) \tag{8.106a}$$

$$H_z^s\Big|_{y=w_i+w_s} = 0 \quad \text{over} \quad g \le z \le (g + d) \tag{8.106b}$$

In this region, the components of magnetic field intensity satisfy eddy current equation, that is,

$$\nabla^2 H_y^s = \eta_s^2 H_y^s \tag{8.107a}$$

where

$$\eta_s^2 = j\omega \cdot \mu_S \cdot \sigma_S \tag{8.107b}$$

Thus, the expression for the peripheral component of magnetic field intensity in the solid part of the polar region, H_y^s, can be given as follows:

$$H_y^s = -\sum_{k-odd}^{\infty} c_k'' \cdot \sin\frac{k\pi}{2w_s} \cdot (y - w_i) \cdot \frac{\sinh\{\beta_k .(z - g - d)\}}{(k\pi/2w_s) \cdot \sinh(\beta_k \cdot d)}$$

$$+\sum_{l-odd}^{\infty} d_l'' \cdot \cos\frac{l\pi}{2d} \cdot (z - g) \cdot \frac{\cosh\{\gamma_l .(y - w_i - w_s)\}}{\gamma_l \cdot \sinh(\gamma_l \cdot w_s)} \tag{8.108a}$$

where c_k'' and d_l'' indicate two sets of arbitrary constants:

$$\beta_k = \sqrt{\left(\frac{k\pi}{2w_s}\right)^2 + \eta_s^2} \tag{8.108b}$$

and

$$\gamma_l = \sqrt{\left(\frac{l\pi}{2d}\right)^2 + \eta_s^2} \tag{8.108c}$$

This expression for H_y^s satisfies the Equation 8.106a identically. Now, since

$$\nabla \bullet \boldsymbol{H}^s = 0 \tag{8.109}$$

and

$$H_x^s = 0, \tag{8.110}$$

we get

$$H_z^s = \sum_{k-odd}^{\infty} c_k'' \cdot \cos\left\{\frac{k\pi}{2w_s} \cdot (y - w_i)\right\} \cdot \frac{\cosh\{\beta_k \cdot (z - g - d)\}}{\beta_k \cdot \sinh(\beta_k \cdot d)}$$
$$-\sum_{l-odd}^{\infty} d_l'' \cdot \sin\left\{\frac{l\pi}{2d} \cdot (z - g)\right\} \cdot \frac{\sinh\{\gamma_l \cdot (y - w_i - w_s)\}}{(l\pi/2d) \cdot \sinh(\gamma_l \cdot w_s)} \tag{8.111}$$

This expression for H_z^s satisfies the Equation 8.106b identically.

On the boundary surfaces, the expressions for this field are as follows:

$$H_y^s\Big|_{y=w_i} = \sum_{l-odd}^{\infty} d_l'' \cdot \cos\left\{\frac{l\pi}{2d}\cdot(z-g)\right\}\cdot\frac{\coth(\gamma_l\cdot w_s)}{\gamma_l} \tag{8.112a}$$

$$H_y^s\Big|_{z=g} = \sum_{k-odd}^{\infty} c_k'' \cdot \sin\left\{\frac{k\pi}{2w_s}\cdot(y-w_i)\right\}\cdot\frac{2w_s}{k\pi} + \sum_{l-odd}^{\infty} d_l'' \cdot \frac{\cosh\{\gamma_l(y-w_i-w_s)\}}{\gamma_l\cdot\sinh(\gamma_l\cdot w_s)} \tag{8.112b}$$

$$H_z^s\Big|_{y=w_i} = \sum_{k-odd}^{\infty} c_k'' \cdot \frac{\cosh\{\beta_k(z-g-d)\}}{\beta_k\cdot\sinh(\beta_k\cdot d)} + \sum_{l-odd}^{\infty} d_l'' \cdot \sin\left\{\frac{l\pi}{2d}\cdot(z-g)\right\}\cdot\frac{2d}{l\pi} \tag{8.112c}$$

and

$$H_z^s\Big|_{z=g} = \sum_{k-odd}^{\infty} c_k'' \cdot \cos\left\{\frac{k\pi}{2w_s}\cdot(y-w_i)\right\}\cdot\frac{\coth(\beta_k\cdot d)}{\beta_k} \tag{8.112d}$$

Field expressions given above involve a number or arbitrary constants. These can be evaluated by numerically solving a number of linear algebraic simultaneous equations developed using various boundary conditions shown below

At $y = w_i$:

$$H_z^i\Big|_{y=w_i} = H_z^s\Big|_{y=w_i} \quad \text{over} \quad g \le z \le (g+d) \tag{8.113a}$$

$$\mu_o \cdot H_y^i\Big|_{y=w_i} = \mu_s \cdot H_y^s\Big|_{y=w_i} \quad \text{over} \quad g \le z \le (g+d) \tag{8.113b}$$

At $z = g$:

$$H_z^i\Big|_{z=g} = H_z^a\Big|_{z=g} \quad \text{over} \quad 0 \le y \le w_i \tag{8.113c}$$

$$\mu_s \cdot H_z^s\Big|_{z=g} = \mu_o \cdot H_z^a\Big|_{z=g} \quad \text{over} \quad w_i \le y \le (w_i + w_s) \tag{8.113d}$$

$$H_y^a\Big|_{z=g} = \begin{cases} H_y^i\Big|_{z=g}, & \text{over} \quad 0 \le y \le w_i \\ H_y^s\Big|_{z=g}, & \text{over} \quad w_i \le y \le (w_i + w_s) \\ 0, & \text{over} \quad (w_i + w_s) \le y \le \lambda \end{cases} \tag{8.113e}$$

Consider Equations 8.113a, 8.105b and 8.112c. Multiply both sides of Equation 8.113a by $\sin\{(L\pi/2d)\cdot(z-g)\}$, L being any odd integer. Then integrate over $g \le z \le (g+d)$. Now, if L is replaced by l, we get d_l'' in terms of J_x, d_q' and c_k''. The last two are in the summed up forms, while J_x is given by Equation 8.99.

Next, consider Equations 8.113b, 8.105c and 8.112a. Now, multiply both sides of Equation 8.113b by $\sin\{(Q\pi/d)\cdot(z-d-g)\}$, for any integer value of Q. Then integrate over $g \le z \le (g+d)$. Now, if Q is replaced by q, we get d_q' in terms of c_p' and d_l''; both in summed up forms.

Consider Equations 8.113c, 8.105d and 8.98b. Multiply both sides of Equation 8.113c by $\sin((P\pi/w_i)\cdot y)$, for any integer value of P. Then integrate over $0 \le y \le w_i$. Now, if P is replaced by p, we get c_p' in terms of J_x, d_q', a_m and b_m; the last three are in summed up forms.

Now, let us consider Equations 8.113d, 8.112d and 8.98b. Multiply both sides of Equation 8.113d by $\cos\{(K\pi/2w_s)\cdot(y-w_i)\}$, for any odd integer value of K. Then integrate over $w_i \le y \le (w_i+w_s)$. Now, if K is replaced by k, we get c_k'' in terms of a_m and b_m; both in summed up forms. Lastly, consider Equations 8.113e, 8.98a, 8.105a and 8.112b. This boundary condition yields two different equations as indicated below.

Let us multiply both sides of the Equations 8.113e by $\cos((M\pi/\lambda)y)$, for any integer value of M. Then integrate over $0 \le y \le \lambda$. Thus,

$$\int_0^{\lambda}\left[H_y^a\Big|_{z=g}\cdot\cos\left(\frac{M\pi}{\lambda}y\right)\right]\cdot dy = \int_0^{w_i}\left[H_y^i\Big|_{z=g}\cdot\cos\left(\frac{M\pi}{\lambda}y\right)\right]\cdot dy + \int_{w_i}^{(w_i+w_s)}\left[H_y^s\Big|_{z=g}\cdot\cos\left(\frac{M\pi}{\lambda}y\right)\right]\cdot dy \tag{8.113f}$$

To generate the second equation, multiply both sides of Equation 8.113e by $\sin((M\pi/\lambda)y)$, for any integer value of M. Then integrate over $0 \le y \le \lambda$. Thus,

$$\int_0^{\lambda}\left[H_y^a\Big|_{z=g}\cdot\sin\left(\frac{M\pi}{\lambda}y\right)\right]\cdot dy = \int_0^{w_i}\left[H_y^i\Big|_{z=g}\cdot\sin\left(\frac{M\pi}{\lambda}y\right)\right]\cdot dy + \int_{w_i}^{(w_i+w_s)}\left[H_y^s\Big|_{z=g}\cdot\sin\left(\frac{M\pi}{\lambda}y\right)\right]\cdot dy \tag{8.113g}$$

In these equations, after performing the integrations as indicated above, if we replace M by m, the first equation gives a_m in terms of c''_p, c''_k and d''_l, all the three in summed up forms. While, the second equation gives b_m, again in terms of c'_p, c''_k and d''_l, all the three in summed up forms.

Numerical solution of these equations gives values for a_m, b_m, c'_p, d'_q, c''_k and d''_l. By back substitution in Equations 8.97b and 8.97c, we find the values for c_m and d_m.

Having thus determined the field distributions, eddy currents in the solid rotor and the torque developed can be found.

8.6 Transient Fields in Plates due to Type 2 Impact Excitations

Nonperiodic eddy currents are induced in conductors subjected to electromagnetic transients. These transients can be broadly classified into the following two types.

Type 1 transients: These are defined as those caused by sudden interruption of dc current or due to sudden application of a constant voltage source to an electromagnetic system. The former is called *current impact excitation* and the latter is termed as *voltage impact excitation*. Consider a passive electromagnetic system connected to a dc current source. If its terminals are suddenly short-circuited, the terminal voltage instantly becomes zero. Accidental short-circuiting of input terminals of live electromagnetic loads driven by constant voltage is often encountered in practice. Although the voltage source is quickly isolated by a high-speed circuit-breaker, the short-circuit fault may persist for some time.

Type 2 transients: The sudden change of terminal voltage is defined as type 2 *voltage impact excitation*. If a relaxed network is suddenly connected to a constant current source, a sudden change in the input current occurs. The resulting electromagnetic transient is said to be due to type 2 *current impact excitation*.

The electromagnetic transient in large conducting plates due to type 1 impact excitations is described in the literature.[13,14] In the present treatment, Maxwell's equations are solved for transient fields in a large conducting plate with constant values for permeability (μ), permittivity (ε) and conductivity (σ). Transient fields considered are those produced due to type 2 impact excitations.

Consider Figure 8.7 showing a large conducting plate of thickness, W, carrying uniform current sheets of density, K_y, on its surfaces. These current sheets simulate the current-carrying excitation winding. In view of the symmetry, only y-component of electric field and only z-component of magnetic field exist. Further, both transient fields vanish as t tends to infinity. These fields satisfy Maxwell's equations in one dimension.

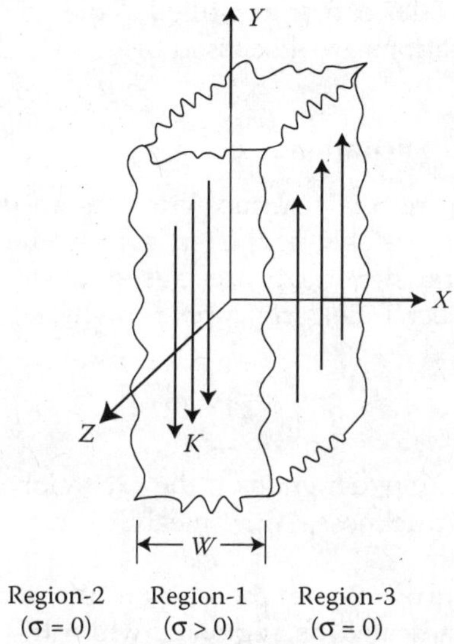

FIGURE 8.7
Large conducting plate with surface current sheets. (Courtesy of The Electromagnetics Academy.)

$$\frac{\partial E_y}{\partial x} = -\mu \cdot \frac{\partial H_z}{\partial t} \tag{8.114}$$

$$-\frac{\partial H_z}{\partial x} = \sigma \cdot E_y + \varepsilon \cdot \frac{\partial E_y}{\partial t} \tag{8.115}$$

Therefore, electromagnetic fields obey the following equations:

$$\frac{\partial^2 E_y}{\partial x^2} = \mu\sigma \cdot \frac{\partial E_y}{\partial t} + \mu\varepsilon \cdot \frac{\partial^2 E_y}{\partial t^2} \tag{8.116}$$

$$\frac{\partial^2 H_z}{\partial x^2} = \mu\sigma \cdot \frac{\partial H_z}{\partial t} + \mu\varepsilon \cdot \frac{\partial^2 H_z}{\partial t^2} \tag{8.117}$$

For free space

$$\sigma = 0 \tag{8.118a}$$

$$\mu = \mu_o \tag{8.118b}$$

$$\varepsilon = \varepsilon_o \tag{8.118c}$$

Equations 8.116 and 8.117 can be identified as damped wave equations. The solutions of these equations are discussed below.

8.6.1 Current Impact Excitation

With reference to Figure 8.7, if surface currents on the surfaces located at $x = \pm W/2$ are instantly established from zero value to a constant value $K_y = \pm K_o$, say at $t = 0$, transient fields are caused due to current impact excitation. At this instant, electric field $\pm E_o$ appears on the surfaces $x = \pm W/2$, where

$$E_o = \rho_s \cdot K_o \tag{8.119}$$

The surface resistivity ρ_s depends on the excitation winding resistance. It is independent of the thickness, W, permeability, μ, permittivity, ε_o and conductivity, σ, of the plate.

As shown in Figure 8.7, let the region be occupied by the plate, $-W/2 < x < W/2$, be indicated as region 1, while the regions $x > W/2$ and $x < -W/2$ be indicated as regions 2 and –3, respectively. In view of the symmetry, it will be sufficient to consider the field distributions in regions 1 and –2 only. At the boundary between these two regions, that is, at $x = W/2$, we have, for $t \geq 0$:

$$H_{1z}\Big|_{x=W/2} = H_{2z}\Big|_{x=W/2} + K_o \tag{8.120a}$$

and

$$E_{1y}\Big|_{x=W/2} = E_{2y}\Big|_{x=W/2} \tag{8.120b}$$

where, suffix 1 indicates fields in region 1 and suffix 2 indicates fields in region 2.

Before the onset of transient, that is, for $t < 0$, the fields are

$$H_{1z}\Big|_{t<0} = H_{2z}\Big|_{t<0} = 0 \tag{8.121a}$$

$$E_{1y}\Big|_{t<0} = E_{2y}\Big|_{t<0} = 0 \tag{8.121b}$$

In the conducting region 1, the magnetic field does not appear instantly; therefore, boundary conditions for $t = 0$ are

$$H_{1z}\Big|_{x=W/2} = 0 \tag{8.122a}$$

$$H_{2z}\big|_{x=W/2} = -K_o \tag{8.122b}$$

$$E_{1y}\big|_{x=W/2} = E_{2y}\big|_{x=W/2} = E_o \tag{8.122c}$$

After the transient is over, the steady-state values will be

$$H_{1z}\big|_{t\to\infty} = K_o \tag{8.123a}$$

$$H_{2z}\big|_{t\to\infty} = 0 \tag{8.123b}$$

$$E_{1y}\big|_{t\to\infty} = 0 \tag{8.123c}$$

$$E_{2y}\big|_{t\to\infty} = 0 \tag{8.123d}$$

Consider the electromagnetic fields inside the plate. These fields must satisfy the initial and final conditions given by Equations 8.121a, 8.121b, 8.122a, 8.122c, 8.123a and 8.123c. Because of geometrical symmetry, in the conducting plate, the magnetic field is an even function and the electric field is an odd function of x. Therefore, in view of Equations 8.114 and 8.115, solutions for Equations 8.116 and 8.117, for $t \geq 0$, are tentatively taken as

$$H_{1z} = K_o - K_o \cdot e^{-t/\tau} - K_o \cdot \left(\rho_s + \frac{\mu}{\tau} \cdot \frac{W}{4}\right) \cdot \left(\frac{\theta}{\mu} \cdot \frac{2}{W}\right) \cdot \frac{\cos(\theta \cdot (2x/W))}{\sin(\theta)} \cdot t \cdot e^{-t/2\tau}$$
$$+ \sum_{m=1}^{\infty} \frac{c_m}{(\beta_m - \alpha_m)} \cdot \frac{m\pi}{\mu} \cdot \frac{2}{W} \cdot \cos\left(m\pi \cdot \frac{2x}{W}\right) \cdot \{e^{-\alpha_m t} - e^{-\beta_m t}\} \tag{8.124a}$$

and

$$E_{1y} = -K_o \cdot \frac{\mu}{\tau} \cdot \frac{W}{4} \cdot \frac{2x}{W} \cdot e^{-t/\tau} + K_o \cdot \left(\rho_s + \frac{\mu}{\tau} \cdot \frac{W}{4}\right) \cdot \frac{\sin(\theta \cdot (2x/W))}{\sin(\theta)}$$
$$\times \left\{1 - \frac{t}{2\tau}\right\} \cdot e^{-t/2\tau} + \sum_{m=1}^{\infty} \frac{c_m}{(\beta_m - \alpha_m)} \cdot \sin\left(m\pi \cdot \frac{2x}{W}\right) \cdot \{\alpha_m \cdot e^{-\alpha_m t} - \beta_m \cdot e^{-\beta_m t}\}$$
$$\tag{8.124b}$$

where c_m indicates a set of arbitrary constants, while

$$\tau = \varepsilon/\sigma, \quad \text{the relaxation time for the plate} \tag{8.125a}$$

$$\theta = \frac{W}{4} \cdot \sqrt{\frac{\mu\sigma}{\tau}}, \quad \text{the plate parameter} \tag{8.125b}$$

$$\alpha_m, \beta_m = \frac{1}{2\mu\varepsilon} \cdot \left[\mu\sigma \pm \sqrt{(\mu\sigma)^2 - 4\mu\varepsilon \cdot \left(m\pi \frac{2}{W} \right)^2} \right] \tag{8.125c}$$

for $m = 1, 2, 3, \ldots$.

The above field expressions satisfy Equations 8.121a, 8.123a and 8.123c identically.

In view of Equation 8.126b and the initial condition defined by Equation 8.123b, the following equation is found:

$$\sum_{m=1}^{\infty} c_m \cdot \sin\left(m\pi \cdot \frac{2x}{W} \right) = -K_o \cdot \frac{\mu}{\tau} \cdot \frac{W}{4} \cdot \frac{2x}{W} + K_o \cdot \left(\rho_s + \frac{\mu}{\tau} \cdot \frac{W}{4} \right) \cdot \frac{\sin(\theta \cdot (2x/W))}{\sin(\theta)} \tag{8.126a}$$

over $-(W/2) < x < (W/2)$.

Therefore, c_m is identified as a set of coefficients in the Fourier series expansions for the expression on the RHS of this equation:

$$c_m = K_o \cdot \frac{\mu}{\tau} \cdot \frac{W}{2} \cdot \frac{\cos(m\pi)}{(m\pi)} \cdot \frac{\theta^2}{\theta^2 - (m\pi)^2} + K_o \cdot 2\rho_s \cdot \cos(m\pi) \cdot \frac{(m\pi)}{\theta^2 - (m\pi)^2} \tag{8.126b}$$

for $m = 1, 2, 3, \ldots, \infty$.

On the boundary $x = W/2$, we get from Equations 8.124a and 8.124b, for $t \geq 0$:

$$H_{1z}\big|_{x=W/2} = K_o - K_o \cdot e^{-t/\tau} - K_o \cdot \left(\rho_s + \frac{\mu}{\tau} \cdot \frac{W}{4} \right) \cdot \left(\frac{\theta}{\mu} \cdot \frac{2}{W} \right) \cdot \cot(\theta) \cdot t \cdot e^{-t/2\tau}$$
$$+ \sum_{m=1}^{\infty} \frac{c_m}{(\beta_m - \alpha_m)} \cdot \frac{m\pi}{\mu} \cdot \frac{2}{W} \cdot \cos(m\pi) \cdot \{e^{-\alpha_m t} - e^{-\beta_m t}\} \tag{8.127a}$$

and

$$E_{1y}\big|_{x=W/2} = -K_o \cdot \frac{\mu}{\tau} \cdot \frac{W}{4} \cdot e^{-t/\tau} + K_o \cdot \left(\rho_s + \frac{\mu}{\tau} \cdot \frac{W}{4} \right) \cdot \left\{ 1 - \frac{t}{2\tau} \right\} \cdot e^{-t/2\tau} \tag{8.127b}$$

Next, consider the electromagnetic fields outside the plate, for $z \geq W/2$, shown as region 2 in Figure 8.7. These fields must satisfy the initial and final conditions given by Equations 8.121a, 8.121b, 8.122b and 8.122c. Both magnetic and electric fields, as given by Equations 8.121a and 8.121b, are zero till the current sheets on the plate surfaces are suddenly switched on at the instant $t = 0$. For $t \geq 0$, the magnetic field is discontinuous on the plate surfaces as shown by Equation 8.120a. Since both eddy currents and displacement currents oppose any sudden change of magnetic flux in the plate,[4] H_{1z} cannot change instantly from its zero value. Consequently, on the plate surface in region 2, the magnetic field H_{2z} suddenly changes from its original zero value to $-K_o$, at $t = 0$, vide Equation 8.122b. At a distant point, the effect of sudden appearance of the current sheets on the plate surfaces is an abrupt change in the magnitude of electromagnetic field at that point. This change, however, takes place at a later instance as the electromagnetic disturbance propagates in free space with a finite velocity c. The sudden appearance of current sheets causes travelling waves in free space that vanish as t tends to infinity. Wavefront of each wave undergoes damping while moving away from the plate surfaces with the velocity c.

The electromagnetic fields in region 2 satisfy wave equation in free space. Thus,

$$\frac{\partial^2 E_{2y}}{\partial x^2} = \mu_0 \varepsilon_0 \cdot \frac{\partial^2 E_{2y}}{\partial t^2} \tag{8.128a}$$

and

$$\frac{\partial^2 H_{2z}}{\partial x^2} = \mu_0 \varepsilon_0 \cdot \frac{\partial^2 H_{2z}}{\partial t^2} \tag{8.128b}$$

Further, we have

$$E_{2y}(t \pm \sqrt{\mu_0 \varepsilon_0} \cdot x') = \mp \sqrt{\frac{\mu_0}{\varepsilon_0}} \cdot H_{2z}(t \pm \sqrt{\mu_0 \varepsilon_0} \cdot x') \tag{8.129a}$$

or

$$H_{2z}(t \pm \sqrt{\mu_0 \varepsilon_0} \cdot x') = \mp \sqrt{\frac{\varepsilon_0}{\mu_0}} \cdot E_{2y}(t \pm \sqrt{\mu_0 \varepsilon_0} \cdot x') \tag{8.129b}$$

where

$$x' \overset{def}{=} x - W/2 \tag{8.130}$$

Now, in view of boundary conditions, vide Equations 8.120a and 8.120b, the expressions for magnetic and electric fields in region 2 found at the boundary $x = W/2$, for $t \geq 0$ are

$$H_{2z}\big|_{x=W/2} = -K_o \cdot e^{-t/\tau} - K_o \cdot \left(\rho_s + \frac{\mu}{\tau} \cdot \frac{W}{4}\right) \cdot \left(\frac{\theta}{\mu} \cdot \frac{2}{W}\right) \cdot \cot(\theta) \cdot t \cdot e^{-t/2\tau}$$
$$+ \sum_{m=1}^{\infty} \frac{c_m}{(\beta_m - \alpha_m)} \cdot \cos(m\pi) \cdot \{e^{-\alpha_m t} - e^{-\beta_m t}\} \tag{8.131a}$$

$$E_{2y}\big|_{x=W/2} = -K_o \cdot \frac{\mu}{\tau} \cdot \frac{W}{4} \cdot e^{-t/\tau} + K_o \cdot \left(\rho_s + \frac{\mu}{\tau} \cdot \frac{W}{4}\right) \cdot \left\{1 - \frac{t}{2\tau}\right\} \cdot e^{-t/2\tau} \tag{8.131b}$$

Therefore, in view of Equations 8.129a and 8.129b, the expressions for magnetic and electric fields in region 2 can be given as

$$H_{2z} = -K_o \cdot \frac{1}{2} \cdot (e^{-t_-/\tau} + e^{-t_+/\tau}) - K_o \cdot \frac{\mu}{\tau} \cdot \frac{W}{8} \cdot \sqrt{\frac{\varepsilon_0}{\mu_0}} \cdot (e^{-t_-/\tau} - e^{-t_+/\tau})$$
$$- K_o \cdot \left(\rho_s + \frac{\mu}{\tau} \cdot \frac{W}{4}\right) \cdot \left(\frac{\theta}{\mu} \cdot \frac{1}{W}\right) \cdot \cot(\theta) \cdot (t_- \cdot e^{-t_-/2\tau} + t_+ \cdot e^{-t_+/2\tau})$$
$$+ \sum_{m=1}^{\infty} \frac{c_m}{(\beta_m - \alpha_m)} \cdot \cos(m\pi) \cdot \frac{1}{2} \cdot \{(e^{-\alpha_m t_-} - e^{-\beta_m t_-}) + (e^{-\alpha_m t_+} - e^{-\beta_m t_+})\}$$
$$+ K_o \cdot \frac{1}{2} \cdot \sqrt{\frac{\varepsilon_0}{\mu_0}} \cdot \left(\rho_s + \frac{\mu}{\tau} \cdot \frac{W}{4}\right) \cdot \left\{\left(1 - \frac{t_-}{2\tau}\right) e^{-e^{-t_-/2\tau}} - \left(1 - \frac{t_+}{2\tau}\right) e^{-e^{-t_+/2\tau}}\right\} \tag{8.132a}$$

$$E_{2y} = -K_o \cdot \frac{\mu}{\tau} \cdot \frac{W}{8} \cdot \left(e^{-(t_-/\tau)} + e^{-(t_+/\tau)}\right) + K_o \cdot \frac{1}{2} \cdot \left(\rho_s + \frac{\mu}{\tau} \cdot \frac{W}{4}\right) \cdot \left\{\left(1 - \frac{t_-}{2\tau}\right) \cdot e^{-t_-/2\tau}\right.$$
$$\left. + \left(1 - \frac{t_+}{2\tau}\right) \cdot e^{-t_+/2\tau}\right\} + \sum_{m=1}^{\infty} \frac{c_m}{(\beta_m - \alpha_m)} \cdot \cos(m\pi) \cdot \sqrt{\frac{\mu_0}{\varepsilon_0}} \cdot \frac{1}{2} \cdot \left\{\left(e^{-\alpha_m t_-} - e^{-\beta_m t_-}\right)\right.$$
$$\left. - \left(e^{-\alpha_m t_+} - e^{-\beta_m t_+}\right)\right\}$$
$$- K_o \cdot \left(\rho_s + \frac{\mu}{\tau} \cdot \frac{W}{4}\right) \cdot \left(\frac{\theta}{\mu} \cdot \frac{1}{W}\right) \cdot \cot(\theta) \cdot \sqrt{\frac{\mu_0}{\varepsilon_0}} \cdot (t_- \cdot e^{-t_-/2\tau} - t_+ \cdot e^{-t_+/2\tau})$$
$$- K_o \cdot \sqrt{\frac{\mu_0}{\varepsilon_0}} \cdot \frac{1}{2} \cdot (e^{-t_-/\tau} - e^{-t_+/\tau}) \tag{8.132b}$$

where the retarded time, t_-, and the accelerated time, t_+, are defined as

$$t_- \overset{def}{=} t - \sqrt{\mu_0 \varepsilon_0} \cdot x' \tag{8.133a}$$

$$t_+ \overset{def}{=} t + \sqrt{\mu_0 \varepsilon_0} \cdot x' \tag{8.133b}$$

In order to satisfy the initial conditions, vide Equations 8.120a and 8.120b, we multiply the RHS of Equations 8.132a and 8.132b by the unit step function $u(t_-)$. Field expressions thus modified are consistent with the observation made earlier in this section.

8.6.2 Voltage Impact Excitation

As an example of voltage impact excitation of type 2, consider a large conducting plate, with surface currents of density $\pm K_o$ simulating the uniformly distributed excitation winding on its surfaces carrying dc current. When the terminals of this winding are suddenly short-circuited, say at $t = 0$, the terminal voltage instantly becomes zero. Consequently, the dc current flowing in the winding eventually ceases. As the excitation current tends to decrease, the magnetic field inside the plate also tends to reduce. The rate of change of this magnetic field induces electric field supporting the original electric field E_o (i.e. original terminal voltage per unit core length), along the excitation winding on the plate surface. Therefore, there is no sudden change in the excitation current. However, a sudden change in the excitation current is possible in the case of nonconducting plates without eddy currents.[14]

Consider Figure 8.7, showing a large conducting plate of thickness, W, carrying uniform current sheets of density, $K_y = \pm K_o$, on its surfaces located at $x = \pm W/2$. These current sheets simulate the excitation winding carrying the steady-state dc current for $t \leq 0$, and the transient current for $t \geq 0$. Before the onset of transient, the initial fields are

$$H_{1z}\big|_{t\leq 0} = K_o \tag{8.134a}$$

$$H_{2z}\big|_{t\leq 0} = 0 \tag{8.134b}$$

$$E_{1y}\big|_{t<0} = E_{2y}\big|_{t<0} = 0 \tag{8.134c}$$

The steady-state values for the electromagnetic fields after the transients are over are

$$H_{1z}\big|_{t\to\infty} = H_{2z}\big|_{t\to\infty} = 0 \tag{8.135a}$$

$$E_{1y}\Big|_{t\to\infty} = E_{2y}\Big|_{t\to\infty} = 0 \tag{8.135b}$$

The boundary conditions for $t > 0$ are as follows:

$$H_{2z}\Big|_{x=W/2} = H_{1z}\Big|_{x=W/2} - K_y \tag{8.136a}$$

and

$$E_{2y}\Big|_{x=W/2} = E_{1y}\Big|_{x=W/2} \tag{8.136b}$$

where K_y indicates the transient current density on the plate surface, $x = W/2$ simulating the transient current in the short circuited excitation winding.

The boundary conditions for $t = 0$ are given below

$$H_{1z}\Big|_{x=W/2} = K_o \tag{8.137a}$$

$$E_{1y}\Big|_{x=W/2} = E_{2y}\Big|_{x=W/2} = E_o \tag{8.137b}$$

Consider the electromagnetic fields inside the plate. These fields must satisfy the initial and final conditions given by Equations 8.134a through 8.135b. Because of symmetry, the magnetic field is an even function and the electric field is an odd function of x. Therefore, in view of Equations 8.114 and 8.115, solutions for Equations 8.116 and 8.117, for $t \geq 0$, are tentatively taken as

$$H_{1z} = K_o \cdot e^{-t/\tau} - K_o \cdot \left(\rho_s - \frac{\mu}{\tau}\cdot\frac{W}{2}\right)\cdot\left(\frac{\theta}{\mu}\cdot\frac{2}{W}\right)\cdot\frac{\cos(\theta\cdot(2x/W))}{\sin(\theta)}\cdot t\cdot e^{-t/2\tau}$$
$$+\sum_{m=1}^{\infty}\frac{d_m}{(\beta_m-\alpha_m)}\cdot\frac{m\pi}{\mu}\cdot\frac{2}{W}\cdot\cos\left(m\pi\cdot\frac{2x}{W}\right)\cdot\{e^{-\alpha_m t}-e^{-\beta_m t}\} \tag{8.138a}$$

$$E_{1y} = K_o\cdot\frac{\mu}{\tau}\cdot x\cdot e^{-t/\tau} + K_o\cdot\left(\rho_s - \frac{\mu}{\tau}\cdot\frac{W}{2}\right)\cdot\frac{\sin(\theta\cdot(2x/W))}{\sin(\theta)}\cdot\left\{1-\frac{t}{2\tau}\right\}\cdot e^{-t/2\tau}$$
$$+\sum_{m=1}^{\infty}\frac{d_m}{(\beta_m-\alpha_m)}\cdot\sin\left(m\pi\cdot\frac{2x}{W}\right)\cdot\{\alpha_m\cdot e^{-\alpha_m t}-\beta_m\cdot e^{-\beta_m t}\} \tag{8.138b}$$

where d_m indicate arbitrary constants.

Considering Equations 8.134c and 8.138b, we have

$$\sum_{m=1}^{\infty} d_m \cdot \sin\left(m\pi \cdot \frac{2x}{W}\right) = K_o \cdot \left[\frac{\mu}{\tau} \cdot x + \left(\rho_s - \frac{\mu}{\tau} \cdot \frac{W}{2}\right) \cdot \frac{\sin(\theta \cdot (2x/W))}{\sin(\theta)}\right]$$

$$\text{over} \quad -\frac{W}{2} < x < \frac{W}{2} \qquad (8.139\text{a})$$

The term d_m can be identified as the coefficient of the Fourier series expansion for the function given on the RHS of Equation 8.139a

$$d_m = -K_o \cdot \frac{\mu}{\tau} \cdot W \cdot \frac{\cos(m\pi)}{(m\pi)} \cdot \frac{\theta^2}{\theta^2 - (m\pi)^2} + K_o \cdot 2\rho_s \cdot \cos(m\pi) \cdot \frac{(m\pi)}{\theta^2 - (m\pi)^2} \qquad (8.139\text{b})$$

for $m = 1, 2, 3, \ldots, \infty$.

On the boundary, $x = W/2$, we get from Equations 8.138a and 8.138b, for $t > 0$

$$H_{1z}\big|_{x=W/2} = K_o \cdot e^{-t/\tau} - K_o \cdot \left(\rho_s - \frac{\mu}{\tau} \cdot \frac{W}{2}\right) \cdot \left(\frac{\theta}{\mu} \cdot \frac{2}{W}\right) \cdot \cot(\theta) \cdot t \cdot e^{-t/2\tau}$$

$$+ \sum_{m=1}^{\infty} \frac{d_m}{(\beta_m - \alpha_m)} \cdot \frac{m\pi}{\mu} \cdot \frac{2}{W} \cdot \cos(m\pi) \cdot \{e^{-\alpha_m t} - e^{-\beta_m t}\} \qquad (8.140\text{a})$$

$$E_{1y}\big|_{x=W/2} = K_o \cdot \frac{\mu}{\tau} \cdot \frac{W}{2} \cdot e^{-t/\tau} + K_o \cdot \left(\rho_s - \frac{\mu}{\tau} \cdot \frac{W}{2}\right) \left\{1 - \frac{t}{2\tau}\right\} \cdot e^{-t/2\tau} \qquad (8.140\text{b})$$

If the terminals of a coil are suddenly short-circuited, the resulting current in the coil is time dependent. At any instant, the resistance drop in the coil must equal the induced voltage due to the decay of coil flux. This leads to an ordinary differential equation. The magnetic field in the core of the coil varies with time as well as space coordinates. Therefore, the system involves both ordinary as well as partial differential equations. The magnetic flux, φ, per unit length in the y-direction is

$$\varphi = \mu \cdot \int_{-W/2}^{W/2} H_{1z} \cdot dx \qquad (8.141\text{a})$$

Therefore, using Equation 8.138a

$$\varphi = \mu \cdot K_o \cdot W \cdot e^{-t/\tau} - K_o \cdot \left(\rho_s - \frac{\mu}{\tau} \cdot \frac{W}{2}\right) \cdot \left(\frac{2}{\mu}\right) \cdot t \cdot e^{-t/2\tau} \tag{8.141b}$$

Hence, the voltage drop in the winding resistance per unit length in the y-direction is

$$\rho_s \cdot K_y = -\frac{d\varphi}{dt} = K_o \cdot \frac{\mu}{\tau} \cdot W \cdot e^{-t/\tau} + K_o \cdot \left(\rho_s - \frac{\mu}{\tau} \cdot W/2\right) 2 \left\{1 - \frac{t}{2\tau}\right\} \cdot e^{-t/2\tau} \tag{8.142a}$$

Thus,

$$K_y = K_o \cdot \left[\frac{\mu}{\tau} \cdot \frac{W}{\rho_s} \cdot e^{-t/\tau} + \left(2 - \frac{\mu}{\tau} \cdot \frac{W}{\rho_s}\right) \cdot \left\{1 - \frac{t}{2\tau}\right\} \cdot e^{-t/2\tau}\right] \tag{8.142b}$$

In view of boundary conditions, vide Equations 8.136a, 8.136b and 8.137b, the expressions for magnetic and electric fields in region 2 at $x = W/2$, found using Equations 8.140a, 8.140b and 8.142b, are as follows:

$$\begin{aligned} H_{2z}\big|_{x=W/2} &= K_o \cdot \left(1 - \frac{\mu}{\tau} \cdot \frac{W}{\rho_s}\right) \cdot e^{-t/\tau} - K_o \cdot \left(\rho_s - \frac{\mu}{\tau} \cdot \frac{W}{2}\right) \cdot \left\{\left(\frac{\theta}{\mu} \cdot \frac{2}{W}\right) \cdot \cot(\theta) \cdot t \right. \\ &\quad \left. + \frac{2}{\rho_s} \cdot \left(1 - \frac{t}{2\tau}\right)\right\} \cdot e^{-t/2\tau} \\ &\quad + \sum_{m=1}^{\infty} \frac{d_m}{(\beta_m - \alpha_m)} \cdot \frac{m\pi}{\mu} \cdot \frac{2}{W} \cdot \cos(m\pi) \cdot (e^{-\alpha_m t} - e^{-\beta_m t}) \end{aligned} \tag{8.143a}$$

$$E_{2y}\big|_{x=W/2} = K_o \cdot \frac{\mu}{\tau} \cdot \frac{W}{2} \cdot e^{-t/\tau} + K_o \cdot \left(\rho_s - \frac{\mu}{\tau} \cdot \frac{W}{2}\right) \cdot \left(1 - \frac{t}{2\tau}\right) \cdot e^{-t/2\tau} \tag{8.143b}$$

The electromagnetic fields in region 2 satisfy wave equation in free space, as given by Equations 8.128a and 8.128b. The relations between H_{2z} and E_{2y}

are given by Equations 8.129a and 8.129b. Field expressions for region 2, therefore, are given by

$$
\begin{aligned}
H_{2z} &= K_o \cdot \left(1 - \frac{\mu}{\tau} \cdot \frac{W}{\rho_s}\right) \cdot \frac{1}{2} \cdot \left\{(e^{-t_-/\tau} + e^{-t_+/\tau}) - 2.(e^{-t_-/2\tau} + e^{-t_+/2\tau})\right\} \\
&\quad - K_o \cdot \left\{\left(\rho_s - \frac{\mu}{\tau} \cdot \frac{W}{2}\right) \cdot \left(\frac{\theta}{\mu} \cdot \frac{1}{W}\right) \cdot \cot(\theta) - \frac{1}{2\tau} \cdot \left(1 - \frac{\mu}{\tau} \cdot \frac{W}{\rho_s}\right)\right\} \\
&\quad \cdot (t_-.e^{-t_-/2\tau} + t_+.e^{-t_+/2\tau}) \\
&\quad + \sum_{m=1}^{\infty} \frac{d_m}{(\beta_m - \alpha_m)} \cdot \frac{m\pi}{\mu} \cdot \frac{1}{W} \cdot \cos(m\pi) \cdot \left\{(e^{-\alpha_m t_-} + e^{-\alpha_m t_+}) - (e^{-\beta_m t_-} + e^{-\beta_m t_+})\right\} \\
&\quad + \sqrt{\frac{\epsilon_0}{\mu_0}} \cdot K_o \cdot \left[\left(\rho_s - \frac{\mu}{\tau} \cdot \frac{W}{2}\right) \cdot \frac{1}{2} \cdot \left\{(e^{-t_-/2\tau} - e^{-t_+/2\tau})\right.\right. \\
&\quad \left.\left. - \frac{1}{2\tau}(t_-.e^{-t_-/2\tau} - t_+.e^{-t_+/2\tau})\right\}\right] \\
&\quad + \sqrt{\frac{\epsilon_0}{\mu_0}} \cdot K_o \cdot \left(\frac{\mu}{\tau} \cdot \frac{W}{4}\right) \cdot (e^{-t_-/\tau} - e^{-t_+/\tau})
\end{aligned}
\tag{8.144a}
$$

$$
\begin{aligned}
E_{2y} &= \sqrt{\frac{\mu_0}{\epsilon_0}} \cdot K_o \cdot \left[\left(1 - \frac{\mu}{\tau} \cdot \frac{W}{\rho_s}\right) \cdot \frac{1}{2} \cdot \left\{(e^{-t_-/\tau} - e^{-t_+/\tau}) - 2.(e^{-t_-/2\tau} - e^{-t_+/2\tau})\right\}\right. \\
&\quad - \left\{\left(\rho_s - \frac{\mu}{\tau} \cdot \frac{W}{2}\right) \cdot \left(\frac{\theta}{\mu} \cdot \frac{1}{W}\right) \cdot \cot(\theta)\right. \\
&\quad \left.\left. - \frac{1}{2\tau} \cdot \left(1 - \frac{\mu}{\tau} \cdot \frac{W}{\rho_s}\right)\right\} \cdot (t_-.e^{-t_-/2\tau} - t_+.e^{-t_+/2\tau})\right] \\
&\quad + \sqrt{\frac{\mu_0}{\epsilon_0}} \cdot \sum_{m=1}^{\infty} \frac{d_m}{(\beta_m - \alpha_m)} \cdot \frac{m\pi}{\mu} \cdot \frac{1}{W} \cdot \cos(m\pi) \cdot \left\{(e^{-\alpha_m t_-} - e^{-\alpha_m t_+})\right. \\
&\quad \left. - (e^{-\beta_m t_-} - e^{-\beta_m t_+})\right\} \\
&\quad + K_o \cdot \left[\left(\rho_s - \frac{\mu}{\tau} \cdot \frac{W}{2}\right) \cdot \frac{1}{2} \cdot \left\{(e^{-t_-/2\tau} + e^{-t_+/2\tau})\right.\right. \\
&\quad \left.\left. - \frac{1}{2\tau}(t_-.e^{-t_-/2\tau} + t_+.e^{-t_+/2\tau})\right\}\right] \\
&\quad + K_o \cdot \left(\frac{\mu}{\tau} \cdot \frac{W}{4}\right) \cdot (e^{-t_-/\tau} + e^{-t_+/\tau})
\end{aligned}
\tag{8.144b}
$$

where the retarded time t_- and the accelerated time t_+ are defined by Equations 8.133a and 8.133b, respectively.

In order to satisfy the initial conditions, vide Equations 8.134b and 8.134c, we multiply the RHS of Equations 8.144a and 8.144b by the unit step function, $u(t_-)$. Field expressions thus modified are consistent with the observation made earlier in this section.

PROJECT PROBLEMS

1. Sum up the infinite series in Equation 8.48, and then obtain the expression for the leakage inductance for a coil group of the single-phase transformer with interleaved windings.
2. Develop a field theory for hysteresis machines with rotor eddy currents.
3. Evaluate the arbitrary constants involved in the expressions for a rotor field of a single-phase induction motor with composite stator poles and solid iron rotor. Develop the expression for the starting torque produced in this motor.
4. Develop the expressions for magnetic field distributions in the solid part of the composite stator poles, in the air gap and in the solid iron rotor of the single-phase induction machine with composite stator poles, for an arbitrary value of slip s.
5. Develop the expression for the torque produced in the single-phase induction machine with composite stator poles and solid iron rotor for an arbitrary value of slip s.
6. Deduce and discuss transient fields in a large nonconducting plate due to the type 2 impact excitation.[14]
7. Find the leakage inductance for a conductor placed in an open rectangular slot without assuming the flux to go straight across the slot opening. Consider the cases when (a) the air gap is infinitely large; (b) the slot opening is facing a smooth iron surface across a finite air-gap length.

References

1. Roth, E., Etude analytique du champ propre d'une encoche, *Rev.Gin. de l'El.*, XXII, 417–424, 1927.
2. Bewley, L. V., *Two-Dimensional Fields in Electrical Engineering*, Dover, New York, pp. 81–83, 1963.
3. Rogowski, W., Ueber das Streufeld und denStreuinduktionskoeffizienten eines Transformators mit Schiebenwicklung und geteilten Endspulen, V. D. I. Heft 71, 1–36, 1909.

4. Sharma, N. D. and Bedford, R. E., *Hysteresis Machines*, Perfect Prints, Thane, India, 2003.
5. Wood, A. J. and Concordia, C., An analysis of solid rotor machines. Part-II The effect of curvature, *Transactions of the AIEE*, 78 (Pt. III B), 1666–1671, 1959 (Feb. 1960 Section).
6. Wood, A. J. and Concordia, C., An analysis of solid rotor machines. Part III. Finite length effects, *Transactions of the AIEE*, 79 (Pt. III), 21–27, 1960.
7. Mukerji, S. K., Linear electromagnetic field analysis of the solid rotor induction machine (tooth-ripple phenomena and end-effects), PhD thesis, IIT Bombay, 1967.
8. Zaher, F. A. A., An analytical solution for the field of a hysteresis motor based on complex permeability, *IEEE Transaction on Energy Conversion*, 5 (1), 156–163, March 1990.
9. Mukerji, S. K., Goel, S. K., Bhooshan, S. and Basu, K. P., Electromagnetic fields theory of electrical machines, Part-I: Poynting theorem for electromechanical energy conversion, *International Journal Electrical Engineering Education*, 41 (2), 137–145, April 2004.
10. Mukerji, S. K., Bhardwaj, M. K. and Shukla, S., A simplified theory of hysteresis machines, *Proceedings of National Conference on Recent Advances in Technology and Engineering (RATE 2012)*, Mangalayatan University, January 20–22, 2012.
11. Mukerji, S. K., Bhardwaj, M. K., George, M. and Sharma, G. K., Asynchronous operation of hysteresis machines, *International Journal of Applied Engineering and Computer Science*, 1 (1), 1–12, January 2014.
12. Wakui, G., Alternating hysteresis and rotational hysteresis in the hysteresis motor, *Transactions of the IEE Japan*, 45, 1558–1567, 1970.
13. Mukerji, S. K., Singh, G. K., Goel, S. K. and Manuja, S., A theoretical study of electromagnetic transients in a large conducting plate due to current impact excitation, *Progress in Electromagnetics Research, 76*, 15–29, 2007.
14. Mukerji, S. K., Singh, G. K., Goel, S. K. and Manuja, S., A theoretical study of electromagnetic transients in a large conducting plate due to voltage impact excitation, *Progress in Electromagnetics Research, 78*, 377–392, 2008.

9

Numerical Computation

9.1 Introduction

This chapter describes the procedures used for the evaluation of arbitrary constants, arbitrary functions, field parameters and other quantities discussed in Chapters 4–8. Only one sample case, out of the similar cases, has been discussed in detail. The described procedures are mainly based on the techniques of numerical analysis. In view of the essence of the problems, the meaning and sphere of numerical analysis and the techniques falling under its domain are briefly described.

9.2 Numerical Analysis

Numerical analysis is the study of algorithms that use numerical approximation for the mathematical problems that evolve out of some physical systems or processes. Its overall goal is the design and analysis of techniques to give approximate but acceptable solutions to the complicated problems. These problems may be related to weather predictions, computation of the trajectories of spacecraft, the crash safety of cars, stresses developed in physical structures or the distribution of fields and so on. For estimating trajectories, the accurate numerical solution of a system of ordinary differential equations may be required, whereas car safety may require numerical solutions of partial differential equations. The problem of structure or that of fields may also involve ordinary or partial differential equations, integral equations and so on.

In numerical analysis, the process of interpolation, extrapolation and regression are quite frequently employed. In case of *interpolation*, the value of some unknown function can be evaluated in between the two given values of the function. In *extrapolation*, the value of some unknown function is to be evaluated, which falls outside the given points. This process first assesses the nature of variation of previous values and based on this trend estimates the new values. *Regression* is also a similar process, but it takes into account that the data are imprecise. Given some points, and a measurement of the value of some function at these points (with an error), it determines the unknown function. It mostly relies on the least square error to achieve the goal.

9.2.1 Computational Errors

No technique, which falls in the domain of numerical analysis, is error free. These errors creep in mainly due to the following reasons:

1. In general, all practical computers have a finite memory and it is impossible to exactly represent all the real numbers on such a computing machine. Thus, a class of error referred to as the *round-off errors* are bound to occur.
2. When an iterative method is terminated or a mathematical procedure is approximated, the error due to which the approximate solution differs from the exact solution is referred to as *truncation errors*.
3. Similarly, the discretisation induces a *discretisation error* because the solution of the discrete problem does not coincide with the solution of the continuous problem.

It may be noted that once an error is generated, it generally propagates through subsequent calculations.

9.2.2 Numerical Stability

An algorithm is said to be *numerically stable* if an error, whatever its cause, does not grow to be much larger during the calculations. In view of this notion, the problems can be classified into two categories. These categories are termed as well conditioned and ill conditioned. The problem is called *well conditioned* if there is only a small change in the solution when the problem data are changed by a small amount. In contrast, in *ill-conditioned* problem any small error in the data will grow to become a large error.

9.3 Domain of Numerical Analysis

The field of numerical analysis includes many subdisciplines and encompasses problems of multi-facial nature. These may include the following.

9.3.1 Values of Functions

The evaluation of a function at a given point is one of the simplest problems. In the case of polynomials, the *Horner scheme* is a better approach, since it requires a lesser number of multiplications and additions. In this case, the estimation and control of round-off errors due to the use of floating point arithmetic is of immense importance.

9.3.2 Equations and Systems of Equations

These can be further classified into linear and nonlinear forms. *Linear equations* are an important class of the numerical analysis. There are many methods for solving the systems of linear equations. Some standard methods employ matrix decomposition techniques. These include Gaussian elimination, LU (lower–upper) decomposition, Cholesky decomposition for symmetric (or Hermitian) and positive-definite matrix and QR decomposition for nonsquare matrices. For large systems preference is given to iterative methods, which include Jacobi method, Gauss–Seidel method, successive over relaxation method and conjugate gradient method. General iterative methods can be developed by using a matrix splitting.

Nonlinear equations are solved by using root-finding algorithms. In this case, if the function is differentiable and the derivative is known it can be solved by using Newton's method. The technique referred to as linearisation can also be employed for solving nonlinear equations.

9.3.2.1 Linear Algebraic Equations

In context to the system of equations, it seems to be appropriate to describe the formation of matrices from linear algebraic equations. An *algebraic equation* in which each term is either a constant or the product of a constant and (the first power of) a single variable is referred to as a *linear equation.* A linear equation can involve a number of variables but does not include exponents. An equation involving n variables can be written in the following form:

$$a_1x_1 + a_2 x_2 + \cdots + a_n x_n = b \tag{9.1}$$

where a_1, a_2, …, a_n represent numbers and are called the *coefficients. The parameters* x_1, x_2, …, x_n are the unknowns and b is called the *constant term.* The present analysis gives rise to a set of such equations, which can be written as

$$A_{11}x_1 + a_{12} x_2 + \cdots + a_{1N} x_N = b_1$$

$$A_{21}x_1 + a_{22} x_2 + \cdots + a_{2N} x_N = b_2$$

$$\cdots$$

$$A_{M1x1} + a_{M2} x_2 + \cdots + a_{MN} x_N = b_M \tag{9.2}$$

Equation 9.2 can be written in the form of a matrix equation as

$$[A]\,[x] = [B] \tag{9.3}$$

where

$$[A] = \begin{bmatrix} a_{11} & a_{12} & .. & a_{1N} \\ a_{21} & a_{22} & .. & a_{2N} \\ .. & .. & .. & .. \\ a_{M1} & a_{M2} & .. & a_{MN} \end{bmatrix} [x] = \begin{bmatrix} x_1 \\ x_2 \\ .. \\ x_N \end{bmatrix} \text{ and } [B] = \begin{bmatrix} b_1 \\ b_2 \\ .. \\ b_N \end{bmatrix} \tag{9.4}$$

As can be seen in Equation 9.4, matrix $[A]$ contains coefficients, matrix $[x]$ contains unknown parameters and matrix $[B]$ contains constant known terms. In this equation, both M and N are same positive real integers.

Equation 9.3 can be manipulated to the form:

$$[x] = [A]^{-1}\,[B] \tag{9.5}$$

where matrix $[A]^{-1}$ represents the inverse of matrix $[A]$.

If the numerical values of the elements involved in matrices $[A]$ and $[B]$ are known the elements of matrix $[x]$ can be obtained from Equation 9.5.

9.3.3 Eigenvalue or Singular Value Problems

This is another important area of numerical analysis since many problems can be phrased in terms of eigenvalue or singular value decomposition. Its application in the area of image compression is quite frequent. The spectral image compression algorithm is mainly based on the singular value decomposition and the corresponding tool in statistics is referred to as principal component analysis.

9.3.4 Optimisation Problem

The problem of maximising (or minimising) a given function at a point is referred to as an optimisation problem. Such a function at the identified point may be subjected to some constraints. Depending on the forms of the objective function and the constraint, the field of optimisation is split into several subfields. *Linear programming* is one such field. It deals with the case when both the objective function and the constraints are linear and is commonly dealt with simplex method. The method of Lagrange multiplier, which reduces the problems with constraints to unconstrained optimisation, is also employed.

9.3.5 Differential Equations

Solutions of both, ordinary and partial differential equations, fall in the domain of numerical analysis. Partial differential equations can be solved by first discretising an equation and then bringing it into a finite-dimensional subspace. This can be achieved either through finite difference method (FDM) or finite element method (FEM). Each of these has its own merits and demerits. For some of the engineering problems finite volume method, a version of FEM is more suited. The theoretical justification of these methods often involves theorems from function analysis. This reduces the problem to the solution of an algebraic equation.

Some of the salient features of FDM and FEM are listed below.

9.3.5.1 Finite Difference Method

- In this method the derivatives appearing in the differential equation are replaced by finite differences that approximate them.
- It is used to obtain numerical solutions of ordinary and partial differential equations.
- It is commonly used in computational science and engineering disciplines.
- It is easier to implement.
- Its basic form is restricted to handle rectangular shapes and their simple variations.
- Quality of approximation between grid points is relatively poor.

9.3.5.2 Finite Element Method

- It is a numerical technique for finding approximate solutions to the boundary value problems for differential equations.
- It uses the calculus of variations to minimise an error function and produce a stable solution.
- It encompasses all the methods for connecting simple element equations over many small subdomains, named finite elements, to approximate a more complex equation over a larger domain.
- The subdivision of a whole domain into simpler parts facilitates (i) accurate representation of complex geometries, (ii) inclusion of dissimilar material properties, (iii) easy representation of the total solution and (iv) capture of local effects.
- It can handle much complex geometries and boundaries.
- The quality of approximation is relatively much better than in FDM.

9.3.6 Numerical Integration

Numerical integration, sometimes called numerical quadrature, is used to evaluate definite integrals. The methods used for evaluating a definite integral include the Newton–Cote's formulae (like the midpoint or Simpson's rule) or Gaussian quadrature. In these methods, an integral on a relatively large set is broken into integrals on smaller sets. In higher dimensions, where these methods become prohibitively expensive in terms of computational effort, the Monte Carlo or quasi-Monte Carlo methods may be used. In cases of modestly large dimensions, method of sparse grids provides a better alternative.

The standard technique for approximate evaluation of integration over unbounded intervals involves specially derived quadrature rules. These include (i) the Gauss–Hermite quadrature for integrals on the whole real line and (ii) the Gauss–Laguerre quadrature for integrals on the positive real line. Monte Carlo methods can also be used. Besides, the method of a change of variables to a finite interval for the given line may also be adopted.

This section includes relations for some of the above-referred methods, which may be used for numerical integration. The types of integrations involved in the analysis may fall under any of following three categories.

9.3.6.1 Integration over Bounded Intervals

As noted above, there are many numerical techniques that may be used to solve this category of integrals. Relations for only two of these are given below.

9.3.6.1.1 Simpson's General Formula (Trapezoidal Rule)

In this relation, the x-axis (within given finite limits) is divided into n equal segments, each of length h. The x_0 corresponds to the lower limit of integration a and x_n to its upper limit b. At x_0 and x_n the values of y are taken to be y_0 and y_n, respectively. Thus, $h = (b - a)/n$ and the value of integral can be given as

$$\int_a^b y \cdot dx = h \cdot \left(\frac{y_0}{2} + y_1 + y_2 + \cdots + y_{n-2} + y_{n-1} + y_n \right) \tag{9.6}$$

9.3.6.1.2 Simpson's General Formula (Parabolic Rule)

In this case, the x-axis (within given finite limits) is divided into $2m$ equal segments, each of length h. As illustrated in Figure 9.1, the x_0 corresponds to the lower limit of integration a and x_{2m} to its upper limit b. At x_0 and x_{2m} the

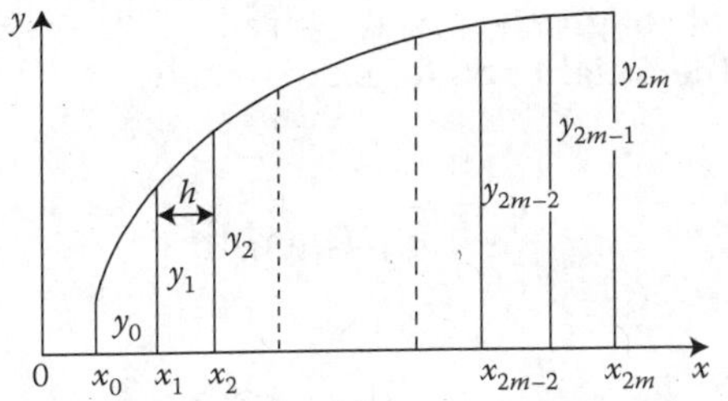

FIGURE 9.1
Division depicting the parabolic rule.

values of y are taken to be y_0 and y_{2m}, respectively. In this case, $h = (b - a)/2m$ and the value of integral can be given as

$$\int_a^b y dx = \frac{h}{3}[(y_0 + y_{2m}) + 4(y_1 + y_3 + \cdots + y_{2m-1}) + 2(y_2 + y_4 + \cdots + y_{2m-2})] \tag{9.7}$$

9.3.6.2 *Integration over Unbounded Intervals*

The integrals belonging to this category are also referred to as improper integrals. These may further be classified into two categories. In the first category, the integration is to be carried out over the entire real axis. Such integrals can be solved by using the Gauss–Hermite quadrature method. In the second category, the integration is to be carried out over half of the real axis. Such integrals can be solved by using the Gauss–Laguerre quadrature method. These two methods are given in the following two sections.

9.3.6.2.1 Gauss–Hermite Quadrature

$$\int_{-\infty}^{+\infty} e^{-x^2} f(x) dx = \sum_{i=1}^{n} w_i f(x_i) + R_n \tag{9.8a}$$

On neglecting the remainder term, it can be written as

$$\int_{-\infty}^{+\infty} e^{-x^2} f(x) dx = \sum_{i=1}^{n} w_i f(x_i) \tag{9.8b}$$

The alternative form of the above equation is

$$\int_{-\infty}^{+\infty} g(x) dx = \sum_{i=1}^{n} w_i e^{x_i^2} g(x_i) \tag{9.8c}$$

In Equations 9.8a through 9.8c, x_i is the ith zero of $H_n(x)$, $H_n(x)$ is the Hermite polynomials, w_i is the weight and R_n is the remainder

$$w_i = \frac{2^{n-1} n! \sqrt{\pi}}{n^2 [H_{n-1}(x_i)]^2} \tag{9.9a}$$

$$R_n = \frac{n!\sqrt{\pi}}{2^n (2n)!} f^{2n}(\xi) \quad (-\infty < \xi < \infty) \tag{9.9b}$$

The weight factors (w_i) and the product $w_i e^{x_i^2}$ for the values of abscissas (x_i) representing zeros of Hermite polynomials are available[14] for $n = 2, 3, 4, 5, 6, 7, 8, 9, 10, 12, 16$ and 20. Table 9.1 gives these values for an arbitrarily selected n (= 9).

9.3.6.2.2 Gauss–Laguerre Quadrature

$$\int_0^{+\infty} e^{-x} f(x)\,dx = \sum_{i=1}^{n} w_i f(x_i) + R_n \tag{9.10a}$$

On neglecting the remainder term, it can be written as

$$\int_0^{+\infty} e^{-x} f(x)\,dx = \sum_{i=1}^{n} w_i f(x_i) \tag{9.10b}$$

The above equation can be written in the following alternative form:

$$\int_0^{+\infty} g(x)\,dx = \sum_{i=1}^{n} w_i e^{x_i} g(x_i) \tag{9.10c}$$

TABLE 9.1

Abscissas and Weight Factors for Hermite Integration for $n = 9$

$\pm x_i$	w_i **(−m) Represents** 10^{-m}	$w_i e^{x_i^2}$
0.00000	7.20235 (−1)	0.72023
0.72355	4.32651 (−1)	0.73030
1.46855	8.84745 (−2)	0.76460
2.26658	4.94362 (−3)	0.84175
3.19099	3.96069 (−5)	1.04700

TABLE 9.2

Abscissas and Weight Factors for Laguerre Integration for $n = 9$

x_i	w_i $(-m)$ Represents 10^{-m}	$w_i e^{x_i}$
0.15232	3.36126 (−1)	0.39143
0.80722	4.11213 (−1)	0.92180
2.00513	1.99287 (−1)	1.48012
3.78347	4.74605 (−2)	2.08677
6.20495	5.59962 (−3)	2.77292
9.37298	3.05249 (−4)	3.59162
13.46623	6.59212 (−6)	4.64876
18.83359	4.11076 (−8)	6.21227
26.37407	3.29087 (−11)	9.36321

In Equations 9.10 through 9.10c, x_i is the ith zero of $L_n(x)$, $L_n(x)$ is the Laguerre polynomials, w_i is the weight and R_n is the remainder

$$w_i = \frac{(n!)^2 x_i}{(n+1)^2[L_{n+1}(x_i)]^2} \tag{9.11a}$$

$$R_n = \frac{(n!)^2}{(2n)!} f^{2n}(\xi) \quad (0 < \xi < \infty) \tag{9.11b}$$

Weight factors (w_i) and the product $w_i e^{x_i}$ for some selected values of abscissas (x_i) representing zeros of Laguerre polynomials are available[14] for $n = 2, 3, 4, 5, 6, 7, 8, 9, 10, 12$ and 15. Table 9.2 gives these values for an arbitrarily selected n (= 9).

9.3.6.2.3 Change of Variable for Infinite Intervals

If, in Equation 9.8a, x is replaced by $t/(1 - t^2)$, then $dx = ((1 + t^2)/(1 - t^2)^2)$. In view of this replacement the limits of '$-\infty$' to '$+\infty$' change to '−1' to '+1'. Thus, the integral of infinite interval reduces to that of finite interval

$$\int_{-\infty}^{+\infty} f(x)\,dx = \int_{-1}^{+1} f\left(\frac{t}{1-t^2}\right)\frac{1+t^2}{(1-t^2)^2}\,dt \tag{9.12}$$

9.3.6.2.4 Change of Variable for Semi-Infinite Intervals

In this case, x is replaced by $a + (t/(1 - t))$ then $dx = dt/(1 - t)^2$ and the limits 'a' to '∞' change from '0' to '1'. Thus, the integral becomes

$$\int_{a}^{+\infty} f(x)\,dx = \int_{0}^{1} f\left(a + \frac{t}{1-t}\right)\frac{dt}{(1-t)^2} \tag{9.13}$$

Here x is replaced by $a - ((1 - t)/t)$ then $dx = dt/t^2$ and the limits '∞' to 'a' change to '0' to '1'. The integral thus becomes

$$\int_{-\infty}^{a} f(x)dx = \int_{0}^{1} f\left(a - \frac{1-t}{t}\right)\frac{dt}{t^2} \quad (9.14)$$

9.3.6.2.5 Multi-Dimensional Integrals

The quadrature rules as such are designed to compute one-dimensional integrals. The multi-dimensional integrals can, however, also be evaluated by repeating one-dimensional integrals. In this approach, the function evaluations exponentially grow with the number of dimensions and some methods to overcome this effect are to be used. Monte Carlo or quasi-Monte Carlo methods provide better alternatives. These methods are easy to apply to multi-dimensional integrals. Besides, these may yield greater accuracy for the same number of function evaluations than repeated integrations using one-dimensional methods. Markov chain Monte Carlo algorithms, which include Metropolis–Hestings algorithm and Gibbs sampling, belong to a large class of useful Monte Carlo methods. Besides, sparse grids are developed by Smolyak for the quadrature of high-dimensional functions. Although it is based on a one-dimensional quadrature rule, it performs more sophisticated combination of univariate results.

9.3.7 Software

There are a number of software routines for numerical problems based on different computer languages. Commercial products implementing different numerical algorithms include the IMSL, NAG and GNU libraries. The other numerical computing applications include MATLAB, S-PLUS, LabVIEW, IDL, FreeMat, Scilab, GNU Octave, IT++, R and some variants of Python. There is wide variation in the performance of these algorithms. The vector and matrix operations are usually fast but in case of scalar loops their speeds may vary.

The arbitrary precision arithmetic is often utilised in many computer algebra systems such as Mathematica to provide greater accuracy. Besides, simple numerical analysis problems can be handled by using any spreadsheet software.

Numerical integration is one of the most intensively studied problems in numerical analysis. It can be implemented through a number of software packages. Some of the free and open source softwares include (i) QUADPACK a collection of algorithms for numerical integration, in Fortran, and is based on Gaussian quadrature, (ii) interalg a solver from OpenOpt/FuncDesigner frameworks, based on interval analysis with guaranteed precision, (iii) the GNU Scientific Library (GSL) is a numerical library that, in C, provides a wide range of mathematical routines, like Monte Carlo integration, (iv) ALGLIB is a collection of algorithms, in C#/C++/Delphi/Visual Basic/and so on, for numerical integration, includes Bulirsch–Stoer and Runge–Kutta integrators, (v) Cuba is a free software library of multi-dimensional integration algorithms, (vi) Cubature code is usable for adaptive multi-dimensional integration and (vii) Scilab is an open source software that provides many powerful features including numerical integration.

9.4 Types of Equations

Before discussing the types of equations involved, it needs to be mentioned that the analysis presented in this text involves a number infinite series and definite (proper and improper) integrals with varied limits. To simplify the analysis, many of these series are summed by using Poisson's sum formula. Some of such series are provided by Equations 4.14, 4.18b, 4.32b and 4.43j. Similarly, the involved integrals are either available in the form of identities or are evaluated by employing conventional methods. Some of such integrals that are finally evaluated are given by Equations A2.1a, A2.1b, A2.5b, A3.2b, A3.2c, A4.1, A4.3, A4.8, A4.13a, A4.13b, A4.13c, 4.7b, 4.8b, 4.23a, 4.23b, 4.25a through 4.25e, 4.45, 5.13a, 5.14a, 5.15, 5.16, 5.74, 5.86, 5.101a through 5.101d, 5.118a, A6.4a, A6.4b, A6.12, 7.13, 7.16a through 7.16e, 7.21a through 7.21n, 7.57, 7.58b, 7.96b, 7.99b, 7.101b, 7.104b, 7.106b, 7.109a through 7.109j, 7.110a through 7.110c, 7.114, 8.4, 8.16, 8.46a through 8.46c and 8.141b.

In view of the evaluation of the above-referred summations and integrations, the analysis is greatly simplified and in some of the cases the final expressions are quite near to the closed forms.

The analysis presented in Chapters 4–8 has ultimately resulted in a number of mathematical expressions. The field and other parameters can be readily evaluated from these expressions. A cursory look at these expressions reveals that these may fall into the following four distinct categories:

- In the first category, there is no involvement of any arbitrary constant, arbitrary function, summation or integration.
- In the second category, there is no involvement of any arbitrary constant or arbitrary function but summations are to be performed to evaluate some field quantities.
- The third category of relations involves some arbitrary constants. These arbitrary constants are to be evaluated by solving some simultaneous linear algebraic equations. Once arbitrary constants are evaluated, the estimation of field parameters will further require summation of involved series.

- The fourth category involves a number of improper integrals and some arbitrary functions. The evaluation of these functions will require solutions of simultaneous linear algebraic equations. Once these functions are evaluated the estimation of field parameters will further require evaluation of involved improper integrals.

In the following sections, the relations belonging to each of the above categories are identified and wherever necessary the methods to evaluate the required quantities are briefly described. Here it needs to be mentioned that while discussing cases falling under the above four categories, the repetition of equations given in chapters is avoided to save the pages. However, the modified forms of the required equations are given. The readers may, therefore, refer to the chapter in accordance with the referred equation numbers.

9.4.1 Relations without Summation or Integration

This section identifies all such cases wherein neither summations nor integrations are needed.

9.4.1.1 Fringing Flux

Section 4.4 deals with the fringing flux for tooth-opposite-tooth orientation with small air gap (g). This is a two-dimensional problem involving solution of Laplace equation and is solved by using Schwarz–Christoffel transformation. The transformation is accomplished in two steps, that is, from z-plane to w-plane (Equation 4.90) and from χ-plane to w-plane (Equation 4.95). The z $(x + jy)$ and χ $(= \varphi + j\psi)$ are the two complex planes. The parameters x and y represent axes of the z-plane and φ and ψ indicate potential and flux. Equation 4.96 can be used to get the flux lines and equipotential lines by using the following steps after assigning a suitable value to g.

A flux line can be obtained by assigning a constant value to φ and different values to ψ, similarly, an equipotential line can be obtained by assigning a constant value to ψ and different values to φ. The corresponding values of x and y can thus be computed. The process can be repeated for another value of φ or ψ to get more and more values of x and y. If the points of coordinates in the z-plane are joined, the resulting plot will represent either a flux line or an equipotential line. In this analysis, no summation or integration is involved.

9.4.1.2 Air-Gap Field of a Conductor Deep inside an Open Slot

Section 4.5 deals with the air-gap field of a conductor that is located deep inside an open slot. Here again g represents air gap and s slot width, which may be assigned suitable numerical values. In this case, the transformation is accomplished in three stages, which include $z = x + jy$, $w = u + jv$ and $\chi = \varphi + j\psi$ planes. As before, the parameters φ and ψ represent values of potentials and fluxes. Equation 4.104 relates z-plane to w-plane and Equation 4.107 relates w-plane with χ-plane. The plots for flux lines and equipotential lines can be obtained either (i) by substituting w from Equation 4.107 in Equation 4.104 for obtaining a new expression for z in terms of φ and ψ or (ii) by evaluating values of w for different values of φ and ψ from Equation 4.107 and substituting these values of w in the expression of z given by Equation 4.104. As in Section 4.4, different values are to be assigned to φ and ψ in turn. In this case, no summation or integration is involved.

9.4.1.3 Analysis of Eddy Current Induction Machines

Section 5.2 presents the analysis of eddy current induction machines. In Section 5.2.1, a simplified two-dimensional model of a polyphase solid rotor induction machine is considered. This model is shown in Figure 5.1, wherein the armature winding is simulated by a surface current sheet. This analysis ultimately leads to the expressions of the eddy current density (Equation 5.11), eddy current loss (Equation 5.13a), force density (Equation 5.14a), mechanical power developed (Equation 5.15), rotor power input (Equation 5.16) and the slip–power relation (Equation 5.17). These are all simple relations without the involvement of any series or integration. In these relations, s $[= 1 - \{ v/(\omega/\ell)\}]$ is the slip, ω is the supply frequency, ω_0 $(= s \cdot \omega)$ is the frequency of sinusoidal field (in the reference frame fixed at the rotor), μ is the permeability, σ is the conductivity, both for rotor iron, while P is the number of stator poles, τ is pole pitch, $\ell = \pi/\tau$, L_R is the rotor length, K_0 is the amplitude of current density with current flowing in axial (x) direction and H_{2y} and H_{2z} are the components of magnetic field intensity given by Equations 5.8a and 5.8b. Further parameters η and α are given by Equations 5.2 and 5.9, respectively. The only arbitrary constant K_1 involved in these relations is given by Equation 5.10b. In these, x, y, z are the coordinates and g is the air-gap length. The values of the above-referred quantities can readily be computed by assigning suitable numerical values to different involved parameters. In this case, also there is no involvement of any infinite series or improper integration.

9.4.1.4 Eddy Currents in a Conducting Plate

Section 5.3 deals with the eddy currents in a conducting plate shown in Figure 5.2. It includes cases of single-phase and polyphase alternating current excitations. The excitation winding carrying ac current is simulated by current sheets on its two surfaces. In view of this analysis, it is observed that in both cases only y-component of eddy current density exists. In these cases given in two subsequent sections, there is no involvement of any summation or integration.

9.4.1.4.1 Single-Phase Excitation

In case of single-phase excitation (Section 5.3.1), the distribution of eddy current density inside the plate is given by Equation 5.22a. In this relation, T represents thickness of the plate, θ is defined by Equation 5.21a and the depth of penetration (d) given by Equation 2.53c.

9.4.1.4.2 Polyphase Excitation

In polyphase excitation (Section 5.3.2), the only surviving component of eddy current density is given by Equation 5.41b. The parameters involved in this relation are given by Equations 5.29, 5.31 and 5.38. Besides, all the other parameters involved are defined in the text. The eddy current density for these cases can readily be obtained by assigning suitable values to involved parameters.

9.4.1.5 Eddy Currents within a Circular Core

The analysis of eddy currents within a circular core is described in Section 5.7. The core of radius R shown in Figure 5.9 leads to Equation 5.113, which spells the z component of magnetic field intensity. It involves Kelvin's functions ber and bei of known arguments $l \cdot \rho$. The parameters l and ρ are shown in the figure. The other parameters include the amplitude of current density K_0, the frequency ω and those given by Equations 5.108c and 5.108d. The eddy current density can now be obtained by using Equation 5.115. In this case, there is no involvement of any summation or integration.

9.4.1.6 Distribution of Current Density in a Circular Conductor

This case is described in Section 5.8. It deals with the distribution of current density in circular conductor with radius R. The configuration is shown in Figure 5.10 along with geometrical parameters. In this case the magnitude of the current density, as a function of the radial distance ρ from the conductor axis, is given by Equation 5.123. In this relation, I is the current flowing at power frequency ω. This expression also involves Kelvin's functions of known arguments. Here too there is no involvement of any summation or integration.

9.4.1.7 Two-Dimensional Fields in Anisotropic Media

Section 6.2 deals with the two-dimensional fields in an anisotropic homogeneous medium, which is characterised by conductivity [$\sigma = (\sigma_x, \sigma_y, \sigma_z)$], permeability [$\mu = (\mu_x, \mu_y, \mu_z)$] and permittivity [$\varepsilon = (\varepsilon_x, \varepsilon_y, \varepsilon_z)$]. The components of complex conductivity are defined by Equations 6.2a through 6.2c. As the field is assumed to be independent of x, and periodic in y and t, its variation is given by $e^{j(\omega t - \ell y)}$, where the time period is $2\pi/\omega$ and the wave length is

$2\pi/\ell$, that is, two pole pitches. In view of Maxwell's equations, the expressions of different components of E and H are given by Equations 6.8 (E_x), 6.9 (H_y), 6.10 (H_z), 6.12 (H_x), 6.13a (E_y) and 6.13b (E_z). The two parameters (viz. K_e and K_h) involved in these relations are given by Equations 6.7a and 6.11a, respectively. Besides, these relations involve four arbitrary constants viz. c_1, c_2, d_1, and d_2. The method for evaluation of these will differ from problem to problem. Two such cases are given in Sections 9.4.1.7 and 9.4.1.8, which in fact describe Sections 6.3 and 6.4. In this case also there is no involvement of any summation or integration.

9.4.1.8 Field in the Cage or Wound Rotor Machine

Section 6.3 basically deals with the evaluation of different components of magnetic field intensity (H) and electric field intensity (E) in the cage or wound rotor induction machine. Its sectional view along with some relevant parameters is shown in Figure 6.1. Some more parameters viz. peak surface current density, slip, angular frequency, pole pitch and so on are defined in the text. In view of the assigned nomenclature x represents axial, y peripheral and z radial directions. Since E and H are to be obtained in the air gap and the slotted rotor regions, these are correspondingly assigned super-fixes (a) and (r) and suffixes x, y and z. This analysis ultimately leads to Equations 6.19 (for $H_y^{(a)}$), 6.20 (for $H_x^{(a)}$), 6.21 (for $H_z^{(a)}$), 6.22a (for $E_x^{(a)}$), 6.22b (for $E_y^{(a)}$) and 6.22c (for $E_z^{(a)}$) in the air gap and to Equations 6.25 (for $H_x^{(r)}$), 6.28 (for $H_y^{(r)}$), 6.30 (for $H_z^{(r)}$), 6.32a, (for $E_x^{(r)}$), 6.32b (for $E_y^{(r)}$) and 6.32c (for $E_z^{(r)}$) in the slotted rotor region. The parameters involved in these equations are given by Equations 6.17a (k_o), 6.25b ($\bar{\sigma}_y$), 6.25c ($\bar{\sigma}_z$), 6.25d (k_h), 6.26a (k_e) and 6.26b ($\bar{\sigma}_x$). Besides, the sole surviving arbitrary constant (a) involved in the expressions of field components is given by Equation 6.40. In view of this analysis, it can be noted that there is no involvement of any summation or integration. For obtaining parameters with the involvement of anisotropic homogeneous region, the media parameter can be replaced by the relations given by Equations 6.41a through 6.41h.

9.4.1.9 Induction Machine with Skewed Rotor Slots

Section 6.4 deals with the evaluation of different components of H and E in the induction machine when its rotor slots are skewed. Figure 6.2 shows the skew angle and the axis of skewed rotor slots. Correspondingly a parameter L, related to the pole pitch (τ) and the skew angle (θ), is given by Equation 6.42. In view of this skew, the relations defining new X (axial), Y (peripheral) and Z (radial) directions are given by Equations 6.43a through 6.43c. Further, the two components of the stator current density viz. K_x and K_y obtained in terms of K_0 are given by Equations 6.46a and 6.46b. The components of H and E are finally given by Equations 6.50a ($H_x^{(a)}$), 6.50b ($H_y^{(a)}$), 6.52 ($H_z^{(a)}$), 6.55b

($E_x^{(a)}$), 6.56b ($E_y^{(a)}$) and 6.57b ($E_z^{(a)}$) in the air gap and by Equations 6.59a ($H_x^{(r)}$), 6.59b ($H_y^{(r)}$), 6.64 ($H_z^{(r)}$), 6.70b ($E_x^{(r)}$), 6.71b ($E_y^{(r)}$) and 6.72b ($E_z^{(r)}$) in the anisotropic rotor region. Some other parameters involved in these relations are given by Equations 6.51 (K), 6.61a ($\bar{\sigma}$), 6.64a (K_0), 6.69a (K_1) and 6.69b (K_2). The involved arbitrary constants a_1, a_2, b_1, b_2, a_0 and b_0 can be approximately obtained in view of the Equations 6.79, 6.80a, 6.80b, 6.81a and 6.81b. For estimating exact values of these arbitrary constants, Equations 6.82a and 6.82b must be kept in mind. This analysis shows that there is no involvement of any summation or integration.

9.4.1.10 Field Theory of Hysteresis Machines

Section 8.4 deals with the field theory of hysteresis machines. Figure 8.4 illustrates an idealised machine along with the coordinate system and some relevant parameters. Accordingly, X represents axial, Y peripheral and Z radial directions. The other required parameters include the amplitude of current density $|K_o|$, supply frequency ω, $\ell = \pi/\lambda$, pole pitch λ and different values of the permeability for the stator (∞ for $z > g$), air gap (μ_o for $g > z > 0$), rotor ring ($\alpha \cdot e^{-j}\beta$ for $0 > z > -d$) and the rotor base (μ_2 over $-d > z > -\infty$). Besides, the velocity in the peripheral direction is given as u_y and the slip is given by Equation 8.51. The H field in the air-gap region is given by Equations 8.55a and 8.55b. The rotor region is divided into two segments referred to as rotor ring (or region 1) and the rotor base (or region 2); the H fields in these two regions are indicated by H_1 and H_2. The H_{2y}, H_{2z}, H_{1y}, H_{1z} components of these two fields are given by Equations 8.57a, 8.57b, 8.58a and 8.58b, respectively. The expressions of these field components involve parameters γ_2 and γ_1, which are given by Equations 8.57c and 8.58c. These field expressions also involve two arbitrary constants a and b, which are given by Equations 8.60a and 8.60b. The expressions of 'a and b' involve parameters C_1 and C_2 and these are defined by Equations 8.61a and 8.61b.

9.4.1.10.1 Induction Machine Action

The induction machine action discussed in Section 8.4.3 relies on the above-evolved parameters. Thus, the eddy current density for two regions (J_{1x} and J_{2x}) is given by Equations 8.63a and 8.63b. The eddy current loss in two regions can now be obtained from Equations 8.64a and 8.64b. Also, the y component of force per unit volume on eddy currents is given by Equations 8.66a and 8.66b. Similarly, the mechanical power developed per unit volume for two regions is given by Equations 8.67a and 8.67b. Lastly, the slip–power relation is given by Equation 8.68. Yet another slip–power relation (Equation 8.69) is obtained in terms of total mechanical power developed due to induction machine action, and the total eddy current loss in the rotor of the machine.

9.4.1.10.2 Hysteresis Machine Action

The hysteresis machine action described in Section 8.4.4 also relies on the relations obtained earlier in Section 8.4. For this case the hysteresis loss per unit volume of the rotor ring is given by Equation 8.76b, the eddy current loss per unit rotor ring volume by Equation 8.76c, the mechanical power due to induction machine action by Equation 8.76d and the mechanical power due to hysteresis machine action by Equation 8.76e. These referred relations involve either the current densities given by Equations 8.63a and 8.63b or the components of magnetic field intensity spelled by Equations 8.58a and 8.58b. As before, the parameters a, b, C_1 and C_2 involved in these referred relations bear the same meanings as spelled earlier. The slip–power relation for this case is spelled by Equations 8.77 and 8.80. In the analysis of Section 8.4, there is no involvement of any series or integrals.

9.4.2 Relations Involving Simple Summations

This section identifies the cases wherein no arbitrary constant is to be evaluated but simple summation is involved at some stage.

9.4.2.1 Eddy Currents in Rectangular and Square Cores

Section 5.4 deals with eddy currents in a core shown in Figure 5.3. The expression of H_Z obtained for rectangular and square cores is given by Equations 5.45 and 5.48, respectively. These expressions are obtained by neglecting the displacement currents and by simulating the excitation winging by a current sheet with current density amplitude K_0. The relation for eddy current density can be readily obtained by using the relation $\boldsymbol{J} = \nabla \times \boldsymbol{H}$. In these relations, the parameters involved can directly be evaluated by using Equations 5.46a (for α_m), 5.46b (for β_n), 5.48a (for γ_p), 5.47a (for a_m) and 5.47b (for b_n). All other parameters are defined in the text or in the referred figure. It is to be noted that none of these parameters involve any summation or integration. However, for computing the eddy current density summation is to be carried out as expressions for H_Z given by Equations 5.45 and 5.48.

9.4.2.2 Eddy Currents in Triangular Core

Section 5.5 deals with the eddy currents in a core of triangular cross-section shown in Figure 5.4. Here too the final expression for H_Z is given by Equation 5.59 from which the components of eddy current density in the triangular core can be readily found by using the relation $\boldsymbol{J} = \nabla \times \boldsymbol{H}$. The parameters involved in this relation are given by Equations 5.54a (for α_{1m}), 5.54b (for η^2) and 5.57a (for α_{2m}). The remaining parameters are given in the text or in

Figures 5.4 and 5.5. Here, except in Equation 5.59, no summation is involved in any of the above-referred equations.

9.4.2.3 Slot Leakage Inductance

Section 8.2 deals with the slot leakage inductance for conductors in open slots. Figure 8.1 illustrates a configuration wherein a slot with width w and depth d contains a current-carrying conductor of thickness c with its bottom and top surfaces located at $y = h_1$ and h_2, respectively. This analysis leads to Equation 8.17 from which leakage inductance can be obtained in terms of total flux linkage given by Equation 8.16. Equation 8.16 involves double summation over odd integer values of m and n. This equation also involves a known parameter $A_{m,n,}$ which is given by Equation 8.7. In general, vector magnetic potential comprises two components A_{Z1} (Equation 8.6) and A_{Z2} (Equation 8.13). While deriving Equation 8.17, A_{Z2}, which depends on external sources, is neglected. The consideration of external sources will require evaluation of two sets of arbitrary constants (b_p and c_q) involved in Equation 8.13 and a fresh expression for inductance is to be obtained from Equation 8.15, wherein A_Z is the sum of A_{Z1} and A_{Z2}.

9.4.2.4 Leakage Inductance of Transformers

Section 8.3 deals with the leakage inductance of transformers. A configuration of a single-phase shell-type transformer with sandwich winding along with various parameters, physical arrangement of coils and regions are shown in Figure 8.2. The constant vector potential surfaces are illustrated in Figure 8.3. The analysis ultimately leads to Equation 8.49, which gives the leakage inductance referred to low voltage side for one coil group. The flux linking the current can be calculated from Equation 8.47 for exact and from Equation 8.48 for approximate values. The parameters involved in the expression of flux are given by Equations 8.47a, 8.47b, 8.18a, 8.39 and 8.41. In this analysis, summation over all values of m is involved only in Equations 8.47 and 8.48 and no evaluation of arbitrary constants is needed.

9.4.2.5 Transient Fields in Plates

Section 8.6 deals with the transient fields in plates due to type 2 impact excitations. Figure 8.7 shows a large conducting plate of thickness W with surface current sheets of density K_y (with magnitude K_0) on its surfaces. The configuration is divided into three regions of which only two (1 and 2) are considered due to symmetry.

9.4.2.5.1 Current Impact Excitation

In case of current impact excitation (Section 8.6.1), the involved parameters include the surface resistivity, ρ_s, permeability, μ, permittivity, ε_o, and conductivity, σ, of the plate. The field components in region 1 are given by Equations 8.124a (H_{1z}) and 8.124b (E_{1y}). In these equations, the involved parameters include the relaxation time (τ), the plate parameter (θ) and parameters (α_m, β_m) that are given by Equations 8.125a, 8.125b and 8.125c, respectively. Besides, the only set of arbitrary constants (c_m) is given by Equation 8.126b. Similarly, the field components in region 2 are given by Equations 8.132a and 8.132b. The other parameters (t_- and t_+) involved in this last set of equations are defined by Equations 8.133a and 8.133b. In this part of the analysis, summation is involved only at the stage of evaluation of field components and that too when c_m is involved in the expressions.

9.4.2.5.2 Voltage Impact Excitation

In case of voltage impact excitation (Section 8.6.2), the field components in region 1 are given by Equations 8.138a and 8.138b. The only set of arbitrary constants (d_m) is given by Equation 8.139a. The magnetic flux (φ) and the voltage drop in the winding resistance, both in per unit length in the y-direction, are given by Equations 8.141b and 8.142a, respectively. The expression for K_y, which emerges from these relations, is given by Equation 8.142b. Lastly, the field components in region 2 are given by Equations 8.144a and 8.144b. It is to be noted that the last two referred equations involve C_m as a set of arbitrary constants. In this case too, the summation is involved only at the stage of evaluation of field components and that too when c_m or d_m is involved in the expressions.

9.4.3 Summations Leading to Simultaneous Linear Algebraic Equations

In subsequent sections many cases of the involvement of infinite series in the field relations are described. The relations lead to the sets of simultaneous linear algebraic equations, which in turn take the form of matrix equations. The manipulation of these matrices ultimately gives the required parameters.

9.4.3.1 Potential Distributions in Tooth-Opposite-Tooth Orientation Case

Section 4.2.1 deals with potential distributions in the tooth-opposite-tooth configuration shown in Figure 4.1. The parameters involved include slot width (s), tooth width (t), tooth pitch (λ, where $\lambda = s + t$), air gap (g) and M and N, the maximum values of m and n, respectively. The evaluation of potential distribution in different regions may involve the following steps:

Step 1: Assign suitable values to slot width (s), gap length (g) and tooth pitch (λ).

Step 2: Evaluate $\mathcal{S}(m, M)$ from Equation 4.18b for $m = 1, 3, 5, \ldots$.

Step 3: Substitute these values of $\mathcal{S}(m, M)$ in Equation 4.11 to get the values of a_m for $m = 1, 3, 5, \ldots$. The modified form of Equation 4.11 is given below

$$a_m + \frac{s}{\lambda}\cdot\frac{s}{g}\cdot\sin\left(\frac{m\pi}{2}\right)\sum_{M-odd}^{\infty} a_M \cdot \frac{\sin(M\pi/2)}{(M\pi/2)} + 2\cdot\left(\frac{\lambda}{\pi S}\right)^2\cdot\sin\left(\frac{m\pi}{2}\right)\cdot\sum_{M-odd}^{\infty}\left[a_M\cdot M\cdot\sin\left(\frac{M\pi}{2}\right)\cdot\mathcal{S}(m,M)\right] = -\frac{1}{2}\cdot\frac{s}{g}\cdot\sin\left(\frac{m\pi}{2}\right) \tag{9.15}$$

or

$$a_m + A_m\cdot\sum_{M-odd}^{\infty} a_M\cdot C_M + D_m\sum_{M-odd}^{\infty}\left[a_M\cdot M\cdot S_M\cdot\mathcal{S}(m,M)\right] = B_m \tag{9.16a}$$

where

$$A_m = \frac{s}{\lambda}\cdot\frac{s}{g}\cdot\sin\left(\frac{m\pi}{2}\right),\quad C_M = \sin\left(\frac{M\pi}{2}\right)\Big/\left(\frac{M\pi}{2}\right)\quad S_M = \sin\left(\frac{M\pi}{2}\right)$$
$$D_m = 2\cdot\left(\frac{\lambda}{\pi S}\right)^2\cdot\sin\left(\frac{m\pi}{2}\right)\quad \text{and}\quad B_m = -\frac{1}{2}\cdot\frac{s}{g}\cdot\sin\left(\frac{m\pi}{2}\right) \tag{9.16b}$$

On assigning different values to m, Equation 9.16a can be written in the following form:

$$a_1 + A_1\cdot\sum_{M-odd}^{\infty} a_M\cdot C_M + D_1\sum_{M-odd}^{\infty}\left[a_M\cdot(M)\cdot S_M\cdot\mathcal{S}(1,M)\right] = B_1$$
$$a_3 + A_3\cdot\sum_{M-odd}^{\infty} a_M\cdot C_M + D_3\sum_{M-odd}^{\infty}\left[a_M\cdot(M)\cdot S_M\cdot\mathcal{S}(3,M)\right] = B_3$$
$$\ldots$$
$$a_M + A_M\cdot\sum_{M-odd}^{\infty} a_M\cdot C_M + D_M\sum_{M-odd}^{\infty}\left[a_M\cdot(M)\cdot S_M\cdot\mathcal{S}(M,M)\right] = B_M \tag{9.17a}$$

Equation 9.17a can further be expanded as

$$
\begin{aligned}
&a_1 + A_1 \cdot (a_1 \cdot C_1 + a_3 \cdot C_3 + \cdots + a_M \cdot C_M) \\
&\quad + D_1\{1 \cdot a_1 \cdot S_1 \cdot \mathcal{S}(1,1) + 3 \cdot a_3 \cdot S_3 \cdot \mathcal{S}(1,3) + \cdots + M \cdot a_M \cdot S_M \cdot \mathcal{S}(1,M)\} = B_1 \\
&a_3 + A_3 \cdot (a_1 \cdot C_1 + a_3 \cdot C_3 + \cdots + a_M \cdot C_M) \\
&\quad + D_3\{1 \cdot a_1 \cdot S_1 \cdot \mathcal{S}(1,1) + 3 \cdot a_3 \cdot S_3 \cdot \mathcal{S}(1,3) + \cdots + M \cdot a_M \cdot S_M \cdot \mathcal{S}(3,M)\} = B_3 \\
&\qquad \cdots \\
&a_M + A_M \cdot (a_1 \cdot C_1 + a_3 \cdot C_3 + \cdots + a_M \cdot C_M) \\
&\quad + D_M\{1 \cdot a_1 \cdot S_1 \cdot \mathcal{S}(1,1) + 3 \cdot a_3 \cdot S_3 \cdot \mathcal{S}(1,3) + \cdots + M \cdot a_M \cdot S_M \cdot \mathcal{S}(M,M)\} = B_M
\end{aligned}
\tag{9.17b}
$$

Equation 9.17b can be written as

$$
\begin{aligned}
&a_1(1 + A_1 \cdot C_1 + D_1 S_1\ \mathcal{S}(1,1)) + a_3\{A_1 \cdot C_3 + 3 \cdot D_1 S_3 \cdot \mathcal{S}(1,3)\} \\
&\quad + \cdots + a_M\{A_1 \cdot C_M + M \cdot D_1 \cdot S_M \cdot \mathcal{S}(1,M)\} = B_1 \\
&a_1(A_3 \cdot C_1 + D_3 \cdot S_1\ \mathcal{S}(1,1)) + \{1 + A_3 \cdot C_3 + 3 \cdot D_3 \cdot S_3 \cdot \mathcal{S}(1,3)\} \\
&\quad + \cdots + a_M\{A_1 \cdot C_M + M \cdot D_3 \cdot S_M \cdot \mathcal{S}(3,M)\} = B_3 \\
&\qquad \cdots\cdots \\
&a_1(A_M \cdot C_1 + D_M \cdot S_1\ \mathcal{S}(1,1)) + a_3\{A_M \cdot C_3 + 3 \cdot D_M \cdot S_3 \cdot \mathcal{S}(1,3)\} \\
&\quad + \cdots + a_M\{1 + A_M \cdot C_M + M \cdot D_M \cdot S_M \cdot \mathcal{S}(M,M)\} = B_M
\end{aligned}
\tag{9.17c}
$$

Equation 9.17c can be written in the form of a matrix equation

$$[A]\ [a] = [B] \tag{9.18a}$$

where

$$
[A] = \begin{bmatrix}
1 + A_1 \cdot C_1 + D_1 S_1\ \mathcal{S}(1,1) & A_1 \cdot C_3 + 3 \cdot D_1\ S_3 \cdot \mathcal{S}(1,3) & .. & A_1 C_M + M \cdot D_1 \cdot S_M \cdot \mathcal{S}(1,M) \\
A_3 \cdot C_1 + D_3 \cdot S_1\ \mathcal{S}(1,1) & 1 + A_3 \cdot C_3 + 3 \cdot D_3 \cdot S_3 \cdot \mathcal{S}(1,3) & .. & A_1 C_M + M \cdot D_3 \cdot S_M \cdot \mathcal{S}(3,M) \\
.. & .. & .. & .. \\
A_M \cdot C_1 + D_M \cdot S_1\ \mathcal{S}(1,1) & A_M \cdot C_3 + 3 \cdot D_M \cdot S_3 \cdot \mathcal{S}(1,3) & .. & 1 + A_M C_M + M D_M \cdot S_M \cdot \mathcal{S}(1,M)
\end{bmatrix}
$$

$$
[a] = \begin{bmatrix} a_1 \\ a_3 \\ \ldots \\ a_M \end{bmatrix} \quad \text{and } [B] = \begin{bmatrix} B_1 \\ B_3 \\ \ldots \\ B_M \end{bmatrix} \tag{9.18b}
$$

In Equation 9.18a, elements of matrix $[A]$ are known coefficients, elements of matrix $[B]$ are also known and the elements of matrix $[a]$ are unknowns, which are to be determined.

In view of Equation 9.18a, we get

$$[a] = [A]^{-1}\,[B] \tag{9.19}$$

where $[A]^{-1}$ is the inverse of matrix $[A]$.

From Equation 9.19, we finally get a_m. The values of all the required elements of Equation 9.18b are given in Equation 9.16b.

Step 4: Substitute values of a_m obtained from Equation 9.19 in Equations 4.7b and 4.8b to get b_0 and b_n for $n = 1, 2, 3, \ldots$.

Step 5: Substitution of values of s, g, λ, a_m and b_n in Equations 4.1 through 4.3 will give $\mathcal{V}_1$, $\mathcal{V}_2$ and $\mathcal{V}_0$. To get values of these potentials, different values of y and z are to be assigned in accordance with regions illustrated in Figure 4.1. Accordingly, the values of z lie between 0 and $>-\infty$ for stator slot, g and $<\infty$ for rotor slot and between 0 and g for the air gap. Similarly, the values of y may lie between $(\lambda - s)/2$ and $(\lambda + s)/2$ for both stator and rotor slots. The potentials in air gap may, however, be computed for values of y from 0 to λ.

To select the maximum values of m and n, the following steps may be adopted:

1. Set a tolerance limit say Δ.
2. Assign some suitable value to integer m (say M_1) and evaluate the arbitrary constants. Let these values be termed as a_{m1} for $m = 1, 2, \ldots, M_1$.
3. Increase the value of integer m to M_2 and again evaluate the arbitrary constants. Let these values be termed as a_{m2} for $m = 1, 2, \ldots, M_2$.
4. If $a_{m1} \sim a_{m2}$ lie within the assigned tolerance limit Δ, then $m = M_2$ is the finally selected value for m. If not, further increase the value of m to M_3 and check the tolerance. Repeat the procedure till tolerance criterion is met.
5. Follow the similar procedure for selecting value of N to be assigned to integer n.

9.4.3.2 *Potential Distributions in Tooth-Opposite-Slot Orientation*

Section 4.2.2 deals with the potential distributions in tooth-opposite-slot configuration shown in Figure 4.2. The involved parameters include slot width (s), tooth width (t), tooth pitch (λ, where $\lambda = s + t$), air gap (g) and M and M the maximum value of m and n, respectively. The potential distribution in different regions can be obtained by using the following steps:

Step 1: Assign suitable values to s, g, λ and t as and where they appear in the equations.

Step 2: Evaluate $\mathcal{S}_1(m, M)$ from Equation 4.28 for values $n = 1 - N$ and $\mathcal{S}_2(m, M)$ from Equation 4.32a for $m \neq M$ and from Equation 4.32b for $m = M$.

Step 3: Substitute these values in Equation 4.27 and evaluate values of c_m for m, $M = 1, 3, 5, \ldots$.

The modified form of Equation 4.27 is written as

$$c_m + S_m \sum_{M-odd}^{\infty} c_M \cdot F_M + D_m \cdot \sum_{M-odd}^{\infty} c_M \cdot K_{mM} = B_m \qquad (9.20a)$$

where

$$B_m = -\frac{s}{g} \cdot \left(\frac{2}{m\pi}\right)^2 \cdot \sin\left(\frac{m\pi}{2}\right) \quad F_M = \frac{\sin(M\pi/2)}{(M\pi/2)} \quad D_m = \left(\frac{\lambda}{s}\right)^2 \cdot \frac{2}{\pi^2} \cdot \sin\left(\frac{m\pi}{2}\right)$$

$$S_m = 2 \cdot \frac{s}{g} \cdot \frac{s}{\lambda}\left(\frac{2}{m\pi}\right)^2 \cdot \sin\left(\frac{m\pi}{2}\right) \quad K_{m\,M} = M \cdot \sin\left(\frac{M\pi}{2}\right)[\mathcal{S}_1(m, M) + \mathcal{S}_2(m, M)] \qquad (9.20b)$$

The expansion steps of Equation 9.20a are given as

$$\begin{aligned} &c_1 + S_1 \sum_{M-odd}^{\infty} c_M \cdot F_M + D_1 \cdot \sum_{M-odd}^{\infty} c_M \cdot K_{1M} = B_1 \\ &c_3 + S_3 \sum_{M-odd}^{\infty} c_M \cdot F_M + D_3 \cdot \sum_{M-odd}^{\infty} c_M \cdot K_{3M} = B_3 \\ &\ldots\ldots \\ &c_M + S_M \sum_{M-odd}^{\infty} c_M \cdot F_M + D_M \cdot \sum_{M-odd}^{\infty} c_M \cdot K_{MM} = B_M \end{aligned} \qquad (9.20c)$$

$$\begin{aligned} &c_1 + S_1(c_1 \cdot F_1 + c_3 \cdot F_3 + \cdots + c_M \cdot F_M) + D_1 \cdot (c_1 \cdot K_{11} + c_3 \cdot K_{13} + \cdots + c_M \cdot K_{1M}) = B_1 \\ &c_3 + S_3(c_1 \cdot F_1 + c_3 \cdot F_3 + \cdots + c_M \cdot F_M) + D_3 \cdot (c_1 \cdot K_{11} + c_3 \cdot K_{13} + \cdots + c_M \cdot K_{1M}) = B_3 \\ &\ldots\ldots \\ &c_M + S_M(c_1 \cdot F_1 + c_3 \cdot F_3 + \cdots + c_M \cdot F_M) + D_M \cdot (c_1 \cdot K_{11} + c_3 \cdot K_{13} + \cdots + c_M \cdot K_{1M}) = B_M \end{aligned} \qquad (9.20d)$$

In view of the procedure followed in Section 9.4.3.1, Equation 9.20d reduces to the following matrix equation from which the values of c_m can be evaluated:

$$\begin{bmatrix} c_1 \\ c_3 \\ \cdots \\ c_M \end{bmatrix} = \begin{bmatrix} 1+S_1F_1+D_1K_{11} & S_1F_3+D_1K_{13} & .. & S_1F_M+D_1S_{1M} \\ S_3F_1+D_3K_{31} & 1+S_3F_3+D_3S_{33} & .. & S_3F_M+D_3S_{3M} \\ .. & .. & .. & .. \\ S_MF_1+D_MK_{M1} & S_MF_3+D_MS_{M3} & .. & 1+S_MF_M+D_MK_{MM} \end{bmatrix}^{-1} \cdot \begin{bmatrix} B_1 \\ B_3 \\ \cdots \\ B_M \end{bmatrix} \tag{9.21}$$

All the required elements for Equation 9.21 are given in Equation 9.20b.

Step 4: Substitute values of c_m obtained from Equation 9.21 in Equations 4.23a and 4.23b to get d_0 and d_n for $n = 1, 2, 3, \ldots$.

Step 5: Substitute values of s, g, λ, c_m and d_n in Equations 4.19, 4.20 and 4.22 to get $\mathcal{V}_1$, $\mathcal{V}_2$ and $\mathcal{V}_0$. To get values of these potentials, different values of y and z are to be assigned in accordance with regions illustrated in Figure 4.2. Accordingly, the values of z lie between 0 and $> -\infty$ for stator slot, between g and $<\infty$ for rotor slot and between 0 and g for the air gap.

In this case, the values of y may lie between $(\lambda - s)/2$ and $(\lambda + s)/2$ for stator slot, and between $(\lambda - s/2)$ and $(\lambda + s/2)$ for rotor slot. The potentials in air gap may be computed for y between $\lambda - t$ and $\lambda + t$. The maximum integer values to m and n may be assigned in accordance of the procedure indicated in Section 9.4.3.1.

9.4.3.3 Potential Distributions in Arbitrary Tooth Orientation

Section 4.2.3 deals with the potential distributions in arbitrary tooth orientation configuration shown in Figure 4.3. The parameters involved include slot width (s), tooth width (t), tooth pitch (λ, where $\lambda = s + t$), air-gap length (g), displacement between stator and rotor teeth ($\delta/2$, where $0 \le \delta \le \lambda$) and N, M, P and Q the maximum value of n, m, p and q, respectively. The estimation of potentials in different regions may require the following steps:

Step 1: Assign suitable values to the dimensional parameters s, g, λ, t and δ.

Step 2: Evaluate $\mathcal{S}_3(m, M)$, $\mathcal{S}_4(m, M)$ and $\mathcal{S}_5(m, M)$ from Equations 4.43d, 4.43e and 4.43j, respectively, for $m, M = 1, 2, 3, \ldots$.

Step 3: Substitute these values in Equation 4.43 to get the values of p_m. This equation is modified as below

$$p_m + C_m \sum_{M=1}^{\infty} p_M \cdot D_M + \sum_{M=1}^{\infty} p_M \cdot G_M\, S_{mM} = B_m \tag{9.22a}$$

where

$$B_m = \left(\frac{s}{2g}\right) \cdot \left(\frac{2}{m\pi}\right)^2 \cdot \{1 - \cos(m\pi)\} \quad D_M = \frac{[1 - \cos(M\pi)]}{M\pi}$$

$$C_m = \times \left(\frac{s}{\lambda}\frac{s}{g}\right) \cdot \left(\frac{2}{m\pi}\right)^2 \cdot \{1 - \cos(m\pi)\} \quad G_M = \left(\frac{1}{\pi}\right) \cdot \left(\frac{\lambda}{s}\right) \cdot \frac{1}{2\pi} \cdot \left(\frac{\lambda}{s}\right)$$

$$\times (M) \cdot \{1 - \cos(M\pi)\}$$

$$S_{mM} = \mathcal{S}_3(m, M) - \cos(m\pi) \cdot \mathcal{S}_4(m, M) + \{1 - \cos(m\pi)\} \cdot \mathcal{S}_5(m, M) \tag{9.22b}$$

The stepwise expansion of Equation 9.22a is given below:

$$\begin{aligned}
p_1 + C_1 \sum_{M=1}^{\infty} p_M \cdot D_M + \sum_{M=1}^{\infty} p_M \cdot G_M S_{1M} &= B_1 \\
p_3 + C_3 \sum_{M=1}^{\infty} p_M \cdot D_M + \sum_{M=1}^{\infty} p_M \cdot G_M S_{3M} &= B_3 \\
\cdots\cdots \\
p_M + C_M \sum_{M=1}^{\infty} p_M \cdot D_M + \sum_{M=1}^{\infty} p_M \cdot G_M S_{MM} &= B_M
\end{aligned} \tag{9.22c}$$

$$\begin{aligned}
&p_1 + C_1(p_1 \cdot D_1 + p_2 \cdot D_{2+\cdots+} p_M \cdot D_M) + (p_1 \cdot G_1 S_{11} + p_2 \cdot G_2 S_{12} + \cdots + p_M \cdot G_M S_{1M})\, S_{1M} = B_1 \\
&p_2 + C_2(p_1 \cdot D_{1+} p_2 \cdot D_{2+\cdots+} p_M \cdot D_M) + (p_1 \cdot G_1 S_{21} + p_2 \cdot G_2 S_{22} + \cdots + p_M \cdot G_M S_{3M}) = B_2 \\
&\cdots \\
&p_M + C_M(p_1 \cdot D_{1+} p_2 \cdot D_{2+\cdots+} p_M \cdot D_M) + (p_1 \cdot G_1 S_{M1} + p_2 \cdot G_2 S_{M2} + \cdots + p_M \cdot G_M S_{MM}) = B_M
\end{aligned} \tag{9.22d}$$

In view of Equation 9.22d, we get the matrix relation $[A]\,[x] = [B]$ where different matrices obtained are as follows:

$$[A] = \begin{bmatrix} 1 + C_1 D_1 + G_1 S_{11} & C_1 D_2 + G_2 S_{12} & .. & C_1 D_M + G_M S_{1M} \\ C_2 D_1 + G_1 S_{21} & 1 + C_2 D_2 + G_2 S_{22} & .. & C_2 D_M + G_M S_{2M} \\ .. & .. & .. & .. \\ C_M D_1 + G_1 S_{M1} & C_M D_2 + G_2 S_{M2} & .. & 1 + C_M D_M + G_M S_{MM} \end{bmatrix}$$

$$[x] = \begin{bmatrix} p_1 \\ p_3 \\ \cdots \\ p_M \end{bmatrix} \text{ and } [B] = \begin{bmatrix} B_1 \\ B_3 \\ \cdots \\ B_M \end{bmatrix} \tag{9.23}$$

The values of arbitrary constants can be evaluated from the relation $[x] = [A]^{-1} [B]$.

Step 3: Substitute values of arbitrary constants in Equations 4.41a and 4.41b to get q_0 and q_n.

Step 4: Substitute values of s, g, λ, δ, q_0, q_n and p_M in Equations 4.33, 4.34 and 4.36 to get V_1, V_2 and V_0. For $\delta = 0$, the results must be the same as obtained for tooth-opposite-tooth case and $\delta = \lambda/4$ must lead to the same results as obtained for tooth-opposite-slot case. To get values of these potentials, different values of y and z are to be assigned in accordance with regions illustrated in Figure 4.3. Accordingly, the values of z has to vary between $-g/2$ and $>-\infty$ for stator slot, $g/2$ and $<\infty$ for rotor slot and between $-g/2$ and $g/2$ for the air gap. Here, y varies over $(t/2 - \delta) \leq y \leq (s + t/2 - \delta)$ for stator slot and $(-s - t/2 + \delta) \leq y \leq (-t/2 + \delta)$ for rotor slot. The potentials in air gap may be computed for y between 0 and λ. Further, the maximum values of M and N may be assigned to m and n in accordance with the procedure indicated in Section 9.4.3.1.

9.4.3.4 Air-Gap Permeance

The air-gap permeance given in Section 4.2.4 can be estimated by taking δ as $0 \leq \delta \leq \lambda/2$, $\mu_0 = 4\pi \times 10^{-7}$ and q_0 given by Equation 4.41a in Section 4.2.3. The value of q_0 is to be substituted in Equations 4.45, 4.46a through 4.46c to get permeance (P), lost flux ($P_\lambda - P$), generalised Carter's coefficient (σ) and the gap-extension factor ($K_{g\delta}$), respectively. The summation of series will be required to estimate the value of q_0 for which the linear simultaneous equations are to be solved as given by Equation 9.23.

9.4.3.5 Magnetic Field near Armature Winding Overhang

Section 4.6 deals with the magnetic field near armature winding overhang. The configuration used is shown in Figure 4.13. In this case, as a first step, select suitable values for supply frequency (ω), slot depth (d), tooth pitch (λ) and gap length (g). Obtain $\ell_{m\cdot n}$ from armature winding details. Find a_m, $\alpha_{m,p}$, $\beta_{m,q}$ and $\gamma_{m,n}$ from Equations 4.118, 4.112, 4.114 and 4.115, respectively, for m, p, $q = 1, 2, 3, \ldots, \infty$ and $n = 1, 3, 5, \ldots, \infty$. Here, it needs to be noted that the application of various boundary conditions results in nine equations, which are numbered as 4.119a through 4.119i. Since x-derivative of Equations 4.119d and 4.119e results in Equations 4.119a and 4.119b, respectively and z-derivative of Equations 4.119f results in Equations 4.119i, Equations 4.119a, 4.119b and 4.119i can be ignored. Also, Equation 4.119c directly relates $A_{m,q}$ with $b_{m,q}$ and can also be ignored. The relations that are to be used for evaluating arbitrary constants are given by Equations 4.119d through 4.119h. These relations can now be manipulated in the steps given below:

Step 1: Multiply Equations 4.119d and 4.119e by $\cos((N\pi/2d) \cdot x)$, where N is any integer, and then integrate over $0 \leq x \leq d$. This gives $A'_{m,n}$ and $A''_{m,n}$ in terms of $P'_{m,p}$ and $P''_{m,p}$ (summed over p), respectively, provided that N is replaced by n after the integrations.

Step 2: In Equation 4.119f, substitute $A'_{m,n}$ and $A''_{m,n}$ in terms of $P'_{m,p}$ and $P''_{m,p}$ (summed over p). Then multiply the resulting equation by sin $((Q\pi/g)\cdot z)$, where Q is any integer, and integrate over $0 \le x \le d$. This gives $A_{m,q}$ in terms of a_m, $P'_{m,p}$ and $P''_{m,p}$ (both, summed over p), provided that Q is replaced by q after the integrations.

Step 3: In Equations 4.119g and 4.119h, substitute $A_{m,q}$, $A'_{m,n}$ and $A''_{m,n}$ in terms of $P'_{m,p}$ and $P''_{m,p}$ (summed over p). Multiply the modified Equations 4.119g and 4.119h by $\sin((P\pi/d)\cdot z)$, where P is any integer, and then integrate over $0 \le x \le d$. This gives two sets of equations (one from the modified Equation 4.119g and the other from the modified Equation 4.119h) relating $P'_{m,p}$ and $P''_{m,p}$, provided that after the integration P is replaced by p.

Step 4: Numerical solutions of the two sets of simultaneous equations thus obtained give $P'_{m,p}$ and $P''_{m,p}$ for all integer values of p.

Step 5: Back substitutions give $A'_{m,n}$ and $A''_{m,n}$ (from step 1), $A_{m,q}$ (from step 2) and $b_{m,q}$ from Equation 4.119c.

Step 6: After obtaining all the arbitrary constants, the field components can be obtained for regions 1, 2, 3 and 4 from Equations 4.112 through 4.115, respectively.

9.4.3.6 Eddy Currents in Regular Polygonal Cross-Section Cores

Section 5.6 describes a pseudoanalytical method for determination of approximate magnetic field intensity in the cores with regular polygonal cross-sections. Once the magnetic intensity is obtained, eddy currents can be readily found. The three regular polygonal cross-sections considered include triangular, hexagonal and octagonal.

9.4.3.6.1 Triangular Cross-Section

In case of triangular cross-section shown in Figure 5.6, the expression for magnetic field intensity is given by Equation 5.69. This expression involves a Fourier coefficient T_m. The function represented by this Fourier series is referred to as torch function. This function in the final form is given by Equation 5.24

$$\left[K\cdot\frac{2}{\pi}\cdot\frac{1}{n}\right] \cong -\sum_{m-odd}^{(2P-1)} T_m \cdot \left[\frac{(m\pi/2)-(n\pi/1)}{\left\{((m\pi/2)-(n\pi/1))^2+\left(\alpha_m\cdot L\left(\sqrt{3}/2\right)\right)^2\right\}} + \frac{(m\pi/2)+(n\pi/1)}{\left\{((m\pi/2)+(n\pi/1))^2+\left(\alpha_m\cdot L\left(\sqrt{3}/2\right)\right)^2\right\}}\right]$$

for $n = 1,3,5,\ldots,(2P-1)$. (5.24)

Equation 5.24 can be written in the following modified form

$$B_n = -\sum_{m-odd}^{2p-1} T_m \cdot A_{m,n} \tag{9.24a}$$

where

$$A_{m,n} = \left[\frac{((m\pi/2) - (n\pi/1))}{\left\{((m\pi/2) - (n\pi/1))^2 + \left(\alpha_m \cdot L\left(\sqrt{3}/2\right)\right)^2\right\}} + \frac{((m\pi/2) + (n\pi/1))}{\left\{((m\pi/2) + (n\pi/1))^2 + \left(\alpha_m \cdot L\left(\sqrt{3}/2\right)\right)^2\right\}} \right]$$

and

$$B_n = K \cdot \frac{2}{\pi} \cdot \frac{1}{n} \tag{9.24b}$$

Equation 9.24, can be written in the form of a matrix equation

$$[A]_{n\times m}[T]_{m\times 1} = [B]_{n\times 1} \quad \text{Or} \quad [T]_{m\times 1} = [A]^{-1}_{m\times n}[B]_{n\times 1} \tag{9.25}$$

Numerical solution of this equation gives values of T_m, which can be substituted in Equation 5.69 to get the magnetic field intensity. Equation 5.69 also involves some other parameters, which are given in the text by Equations 5.61a through 5.61d and 5.64a. The geometrical parameters are given in Figure 5.6. Once H is evaluated, eddy current density can be obtained by using the curl relation.

9.4.3.6.2 Hexagonal Cross-Section

The hexagonal cross-section is shown in Figure 5.7. This case is similar to that of Section 9.4.3.5.1. Equation 5.86 can be converted into matrix relation and the Fourier coefficient T_m can be obtained in the similar manner. The numerical values of T_m can be substituted in Equation 5.81 to get the magnetic field intensity. This equation also involves some other parameters, which are indicated in the text and Figure 5.7. Once H is evaluated, eddy current density J can be obtained as before.

9.4.3.6.3 Octagonal Cross-Section

In case of octagonal cross-section shown in Figure 5.8, the Fourier coefficient T_m can be obtained from Equation 5.99 in the way used in Section 9.4.3.6.1. The values of integrals I_{1m}, I_{2m}, I_{3m} and I_{4m} involved in Equation 5.99 are given by Equations 5.101a through 5.101d. The distribution of the magnetic field in the core can be obtained from Equation 5.93. The eddy current density, thereafter, can be obtained from the magnetic field in the core.

9.4.3.7 Eddy Currents in Laminated Rectangular Cores

Section 5.9 deals with the eddy currents in laminated rectangular cores. Figure 5.11 shows such a core with all necessary parameters. The exciting coil wound around the long rectangular core carrying an alternating current i is simulated by a surface current density K_o given by Equation 5.125 wherein N is the number of turns per unit length of the coil. The current-carrying coil produces time-varying magnetic field H_z, in the core, eddy current density with components J_x and J_y, in the conducting regions and displacement currents in the insulation regions. The magnetic field outside the coil is neglected. For the long rectangular core with uniformly distributed current sheet, the magnetic field is entirely axial and independent of z. The other assumed constant parameters include permeability μ for iron regions, permittivity ε for insulation regions and conductivity (σ, σ') for both types of regions. In view of Maxwell's equations for harmonic fields, the expressions for the components of H are given by Equations 5.131 ($H_z^{0'}$), 5.134 ($H_z^{m'}$), 5.137a ($H_z^{n'}$) and 5.140 (H_z^{m}). Similarly, the components of eddy current density J are given by Equations 5.132a ($J_x^{0'}$), 5.132b ($J_y^{0'}$), 5.135a ($J_x^{m'}$), 5.135b ($J_y^{m'}$), 5.138a ($J_x^{n'}$) 5.138b ($J_y^{n'}$), 5.141a (J_x^{m}) and 5.141b (J_y^{m}). The known parameters involved in these relations are given by Equations 5.126a (γ), 5.127a (γ'), 5.128c (δ), 5.129c (δ'), 5.131a (α'_p), 5.131b ($b_p^{0'}$), 5.137b ($a_p^{n'}$) and 5.140a (α_p). Besides, these expressions also involve a few sets of arbitrary constants (viz. $a_p^{0'}$, $a_p^{m'}$, $b_p^{m'}$, $b_p^{n'}$, a_p^{m} and b_p^{m}). These arbitrary constants can be evaluated in view of the continuity of tangential components of H and E at the boundaries between different adjacent regions.

9.4.3.8 Tooth-Ripple Harmonics in Solid Rotor Induction Machines

Section 7.2 deals with the tooth-ripple harmonics in solid rotor induction machines containing three-phase stator winding housed in open rectangular stator slots. The configuration of idealised machine is illustrated in Figure 7.2. In this figure, X-axis indicates axial, Y-axis peripheral and Z-axis radial directions. The rotor air-gap surface lies at $z = 0$ and stator air-gap surface at $z = -g$. This figure also shows the geometrical locations of shaft, rotor core, air gap, stator and conductors belonging to red (R), blue (B) and yellow (Y) phases located in stator slots, in two layers and the assumed permeabilities in different parts. Figure 7.3 illustrates the conductors belonging to only red phase and Figure 7.4 illustrates location of different stator slots along with the parameters defining the geometrical location and the dimensions of a conductor in a slot.

Step 1: With reference to Figure 7.4, the vector magnetic potential in the current-carrying stator slot and for the rth current-free slot are given by Equations 7.1 and 7.2, respectively. Equation 7.1 involves a parameter $A_{m,n}$, which is given by Equation 8.7 (Chapter 8). Besides, each of these equations also involve, two sets of complex arbitrary constants a_p^o and b_q^o (in Equation 7.1) and a_p^r and b_q^r for each value of r (in Equation 7.2). The components of magnetic field intensity in stator slots obtained in view of Equation 7.1 are given by Equations 7.3a (H_{sy}^o) and 7.3b (H_{sz}^o). Similarly, the components of magnetic field intensity in the rth slot obtained in view of Equation 7.2 are given by Equations 7.4a (H_{sy}^r) and 7.4b (H_{sz}^r). The components of air-gap field are given by Equations 7.9a (H_{ay}) and 7.9b (H_{az}). In these relations, τ is the pole pitch and ω is the supply frequency. These equations involve four sets of complex Fourier coefficients (h_{1u}^R, h_{2u}^R, h_{1u}^S and h_{2u}^S), which define tangential component of the magnetic field intensity (i.e. H_{ay}) on rotor and stator air-gap surfaces. Thus, there are eight sets of complex arbitrary constants (a_p^o, b_q^o, a_p^r, b_q^r, h_{1u}^R, h_{2u}^R, h_{1u}^S and h_{2u}^S), which need to be evaluated.

Equations 7.12a through 7.12c relate h_{1u}^S and h_{2u}^S to a_p^o, b_q^o, a_p^r and b_q^r. Further, if the real and imaginary parts are separately equated, Equations 7.12a through 7.12c can be split into two sets of equations. Thus, in view of Equations 7.12a through 7.12c and Equations 7.15a and 7.15b, on replacing U by u, the value for h_{1u}^S for all odd values of u is obtainable in terms of a_p^o, a_p^r and b_q^r. Similarly, for any odd integer U, multiplying the left-hand side (LHS) of Equations 7.12a through 7.12c by $\sin((U\pi/\tau)\cdot y)$ and then integrating over half the pole pitch, that is, $\tau/2$, the value for h_{2u}^S for all odd values of u can be found in terms of a_p^o, a_p^r and b_q^r, provided that after the integration U is replaced by u. Besides, Equations 7.22 through 7.24 also relate a_p^o, b_q^o, a_p^r and b_q^r to h_{1u}^R, h_{1u}^S, h_{2u}^R and h_{2u}^S.

Step 2: In this step, the components of the magnetic field intensity in the air gap are obtained (i) *in a reference frame stationary with respect to the stator,* which are given by Equations 7.26a (H_{ay}) and 7.26b (H_{az}) and (ii) *in a reference frame moving with the rotor,* which are given by Equations 7.32a (H_{ay}) and 7.32b (H_{az}). These equations involve parameters s_u^+ and s_u^-, which are given by Equations 7.31a and 7.31b. The parameters s_u^+ and s_u^-, in fact, are related to the slip of the motor.

As noted earlier the arbitrary constants, h_{1u}^R, h_{2u}^R, h_{1u}^S and h_{2u}^S are, in general, complex quantities. For simplification, these can be related to two new parameters g_u^+ and g_u^- by Equations 7.36 and 7.37. The peripheral component of magnetic field intensity in the rotor region is now given in terms of these new coefficients by Equations 7.38 (H_{ry}) and 7.41 (H_{rz}). These relations involve a set of new parameter ($k_u^{\pm}$) which is given by Equation (7.39a). Equations 7.44a and 7.44b further relate g_u^+, g_u^- and h_{1u}^R, h_{1u}^S, h_{2u}^R and h_{2u}^S. If in these equations g_u^+ and g_u^- are substituted from Equations 7.36 and 7.37 and the real and imaginary parts are equated on the two sides separately, this will result in four (two each for real and imaginary) linearly independent equations. Between six sets of unknowns, viz. g_u^+, g_u^-, h_{1u}^R, h_{2u}^R, h_{1u}^S and h_{2u}^S, we have four sets of equations, namely, Equations 7.36, 7.37, 7.44a and 7.44b. As discussed above, two more sets of equations can be generated using boundary conditions given by Equations 7.11a and 7.11b and all the arbitrary coefficients involved can be evaluated.

The equations referred to above can be numerically solved to obtain the values of unknown arbitrary coefficients. These coefficients can be substituted in relevant equations for the determination of field distributions in different regions due to a series of coils belonging to the same phase and same layer.

Step 3: For estimating the field distribution in the solid rotor due to all the coils of the same phase and same layer, the right-hand side (RHS) of Equations 7.38 and 7.41 can be multiplied by the number of slots per pole per phase, η, and the *distribution factor* (k_{du}) given by Equation 7.45, for the harmonic field of order u. The revised expressions for the field distribution in the solid rotor due to the entire red-phase winding of first layer are given by Equations 7.47a ($H_{ry}^{(1)}$) and 7.47b ($H_{rz}^{(1)}$) and for the second layer by Equations 7.48a ($H_{ry}^{(2)}$) and 7.48b ($H_{rz}^{(2)}$). The components of the magnetic field in the solid rotor due to these two layers are given by Equations 7.49a (H_{ry}) and 7.49b (H_{rz}). These equations now involve four sets of arbitrary constants (viz. $g_u^{+(1)}$, $g_u^{-(1)}$, $g_u^{+(2)}$ and $g_u^{-(2)}$) which are to be numerically evaluated. In view of the approximation given by Equation 7.50, the volume of numerical computations considerably reduces. In view of these approximations, Equations 7.49a and 7.49b are replaced by the simplified forms of Equations 7.51a and 7.51b. In these relations, a new quantity called the *pitch factor* (k_{pu}) for the harmonic field of order u is introduced which is given by Equation 7.51c.

Step 4: The expressions of magnetic field due to the three-phase stator winding currents, in a reference frame fixed on the rotor, are given by Equations 7.52a (H_{ry}) and 7.52b (H_{rz}). Similarly, the expression for eddy current density in the solid rotor is given by Equations 7.54 (J_{rx}) and 7.55 (E_{rx}).

Step 5: In view of the above analysis, the following machine performance parameters with poles (P), rotor axial length (L_R), rotor radial depth (D_R) and pole pitch (τ) can be computed from the equation given against each of these. (i) Eddy current loss (W_E) in the solid rotor from Equation 7.57, (ii) time average of the peripheral force (F_y) from Equation 7.58b, (iii) mechanical power developed (P_M) from Equation 7.60, (iv) power input to the rotor (P_R) from Equation 7.61, (v) torque developed (T) from Equation 7.63 and (vi) well-known relation for all induction and hysteresis machines for each order of harmonic u from Equation 7.62.

9.4.3.9 Three-Dimensional Fields in Solid Rotor Induction Machines

Section 7.3 deals with the three-dimensional fields in solid rotor induction machines. Figure 7.5 illustrates a sectional view of an idealised machine and the current sheet simulating the stator winding. The configuration is divided into six regions labelled as 1, 2, …, 6. The rotor region (region 1) is the only conducting region and all the remaining regions are air regions.

In this figure, X is parallel to axial, Y to peripheral and Z to the radial directions. The rotor length is taken as L_R, stator length as L_S and L_0 is the total length of stator core and the two adjacent air regions. Similarly, rotor core depth is taken as D_R, stator core depth as D_S and air gap as g. The surfaces at the air gap are taken as that at $z = 0$ and $z = -g$ while $z = D_R$ represents the shaft surface. The middle of the axial length of the machine is taken as the surface $x = 0$ and $x = \pm L_R/2$ and $x = \pm L_S/2$ represents the rotor core and the stator core end surfaces, respectively. The two end covers are presumed to be located at $x = \pm L_O/2$. In the analysis, P is the number of stator poles and τ is the pole pitch.

The stator current sheet simulating the stator winding, with balanced three-phase currents, is the primary source for the magnetic field in all regions. Thus, the distribution of magnetic field in each region is characterised by the exponential factor exp $\cdot j(\omega t - \ell y)$, where $\ell = \pi/\tau$, τ being the pole pitch. Field expressions are written in complex form without exponential factor, which is to be inserted while selecting the real part to obtain the expressions for instantaneous field.

In rotor (or conducting) region (labelled as region 1), the components of current density are given by Equations 7.70 (J_{z1}), 7.71 (J_{x1}) and 7.72 (J_{y1}). Similarly, in this region different H components are given by Equations 7.73 (H_{x1}), 7.74 (H_{y1}) and 7.75 (H_{z1}). In the air space adjacent to the rotor end surface, labelled as region 2, the different field components are given by Equations 7.81 ($\mathcal{V}_2$), 7.82a (H_{x2}), 7.82b (H_{y2}) and 7.82c (H_{z2}). In the air space adjacent to the stator end surface labelled as region 3, the different field components are given by Equations 7.88 ($\mathcal{V}_3$), 7.89a (H_{x3}), 7.89b (H_{y3}) and 7.89c (H_{z3}). Lastly, in the air-gap region labelled as region 4, different field components are given by Equations 7.91 ($\mathcal{V}_4$), 7.92a (H_{x4}), 7.92b (H_{y4}) and 7.92c (H_{z4}). In the above-referred equations, the involved parameters are given by Equations 7.70a (β_q), 7.71a (α_p), 7.65a (d^2), 7.81a (γ_q), 7.81b (δ_m), 7.88a (ζ_n), 7.91a (η_r) and 7.94d (K_x). The performance parameters and the corresponding equations which can be obtained in view of the above-noted equations are (i) eddy current loss W_E in the solid rotor (Equation 7.111), (ii) time average of the peripheral force F_y developed in the rotor (Equation 7.115), (iii) mechanical power developed P_M (Equation 7.119a), (iv) rotor power input P_R (Equation 7.119b) and (v) the slip in terms of the eddy current loss, mechanical power developed and the rotor input power (Equation 7.120).

In the relations of field components and the performance parameters, noted above, a number of arbitrary constants (viz. b_q, a_r', a_p, c_m, a_r'' and d_n) are involved. These can be evaluated in view of Equations 7.84, 7.96b, 7.99b, 7.101b, 7.104b and 7.106b, which in their modified forms, in the same sequence, are given below

$$b_q \cdot A_0(q) - \sum_{p-odd}^{\infty} a_p \cdot A_1(q,p) - \sum_{m=1}^{\infty} c_m \cdot A_2(q,m) = 0 \tag{9.26a}$$

$$a'_r \cdot k_0 + \sum_{p-odd}^{\infty} a_p \cdot A_3(r,p) + \sum_{m=1}^{\infty} c_m \cdot A_4(r,m) = 0 \tag{9.26b}$$

$$a_p \cdot A_5(p) - \sum_{q=1}^{\infty} b_q \cdot A_6(p,q) - \sum_{r-odd}^{\infty} a'_r \cdot A_7(p,r) + \sum_{r-odd}^{\infty} a''_r \cdot A_8(p,r) = 0 \tag{9.26c}$$

$$c_m \cdot A_9(m) - \sum_{q-1}^{\infty} b_q \cdot A_{10}(m,q) - \sum_{r-odd}^{\infty} a'_r \cdot A_{11}(m,r) + \sum_{r-odd}^{\infty} a''_r \cdot A_{12}(m,r) = 0 \tag{9.26d}$$

$$a''_r - \sum_{n=1}^{\infty} d_n \cdot A_{13}(r,n) = K_1(r) \tag{9.26e}$$

$$d_n \cdot K_2(n) - \sum_{r-odd}^{\infty} a'_r \cdot A_{14}(r) + \sum_{r-odd}^{\infty} a''_r \cdot A_{15}(n,r) = 0 \tag{9.26f}$$

The elements involved in any modified equation can be obtained just by comparing the original equation with its modified form. These modified equations can be written in the form of a matrix equation $[A] \cdot [x] = [B]$ where $[A]$, $[B]$ and $[x]$ matrices are as given below:

Rows↓......q...... ....p........ ....m........ ....r....... ......r....... ......n.......
→ Columns

$$[A] = \begin{array}{c} q \\ r \\ p \\ m \\ r \\ n \end{array} \begin{bmatrix} [A_0(q)] & [A_1(q,p)] & [A_2(q,m)] & [0] & [0] & [0] \\ [0] & [A_3(r,p)] & [A_4(r,m)] & [K_0] & [0] & [0] \\ [A_6(p,q)] & [A_5(p)] & [0] & [A_7(p,r)] & [A_8(p,r)] & [0] \\ [A_{10}(m,q)] & [0] & [A_9(m)] & [A_{11}(m,r)] & [A_{12}(m,r)] & [0] \\ [0] & [0] & [0] & [0] & [I] & [A_{13}(r,n)] \\ [0] & [0] & [0] & [A_{14}(r)] & [A_{15}(n,r)] & [K_2(n)] \end{bmatrix}$$

$$[x] = \begin{bmatrix} [b_q] \\ [a_p] \\ [c_m] \\ [a'_r] \\ [a''_r] \\ [d_n] \end{bmatrix} \quad \text{and} \quad [B] = \begin{matrix} q \\ p \\ m \\ r \\ r \\ n \end{matrix} \begin{bmatrix} [0] \\ [0] \\ [0] \\ [0] \\ [K_1(r)] \\ [0] \end{bmatrix} \tag{9.27}$$

In matrices [*A*], [*B*] and [*x*] each element in itself is a matrix. Matrix [0] represents a null matrix and [*I*] an identity matrix. The other matrices will have nonzero elements corresponding to its element and order indicated by the involvement of integers in parentheses. It is to mention that [*A*] is a square matrix and [*B*] and [*x*] are column vectors. Once [*A*] and [*B*] are generated, [*x*] can be obtained by using inverting [*A*] and multiplying it with [*B*] as before.

9.4.3.10 Single-Phase Induction Motors with Composite Poles Being Considered

Section 8.5 deals with the single-phase induction motor with composite poles without shading ring, as an alternative to the shaded pole-type machine. The analysis is based on certain simplifying assumptions enumerated in the text. An idealised machine shown in Figure 8.6 contains solid and laminated parts of stator poles, coil sides of excitation windings with currents flowing in opposite directions; laminated stator core, air gap, solid rotor and Cartesian system of space coordinates with the rotor air-gap surface on the *X*–*Y*-plane. The *X*-axis indicates axial, *Y*-axis peripheral and *Z*-axis radial directions. The conductivity and permeability for the solid rotor iron (σ_R and μ_R) are constants. Also, λ indicates the pole pitch and g indicates air-gap length. The air gap extends over $0 \le z \le g$.

The field varies periodically in the peripheral (y) direction with a wavelength of 2λ, where λ indicates the pole pitch. The rotor and the stator air-gap surfaces lie along $z = 0$ and $z = g$, respectively. Along the two air-gap surfaces, the distribution of peripheral and axial components of H field are given by Equations 8.83 and 8.86. Similarly, the peripheral and axial components of H field in the rotor region are given by Equations 8.89 and 8.94. In the expressions of these field components parameters α_m and η_r^2 and four sets of arbitrary constants (viz. a_m, b_m, c_m and d_m) are involved. The first two parameters are given by Equations 8.91a and 8.91b. In view of Equation 8.97b, a_m is related to c_m. Similarly, Equation 8.97c relates b_m to d_m. In both of these relations, the parameter f_m is given by Equation 8.97d and there is no involvement of any summation. At this stage, only two arbitrary constants remain to be evaluated.

In view of Figure 8.6, the interpolar region with peripheral width (w_i) and radial depth (d) is occupied by current-carrying conductors with total ampere-turn ι. The density of excitation current is given by Equation 8.99. The vector magnetic potential comprises particular integral and complementary function. These are given by Equations 8.102a and 8.103a, respectively. Equation 8.103a again involves two sets of arbitrary constants c'_p and d'_q. In this region, the components of magnetic field intensity are given by Equations 8.103b (H''_y) and 8.103c (H''_z) and their final expressions by Equations 8.104a and 8.104b. Similarly, the peripheral components of magnetic field intensity in the solid part of the polar region are given by Equations 8.108a (H^s_y) and 8.111 (H^s_z). Some of the parameters involved in these expressions are given by Equations 8.107b (η_s^2), 8.108b (β_k) and 8.108c (γ_l). These field expressions involve a number of arbitrary constants (viz. c'_p and d'_q in Equations 8.104a and 8.104b and c''_k and d''_l in Equations 8.108a and 8.111). These can be evaluated by numerically solving a number of linear algebraic simultaneous equations developed by using boundary conditions involving the continuity of E and H fields. These boundary conditions are given by Equations 8.113a, 8.113b (at $y = w_i$), 8.113c, 8.113d and 8.113e and 8.111f (at $z = g$). In view of these boundary conditions, d''_l can be obtained in terms of J_x, d'_q and c''_k, d'_q in terms of c'_p and d''_1, c'_p in terms of J_x, d'_q, a_m and b_m, c''_k in terms of a_m and b_m, a_m in terms of c'_p, c''_k and d''_1 and b_m in terms of c'_p, c''_k and d''_l. As mentioned in the text, some of these relations will involve summations. Numerical solution of these resulting equations will give values for a_m, b_m, c'_p, d'_q, c''_k and d''_l. By back substitution in Equations 8.97b and 8.97c, the values for c_m and d_m can be found. The field distributions in different regions can be evaluated by substitution of these arbitrary constants in relevant equations. Having thus determined the field distributions, eddy currents in the solid rotor and the torque developed can be found.

9.4.4 Integrations Leading to Simultaneous Algebraic Equations

Section 4.3 deals with the modelling for aperiodical field distributions. Under this section, three cases of slotted configurations are considered. These include tooth-opposite-tooth orientation (Section 4.3.1), tooth-opposite-slot orientation (Section 4.3.2) and arbitrary teeth orientation (Section 4.3.3). The geometrical parameters involved in these configurations include s, t, g and λ where s is the slot width ($\to\infty$), t is the tooth width, g is the air-gap length and λ is the tooth pitch ($\to\infty$). The expressions for potentials involve two space coordinate parameters y and z. The expressions of potential distributions also involve a number of arbitrary constants (or Fourier coefficients). The analysis, in all these cases, involves improper integrals. To evaluate these arbitrary coefficients, the involved integral equations are manipulated to form sets of simultaneous linear algebraic equations. These equations can further be manipulated into the equations involving matrices.

The translation of the involved integral equations into simultaneous linear algebraic equations can be accomplished by using the trapezoidal rule, parabolic rule or Gauss–Laguerre quadrature method. All these integral equations involve improper integrals. As the trapezoidal and parabolic rules are basically meant for the integrals with finite limits, their application to improper integrals will obviously lead to approximate results. Besides, the Gauss–Laguerre quadrature method is inherently approximate as it converts an improper integral into the series summation wherein the values are obtained at selected values of abscissas (x_i) representing zeros of Laguerre polynomials. Thus, solutions with all these methods will yield approximate results.

It needs to be mentioned that no mathematician will allow such a luxury of replacing improper integrals with those with finite limits but engineers do take such a risk for arriving at a fruitful outcome. In the following sections for each of the three cases cited above, a different method is used. Thus, for the case of tooth-opposite-tooth orientation, the trapezoidal rule is used, for tooth-opposite-slot orientation parabolic rule is employed, whereas in case of arbitrary teeth orientation the Gauss–Laguerre quadrature method is used.

9.4.4.1 Tooth-Opposite-Tooth Orientation Case

For this case, the trapezoidal rule given in Section 9.3.6.1.1 is used. In this case, two arbitrary functions $F(u)$ and $f(w)$ are to be evaluated in view of Equations 4.55 and 4.56. The modified form of Equation 4.55 is

$$F(v) + f(v)\cdot K_1(v) - \int_0^\infty f(w)\cdot K(w,v)\cdot dw = 0 \tag{9.28}$$

where

$$K_1(v) = \frac{\pi}{2}\cdot \sin(v\cdot t/2) \tag{9.29a}$$

and

$$K(w,v) = [K_2(w,v) - K_3(w,v) + K_4(w,v)] \tag{9.29b}$$

In Equation 9.29b

$$K_2(w,\ v) = \cos(w\cdot t/2)\cdot \frac{1}{2}\left\{\frac{1}{(w+v)} + \frac{1}{(w-v)}\right\} \tag{9.29c}$$

$$K_3(w,\ v) = \cos(w\cdot t/2)\frac{1}{2}\left\{\frac{1-\cos(w+v)T/2}{(w+v)} + \frac{1-\cos(w-v)T/2}{(w-v)}\right\} \tag{9.29d}$$

$$K_4(w,\ v) = \sin(w\cdot t/2)\cdot \frac{1}{2}\left\{\frac{\sin(w+v)T/2}{(w+v)} + \frac{\sin(w-v)T/2}{(w-v)}\right\} \tag{9.29e}$$

The modified form of Equation 4.56 is

$$K_5(v)\cdot F(v) - K_6(v)\cdot f(v) - \int_0^\infty K_8(u,v)\cdot F(u,v)\cdot du = K_7(v) \tag{9.30}$$

where

$$K_5(v) = v\cdot \coth\left(v\cdot\frac{g}{2}\right)\cdot\cos\left(v\cdot\frac{t}{2}\right) \tag{9.31a}$$

$$K_6(v) = \frac{\pi}{2}\cdot v \tag{9.31b}$$

$$K_7(v) = \frac{1}{g}\cdot\frac{1}{v} - \frac{1}{2} \tag{9.31c}$$

$$K_8(u,\,v) = \frac{1}{\pi}u\cdot\coth\left(u\cdot\frac{g}{2}\right)\cdot\cos\left(u\cdot\frac{t}{2}\right)\cdot\left\{\frac{1}{(v+u)} - \frac{1}{(v-u)}\right\} \tag{9.31d}$$

Let v takes value of $v_0, v_1, \ldots, v_p$. Equation 9.28 can now be written as

$$F(v_0) + f(v_0)\cdot K_1(v_0) - \int_0^\infty f(w)\cdot K(w,v_0)\cdot dw = 0$$

$$F(v_1) + f(v_1)\cdot K_1(v_1) - \int_0^\infty f(w)\cdot K(w,v_1)\cdot dw = 0$$

$$\ldots\ldots$$

$$F(v_p) + f(v_p)\cdot K_1(v_p) - \int_0^\infty f(w)\cdot K(w,v_p)\cdot dw = 0 \tag{9.32}$$

Similarly, Equation 9.30 can be written as

$$K_5(v_0)\cdot F(v_0) - K_6(v_0)\cdot f(v_0) - \int_0^\infty K_8(u,v_0)\cdot F(u,v_0)\cdot du = K_7(v_0)$$

$$K_5(v_1)\cdot F(v_1) - K_6(v_1)\cdot f(v_1) - \int_0^\infty K_8(u,v_1)\cdot F(u,v_1)\cdot du = K_7(v_1)$$

$$\ldots\ldots$$

$$K_5(v_q)\cdot F(v_q) - K_6(v_q)\cdot f(v_q) - \int_0^\infty K_8(u,v_q)\cdot F(u,v_q)\cdot du = K_7(v_q) \tag{9.33}$$

Consider one of the integrals involved in Equation 9.32, which is reproduced below:

$$\int_0^\infty f(w)\cdot K(w,v_p)\cdot dw \tag{9.34}$$

As noted above, the integral of Equation 9.34 is to be solved by the trapezoidal rule. For implementing this rule, the limits of integral given by Equation 9.34 are to be restricted to finite limits that is from a to b instead of 0 to ∞. The lower limit a may be taken as small as possible and the upper limit b as large as possible. In this method, the entire u-axis within the given limits is to be divided into n equal segments each of length [that is, $h = \{(b-a)/n\}$]. If at the lower limit a the value of u is taken to be u_0 and at the upper limit b, u is taken to be u_n and the corresponding values of the function y are taken as y_0 and y_n the integral of Equation 9.34, in view of Equation 9.6, can be written as

$$\int_0^\infty f(w)\cdot K(w,v_p)\cdot dw \approx \int_a^b y\cdot du = h\cdot\left(\frac{y_0}{2}+y_1+y_2+\cdots+y_{n-2}+y_{n-1}+y_n\right) \tag{9.35}$$

Since in Equation 9.35 $y = f(w)\cdot K(w,v_p)$, its RHS can be written as

$$h\cdot\{f(w_0)\cdot K(w_0,v_p)/2+f(w_1)\cdot K(w_1,v_p)+\cdots+f(w_{n-1})\cdot K(w_{n-1},v_p)$$
$$+f(w_n)\cdot K(w_n,v_p)\}$$

and Equation 9.35 becomes

$$\int_0^\infty f(w)\cdot K(w,v_p)\cdot dw \approx h\cdot\{f(w_0)\cdot K(w_0,v_p)/2+f(w_1)\cdot K(w_1,v_p)$$
$$+\cdots+f(w_{n-1})\cdot K(w_{n-1},v_p)+f(w_n)\cdot K(w_n,v_p)\} \tag{9.36}$$

In view of Equation 9.36, Equation 9.32 can be written as

$$F(v_0)+f(v_0)\cdot K_1(v_0)-(h/2)f(w_0)\cdot K(w_0,v_0)-h\cdot f(w_1)\cdot K(w_1,v_0)$$
$$-\cdots-h\cdot f(w_{n-1})\cdot K(w_{n-1},v_0)-h\cdot f(w_n)\cdot K(w_n,v_0)=0$$
$$F(v_1)+f(v_1)\cdot K_1(v_1)-(h/2)f(w_0)\cdot K(w_0,v_1)-h\cdot f(w_1)\cdot K(w_1,v_1)$$
$$-\cdots-h\cdot f(w_{n-1})\cdot K(w_{n-1},v_1)-h\cdot f(w_n)\cdot K(w_n,v_1)=0$$
$$\cdots\cdots$$
$$F(v_p)+f(v_p)\cdot K_1(v_p)-(h/2)f(w_0)\cdot K(w_0,v_p)-h\cdot f(w_1)\cdot K(w_1,v_p)$$
$$-\cdots-h\cdot f(w_{n-1})\cdot K(w_{n-1},v_p)-h\cdot f(w_n)\cdot K(w_n,v_p)=0 \tag{9.37}$$

Similarly, Equation 9.33 can be written as

$$
\begin{aligned}
&K_5(v_0)\cdot F(v_0) - K_6(v_0)\cdot f(v_0) - (h/2)F(u_0,v_0)\cdot K_8(u_0,v_0) - h\cdot F(u_1,v_0)\cdot K_8(u_1,v_0)\\
&\quad - \cdots - h\cdot F(u_{n-1},v_0)\cdot K_8(u_{n-1},v_0) + h\cdot F(u_n,v_0)\cdot K_8(w_n,v_0) = K_7(v_0)\\
&K_5(v_1)\cdot F(v_1) - K_6(v_1)\cdot f(v_1) - (h/2)F(u_0,v_1)\cdot K_8(u_0,v_1) - h\cdot F(u_1,v_1)\cdot K_8(u_1,v_1)\\
&\quad - \cdots - h\cdot F(u_{n-1},v_1)\cdot K_8(u_{n-1},v_1) - h\cdot F(u_n,v_1)\cdot K_8(w_n,v_1) = K_7(v_1)\\
&\qquad\qquad\qquad\qquad\cdots\cdots\\
&K_5(v_q)\cdot F(v_q) - K_6(v_q)\cdot f(v_q) - (h/2)F(u_0,v_q)\cdot K_8(u_0,v_q) - h\cdot F(u_1,v_q)\cdot K_8(u_1,v_q)\\
&\quad - \cdots - h\cdot F(u_{n-1},v_q)\cdot K_8(u_{n-1},v_q) - h\cdot F(u_n,v_q)\cdot K_8(w_n,v_q) = K_7(v_q)
\end{aligned}
\tag{9.38}
$$

Equations 9.37 and 9.38 can now be written in the matrix form:

$$
[A] = \begin{bmatrix} [A_1] & [A_2] \\ [A_3] & [A_4] \end{bmatrix} \quad [x] = \begin{bmatrix} [x_1] \\ [x_2] \end{bmatrix} \quad [B] = \begin{bmatrix} [B_1] \\ [B_2] \end{bmatrix} \tag{9.39}
$$

where

$$
[A_1] = \begin{bmatrix}
K_1(v_0) - (h/2)K(w_0,v_0) & -h\cdot K(w_1,v_0) & . & -h\cdot K(w_n,v_0)\\
-(h/2)K(w_0,v_1) & K_1(v_1) - h\cdot K(w_1,v_1) & . & -h\cdot K(w_n,v_1)\\
. & . & . & .\\
-(h/2)K(w_0,v_n) & -h\cdot K(w_1,v_n) & . & K_1(v_p) - h\cdot K(w_n,v_p)
\end{bmatrix}
$$

$$
[A_2] = \begin{bmatrix}
1 & 0 & . & 0\\
0 & 1 & . & 0\\
. & . & . & .\\
0 & 0 & . & 1
\end{bmatrix}
\quad
[A_3] = \begin{bmatrix}
-K_6(v_0) & 0 & . & 0\\
0 & -K_6(v_1) & . & 0\\
. & . & . & .\\
0 & 0 & . & -K_6(v_q)
\end{bmatrix}
$$

$$
[A_4] = \begin{bmatrix}
K_5(v_0) - (h/2)K_8(u_0,v_0) & -h\cdot K_8(u_1,v_0) & . & -h\cdot K_8(n_n,v_0)\\
-h\cdot K_8(u_0,v_1) & K_5(v_1) - h\cdot K_8(u_1,v_1) & . & -h\cdot K_8(u_n,v_1)\\
. & . & . & .\\
-h\cdot K_8(u_0,v_n) & -h\cdot K_8(u_1,v_n) & . & K_5(v_q) - h\cdot K_8(u_n,v_q)
\end{bmatrix}
$$

$$
[x_1] = \begin{bmatrix} f(w_0)\\ f(w_1)\\ .\\ f(w_p) \end{bmatrix}
\quad
[x_2] = \begin{bmatrix} F(v_0)\\ F(v_1)\\ .\\ F(v_q) \end{bmatrix}
\quad
[B_1] = \begin{bmatrix} 0\\ 0\\ 0\\ 0 \end{bmatrix}
\quad \text{and}\ [B_2] = \begin{bmatrix} K_7(v_0)\\ K_7(v_1)\\ .\\ K_7(v_q) \end{bmatrix}
\tag{9.40}
$$

The values of arbitrary functions $f(w)$ and $F(v)$ can be obtained by using the relation

$$[x] = [A]^{-1}[B]$$

If the values of $F(u)$ and $f(w)$ are substituted in Equations 4.47 and 4.49, the values of $\mathcal{V}_0$ and $\mathcal{V}_1$ can be obtained. In the expressions of these potentials, the improper integral will be solved by using the same technique as described above.

9.4.4.2 Tooth-Opposite-Slot Orientation

In this case, the parabolic rule of Section 9.3.6.1.2 is employed. Here also, two arbitrary functions $F(u)$ and $f(w)$ are to be evaluated in view of Equations 4.64 and 4.65.

The modified form of Equation 4.64 is

$$F(v) + f(v) \cdot K_1(v) - \int_0^\infty f(w) \cdot K(w,v) \cdot dw = K_2(v) \tag{9.41}$$

where

$$K_1(v) = \sin(v \cdot t/2) \tag{9.42a}$$

$$\begin{aligned} K(w, v) = \cos(w \cdot t/2) \cdot \left[\frac{\cos\{(w+v)(T/2)\}}{(w+v)} + \frac{\cos\{(w-v)(T/2)\}}{(w-v)} \right] \\ + \sin(w \cdot t/2) \cdot \left[\frac{\sin\{(w+v)(T/2)\}}{(w+v)} + \frac{\sin\{(w-v)(T/2)\}}{(w-v)} \right] \end{aligned} \tag{9.42b}$$

$$K_2(v) = \frac{+\sin(v \cdot T/2)}{v} \tag{9.42c}$$

The modified form of Equation 4.65 is

$$F(v) \cdot K_3(v) - \int_0^\infty F(u) \cdot K_4(u,v) \cdot du - f(v) = K_5(v) \tag{9.43}$$

where

$$K_3(v) = \sin(v \cdot t/2) \tag{9.44a}$$

$$K_4(u, v) = \frac{1}{\pi}(u/v) \cdot \cos(u \cdot t/2) \cdot \left\{ \frac{1}{(v+u)} + \frac{1}{(v-u)} \right\} \tag{9.44b}$$

$$K_5(v) = -1/v \tag{9.44c}$$

Let v take value of $v_0, v_1, \ldots, v_{2p}$. Equation 9.41 can now be written as

$$F(v_0) + f(v_0) \cdot K_1(v_0) - \int_0^\infty f(w) \cdot K(w, v_0) \cdot dw = K_2(v_0)$$

$$F(v_1) + f(v_1) \cdot K_1(v_1) - \int_0^\infty f(w) \cdot K(w, v_1) \cdot dw = K_2(v_1)$$

$$\ldots\ldots$$

$$F(v_{2p}) + f(v_{2p}) \cdot K_1(v_{2p}) - \int_0^\infty f(w) \cdot K(w, v_{2p}) \cdot dw = K_2(v_{2p}) \tag{9.45}$$

Similarly, Equation 9.43 can be written as

$$K_3(v_0) \cdot F(v_0) - f(v_0) - \int_0^\infty K_4(u, v_0) \cdot F(u) \cdot du = K_5(v_0)$$

$$K_3(v_1) \cdot F(v_1) - f(v_1) - \int_0^\infty K_4(u, v_1) \cdot F(u) \cdot du = K_5(v_1)$$

$$\ldots\ldots$$

$$K_3(v_{2q}) \cdot F(v_{2q}) - f(v_{2q}) - \int_0^\infty K_4(u, v_{2q}) \cdot F(u) \cdot du = K_5(v_{2q}) \tag{9.46}$$

Consider one of the integrals involved in Equation 9.45, which is reproduced below

$$\int_0^\infty f(w) \cdot K(w, v_{2p}) \cdot dw \tag{9.47}$$

To evaluate an integral by parabolic rule, first the limits of this integral are to be restricted from a to b instead of 0 to ∞. Again, the lower limit a may be as small as possible and the upper limit b as large as possible. In this case, the entire u-axis within the given limits is to be divided into $2m$ equal segments each of length h, where $h = \{(b - a)/2m\}$. If at the lower limit a the value of u is taken to be u_0 and at the upper limit u is taken to be u_{2m} and the corresponding values of the function y are taken as y_0 and y_{2m} the integral of Equation 9.47, in view of Equation 9.7, can be written as

$$\int_0^\infty f(w)\cdot K(w,v_p)\cdot dw \approx \int_a^b y\cdot du = \frac{h}{3}[(y_0 + y_{2m}) + 4(y_1 + y_3 + \cdots + y_{2m-1}) + 2(y_2 + y_4 + \cdots + y_{2m-2})] \tag{9.48}$$

Since in Equation 9.48 $y = f(w)\cdot K(w,v_{2p})$, its RHS can be written as

$$\begin{aligned}\frac{h}{3}[&\{f(w_0)K(w_0,v_{2p}) + f(w_{2m})K(w_{2m},v_{2p})\}\\ &+ 4\{f(w_1)K(w_1,v_{2p}) + f(w_3)K(w_3,v_{2p}) + \cdots + f(w_{2m-1})K(w_{2m-1},v_{2p})\}\\ &+ 2\{f(w_2)K(w_2,v_{2p}) + f(w_4)K(w_4,v_{2p}) + \cdots + f(w_{2m-2})K(w_{2m-2},v_{2p})\}]\end{aligned} \tag{9.49}$$

and Equation 9.48 becomes

$$\begin{aligned}\int_0^\infty f(w)\cdot K(w,v_{2p})\cdot dw \approx \frac{h}{3}[&\{f(w_0)K(w_0,v_{2p}) + f(w_{2m})K(w_{2m},v_{2p})\}\\ &+ 4\{f(w_1)K(w_1,v_{2p}) + f(w_3)K(w_3,v_{2p}) + \cdots + f(w_{2m-1})K(w_{2m-1},v_{2p})\}\\ &+ 2\{f(w_2)K(w_2,v_{2p}) + f(w_4)K(w_4,v_{2p}) + \cdots + f(w_{2m-2})K(w_{2m-2},v_{2p})\}]\end{aligned} \tag{9.50}$$

In view of Equation 9.50, Equation 9.45 can be written as

$$\begin{aligned}F(v_0) + f(v_0)\cdot K_1(v_0) - \frac{h}{3}[&\{f(w_0)K(w_0,v_0) + f(w_{2m})K(w_{2m},v_0)\}\\ &+ 4\{f(w_1)K(w_1,v_0) + f(w_3)K(w_3,v_0) + \cdots + f(w_{2m-1})K(w_{2m-1},v_0)\}\\ &+ 2\{f(w_2)K(w_2,v_0) + f(w_4)K(w_4,v_0) + \cdots + f(w_{2m-2})K(w_{2m-2},v_0)\}] = K_2(v_0)\end{aligned}$$

$$\begin{aligned}F(v_1) + f(v_1)\cdot K_1(v_1) - \frac{h}{3}[&\{f(w_0)K(w_0,v_1) + f(w_{2m})K(w_{2m},v_1)\}\\ &+ 4\{f(w_1)K(w_1,v_1) + f(w_3)K(w_3,v_1) + \cdots + f(w_{2m-1})K(w_{2m-1},v_1)\}\\ &+ 2\{f(w_2)K(w_2,v_1) + f(w_4)K(w_4,v_1) + \cdots + f(w_{2m-2})K(w_{2m-2},v_1)\}] = K_2(v_1)\end{aligned}$$

$$\cdots\cdots$$

$$\begin{aligned}
&F(v_{2p}) + f(v_{2p}) \cdot K_1(v_{2p}) - \frac{h}{3}[\{f(w_0)K(w_0,v_{2p}) + f(w_{2m})K(w_{2m},v_{2p})\} \\
&\quad + 4\{f(w_1)K(w_1,v_{2p}) + f(w_3)K(w_3,v_{2p}) + \cdots + f(w_{2m-1})K(w_{2m-1},v_{2p})\} \\
&\quad + 2\{f(w_2)K(w_2,v_{2p}) + f(w_4)K(w_4,v_{2p}) + \cdots \\
&\quad + f(w_{2m-2})K(w_{2m-2},v_{2p})\}] = K_2(v_{2p}) \qquad (9.51)
\end{aligned}$$

Similarly, Equation 9.46 can be written as

$$\begin{aligned}
&F(v_0) \cdot K_3(v_0) - f(v_0) - \frac{h}{3}[\{F(u_0)K_4(u_0,v_0) + F(u_{2m})K_4(u_{2m},v_0)\} \\
&+ 4\{F(u_1)K_4(u_1,v_0) + F(u_3)K_4(u_3,v_0) + \cdots + F(u_{2m-1})K_4(u_{2m-1},v_0)\} \\
&+ 2\{F(u_2)K_4(u_2,v_0) + F(u_4)K_4(u_4,v_0) + \cdots + F(u_{2m-2})K_4(u_{2m-2},v_0)\}] = K_5(v_0)
\end{aligned}$$

$$\begin{aligned}
&F(v_1) \cdot K_3(v_1) - f(v_1) - \frac{h}{3}[\{F(u_0)K_4(u_0,v_1) + F(u_{2m})K_4(u_{2m},v_1)\} \\
&+ 4\{F(u_1)K_4(u_1,v_1) + F(u_3)K_4(u_3,v_1) + \cdots + F(u_{2m-1})K_4(u_{2m-1},v_1)\} \\
&+ 2\{F(u_2)K_4(u_2,v_1) + F(u_4)K_4(u_4,v_1) + \cdots + F(u_{2m-2})K_4(u_{2m-2},v_1)\}] = K_5(v_1)
\end{aligned}$$

$$\cdots\cdots$$

$$\begin{aligned}
&F(v_{2q}) \cdot K_3(v_{2q}) - f(v_{2q}) - \frac{h}{3}[\{F(u_0)K_4(u_0,v_{2q}) + F(u_{2m})K_4(u_{2m},v_{2q})\} \\
&+ 4\{F(u_1)K_4(u_1,v_{2q}) + F(u_3)K_4(u_3,v_{2q}) + \cdots + F(u_{2m-1})K_4(u_{2m-1},v_{2q})\} \\
&+ 2\{F(u_2)K_4(u_2,v_{2q}) + F(u_4)K_4(u_4,v_{2q}) + \cdots + F(u_{2m-2})K_4(u_{2m-2},v_{2q})\}] = K_5(v_{2q}) \\
&\qquad (9.52)
\end{aligned}$$

Equations 9.51 and 9.52 can now be written in the matrix form

$$[A] = \begin{bmatrix} [A_1] & [A_2] \\ [A_3] & [A_4] \end{bmatrix} \quad [x] = \begin{bmatrix} [x_1] \\ [x_2] \end{bmatrix} \quad [B] = \begin{bmatrix} [B_1] \\ [B_2] \end{bmatrix} \qquad (9.53)$$

where

$$[A_1] = \begin{bmatrix}
K_1(v_0) - \left(\frac{h}{3}\right)K(w_0,v_0) & -\left(\frac{4h}{3}\right)\cdot K(w_1,v_0) & -\left(\frac{2h}{3}\right)\cdot K(w_2,v_0) & \cdot & -\left(\frac{2h}{3}\right)\cdot K(w_{2m-2},v_0) & -\left(\frac{4h}{3}\right)\cdot K(w_{2m-1},v_0) & -\left(\frac{h}{3}\right)\cdot K(w_{2m},v_0) \\
-\left(\frac{h}{3}\right)K(w_0,v_1) & K_1(v_1) - \left(\frac{4h}{3}\right)\cdot K(w_1,v_1) & -\left(\frac{2h}{3}\right)\cdot K(w_2,v_1) & \cdot & -\left(\frac{2h}{3}\right)\cdot K(w_{2m-2},v_1) & -\left(\frac{4h}{3}\right)\cdot K(w_{2m-1},v_1) & -\left(\frac{h}{3}\right)\cdot K(w_{2m},v_1) \\
-\left(\frac{h}{3}\right)K(w_0,v_2) & \cdot & K_1(v_2) - \left(\frac{2h}{3}\right)\cdot K(w_2,v_2) & \cdot & \cdot & \cdot & -\left(\frac{h}{3}\right)\cdot K(w_{2m},v_2) \\
\cdot & \cdot & \cdot & & \cdot & \cdot & \cdot \\
-\left(\frac{h}{3}\right)K(w_0,v_{2p-2}) & \cdot & \cdot & \cdot & K_1(v_{2p-2}) - \left(\frac{2h}{3}\right)\cdot K(w_{2m-2},v_{2p-2}) & \cdot & -\left(\frac{h}{3}\right)\cdot K(w_{2m},v_{2p-2}) \\
-\left(\frac{h}{3}\right)K(w_0,v_{2p-1}) & \cdot & \cdot & \cdot & \cdot & K_1(v_{2p-1}) - \left(\frac{4h}{3}\right)\cdot K(w_{2m-1},v_{2p-1}) & -\left(\frac{h}{3}\right)\cdot K(w_{2m},v_{2p-1}) \\
-\left(\frac{h}{3}\right)K(w_0,v_{2p}) & -\left(\frac{4h}{3}\right)\cdot K(w_1,v_{2p}) & -\left(\frac{2h}{3}\right)\cdot K(w_2,v_{2p}) & \cdot & -\left(\frac{2h}{3}\right)\cdot K(w_{2m-2},v_{2p}) & -\left(\frac{4h}{3}\right)\cdot K(w_{2m-1},v_{2p}) & K_1(v_{2p}) - \left(\frac{4h}{3}\right)\cdot K(w_{2m},v_{2p})
\end{bmatrix}$$

$$[A_2] = \begin{bmatrix} 1 & 0 & . & 0 \\ 0 & 1 & . & 0 \\ . & . & . & . \\ 0 & 0 & . & 1 \end{bmatrix} \quad [A_3] = \begin{bmatrix} -K_3(v_0) & 0 & . & 0 \\ 0 & -K_3(v_1) & . & 0 \\ . & . & . & . \\ 0 & 0 & . & -K_3(v_{2q}) \end{bmatrix}$$

$$[A_4] = \begin{bmatrix}
K_3(v_0) - \left(\frac{h}{3}\right)K_4(u_0,v_0) & -\left(\frac{4h}{3}\right)\cdot K(u_1,v_0) & -\left(\frac{2h}{3}\right)\cdot K_4(u_2,v_0) & . & -\left(\frac{2h}{3}\right)\cdot K_4(u_{2m-2},v_0) & -\left(\frac{4h}{3}\right)\cdot K_4(u_{2m-1},v_0) & -\left(\frac{h}{3}\right)\cdot K_4(u_{2m},v_0) \\
-\left(\frac{h}{3}\right)K_4(u_0,v_1) & K_3(v_1) - \left(\frac{4h}{3}\right)\cdot K_4(u_1,v_1) & -\left(\frac{2h}{3}\right)\cdot K_4(u_2,v_1) & . & -\left(\frac{2h}{3}\right)\cdot K_4(u_{2m-2},v_1) & -\left(\frac{4h}{3}\right)\cdot K_4(u_{2m-1},v_1) & -\left(\frac{h}{3}\right)\cdot K_4(u_{2m},v_1) \\
-\left(\frac{h}{3}\right)K_4(u_0,v_2) & . & K_3(v_2) - \left(\frac{2h}{3}\right)\cdot K_4(u_2,v_2) & . & . & . & -\left(\frac{h}{3}\right)\cdot K_4(u_{2m},v_2) \\
. & . & . & & . & . & . \\
-\left(\frac{h}{3}\right)K_4(u_0,v_{2q-2}) & . & . & & K_3(v_{2q-2}) - \left(\frac{2h}{3}\right)\cdot K_4(u_{2m-2},v_{2q-2}) & . & -\left(\frac{h}{3}\right)\cdot K_4(u_{2m},v_{2q-2}) \\
-\left(\frac{h}{3}\right)K_4(u_0,v_{2q-1}) & . & . & . & . & K_3(v_{2q-1}) - \left(\frac{4h}{3}\right)\cdot K_4(u_{2m-1},v_{2q-1}) & -\left(\frac{h}{3}\right)\cdot K_4(u_{2m},v_{2q-1}) \\
-\left(\frac{h}{3}\right)K_4(u_0,v_{2q}) & -\left(\frac{4h}{3}\right)\cdot K_4(u_1,v_{2q}) & -\left(\frac{2h}{3}\right)\cdot K_4(u_2,v_{2q}) & . & -\left(\frac{2h}{3}\right)\cdot K_4(u_{2m-2},v_{2q}) & -\left(\frac{4h}{3}\right)\cdot K_4(u_{2m-1},v_{2q}) & K_3(v_{2q}) - \left(\frac{h}{3}\right)\cdot K_4(u_{2m},v_{2q})
\end{bmatrix}$$

$$[x_1] = \begin{bmatrix} f(w_0) \\ f(w_1) \\ \cdot \\ f(w_{2p}) \end{bmatrix} \quad [x_2] = \begin{bmatrix} F(v_0) \\ F(v_1) \\ \cdot \\ F(v_{2q}) \end{bmatrix} \quad [B_1] = \begin{bmatrix} K_2(v_0) \\ K_2(v_1) \\ \cdot \\ K_2(v_{2p}) \end{bmatrix} \quad \text{and } [B_2] = \begin{bmatrix} K_5(v_0) \\ K_5(v_1) \\ \cdot \\ K_5(v_{2q}) \end{bmatrix} \tag{9.54}$$

The values of arbitrary functions $f(w)$ and $F(v)$ can be obtained by using the relation

$$[x] = [A]^{-1}[B]$$

If the values of $F(u)$ and $f(w)$ are substituted in Equations 4.57 and 4.58, the values of $\mathcal{V}_0$ and $\mathcal{V}_1$ can be obtained. In the expressions of these potentials, the improper integral will be solved by using the same technique as described above.

9.4.4.3 Arbitrary Orientation of Teeth

For this case, the Gauss–Laguerre quadrature method described in Section 9.3.6.2.2 is used. This case involves four arbitrary functions viz. $f_1(w)$, $f_2(w)$, $F'(u)$ and $F''(u)$, which are to be evaluated.

The modified form of Equation 4.77a is

$$F'(v) - f_1(v)\cdot S_1(v) + f_2(v)\cdot S_2(v) + C_1(v)\int_0^{\infty} f_1(w)\cdot A(w,v)\cdot dw + C_2(v)\int_0^{\infty} f_2(w)\cdot A(w,v)\cdot dw = B_1(v) \tag{9.55}$$

where

$$S_1(v) = \frac{1}{2}\cdot \sin\{v\cdot(\delta + t/2)\} \quad S_2(v) = \frac{1}{2}\cdot \sin\{v\cdot(\delta - t/2)\}$$

$$C_1(v) = \frac{1}{\pi}\cdot \cos\{v\cdot(\delta + t/2)\} \quad C_2(v) = \frac{1}{\pi}\cdot \cos\{v\cdot(\delta - t/2)\}$$

$$A_1 = \left[\frac{w}{(w^2 - v^2)}\right] \quad B_1(v) = -\left[\frac{\sin\{v\cdot(\delta + t/2)\} - \sin\{v\cdot(\delta - t/2)\}}{2\pi v}\right] \tag{9.56}$$

The modified form of Equation 4.77b is

$$F''(v) + f_1(v) \cdot S_3(v) - f_2(v) \cdot C_3(v) + \int_0^\infty f_1(w) \cdot S_4(w,v) \cdot dw + \int_0^\infty f_2(w) \cdot S_5(w,v) \cdot dw = B_2(v) \tag{9.57}$$

where

$$S_3(v) = \frac{1}{2} \cdot \cos\{v \cdot (\delta - t/2)\} \quad C_3(v) = \frac{1}{2} \cdot \cos\{v \cdot (\delta + t/2)\}$$

$$S_4(w,v) = \frac{1}{2\pi} \cdot \left[\frac{\sin\{(2w - v) \cdot (\delta + t/2)\}}{(w - v)} - \frac{\sin\{(2w + v) \cdot (\delta + t/2)\}}{(w + v)} \right]$$

$$S_5(w,v) = \frac{1}{2\pi} \cdot \left[\frac{\sin\{(2w - v) \cdot (\delta - t/2)\}}{(w - v)} - \frac{\sin\{(2w + v).(\delta - t/2)\}}{(w + v)} \right]$$

$$B_2(v) = \frac{1}{2\pi} \cdot \left[\frac{\cos\{v \cdot (\delta + t/2)\} - \cos\{v \cdot (\delta - t/2)\}}{v} \right] \tag{9.58}$$

The modified form of Equation 4.84a is

$$f_1(v) - \int_0^\infty F'(u) \cdot K_1(u,v) \cdot du - \int_0^\infty F''(u) \cdot K_2(u,v) \cdot du + F'(v) \cdot K_3(v) - F''(v) \cdot K_4(v) = B_3(v) \tag{9.59}$$

where

$$K_1(u,v) = \frac{2}{\pi} \coth(u \cdot g/2) \cdot \cos\{u \cdot (\delta + t/2)\} \left\{ \frac{u}{(v^2 - u^2)} \right\}$$

$$K_2(u,v) = \frac{2}{\pi} \tanh(u \cdot g/2) \cdot \sin\{u \cdot (\delta + t/2)\} \left\{ \frac{u}{(v^2 - u^2)} \right\}$$

$$K_3(v) = \frac{1}{v} \cdot \coth(v \cdot g/2) \cdot \sin\{v \cdot (\delta + t/2)\}$$

$$K_4(v) = \frac{1}{v} \cdot \tanh(v \cdot g/2) \cdot \cos\{u \cdot (\delta + t/2)\} \quad \text{and} \quad B_3(v) = \frac{1}{\pi v} \tag{9.60}$$

The modified form of Equation 4.84b is

$$f_2(v) + \int_0^\infty F'(u) \cdot L_1(u,v) \cdot du - \int_0^\infty F''(u) \cdot L_2(u,v) \cdot du - F'(v) \cdot L_3(v) - F''(v) \cdot L_4(v) = B_3(v) \tag{9.61}$$

where

$$L_1(u,v) = \frac{2}{\pi}\coth(u \cdot g/2) \cdot \cos\{u \cdot (\delta - t/2)\}\left\{\frac{u}{(v^2 - u^2)}\right\}$$

$$L_2(u,v) = \frac{2}{\pi}\tanh(u \cdot g/2) \cdot \sin\{u \cdot (\delta - t/2)\}\left\{\frac{u}{(v^2 - u^2)}\right\}$$

$$L_3(v) = \frac{1}{v} \cdot \coth(v \cdot g/2) \cdot \sin\{v \cdot (\delta - t/2)\}$$

$$L_4(v) = \frac{1}{v} \cdot \tanh(v \cdot g/2) \cdot \cos\{u \cdot (\delta - t/2)\} \tag{9.62}$$

For the application of Gauss–Laguerre quadrature method, consider one of the integrals involved in Equation 9.55 and the relation given by Equation 9.10c

$$\int_0^\infty f_1(w) \cdot A(w,v) \cdot dw = \int_0^\infty g(w) \cdot dw \tag{9.63}$$

$$\int_0^\infty g(x) \cdot dx = \sum_{i=1}^{n} w_i e^{x_i} g(x_i) \tag{9.10c}$$

On comparing Equation 9.63 and LHS of Equation 9.10c, we get

$$g(w) = f_1(w) \cdot A(w,v) \tag{9.64a}$$

On replacing w by x in Equation 9.64a, we get

$$g(x) = f_1(x) \cdot A(x,v) \tag{9.64b}$$

and

$$dx = dw \tag{9.64c}$$

On replacing x by x_i in Equation 9.64b, we get

$$g(x_i) = f_1(x_i)A(x_i,v) \tag{9.64d}$$

Thus, RHS of Equation 9.10c can now be written as

$$\sum_{i=1}^{n} w_i e^{x_i} g(x_i) = \sum_{i=1}^{n} w_i e^{x_i} f_1(x_i)A(x_i,v) \tag{9.65}$$

In view of similar conversions of other integrals into summation, Equations 9.55, 9.57, 9.59 and 9.61 becomes

$$F'(x_j) - f_1(x_j)\cdot S_1(x_j) + f_2(x_j)\cdot S_2(x_j) + C_1(x_j)\sum_{i=1}^{n} w_i e^{x_i} f_1(x_i)A(x_i,x_j)$$
$$+ C_2(x_j)\sum_{i=1}^{n} w_i e^{x_i} f_2(x_i)A(x_i,x_j) = B_1(x_j) \tag{9.66}$$

$$F''(x_j) + f_1(x_j)\cdot S_3(x_j) - f_2(x_j)\cdot C_3(x_j) + \sum_{i=1}^{n} w_i e^{x_i} f_1(x_i)S_4(x_i,x_j)$$
$$+ \sum_{i=1}^{n} w_i e^{x_i} f_2(x_i)S_5(x_i,x_j) = B_2(x_j) \tag{9.67}$$

$$f_1(x_j) - \sum_{i=1}^{n} w_i e^{x_i} F'(x_i)K_1(x_i,x_j) - \sum_{i=1}^{n} w_i e^{x_i} F''(x_i)K_2(x_i,x_j)$$
$$+ F'(x_j)\cdot K_3(x_j) - F''(x_j)\cdot K_4(x_j) = B_3(x_j) \tag{9.68}$$

$$f_2(x_j) + \sum_{i=1}^{n} w_i e^{x_i} F'(x_i)L_1(x_i,x_j) - \sum_{i=1}^{n} w_i e^{x_i} F''(x_i)L_2(x_i,x_j)$$
$$- F'(x_j)\cdot L_3(x_j) - F''(x_j)\cdot L_4(x_j) = B_3(x_j) \tag{9.69}$$

In these equations, it is to be noted that this method will yield the values of functions viz. F', F'', f_1 and f_2 only at the values of abscissas (i.e. at x_i); all other functions involved in the analysis viz. S_1, C_1 and so on are also to be evaluated at these abscissas, which are shown as (x_j).

Equations 9.66 through 9.69 can now be written in the form of a matrix equation $[A]\cdot[x]=[B]$ and the values of coefficients can be obtained from equation $[x]=[A]^{-1}[B]$, where $[A]^{-1}$ indicates inverse of $[A]$. In these relations, the values of matrices $[A]$, $[x]$ and $[B]$ are as given below

$$[A]=\begin{bmatrix}[A_{11}] & [A_{12}] & [A_{13}] & [A_{14}]\\ [A_{21}] & [A_{22}] & [A_{23}] & [A_{24}]\\ [A_{31}] & [A_{32}] & [A_{33}] & [A_{34}]\\ [A_{41}] & [A_{42}] & [A_{43}] & [A_{44}]\end{bmatrix}\quad [x]=\begin{bmatrix}[F'(x_i)]\\ [F''(x_i)]\\ [f_1(x_i)]\\ [f_2(x_i)]\end{bmatrix}\quad [B]=\begin{bmatrix}[B_1(x_j)]\\ [B_2(x_j)]\\ [B_3(x_j)]\\ [B_3(x_j)]\end{bmatrix} \tag{9.70}$$

Matrix A contains 16 submatrices. The submatrices A_{11}, A_{22}, A_{33} and A_{44} are unit matrices, while A_{12}, A_{21}, A_{34} and A_{43} are null matrices. The other matrices are as given below

$$[A_{13}]=\begin{bmatrix}S_1(x_1)+C_1(x_1)w_1e^{x_1}A(x_1,x_1) & C_1(x_1)w_2e^{x_2}A(x_2,x_1) & . & C_1(x_1)w_ne^{x_n}A(x_n,x_1)\\ C_1(x_2)w_1e^{x_1}A(x_1,x_2) & S_1(x_2)+C_1(x_2)w_2e^{x_2}A(x_2,x_2) & . & C_1(x_2)w_ne^{x_n}A(x_n,x_2)\\ . & . & . & .\\ C_1(x_n)w_1e^{x_1}A(x_1,x_n) & C_1(x_n)w_2e^{x_2}A(x_2,x_n) & . & S_1(x_n)+C_1(x_n)w_ne^{x_n}A(x_n,x_n)\end{bmatrix}$$

$$[A_{14}]=\begin{bmatrix}S_2(x_1)+C_2(x_1)w_1e^{x_1}A(x_1,x_1) & C_2(x_1)w_2e^{x_2}A(x_2,x_1) & . & C_2(x_1)w_ne^{x_n}A(x_n,x_1)\\ C_2(x_2)w_1e^{x_1}A(x_1,x_2) & S_2(x_2)+C_2(x_2)w_2e^{x_2}A(x_2,x_2) & . & C_2(x_2)w_ne^{x_n}A(x_n,x_2)\\ . & . & . & .\\ C_2(x_n)w_1e^{x_1}A(x_1,x_n) & C_2(x_n)w_2e^{x_2}A(x_2,x_n) & . & S_2(x_n)+C_2(x_n)w_ne^{x_n}A(x_n,x_n)\end{bmatrix}$$

$$[A_{23}]=\begin{bmatrix}S_3(x_1)+w_1e^{x_1}S_4(x_1,x_1) & w_2e^{x_2}S_4(x_2,x_1) & . & w_ne^{x_n}S_4(x_n,x_1)\\ w_1e^{x_1}S_4(x_1,x_2) & S_3(x_2)+w_2e^{x_2}S_4(x_2,x_2) & . & w_ne^{x_n}S_4(x_n,x_2)\\ . & . & . & .\\ w_1e^{x_1}S_4(x_1,x_n) & w_2e^{x_2}S_4(x_2,x_n) & . & S_3(x_n)+w_ne^{x_n}S_4(x_n,x_n)\end{bmatrix}$$

$$[A_{24}]=\begin{bmatrix}C_3(x_1)+w_1e^{x_1}S_5(x_1,x_1) & w_2e^{x_2}S_5(x_2,x_1) & . & w_ne^{x_n}S_5(x_n,x_1)\\ w_1e^{x_1}S_5(x_1,x_2) & C_3(x_2)+w_2e^{x_2}S_5(x_2,x_2) & . & w_ne^{x_n}S_5(x_n,x_2)\\ . & . & . & .\\ w_1e^{x_1}S_5(x_1,x_n) & w_2e^{x_2}S_5(x_2,x_n) & . & C_3(x_n)+w_ne^{x_n}S_5(x_n,x_n)\end{bmatrix}$$

$$[A_{31}] = \begin{bmatrix} K_3(x_1) + w_1 e^{x_1} K_1(x_1,x_1) & w_2 e^{x_2} K_1(x_2,x_1) & . & w_n e^{x_n} K_1(x_n,x_1) \\ w_1 e^{x_1} K_1(x_1,x_2) & K_3(x_2) + w_2 e^{x_2} K_1(x_2,x_2) & . & w_n e^{x_n} K_1(x_n,x_2) \\ . & . & . & . \\ w_1 e^{x_1} K_1(x_1,x_n) & w_2 e^{x_2} K_1(x_2,x_n) & . & K_3(x_n) + w_n e^{x_n} K_1(x_n,x_n) \end{bmatrix}$$

$$[A_{32}] = \begin{bmatrix} K_4(x_1) + w_1 e^{x_1} K_2(x_1,x_1) & w_2 e^{x_2} K_2(x_2,x_1) & . & w_n e^{x_n} K_2(x_n,x_1) \\ w_1 e^{x_1} K_2(x_1,x_2) & K_4(x_2) + w_2 e^{x_2} K_2(x_2,x_2) & . & w_n e^{x_n} K_2(x_n,x_2) \\ . & . & . & . \\ w_1 e^{x_1} K_2(x_1,x_n) & w_2 e^{x_2} K_2(x_2,x_n) & . & K_4(x_n) + w_n e^{x_n} K_2(x_n,x_n) \end{bmatrix}$$

$$[A_{41}] = \begin{bmatrix} L_3(x_1) + w_1 e^{x_1} L_1(x_1,x_1) & w_2 e^{x_2} L_1(x_2,x_1) & . & w_n e^{x_n} L_1(x_n,x_1) \\ w_1 e^{x_1} L_1(x_1,x_2) & L_3(x_2) + w_2 e^{x_2} L_1(x_2,x_2) & . & w_n e^{x_n} L_1(x_n,x_2) \\ . & . & . & . \\ w_1 e^{x_1} L_1(x_1,x_n) & w_2 e^{x_2} L_1(x_2,x_n) & . & L_3(x_n) + w_n e^{x_n} L_1(x_n,x_n) \end{bmatrix}$$

$$[A_{42}] = \begin{bmatrix} L_4(x_1) + w_1 e^{x_1} L_2(x_1,x_1) & w_2 e^{x_2} L_2(x_2,x_1) & . & w_n e^{x_n} L_2(x_i,x_1) \\ w_1 e^{x_1} L_2(x_1,x_2) & L_4(x_2) + w_2 e^{x_2} L_2(x_2,x_2) & . & w_n e^{x_n} L_2(x_n,x_2) \\ . & . & . & . \\ w_1 e^{x_1} L_2(x_1,x_n) & w_2 e^{x_2} L_2(x_2,x_n) & . & L_4(x_n) + w_n e^{x_n} L_2(x_n,x_n) \end{bmatrix} \tag{9.71}$$

Once the values of $f_1(w)$, $f_2(w)$, $F'(u)$ and $F''(u)$ are evaluated, these can be substituted in Equations 4.67 through 4.70, and 4.72 to obtain potential distributions V_1, V_2, V_3, V_4 and V_0 over the desired ranges. The numerical values of these potentials are again obtained by numerical integration for which the procedure will remain the same as outlined above.

Further Reading

Abramowitz, M. and Segun, I. A., *Handbook of Mathematical Functions*, Dover Publications, Inc., New York, 253–293, 1965.

George, E. F., Michael, A. M., and Cleve, B. M., *Computer Methods for Mathematical Computations*, Prentice-Hall, Englewood Cliffs, NJ, 1977, Chapter 5.

Golub, G. H. and Charles F. V., *Matrix Computations*, Johns Hopkins University Press, Baltimore and London, 3rd Edn., 1996.

Higham, N. J., *Accuracy and Stability of Numerical Algorithms*, Society for Industrial and Applied Mathematics, Philadelphia, 1996.

Hildebrand, F. B., *Introduction to Numerical Analysis*, 2nd Edn., McGraw-Hill, US, 1974.

Josef, S. and Ronald, B., *Introduction to Numerical Analysis*, Springer-Verlag, New York, 1980, Chapter 3.

Leader, J. J., *Numerical Analysis and Scientific Computation*, Addison Wesley, New York, 2004.

Philip J. D. and Philip R., *Methods of Numerical Integration*, Academic Press, New York, 1975. Reprinted Dover Publications, Mineola, New York, 2007.

Press, W. H., Teukolsky, S. A., Vetterling, W. T. and Flannery, B. P. Chapter 4: Integration of functions, *Numerical Recipes: The Art of Scientific Computing*, 3rd Edn., Cambridge University Press, New York, 2007.

Saad, Y., *Numerical Methods for Large Eigenvalue Problems*, Classics edition, SIAM, Philadelphia, PA, 2011.

Appendix 1: Hilbert Transform

The Hilbert transform $\hat{f}(t)$ of the function $f(t)$ is defined as

$$\hat{f}(t) \stackrel{def}{=} \mathcal{H}[f(t)] = \frac{1}{\pi}\int_{-\infty}^{\infty} \frac{f(\tau)}{t-\tau}d\tau = \frac{1}{\pi}\int_{-\infty}^{\infty} \frac{f(t-\tau)}{\tau}d\tau \qquad \text{(A1.1)}$$

The integrand with a singularity at $\tau = t$, results in an improper integral. To evaluate this integral, Cauchy's principal value of the integral should be taken, i.e.

$$\int_{-\infty}^{\infty} \frac{f(\tau)}{t-\tau}d\tau = \lim_{\epsilon \to 0}\left[\int_{-\infty}^{t-\epsilon} \frac{f(\tau)}{t-\tau}d\tau + \int_{t+\epsilon}^{\infty} \frac{f(\tau)}{t-\tau}d\tau\right]$$

The Fourier transform $F(\omega)$ of a real valued function $f(t)$ is defined as

$$F(\omega) \stackrel{def}{=} \Im[f(t)] = \int_{-\infty}^{\infty} f(t)\cdot \exp(-j2\pi\omega t)\cdot dt \qquad \text{(A1.2)}$$

The Hilbert transform $\mathcal{H}[F(\omega)]$ of $F(\omega)$ is denoted by $\hat{F}(\omega)$. The relationship between the two transforms is given as

$$\hat{F}(\omega) \stackrel{def}{=} \mathcal{H}[F(\omega)] = -j\text{sgn}(\omega)\cdot F(\omega) \qquad \text{(A1.3)}$$

In Equation A1.3, the signum (or sgn) function is defined as

$$\text{sgn}(\omega) \stackrel{def}{=} \begin{cases} 1, & \text{for } \omega > 0 \\ 0, & \text{for } \omega = 0 \\ -1 & \text{for } \omega < 0 \end{cases} \qquad \text{(A1.4)}$$

Note: Further details about the Hilbert transform may be seen in *Communication Systems*, Fourth Edition, by Simon Haykin, John Wiley & Sons, pp. 723–725, 2001, New Delhi.

Appendix 2: Evaluation of Integrals Involved in Section 4.3.1

Part A: Integrals Involved in Equation 4.52

For ready reference, Equation 4.52 is reproduced below:

$$\frac{2}{\pi}\int_0^\infty F(u)\cdot\cos(u\cdot y)\cdot du = 0 \quad \text{over } 0 \le y < t/2$$

$$= \frac{2}{\pi}\int_0^\infty f(w)\cdot\sin\{w\cdot(y - t/2)\}\cdot dw - 1/2 \quad \text{over } t/2 \le y < \infty, \tag{4.52}$$

Multiply both sides of Equation 4.52 by cos(vy) and integrate over $0 \le y \le \infty$, to get

$$LHS = \frac{2}{\pi}\int_0^\infty F(u)\cdot du\int_0^\infty \cos(u\cdot y)\cdot\cos(v\cdot y)\cdot dy$$

Using the identities

$$\int_0^\infty \cos(w\cdot t)\cdot dt \equiv \pi\mathcal{S}(w) \tag{A2.1a}$$

$$\int_0^\infty F(w)\cdot\mathcal{S}(w - v)\cdot dw \equiv F(v) \tag{A2.1b}$$

The left-hand side (LHS) of Equation 4.52 can be written as

$$LHS = \int_0^\infty F(u)\cdot\{\mathcal{S}(u - v) + \mathcal{S}(u + v)\}\cdot du = F(v) \tag{A2.2}$$

Similarly, the right-hand side (RHS) of Equation 4.52 can be written as

$$RHS = \frac{2}{\pi}\int_0^\infty f(w)\cdot dw \int_{T/2}^\infty \sin\{w\cdot(y-t/2)\}\cdot\cos(v\cdot y)\cdot dy - 1/\left(2\int_{T/2}^\infty \cos(v\cdot y)\cdot dy\right)$$

For simplification, RHS can be divided into two parts, that is,

$$RHS \stackrel{def}{=} RHS1 + RHS2 \tag{A2.3}$$

*RHS*1 and *RHS*2 are separately evaluated as below

$$RHS1 \stackrel{def}{=} \frac{2}{\pi}\int_0^\infty f(w)\cdot dw\left[\int_{T/2}^\infty \sin\{w\cdot(y-t/2)\}\cdot\cos(v\cdot y)\cdot dy\right] \tag{A2.4a}$$

$$= \frac{2}{\pi}\int_0^\infty f(w)\cdot dw\left[\int_0^\infty \sin\{w\cdot(y-t/2)\}\cdot\cos(v\cdot y)\cdot dy\right]$$

$$- \frac{2}{\pi}\int_0^\infty f(w)\cdot dw\left[\int_0^{T/2} \sin\{w\cdot(y-t/2)\}\cdot\cos(v\cdot y)\cdot dy\right]$$

$$= \frac{2}{\pi}\int_0^\infty f(w)\cdot dw\left[\int_0^\infty \cos(w\cdot t/2)\cdot\{\sin(w\cdot y)\cdot\cos(v\cdot y)\}\cdot dy\right]$$

$$- \frac{2}{\pi}\int_0^\infty f(w)\cdot dw\left[\int_0^\infty \sin(w\cdot t/2)\cdot\{\cos(w\cdot y)\cdot\cos(v\cdot y)\}\cdot dy\right]$$

$$- \frac{2}{\pi}\int_0^\infty f(w)\cdot dw\left[\int_0^{T/2} \cos(w\cdot t/2)\cdot\{\sin(w\cdot y)\cdot\cos(v\cdot y)\}\cdot dy\right]$$

$$+ \frac{2}{\pi}\int_0^\infty f(w)\cdot dw\left[\int_0^{T/2} \sin(w\cdot t/2)\cdot\{\cos(w\cdot y)\cdot\cos(v\cdot y)\}\cdot dy\right]$$

$$= \frac{2}{\pi}\int_0^\infty f(w)\cdot\cos(w\cdot t/2)\cdot dw\left[\int_0^\infty \{\sin(w\cdot y)\cdot\cos(v\cdot y)\}\cdot dy\right]$$

$$- \frac{2}{\pi}\int_0^\infty f(w)\cdot\sin(w\cdot t/2)\cdot dw\left[\int_0^\infty \{\cos(w\cdot y)\cdot\cos(v\cdot y)\}\cdot dy\right]$$

$$-\frac{2}{\pi}\int_0^\infty f(w)\cdot\cos(w\cdot t/2)\cdot dw\left[\int_0^{T/2}\{\sin(w\cdot y)\cdot\cos(v\cdot y)\}\cdot dy\right]$$

$$+\frac{2}{\pi}\int_0^\infty f(w)\cdot\sin(w\cdot t/2)\cdot dw\left[\int_0^{T/2}\{\cos(w\cdot y)\cdot\cos(v\cdot y)\}\cdot dy\right] \qquad \text{(A2.4b)}$$

Now, since

$$\int_0^\infty\{\sin(w\cdot y)\cdot\cos(v\cdot y)\}\cdot dy = \frac{1}{2}\int_0^\infty\{\sin(w+v)y+\sin(w-v)y\}\cdot dy$$

$$= \frac{1}{2}\left\{\frac{1}{(w+v)}+\frac{1}{(w-v)}\right\} \quad \text{for } w \neq v$$

$$= \frac{1}{4}\cdot\frac{1}{v} \quad \text{for } w = v$$

$$\int_0^\infty\{\cos(w\cdot y)\cdot\cos(v\cdot y)\}\cdot dy = \frac{1}{2}\int_0^\infty\{\cos(w+v)y+\cos(w-v)y\}\cdot dy$$

$$= \frac{\pi}{2}\{\mathcal{S}(w+v)+\mathcal{S}(w-v)\}$$

$$\int_0^{T/2}\{\sin(w\cdot y)\cdot\cos(v\cdot y)\}\cdot dy = \frac{1}{2}\int_0^{T/2}\{\sin(w+v)y+\sin(w-v)y\}\cdot dy$$

$$= \frac{1}{2}\left\{\frac{\cos(w+v)y}{(w+v)}+\frac{\cos(w-v)y}{(w-v)}\right\}\Bigg|_{T/2}^{0}$$

$$= \frac{1}{2}\left\{\frac{1-(\cos(w+v)T/2)}{(w+v)}+\frac{1-(\cos(w-v)T/2)}{(w-v)}\right\}$$

and

$$\int_0^{T/2}\{\cos(w\cdot y)\cdot\cos(v\cdot y)\}\cdot dy = \frac{1}{2}\int_0^{T/2}\{\cos(w+v)y+\cos(w-v)y\}\cdot dy$$

$$= \frac{1}{2}\left\{\frac{\sin(w+v)y}{(w+v)} + \frac{\sin(w-v)y}{(w-v)}\right\}\Bigg|_0^{T/2} = \frac{1}{2}\left\{\frac{\sin(w+v)T/2}{(w+v)} + \frac{\sin(w-v)T/2}{(w-v)}\right\}$$

Therefore, from Equation A2.4b, we get for $w \neq v$

$$RHS1 = \frac{2}{\pi}\int_0^\infty f(w)\cdot\cos(w\cdot t/2)\cdot\frac{1}{2}\left\{\frac{1}{(w+v)} + \frac{1}{(w-v)}\right\}\cdot dw$$

$$-\frac{2}{\pi}\int_0^\infty f(w)\cdot\sin(w\cdot t/2)\cdot\frac{\pi}{2}\{\delta(w+v)+\delta(w-v)\}\cdot dw$$

$$-\frac{2}{\pi}\int_0^\infty f(w)\cdot\cos(w\cdot t/2)\frac{1}{2}\left\{\frac{1-(\cos(w+v)T/2)}{(w+v)} + \frac{1-(\cos(w-v)T/2)}{(w-v)}\right\}\cdot dw$$

$$+\frac{2}{\pi}\int_0^\infty f(w)\cdot\sin(w\cdot t/2)\cdot\frac{1}{2}\left\{\frac{\sin(w+v)T/2}{(w+v)} + \frac{\sin(w-v)T/2}{(w-v)}\right\}\cdot dw \tag{A2.5a}$$

And the expression for $RHS1$ for $w = v$ can be obtained from Equation A2.5a on ignoring the term $1/(w-v)$.

Now,

$$RHS2 \stackrel{def}{=} -\frac{1}{2}\int_{T/2}^\infty \cos(v\cdot y)\cdot dy$$

$$= -\frac{1}{2}\int_0^\infty \cos(v\cdot y)\cdot dy + 1/\left(2\int_0^{T/2}\cos(v\cdot y)\cdot dy\right)$$

$$= -\frac{\pi}{2}\cdot\delta(v) + \frac{1}{2}\frac{\sin(v\cdot T/2)}{v} \tag{A2.5b}$$

Therefore, in view of Equations A2.2, A2.5a and A2.5b, we get the following relation between $F(v)$ and $f(w)$ for $0 < v < \infty$:

$$F(v) = \int_0^\infty f(w)\cdot\cos(w\cdot t/2)\cdot\frac{1}{2}\left\{\frac{1}{(w+v)} + \frac{1}{(w-v)}\right\}\cdot dw - f(v)\cdot\frac{\pi}{2}\cdot\sin(v\cdot t/2)$$

$$-\int_0^\infty f(w)\cdot\cos(w\cdot t/2)\frac{1}{2}\left\{\frac{1-(\cos(w+v)T/2)}{(w+v)} + \frac{1-(\cos(w-v)T/2)}{(w-v)}\right\}\cdot dw$$

$$+\int_0^\infty f(w)\cdot\sin(w\cdot t/2)\cdot\frac{1}{2}\left\{\frac{\sin(w+v)T/2}{(w+v)} + \frac{\sin(w-v)T/2}{(w-v)}\right\}\cdot dw \tag{A2.6}$$

Part B: Integrals Involved in Equation 4.54

For a second set of such relation Equation 4.54, which is reproduced below for ready reference, is to be considered

$$-\frac{2}{\pi}\int_0^\infty u\cdot F(u)\cdot\cos\{u\cdot(Y+t/2)\}\cdot\coth(u\cdot g/2)\cdot du-\frac{1}{g}$$
$$=\frac{2}{\pi}\int_0^\infty w\cdot f(w)\cdot\sin(w\cdot Y)\cdot dw-\frac{1}{\pi}\cdot\left(\frac{1}{Y}\right)$$
$$\text{over } 0\le Y<\infty \tag{4.54}$$

Multiply both sides of this equation by $\sin(vY)$ and then integrate over $0 \le Y < \infty$. The first term on the LHS will be

$$LHS1 \stackrel{def}{=} -\frac{2}{\pi}\int_0^\infty u\cdot F(u)\cdot\coth(u\cdot g/2)\cdot du\int_0^\infty \sin(vY)\cdot\cos\{u\cdot(Y+t/2)\}\cdot dY$$
$$\stackrel{def}{=} -\frac{1}{\pi}\int_0^\infty u\cdot F(u)\cdot\coth(u\cdot g/2)\cdot du\int_0^\infty [\sin\{(v+u)Y+u\cdot t/2\}$$
$$+\sin\{(v-u)Y-u\cdot t/2\}]\cdot dY \tag{A2.7a}$$

Or

$$LHS1 \stackrel{def}{=} -\frac{1}{\pi}\int_0^\infty u\cdot F(u)\cdot\coth(u\cdot g/2)\cdot du\int_0^\infty [\cos(u\cdot t/2)\cdot\sin\{(v+u)Y\}$$
$$+\sin(u\cdot t/2)\cdot\cos\{(v+u)Y\}+\cos(u\cdot t/2)\cdot\sin\{(v-u)Y\}$$
$$-\sin(u\cdot t/2)\cdot\cos\{(v-u)Y\}]\cdot dY$$

$$\stackrel{def}{=} -\frac{1}{\pi}\int_0^\infty u\cdot F(u)\cdot\coth(u\cdot g/2)\cdot du \ge \left[\cos(u\cdot t/2)\cdot\left\{\frac{1}{(v+u)}-\frac{1}{(v-u)}\right\}\right.$$
$$\left.+\pi\cdot\sin(u\cdot t/2)\cdot\{\mathcal{S}(v+u)-\mathcal{S}(v-u)\}\right] \tag{A2.7b}$$

$$LHS1 = -\frac{1}{\pi}\int_0^\infty u\cdot F(u)\cdot\coth\big(u\cdot(g/2)\big)\cdot\cos\big(u\cdot(t/2)\big)\cdot\left\{\frac{1}{(v+u)}-\frac{1}{(v-u)}\right\}\cdot du$$
$$+v\cdot F(v)\cdot\coth\big(v\cdot(g/2)\big)\cdot\cos\big(v\cdot(t/2)\big) \tag{A2.7c}$$

and,

$$LHS2 \stackrel{def}{=} -\frac{1}{g}\int_0^\infty \sin(v \cdot Y) \cdot dY = -\frac{1}{g} \cdot \frac{1}{v} \quad \text{for } v \neq 0$$
$$= 0 \quad \text{for } v = 0 \tag{A2.7d}$$

$$LHS \stackrel{def}{=} LHS1 + LHS2 \tag{A2.8}$$

Therefore, for $v \neq 0$

$$LHS = -\frac{1}{\pi}\int_0^\infty u \cdot F(u) \cdot \coth\big(u \cdot (g/2)\big) \cdot \cos\big(u \cdot (t/2)\big) \cdot \left\{\frac{1}{(v+u)} - \frac{1}{(v-u)}\right\} \cdot du$$
$$+ v \cdot F(v) \cdot \coth\big(v \cdot (g/2)\big) \cdot \cos\big(v \cdot (t/2)\big) - \frac{1}{g} \cdot \frac{1}{v} \tag{A2.9}$$

Similarly, let

$$RHS \cdot \stackrel{def}{=} RHS1 + RHS2 \tag{A2.10a}$$

where

$$RHS1 \stackrel{def}{=} \int_0^\infty w \cdot f(w) \cdot dw \int_0^\infty \sin(v \cdot Y) \cdot \sin(w \cdot Y) \cdot dY \tag{A2.10b}$$

or

$$RHS1 \stackrel{def}{=} \frac{1}{2}\int_0^\infty w \cdot f(w) \cdot dw \int_0^\infty [\cos\{(v-w)Y\} - \cos\{(v+w)Y\}] \cdot dY$$

or

$$RHS1 \stackrel{def}{=} \frac{\pi}{2}\int_0^\infty w \cdot f(w) \cdot [\mathcal{S}(v-w) - \mathcal{S}(v+w)] \cdot dw$$

Since $v > 0$, we get

$$RHS1 = \frac{\pi}{2} \cdot v \cdot f(v) \tag{A2.10c}$$

and

$$RHS2 \stackrel{def}{=} -\frac{1}{\pi} \cdot \int_0^{\infty} \frac{\sin(v \cdot Y)}{Y} \cdot dY \tag{A2.11a}$$

or

$$RHS2 = -\frac{1}{2} \cdot [U(v) - U(-v)] \tag{A2.11b}$$

Therefore, using Equations A2.10a, A2.10c and A2.11b, we get

$$RHS = \frac{\pi}{2} \cdot v \cdot f(v) - \frac{1}{2} \tag{A2.12}$$

Thus, in view of Equations A2.9 and A2.12, we get the following relation:

$$\begin{aligned} &-\frac{1}{\pi}\int_0^{\infty} u \cdot F(u) \cdot \coth\big(u \cdot (g/2)\big) \cdot \cos\big(u \cdot (t/2)\big) \cdot \left\{\frac{1}{(v+u)} - \frac{1}{(v-u)}\right\} \cdot du \\ &+ v \cdot F(v) \cdot \coth\big(v \cdot (g/2)\big) \cdot \cos\big(v \cdot (t/2)\big) - \frac{1}{g} \cdot \frac{1}{v} = \frac{\pi}{2} \cdot v \cdot f(v) - \frac{1}{2} \end{aligned} \tag{A2.13}$$

Appendix 3: Evaluation of Integrals Involved in Section 4.3.2

Part A: Integrals Involved in Equation 4.61

For ready reference, Equation 4.61 is reproduced below:

over

$$\frac{2}{\pi}\int_0^\infty F(u)\cdot\cos(u\cdot y)\cdot du = 1 \text{ over } 0 \le y < \frac{T}{2}$$

$$= \frac{2}{\pi}\int_0^\infty f(w)\cdot\sin\{w\cdot(y-t/2)\}\cdot dw \text{ over } \frac{T}{2} \le y < \infty \tag{4.61}$$

Multiply both sides of Equation 4.61 by $\cos(v\cdot y)$ and integrate over $0 < Y < \infty$ to get

$$\int_0^\infty F(u)\cdot du\int_0^\infty 2\cos(u\cdot y)\cdot\cos(v\cdot y)\cdot dy = \pi\int_0^{T/2}\cos(v\cdot y)\cdot dy$$

$$+\int_0^\infty f(w)\cdot dw\int_{T/2}^\infty 2\sin\{w\cdot(y-t/2)\}\cos(v\cdot y)\cdot dy$$

or

$$\int_0^\infty F(u)\cdot du\int_0^\infty 2\cos(u\cdot y)\cdot\cos(v\cdot y)\cdot dy = \pi\int_0^{T/2}\cos(v\cdot y)\cdot dy$$

$$+\int_0^\infty f(w)\cdot dw\left[\int_0^\infty 2\sin\{w\cdot(y-t/2)\}\cos(v\cdot y)\cdot dy\right.$$

$$- \int_0^{T/2} 2\sin\{w \cdot (y - t/2)\}\cos(v \cdot y) \cdot dy \Bigg] \tag{A3.1}$$

Now, since

$$2\sin\{w \cdot (y - t/2)\}\cos(v \cdot y) = \cos(w \cdot t/2) \times \left[\sin\{(w + v)y\} + \sin\{(w - v)y\}\right]$$

$$-\sin(w \cdot t/2) \cdot \left[\cos\{(w + v)y\} + \cos\{(w - v)y\}\right] \tag{A3.2a}$$

$$\int_0^{\infty} \sin(a \cdot y)\, dy = \frac{1}{a}, \quad a \neq 0$$
$$= 0, \quad a = 0 \tag{A3.2b}$$

and

$$\int_0^{\infty} \cos(a \cdot y)\, dy = \pi \cdot \mathcal{S}(a) \tag{A3.2c}$$

Therefore, left-hand side (*LHS*) of Equation A3.1 can be written as

$$LHS = \int_0^{\infty} F(u) \cdot du \int_0^{\infty} \left[\cos(u - v) \cdot y + \cos(u + v)y\right] \cdot dy$$

$$= \int_0^{\infty} F(u) \cdot du\pi\left[\mathcal{S}(u - v) + \mathcal{S}(u + v)\right] = \pi \cdot F(v) \tag{A3.3a}$$

The right-hand side (*RHS*) of Equation A3.1 can be written as

$$RHS = \pi \int_0^{T/2} \cos(v \cdot y) \cdot dy$$

$$+\int_0^\infty f(w)\cdot dw\cos(w\cdot t/2)\cdot\int_0^\infty\left[\sin\{(w+v)y\}+\sin\{(w-v)y\}\right]\cdot dy$$

$$-\sin(w\cdot t/2)\cdot\int_0^\infty\left[\cos\{(w+v)y\}+\cos\{(w-v)y\}\right]\cdot dy$$

$$-\cos(w\cdot t/2)\cdot\int_0^{T/2}\left[\sin\{(w+v)y\}+\sin\{(w-v)y\}\right]\cdot dy$$

$$+\sin(w\cdot t/2)\cdot\int_0^{T/2}\left[\cos\{(w+v)y\}+\cos\{(w-v)y\}\right]\cdot dy$$

$$=\pi\cdot\frac{\sin(v\cdot T/2)}{v}$$

$$+\int_0^\infty f(w)\cdot dw\cos(w\cdot t/2)\cdot\left[\frac{1}{(w+v)}+\frac{1}{(w-v)}\right]$$

$$-\sin(w\cdot t/2)\cdot\pi\cdot\left[\mathcal{S}(w+v)+\mathcal{S}(w-v)\right]$$

$$+\cos(w\cdot t/2)\cdot\left[\frac{\cos\{(w+v)(T/2)\}-1}{(w+v)}+\frac{\cos\{(w-v)(T/2)\}-1}{(w-v)}\right]$$

$$+\sin(w\cdot t/2)\cdot\left[\frac{\sin\{(w+v)(T/2)\}}{(w+v)}+\frac{\sin\{(w-v)(T/2)\}}{(w-v)}\right]$$

$$=\pi\cdot\frac{\sin(v\cdot T/2)}{v}+\int_0^\infty f(w)\cdot dw\cos(w\cdot t/2)$$

$$\times\left[\frac{\cos\{(w+v)(T/2)\}}{(w+v)}+\frac{\cos\{(w-v)(T/2)\}}{(w-v)}\right]$$

$$+\sin(w\cdot t/2)\cdot\left[\frac{\sin\{(w+v)(T/2)\}}{(w+v)}+\frac{\sin\{(w-v)(T/2)\}}{(w-v)}\right]-f(v)\cdot\sin(v\cdot t/2)\cdot\pi \tag{A3.3b}$$

Therefore, for $v \neq w$

$$F(v) = \frac{\sin(v \cdot T/2)}{v} - f(v) \cdot \sin(v \cdot t/2)$$

$$+ \int_0^{\infty} f(w) \cdot dw \cos(w \cdot t/2) \cdot \left[\frac{\cos\{(w+v)(T/2)\}}{(w+v)} + \frac{\cos\{(w-v)(T/2)\}}{(w-v)} \right]$$

$$+ \sin(w \cdot t/2) \cdot \left[\frac{\sin\{(w+v)(T/2)\}}{(w+v)} + \frac{\sin\{(w-v)(T/2)\}}{(w-v)} \right] \quad \text{(A3.4)}$$

Part B: Integrals Involved in Equation 4.63

Consider Equation 4.63, which is reproduced below for ready reference

$$-\frac{2}{\pi} \int_0^{\infty} u \cdot F(u) \cdot \cos\{u \cdot (Y + t/2)\} \cdot du = \frac{2}{\pi} \int_0^{\infty} w f(w) \cdot \sin(w \cdot Y) \cdot dw - \frac{2}{\pi} \cdot \frac{1}{Y}$$

$$\text{Over } 0 < Y < \infty \quad \text{(4.63)}$$

Multiply both sides of Equation 4.63 by $\sin(v \cdot Y)$ and integrate over $0 < Y < \infty$, we get

$$LHS = -\frac{2}{\pi} \int_0^{\infty} u \cdot F(u) \cdot du \int_0^{\infty} \sin(v \cdot Y) \cdot \cos\{u \cdot (Y + t/2)\} \cdot dY$$

$$= -\frac{1}{\pi} \int_0^{\infty} u \cdot F(u) \cdot du \int_0^{\infty} 2\sin(v \cdot Y) \cdot \cos\{u \cdot (Y + t/2)\} \cdot dY$$

$$= -\frac{1}{\pi} \int_0^{\infty} u \cdot F(u) \cdot du \int_0^{\infty} \left[\sin\{(v + u) \cdot Y + u \cdot t/2\} \right.$$

$$+\sin\left\{(v-u)\cdot Y - u\cdot t/2\right\}\Big]\cdot dY$$

$$= -\frac{1}{\pi}\int_0^\infty u\cdot F(u)\cdot du\int_0^\infty \cos(u\cdot t/2)\cdot\left[\sin\left\{(v+u)\cdot Y\right\}+\sin\left\{(v-u)\cdot Y\right\}\right]$$

$$+\sin(u\cdot t/2)\cdot\left[\cos\left\{(v+u)\cdot Y\right\}-\cos\left\{(v-u)\cdot Y\right\}\right]\cdot dY$$

$$= -\frac{1}{\pi}\int_0^\infty u\cdot F(u)\cdot du\left[\cos(u\cdot t/2)\cdot\left\{\frac{1}{(v+u)}+\frac{1}{(v-u)}\right\}-\sin(u\cdot t/2)\cdot\pi\cdot\delta(v-u)\right]$$

$$= -\frac{1}{\pi}\int_0^\infty u\cdot F(u)\cdot\cos(u\cdot t/2)\cdot\left\{\frac{1}{(v+u)}+\frac{1}{(v-u)}\right\}\cdot du + v\cdot F(v)\cdot\sin(v\cdot t/2) \tag{A3.5a}$$

and

$$RHS = \frac{1}{\pi}\int_0^\infty w\cdot f(w)\cdot dw\int_0^\infty 2\cdot\sin(v\cdot Y)\cdot\sin(w\cdot Y)\cdot dY - \frac{2}{\pi}\cdot\int_0^\infty\frac{\sin(v\cdot Y)}{Y}\cdot dY$$

$$= \frac{1}{\pi}\int_0^\infty w\cdot f(w)\cdot dw\int_0^\infty\left[\cos\left\{(v-w)Y\right\}-\cos\left\{(v+w)Y\right\}\right]\cdot dY$$

$$-\frac{2}{\pi}\cdot\int_0^\infty\frac{\sin(v\cdot Y)}{Y}\cdot dY$$

$$= \frac{1}{\pi}\int_0^\infty w\cdot f(w)\cdot dw\cdot\pi\left[\mathcal{S}(v-w)-\mathcal{S}(v+w)\right]-\left[U(v)-U(-v)\right]$$

$$= v\cdot f(v) - U(v) \tag{A3.5b}$$

Therefore,

$$v \cdot F(v) \cdot \sin(v \cdot t/2) - \frac{1}{\pi}\int_0^{\infty} u \cdot F(u) \cdot \cos(u \cdot t/2) \cdot \left\{\frac{1}{(v+u)} + \frac{1}{(v-u)}\right\} \cdot du$$
$$= v \cdot f(v) - 1$$

For,

$$v \neq u \tag{A3.6}$$

Appendix 4: Evaluation of Integrals Involved in Section 4.3.3

Equation 4.76a is reproduced below:

$$\int_0^{\infty}\left\{F'(u)\cdot\cos(u\cdot y)+F''(u)\cdot\sin(u\cdot y)\right\}\cdot du$$

$$=\int_0^{\infty} f_2(w)\cdot\sin(w\cdot y_2)\cdot dw \quad \text{over} \quad -\infty\le y\le(\delta-t/2)$$

$$=-1/2 \quad \text{over } (\delta-t/2)\le y\le(\delta+t/2)$$

$$=-\int_0^{\infty} f_1(w)\cdot\sin(w\cdot y_1)\cdot dw \quad \text{over } (\delta+t/2)\le y\le\infty \tag{4.76a}$$

Part A: First Integration of LHS of Equation 4.76a

Multiply LHS of Equation 4.76a by $\cos(v\cdot y)$, integrate the resulting expression over $-\infty\le y\le\infty$, and use the relation

$$\int_{-\infty}^{\infty}\cos(v\cdot y)\cdot dy = 2\pi\cdot\mathcal{S}(v) \tag{A4.1}$$

where $\mathcal{S}(v)$ is called delta (or impulse) function of the variable v.

This exercise yields

$$\int_0^{\infty}\int_{-\infty}^{\infty}\left[\left\{F'(u)\cdot\cos(u\cdot y)+F''(u)\cdot\sin(u\cdot y)\right\}\cdot\cos(v\cdot y)\cdot dy\right]\cdot du$$

$$=\int_0^{\infty}\frac{1}{2}\cdot F'(u)\cdot\left[\int_{-\infty}^{\infty}\left\{\cos(u-v)y+\cos(u+v)y\right\}\cdot dy\right]\cdot du$$

$$= \int_0^{\infty} \pi \cdot F'(u) \cdot \left[\mathcal{S}(u - v) + \mathcal{S}(u + v)\right] \cdot du$$

$$= \pi \cdot F'(v) \tag{A4.2}$$

Part B: Second Integration of RHS of Equation 4.76a

Multiply RHS of Equation 4.76a by $\cos(v \cdot y)$ and piece-wise integrate over $-\infty \le y \le \infty$, and use the following relation:

$$\int_0^{\infty} \sin(v \cdot Y) \cdot dY = \frac{1}{v}, \quad \text{for } v \neq 0 \tag{A4.3}$$

This exercise yields

$$\int_0^{\infty} f_2(w) \cdot \left[\int_{-\infty}^{(\delta - t/2)} \sin\{w \cdot (y - \delta + t/2)\} \cdot \cos(v \cdot y) dy\right] \cdot dw$$

$$-\frac{1}{2} \int_{(\delta - t/2)}^{(\delta + t/2)} \cos(v \cdot y) dy - \int_0^{\infty} f_1(w) \cdot \left[\int_{(\delta + t/2)}^{\infty} \sin\{w \cdot (y - \delta - t/2\ )\} \cdot \cos(v \cdot y) dy\right] \cdot dw$$

$$= \int_0^{\infty} f_2(w) \cdot \cos\{w \cdot (\delta - t/2)\} \left[\int_{-\infty}^{(\delta - t/2)} \frac{1}{2} \cdot \left[\sin\{(w + v) \cdot y\} + \sin\{(w - v) \cdot y\}\right] \cdot dy\right] \cdot dw$$

$$-\int_0^{\infty} f_2(w) \cdot \sin\{w \cdot (\delta - t/2)\} \left[\int_{-\infty}^{(\delta - t/2)} \frac{1}{2} \cdot \left[\cos\{(w + v) \cdot y\} + \cos\{(w - v) \cdot y\}\right] \cdot dy\right] \cdot dw$$

$$-\frac{1}{2} \int_{(\delta - t/2)}^{(\delta + t/2)} \cos(v \cdot y) dy - \int_0^{\infty} f_1(w) \cdot \cos\{w \cdot (\delta + t/2)\} \int_{(\delta + t/2)}^{\infty} \frac{1}{2} \cdot \left[\sin\{(w + v) \cdot y\}\right.$$

$$\left. + \sin\{(w - v) \cdot y\}\right] \cdot dy \cdot dw + \int_0^{\infty} f_1(w) \cdot \sin\{w \cdot (\delta + t/2)\}$$

$$\left[\int_{(\delta+t/2)}^{\infty}\frac{1}{2}\cdot\left[\cos\{(w+v)\cdot y\}+\cos\{(w-v)\cdot y\}\right]\cdot dy\right]\cdot dw$$

$$=-\int_0^{\infty} f_2(w)\cdot\frac{1}{2}\cdot\cos\{w\cdot(\delta-t/2)\}\left[\int_0^{\infty}\left[\sin\{(w+v)\cdot y\}+\sin\{(w-v)\cdot y\}\right]\cdot dy\right]\cdot dw$$

$$+\int_0^{\infty} f_2(w)\cdot\frac{1}{2}\cdot\cos\{w\cdot(\delta-t/2)\}\left[\int_0^{(\delta-t/2)}\left[\sin\{(w+v)\cdot y\}+\sin\{(w-v)\cdot y\}\right]\cdot dy\right]\cdot dw$$

$$-\int_0^{\infty} f_2(w)\cdot\frac{1}{2}\cdot\sin\{w\cdot(\delta-t/2)\}\left[\int_0^{\infty}\left[\cos\{(w+v)\cdot y\}+\cos\{(w-v)\cdot y\}\right]\cdot dy\right]\cdot dw$$

$$-\int_0^{\infty} f_2(w)\cdot\frac{1}{2}\cdot\sin\{w\cdot(\delta-t/2)\}\int_0^{(\delta-t/2)}\left[\cos\{(w+v)\cdot y\}+\cos\{(w-v)\cdot y\}\right]\cdot dy\cdot dw$$

$$-\frac{1}{2}\int_{(\delta-t/2)}^{(\delta+t/2)}\cos(v\cdot y)dy-\int_0^{\infty} f_1(w)\cdot\frac{1}{2}\cdot\cos\left\{w\cdot\left(\delta+\frac{t}{2}\right)\right\}$$
$$\times\left[\int_0^{\infty}\left[\sin\{(w+v)\cdot y\}+\sin\{(w-v)\cdot y\}\right]\cdot dy\right]\cdot dw$$

$$+\int_0^{\infty} f_1(w)\cdot\frac{1}{2}\cdot\cos\{w\cdot(\delta+t/2)\}\left[\int_0^{(\delta+t/2)}\left[\sin\{(w+v)\cdot y\}+\sin\{(w-v)\cdot y\}\right]\cdot dy\right]\cdot dw$$

$$+\int_0^{\infty} f_1(w)\cdot\frac{1}{2}\cdot\sin\{w\cdot(\delta+t/2)\}\left[\int_0^{\infty}\left[\cos\{(w+v)\cdot y\}+\cos\{(w-v)\cdot y\}\right]\cdot dy\right]\cdot dw$$

$$-\int_0^{\infty} f_1(w)\cdot\frac{1}{2}\cdot\sin\{w\cdot(\delta+t/2)\}\cdot\left[\int_0^{(\delta+t/2)}\left[\cos\{(w+v)\cdot y\}+\cos\{(w-v)\cdot y\}\right]\cdot dy\right]\cdot dw$$

$$=-\int_0^{\infty} f_2(w)\cdot\frac{1}{2}\cdot\cos\{w\cdot(\delta-t/2)\}\cdot\left[\frac{1}{(w+v)}+\frac{1}{(w-v)}\right]\cdot dw$$

$$-\int_0^\infty f_2(w)\cdot\frac{1}{2}\cdot\cos\{w\cdot(\delta-t/2)\}$$
$$\times\left[\frac{\cos\{(w+v)\cdot(\delta-t/2)\}-1}{(w+v)}+\frac{\cos\{(w-v)\cdot(\delta-t/2)\}-1}{(w-v)}\right]\cdot dw$$

$$-\int_0^\infty f_2(w)\cdot\frac{\pi}{2}\cdot\sin\{w\cdot(\delta-t/2)\}\cdot[\mathcal{S}(w+v)+\mathcal{S}(w-v)]\cdot dw$$

$$-\int_0^\infty f_2(w)\cdot\frac{1}{2}\cdot\sin\{w\cdot(\delta-t/2)\}\left[\frac{\sin\{(w+v)\cdot(\delta-t/2)\}}{(w+v)}+\frac{\sin\{(w-v)\cdot(\delta-t/2)\}}{(w-v)}\right]\cdot dw$$

$$-\left[\frac{\sin\{v\cdot(\delta+t/2)\}-\sin\{v\cdot(\delta-t/2)\}}{2v}\right]-\int_0^\infty f_1(w)\cdot\frac{1}{2}\cdot\cos\{w\cdot(\delta+t/2)\}$$
$$\times\left[\frac{1}{(w+v)}+\frac{1}{(w-v)}\right]\cdot dw$$

$$-\int_0^\infty f_1(w)\cdot\frac{1}{2}\cdot\cos\{w\cdot(\delta+t/2)\}$$
$$\times\left[\frac{\cos\{(w+v)\cdot(\delta+t/2)\}-1}{(w+v)}+\frac{\cos\{(w-v)\cdot(\delta+t/2)\}-1}{(w-v)}\right]\cdot dw$$

$$+\int_0^\infty f_1(w)\cdot\frac{\pi}{2}\cdot\sin\{w\cdot(\delta+t/2)\}[\mathcal{S}(w+v)+\mathcal{S}(w-v)]\cdot dw$$

$$-\int_0^\infty f_1(w)\cdot\frac{1}{2}\cdot\sin\{w\cdot(\delta+t/2)\}\left[\frac{\sin\{(w+v)\cdot(\delta+t/2)\}}{(w+v)}+\frac{\sin\{(w-v)\cdot(\delta+t/2)\}}{(w-v)}\right]\cdot dw$$

$$=-\int_0^\infty f_2(w)\cdot\frac{1}{2}\cdot\left[\frac{\cos\{v\cdot(\delta-t/2)\}}{(w+v)}+\frac{\cos\{v\cdot(\delta-t/2)\}}{(w-v)}\right]\cdot dw$$

$$-f_2(v)\cdot\frac{\pi}{2}\cdot\sin\{v\cdot(\delta-t/2)\}-\left[\frac{\sin\{v\cdot(\delta+t/2)\}-\sin\{v\cdot(\delta-t/2)\}}{2v}\right]$$

$$-\int_0^\infty f_1(w)\cdot\frac{1}{2}\cdot\left[\frac{\cos\{v\cdot(\delta+t/2)\}}{(w+v)}+\frac{\cos\{v\cdot(\delta+t/2)\}}{(w-v)}\right]\cdot dw+f_1(v)\cdot\frac{\pi}{2}\cdot\sin\{v\cdot(\delta+t/2)\}$$

$$=-\cos\{v\cdot(\delta-t/2)\}\cdot\int_0^\infty f_2(w)\cdot\left[\frac{w}{(w^2-v^2)}\right]\cdot dw-f_2(v)\cdot\frac{\pi}{2}\cdot\sin\{v\cdot(\delta-t/2)\}$$

$$-\left[\frac{\sin\{v\cdot(\delta+t/2)\}-\sin\{v\cdot(\delta-t/2)\}}{2v}\right]-\cos\{v\cdot(\delta+t/2)\}\cdot\int_0^\infty f_1(w)\left[\frac{w}{(w^2-v^2)}\right]\cdot dw$$

$$+f_1(v)\cdot\frac{\pi}{2}\cdot\sin\{v\cdot(\delta+t/2)\} \tag{A4.4}$$

Thus, in view of Equations A4.2 and A4.4

$$\pi\cdot F'(v)=-\cos\{v\cdot(\delta-t/2)\}\cdot\int_0^\infty f_2(w)\cdot\left[\frac{w}{(w^2-v^2)}\right]\cdot dw$$

$$-f_2(v)\cdot\frac{\pi}{2}\cdot\sin\{v\cdot(\delta-t/2)\}-\left[\frac{\sin\{v\cdot(\delta+t/2)\}-\sin\{v\cdot(\delta-t/2)\}}{2v}\right]$$

$$-\cos\{v\cdot(\delta+t/2)\}\cdot\int_0^\infty f_1(w)\left[\frac{w}{(w^2-v^2)}\right]\cdot dw$$

$$+f_1(v)\cdot\frac{\pi}{2}\cdot\sin\{v\cdot(\delta+t/2)\}\quad(\text{for all } v>0) \tag{A4.5}$$

Part C: Third Integration of LHS of Equation 4.76a

Multiply LHS of Equation 4.76a by $\sin(v\cdot y)$ and integrate over $-\infty\le y\le\infty$, to get

$$\int_0^\infty\left[\int_{-\infty}^\infty\left[\{F'(u)\cdot\cos(u\cdot y)+F''(u)\cdot\sin(u\cdot y)\}\cdot\sin(v\cdot y)\right]\cdot dy\right]\cdot du$$

$$= \int_0^\infty \frac{1}{2} \cdot F''(u) \cdot \left[\int_{-\infty}^{\infty} \{\cos(u - v)y - \cos(u + v)y\} \cdot dy \right] \cdot du$$

$$= \int_0^\infty \pi \cdot F'(u) \cdot \left[\mathcal{S}(u - v) - \mathcal{S}(u + v) \right] \cdot du = \pi \cdot F''(v) \tag{A4.6}$$

Part D: Fourth Integration of RHS of Equation 4.76a

Multiply RHS of Equation 4.76a by $\sin(v \cdot y)$ and piece-wise integrate over $-\infty \leq y \leq \infty$, to get

$$\int_0^\infty f_2(w) \cdot \left[\int_{-\infty}^{(\delta - t/2)} \sin\{w \cdot (y - \delta + t/2)\} \cdot \sin(v \cdot y) dy \right] \cdot dw - \frac{1}{2} \int_{(\delta - t/2)}^{(\delta + t/2)} \sin(v \cdot y) dy$$

$$- \int_0^\infty f_1(w) \cdot \left[\int_{(\delta + t/2)}^{\infty} \sin\{w \cdot (y - \delta - t/2)\} \cdot \sin(v \cdot y) dy \right] \cdot dw$$

$$= \int_0^\infty f_2(w) \cdot \cos\{w \cdot (\delta - t/2)\} \left[\int_{-\infty}^{(\delta - t/2)} \frac{1}{2} \cdot \left[\cos\{(w - v) \cdot y\} - \cos\{(w + v) \cdot y\} \right] \cdot dy \right] \cdot dw$$

$$- \int_0^\infty f_2(w) \cdot \sin\{w \cdot (\delta - t/2)\} \left[\int_{-\infty}^{(\delta - t/2)} \frac{1}{2} \cdot \left[\sin\{(w + v) \cdot y\} \right.\right.$$

$$\left.\left. - \sin\{(w - v) \cdot y\} \right] \cdot dy \right] \cdot dw - \frac{1}{2} \int_{(\delta - t/2)}^{(\delta + t/2)} \sin(v \cdot y) dy$$

$$- \int_0^\infty f_1(w) \cdot \cos\{w \cdot (\delta + t/2)\} \left[\int_{(\delta + t/2)}^{\infty} \frac{1}{2} \cdot \left[\cos\{(w - v) \cdot y\} - \cos\{(w + v) \cdot y\} \right] \cdot dy \right] \cdot dw$$

$$+\int_0^\infty f_1(w)\cdot\sin\{w\cdot(\delta+t/2)\}\left[\int_{(\delta+t/2)}^{\infty}\frac{1}{2}\cdot\left[\sin\{(w+v)\cdot y\}-\sin\{(w-v)\cdot y\}\right]\cdot dy\right]\cdot dw$$

$$=\int_0^\infty f_2(w)\cdot\frac{1}{2}\cdot\cos\{w\cdot(\delta-t/2)\}\left[\int_0^\infty\left[\cos\{(w-v)\cdot y\}-\cos\{(w+v)\cdot y\}\right]\cdot dy\right]\cdot dw$$

$$-\int_0^\infty f_2(w)\cdot\frac{1}{2}\cdot\cos\{w\cdot(\delta-t/2)\}\left[\int_0^{(\delta-t/2)}\left[\cos\{(w-v)\cdot y\}-\cos\{(w+v)\cdot y\}\right]\cdot dy\right]\cdot dw$$

$$+\int_0^\infty f_2(w)\cdot\frac{1}{2}\cdot\sin\{w\cdot(\delta-t/2)\}\left[\int_0^\infty\left[\sin\{(w+v)\cdot y\}-\sin\{(w-v)\cdot y\}\right]\cdot dy\right]\cdot dw$$

$$-\int_0^\infty f_2(w)\cdot\frac{1}{2}\cdot\sin\{w\cdot(\delta-t/2)\}\left[\int_0^{(\delta-t/2)}\left[\sin\{(w+v)\cdot y\}-\sin\{(w-v)\cdot y\}\right]\cdot dy\right]\cdot dw$$

$$-\frac{1}{2}\int_{(\delta-t/2)}^{(\delta+t/2)}\sin(v\cdot y)dy$$

$$-\int_0^\infty f_1(w)\cdot\frac{1}{2}\cdot\cos\{w\cdot(\delta+t/2)\}\left[\int_0^\infty\left[\cos\{(w-v)\cdot y\}-\cos\{(w+v)\cdot y\}\right]\cdot dy\right]\cdot dw$$

$$+\int_0^\infty f_1(w)\cdot\frac{1}{2}\cdot\cos\{w\cdot(\delta+t/2)\}\left[\int_0^{(\delta+t/2)}\left[\cos\{(w-v)\cdot y\}-\cos\{(w+v)\cdot y\}\right]\cdot dy\right]\cdot dw$$

$$-\int_0^\infty f_1(w)\cdot\frac{1}{2}\cdot\sin\{w\cdot(\delta+t/2)\}\left[\int_0^\infty\left[\sin\{(w+v)\cdot y\}-\sin\{(w-v)\cdot y\}\right]\cdot dy\right]\cdot dw$$

$$+\int_0^\infty f_1(w)\cdot\frac{1}{2}\cdot\sin\{w\cdot(\delta+t/2)\}\left[\int_0^{(\delta+t/2)}\left[\sin\{(w+v)\cdot y\}-\sin\{(w-v)\cdot y\}\right]\cdot dy\right]\cdot dw$$

(A4.7)

Therefore, in view of Equations A4.6 and A4.7, and the relation

$$\int_0^{\infty} \cos(v \cdot y) \cdot dy = \pi \cdot \mathcal{S}(v) \tag{A4.8}$$

We get

$$\pi \cdot F''(v) = \int_0^{\infty} f_2(w) \cdot \frac{\pi}{2} \cdot \cos\{w \cdot (\delta - t/2)\} \cdot [\mathcal{S}(w - v) - \mathcal{S}(w + v)] \cdot dw$$

$$-\int_0^{\infty} f_2(w) \cdot \frac{1}{2} \cdot \cos\{w \cdot (\delta - t/2)\} \cdot \left[\frac{\sin\{(w - v) \cdot (\delta - t/2)\}}{(w - v)} - \frac{\sin\{(w + v) \cdot (\delta - t/2)\}}{(w + v)}\right] \cdot dw$$

$$+\int_0^{\infty} f_2(w) \cdot \frac{1}{2} \cdot \sin\{w \cdot (\delta - t/2)\} \cdot \left[\frac{2}{(w + v)} - \frac{1}{(w - v)}\right] \cdot dw$$

$$-\int_0^{\infty} f_2(w) \cdot \frac{1}{2} \cdot \sin\{w \cdot (\delta - t/2)\}$$

$$\times \left[\frac{\cos\{(w - v) \cdot (\delta - t/2)\} - 1}{(w - v)} - \frac{\cos\{(w + v) \cdot (\delta - t/2)\} - 1}{(w + v)}\right] \cdot dw$$

$$-\frac{1}{2} \int_{(\delta - t/2)}^{(\delta + t/2)} \sin(v \cdot y) dy$$

$$-\int_0^{\infty} f_1(w) \cdot \frac{\pi}{2} \cdot \cos\{w \cdot (\delta + t/2)\}[\mathcal{S}(w - v) - \mathcal{S}(w + v)] \cdot dw$$

$$+\int_0^{\infty} f_1(w) \cdot \frac{1}{2} \cdot \cos\{w \cdot (\delta + t/2)\} \cdot \left[\frac{\sin\{(w - v) \cdot (\delta + t/2)\}}{(w - v)} - \frac{\sin\{(w + v) \cdot (\delta + t/2)\}}{(w + v)}\right] \cdot dw$$

$$-\int_0^{\infty} f_1(w) \cdot \frac{1}{2} \cdot \sin\{w \cdot (\delta + t/2)\} \cdot \left[\frac{1}{(w + v)} - \frac{1}{(w - v)}\right] \cdot dw$$

$$+\int_0^\infty f_1(w)\cdot\frac{1}{2}\cdot\sin\{w\cdot(\delta+t/2)\}$$

$$\times\left[\frac{\cos\{(w-v)\cdot(\delta+t/2)\}-1}{(w-v)}-\frac{\cos\{(w+v)\cdot(\delta+t/2)-1\}}{(w+v)}\right]\cdot dw \qquad \text{(A4.9)}$$

On simplification, Equation A4.9 reduces to

$$\pi\cdot F''(v)=f_2(v)\cdot\frac{\pi}{2}\cdot\cos\{v\cdot(\delta-t/2)\}-f_1(v)\cdot\frac{\pi}{2}\cdot\cos\{v\cdot(\delta+t/2)\}$$

$$-\int_0^\infty f_2(w)\cdot\frac{1}{2}\cdot\left[\frac{\sin\{(2w-v)\cdot(\delta-t/2)\}}{(w-v)}-\frac{\sin\{(2w+v)\cdot(\delta-t/2)\}}{(w+v)}\right]\cdot dw$$

$$+\frac{1}{2}\cdot\left[\frac{\cos\{v\cdot(\delta+t/2)\}-\cos\{v\cdot(\delta-t/2)\}}{v}\right]$$

$$+\int_0^\infty f_1(w)\cdot\frac{1}{2}\cdot\left[\frac{\sin\{(2w-v)\cdot(\delta+t/2)\}}{(w-v)}-\frac{\sin\{(2w+v)\cdot(\delta+t/2)\}}{(w+v)}\right]\cdot dw\ (\text{for all } v>0)$$

$$\text{(A4.10)}$$

Part E: Integration of Equations 4.82b and 4.83b

Equations 4.82b and 4.83b are reproduced below for ready reference:

$$-\frac{1}{\pi}\cdot\frac{1}{Y}+\int_0^\infty f_1(w)\cdot w\cdot\sin(w\cdot Y)\cdot dw$$

$$=\int_0^\infty F'(u)\cdot\coth(u\cdot g/2)\cdot\left[\cos(u\cdot Y)\cdot\cos\{u\cdot(\delta+t/2)\}-\sin(u\cdot Y)\cdot\sin\{u\cdot(\delta+t/2)\}\right]$$

$$+ F''(u) \cdot \tanh(u \cdot g/2) \cdot \left[\sin(u \cdot Y) \cdot \cos\{u \cdot (\delta + t/2)\}\right.$$

$$\left. + \cos(u \cdot Y) \cdot \sin\{u \cdot (\delta + t/2)\}\right] \cdot u \cdot du \tag{4.82b}$$

over $0 \leq Y < \infty$

$$\frac{1}{\pi} \cdot \frac{1}{Y} - \int_0^\infty f_2(w) \cdot w \cdot \sin(w \cdot Y) \cdot dw$$

$$= \int_0^\infty \Big[F'(u) \cdot \coth(u \cdot g/2) \cdot \left[\cos(u \cdot Y) \cdot \cos\{u \cdot (\delta - t/2)\} - \sin(u \cdot Y) \cdot \sin\{u \cdot (\delta - t/2)\}\right]$$

$$-F''(u) \cdot \tanh(u \cdot g/2) \cdot \left[\sin(u \cdot Y) \cdot \cos\{u \cdot (\delta - t/2)\}\right.$$

$$\left. + \cos(u \cdot Y) \cdot \sin\{u \cdot (\delta - t/2)\}\right]\Big] \cdot u \cdot du \tag{4.83b}$$

over $0 < Y < \infty$.

Multiply both sides of Equation 4.82b with $\sin(v \cdot Y)$ and integrate over $0 < Y < \infty$, to get

$$-\frac{1}{\pi} \cdot \int_0^\infty \frac{\sin(v \cdot Y)}{Y} \cdot dY + \int_0^\infty f_1(w) \cdot w \cdot \left[\int_0^\infty \sin(v \cdot Y) \cdot \sin(w \cdot Y) \cdot dY\right] \cdot dw$$

$$= \int_0^\infty \Bigg[F'(u) \cdot \coth(u \cdot g/2) \cdot \Bigg[\cos\{u \cdot (\delta + t/2)\} \cdot \int_0^\infty \sin(v \cdot Y) \cdot \cos(u \cdot Y) \cdot dY$$

$$-\sin\{u \cdot (\delta + t/2)\} \cdot \int_0^\infty \sin(v \cdot Y) \cdot \sin(u \cdot Y) \cdot dY \Bigg]$$

$$+F''(u) \cdot \tanh(u \cdot g/2) \cdot \Bigg[\cos\{u \cdot (\delta + t/2)\} \cdot \int_0^\infty \sin(v \cdot Y) \cdot \sin(u \cdot Y) \cdot dY$$

$$+\sin\{u\cdot(\delta+t/2)\}\cdot\int_0^\infty \sin(v\cdot Y)\cdot\cos(u\cdot Y)\cdot dY\Bigg]\Bigg]\cdot u\cdot du \tag{A4.11}$$

Multiply both sides of Equation 4.83b with $\sin(v\cdot Y)$ and integrate over $0<Y<\infty$, to get

$$\frac{1}{\pi}\cdot\int_0^\infty \frac{\sin(v\cdot Y)}{Y}\cdot dY-\int_0^\infty f_2(w)\cdot w\cdot\left[\int_0^\infty \sin(v\cdot Y)\cdot\sin(w\cdot Y)\cdot dY\right]\cdot dw$$

$$=\int_0^\infty F'(u)\cdot\coth(u\cdot g/2)\cdot\Bigg[\cos\{u\cdot(\delta-t/2)\}\cdot\int_0^\infty \sin(v\cdot Y)\cdot\cos(u\cdot Y)\cdot dY$$

$$-\sin\{u\cdot(\delta-t/2)\}\cdot\int_0^\infty \sin(v\cdot Y)\cdot\sin(u\cdot Y)\cdot dY\Bigg]$$

$$-F''(u)\cdot\tanh(u\cdot g/2)\cdot\Bigg[\cos\{u\cdot(\delta-t/2)\}\cdot\int_0^\infty \sin(v\cdot Y)\cdot\sin(u\cdot Y)\cdot dY$$

$$+\sin\{u\cdot(\delta-t/2)\}\cdot\int_0^\infty \sin(v\cdot Y)\cdot\cos(u\cdot Y)\cdot dY\Bigg]\cdot u\cdot du \tag{A4.12}$$

Consider the following identities:

$$\int_0^\infty \frac{\sin(v\cdot Y)}{Y}\cdot dY\equiv-\frac{\pi}{2},\quad \text{for } v<0$$

$$\equiv 0,\quad \textit{for } v=0$$

$$\equiv\frac{\pi}{2},\quad \text{for } v>0 \tag{A4.13a}$$

$$\int_0^\infty \sin(v\cdot Y)\cdot dY\equiv\frac{1}{v},\quad \textit{for } v\neq 0 \tag{A4.13b}$$

$$\int_0^{\infty} \cos(v \cdot Y) \cdot dY \equiv \pi \cdot \mathcal{S}(v) \tag{A4.13c}$$

In view of the identities given by Equation A4.13a–c, Equation A4.11 can be rewritten as

$$-\frac{1}{2} + \int_0^{\infty} f_1(w) \cdot w \cdot \frac{\pi}{2} \cdot \left[\mathcal{S}(w - v) - \mathcal{S}(w + v)\right] \cdot dw$$

$$= \int_0^{\infty} \left[F'(u) \cdot \coth(u \cdot g/2) \cdot \left[\cos\{u \cdot (\delta + t/2)\} \cdot \frac{1}{2} \cdot \left\{ \frac{1}{(v+u)} + \frac{1}{(v-u)} \right\} \right.\right.$$

$$\left. - \sin\{u \cdot (\delta + t/2)\} \cdot \frac{\pi}{2} \cdot \{\mathcal{S}(u - v) - \mathcal{S}(u + v)\} \right]$$

$$+ F''(u) \cdot \tanh(u \cdot g/2) \cdot \left[\cos\{u \cdot (\delta + t/2)\} \cdot \frac{\pi}{2} \cdot \{\mathcal{S}(u - v) - \mathcal{S}(u + v)\} \right.$$

$$\left.\left. + \sin\{u \cdot (\delta + t/2)\} \cdot \frac{1}{2} \cdot \left\{ \frac{1}{(v+u)} + \frac{1}{(v-u)} \right\} \right]\right] \cdot u \cdot du \tag{A4.14}$$

Or

$$-\frac{1}{2} + \frac{\pi}{2} \cdot v \cdot f_1(v) = \int_0^{\infty} \left[F'(u) \cdot \coth(u \cdot g/2) \cdot \cos\{u \cdot (\delta + t/2)\} \right.$$

$$\left. + F''(u) \cdot \tanh(u \cdot g/2) \cdot \sin\{u \cdot (\delta + t/2)\} \right] \cdot \left\{ \frac{v \cdot u}{\left(v^2 - u^2\right)} \right\} \cdot du$$

$$-F'(v) \cdot \coth(v \cdot g/2) \cdot \sin\{v \cdot (\delta + t/2)\} \cdot \frac{\pi}{2}$$

$$+F''(v) \cdot \tanh(v \cdot g/2) \cdot \cos\{u \cdot (\delta + t/2)\} \cdot \frac{\pi}{2} \tag{A4.15}$$

For all $v > 0$.

Similarly, in view of identities given by Equations A4.13a–c. Equation A4.12 can be rewritten as

$$\frac{1}{2} - \frac{\pi}{2} \cdot v \cdot f_2(v) = \int_0^\infty \Big[F'(u) \cdot \coth(u \cdot g/2) \cdot \cos\{u \cdot (\delta - t/2)\}$$

$$- F''(u) \cdot \tanh(u \cdot g/2) \cdot \sin\{u \cdot (\delta - t/2)\}\Big] \cdot \left\{\frac{v \cdot u}{(v^2 - u^2)}\right\} \cdot du$$

$$- F'(v) \cdot v \cdot \coth(v \cdot g/2) \cdot \sin\{v \cdot (\delta - t/2)\} \cdot \frac{\pi}{2}$$

$$- F''(v) \cdot v \cdot \tanh(v \cdot g/2) \cdot \cos\{v \cdot (\delta - t/2)\} \cdot \frac{\pi}{2} \tag{A4.16}$$

For all $v > 0$.

Appendix 5: Evaluation of Arbitrary Constants Involved in Section 5.9

At the boundary, $y = (mT - T_1/2)$

$$H_z^m\Big|_{y=(mT-T_1/2)} = H_z^{m'}\Big|_{y=(mT-T_1/2)} \quad \text{(A5.1)}$$

and

$$\frac{J_x^m}{\sigma}\Bigg|_{y=(mT-T_1/2)} = \frac{J_x^{m'}}{\sigma'}\Bigg|_{y=(mT-T_1/2)} \quad \text{(A5.2)}$$

over $-W/2 \le x \le W/2$ for $m = 1,2,\ldots,n$.

Also, at the boundary, $y = (mT - T_1/2 - T_2)$

$$H_z^m\Big|_{y=(mT-T_1/2-T_2)} = H_z^{(m-1)'}\Big|_{y=(mT-T_1/2-T_2)} \quad \text{(A5.3)}$$

and

$$\frac{J_x^m}{\sigma}\Bigg|_{y=(mT-T_1/2-T_2)} = \frac{J_x^{(m-1)'}}{\sigma'}\Bigg|_{y=(mT-T_1/2-T_2)} \quad \text{(A5.4)}$$

over $-W/2 \le x \le W/2$, for $m = 1,2,\ldots,n$.

Therefore, considering Equations A5.1, 5.140 and 5.134

$$a_p^m - b_p^{m'} = K_o \cdot \frac{4}{\pi} \cdot \left(\frac{W}{\pi}\right)^2 \cdot (\gamma'^2 - \gamma^2) \cdot \frac{P \cdot \sin(p(\pi/2))}{[p^2 - ((w/\pi)\gamma')^2] \cdot [p^2 - ((w/\pi)\gamma)^2]} \quad \text{(A5.5)}$$

for $m = 1,2,\ldots,n$.

While considering Equations A5.2, 5.141a and 5.135a

$$\left[\coth(\alpha_p \cdot T_2) \cdot a_p^m - \text{cosech}(\alpha_p \cdot T_2) \cdot b_p^m\right] \cdot \left(\frac{\delta}{\sigma} \cdot \alpha_p\right)$$

$$= [\text{cosech}(\alpha'_p T_1) \cdot a_p^{m'} - \coth(\alpha'_p \cdot T_1) \cdot b_p^{m'}] \cdot \left(\frac{\delta'}{\sigma'} \cdot \alpha'_p\right) \tag{A5.6}$$

For $m = 1,2,\ldots,n$.

In view of Equations A5.2, 5.141a and 5.138a, for $m = n$, we get

$$[\coth(\alpha_p \cdot T_2) \cdot a_p^n - \text{cosech}(\alpha_p \cdot T_2) \cdot b_p^n] \cdot \left(\frac{\delta}{\sigma} \cdot \alpha_p\right)$$

$$= [\text{cosech}(\alpha'_p T_1/2) \cdot a_p^{n'} - \coth(\alpha'_p \cdot T_1/2) \cdot b_p^{n'}] \cdot \left(\frac{\delta'}{\sigma'} \cdot \alpha'_p\right) \tag{A5.7}$$

Similarly, on considering Equations A5.3, 5.140 and 5.134, we get

$$b_p^m - a_p^{(m-1)'} = K_o \cdot \frac{4}{\pi} \cdot \left(\frac{W}{\pi}\right)^2 \cdot (\gamma'^2 - \gamma^2)$$

$$\times \frac{p \cdot \sin(p\pi/2)}{[p^2 - ((W/\pi) \cdot \gamma')^2] \cdot [p^2 - ((W/\pi) \cdot \gamma)^2]} \tag{A5.8}$$

For $m = 2,3,\ldots,n$.

$$b_p^1 - a_p^{0'} = K_o \cdot \frac{4}{\pi} \cdot \left(\frac{W}{\pi}\right)^2 \cdot (\gamma'^2 - \gamma^2)$$

$$\times \frac{p \cdot \sin(p\pi/2)}{[p^2 - ((W/\pi) \cdot \gamma')^2] \cdot [p^2 - ((W/\pi) \cdot \gamma)^2]} \tag{A5.9}$$

While considering Equations A5.4, 5.141a and 5.135a

$$[\text{cosech}(\alpha_p \cdot T_2) \cdot a_p^m - \coth(\alpha_p \cdot T_2) \cdot b_p^m] \cdot \left(\frac{\delta}{\sigma} \cdot \alpha_p\right)$$

$$= [\coth(\alpha'_p T_1) \cdot a_p^{(m-1)'} - \text{cosech}(\alpha'_p \cdot T_1) \cdot b_p^{(m-1)'}] \cdot \left(\frac{\delta'}{\sigma'} \cdot \alpha'_p\right) \tag{A5.10a}$$

For $m = 2,3,\ldots,n$.

Also, in view of Equations A5.4, 5.141a and 5.132a, for $m = 1$

$$[\text{cosech}(\alpha_p \cdot T_2) \cdot a_p^1 - \coth(\alpha_p \cdot T_2) \cdot b_p^1] \cdot \left(\frac{\delta}{\sigma} \cdot \alpha_p\right)$$

$$= [\coth(\alpha_p' T_1/2) \cdot a_p^{0'} - \text{cosech}(\alpha_p' \cdot T_1/2) \cdot b_p^{0'}] \cdot \left(\frac{\delta'}{\sigma'} \cdot \alpha_p'\right) \quad \text{(A5.10b)}$$

A5.1 Simplified Solution

Although all arbitrary constants can be evaluated by solving the above equations, the resulting expressions for arbitrary constants are quite involved. For simplified solution, noting that cosine and sine hyperbolic functions are usually large, one may assume that

$$\coth(\alpha_p \cdot T_2) \cong \coth(\alpha_p' T_1/2) \cong 1 \quad \text{(A5.11)}$$

and

$$\text{cosech}(\alpha_p \cdot T_2) \cong \text{cosech}(\alpha_p' \cdot T_1/2) \ll 1 \quad \text{(A5.12)}$$

Thus, from Equations A5.6 and A5.7

$$a_p^m \cdot \left(\frac{\delta}{\sigma} \cdot \alpha_p\right) + b_p^{m'} \cdot \left(\frac{\delta'}{\sigma'} \cdot \alpha_p'\right) \cong 0 \quad \text{(A5.13)}$$

For $m = 1,2,3,\ldots,n$.
And, from Equations A5.9 and A5.10

$$b_p^m \cdot \left(\frac{\delta}{\sigma} \cdot \alpha_p\right) + a_p^{(m-1)'} \cdot \left(\frac{\delta'}{\sigma'} \cdot \alpha_p'\right) \cong 0 \quad \text{(A5.14)}$$

For $m = 1,2,3,\ldots,n$.

Solving Equations A5.5, A5.8, A5.13 and A5.14, simultaneously, we get, for $m = 1,2,3,\ldots,n$.

$$a_p^m \cong b_p^m \cong \left[K_o \cdot \frac{4}{\pi} \cdot \left(\frac{W}{\pi}\right)^2 \cdot (\gamma'^2 - \gamma^2)\right] \cdot \frac{\alpha_p' \cdot (\delta'/\sigma')}{((\delta/\sigma) \cdot \alpha_p) + ((\delta'/\sigma') \cdot \alpha_p')}$$

$$\times \frac{p \cdot \sin(p\pi/2)}{[p^2 - ((W/\pi) \cdot \gamma')^2] \cdot [p^2 - ((W/\pi) \cdot \gamma)^2]} \quad \text{(A5.15)}$$

$$a_p^{(m-1)'} \cong b_p^{m'} \cong -\left[K_o \cdot \frac{4}{\pi} \cdot \left(\frac{W}{\pi}\right)^2 \cdot (\gamma'^2 - \gamma^2)\right] \cdot \frac{\alpha_p \cdot (\delta/\sigma)}{((\delta/\sigma) \cdot \alpha_p) + ((\delta'/\sigma') \cdot \alpha'_p)}$$
$$\times \frac{p \cdot \sin(p\pi/2)}{[p^2 - ((W/\pi) \cdot \gamma')^2] \cdot [p^2 - ((W/\pi) \cdot \gamma)^2]} \tag{A5.16}$$

Expressions for $b_p^{0'}$ and $a_p^{n'}$ are given by Equations 5.131b and 5.137b, respectively.

Appendix 6: Current Sheet Simulation of Stator Winding

The stator winding consists of two parts, namely, (i) *'main winding'* or the part of the winding embedded in stator slots, and (ii) *'end winding'* or the overhung parts of the winding. The main winding for machines without skewed stator slots can be simulated by an axial current sheet, whereas the end winding can be simulated by axial, peripheral and radial current sheets. In the following treatment, the radial component of the current sheet is neglected. For the purpose of simulation, each coil-side is replaced by a current ribbon of width W, equal to the stator slot opening. The current sheet simulation presented here is restricted to an integral-slot, short-chorded, double-layer balanced polyphase winding consisting of diamond coils. The phase spread for three-phase winding is assumed to be $\pi/3$ electrical radians.

Figure A6.1 shows a part of the double-layer stator winding and the three regions into which it is divided. Region 1 ($-L''/2 < x < -L'/2$) and region 3 ($L'/2 < x < L''/2$) are the end winding regions and region 2 ($-L''/2 < x < L'/2$) represents the straight part of the winding or main winding.

A6.1 Main Winding: Region 2

Consider one of the phases, say phase R, of the main winding. The phase axis for the bottom layer is $((1/2)\xi\lambda)$ ahead of axis of the entire phase winding and that of the top layer is $((1/2)\xi\lambda)$ behind. The current sheets for the top (K_{xT2}^{R}) and bottom (K_{xB2}^{R}) layer of this phase of the main winding can be given by

$$K_{xT2}^{R} = \sum_{m-odd}^{\infty} K'_m \cdot \sin\left\{m \cdot \ell \cdot \left(y + \frac{1}{2}\xi \cdot \lambda\right)\right\} \cdot \sin(\omega t) \tag{A6.1}$$

$$K_{xB2}^{R} = \sum_{m-odd}^{\infty} K'_m \cdot \sin\left\{m \cdot \ell \cdot \left(y - \frac{1}{2}\xi \cdot \lambda\right)\right\} \cdot \sin(\omega t) \tag{A6.2}$$

where $\ell = \pi/\tau$

τ = pole pitch

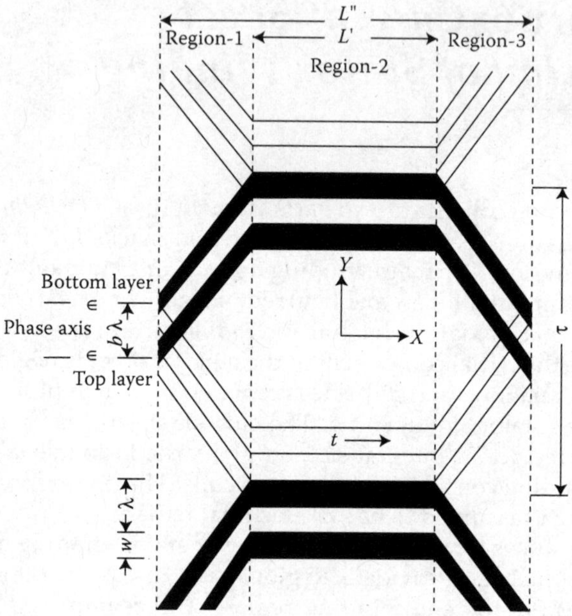

FIGURE A6.1
Phase R of a three-phase double-layer winding for two slots per pole per phase and short pitched by one slot-pitch (the shaded parts represent the top layer).

$$\xi = \text{slots short of full pitch (chording)}$$

$$\lambda = \text{slot pitch}$$

K'_m involved in Equations A6.1 and A6.2 is given as

$$K'_m = \frac{4}{\tau}\cdot\left[\sum_{r=1}^{\eta/2}\int_{[\tau-(2r-1)\cdot\lambda-W]/2}^{[\tau-(2r-1)\cdot\lambda+W]/2}\left(-\frac{1}{W}\right)\cdot(N_t\cdot I\cdot\sqrt{2})\cdot\sin(m\cdot\ell\cdot y)\cdot dy\right] \quad \text{(A6.3a)}$$

for η–even

$$K'_m = \frac{4}{\tau}\cdot\left[\sum_{r=1}^{(\eta-1)/2}\int_{[\tau-(2r-1)\cdot\lambda-W]/2}^{[\tau-(2r-1)\cdot\lambda+W]/2}\left(-\frac{1}{W}\right)\cdot\left(N_t\cdot I\cdot\sqrt{2}\right)\cdot\sin(m\cdot\ell\cdot y)\cdot dy\right]$$
$$+\int_{[\tau-W]/2}^{\tau/2}\left(-\frac{1}{W}\right)\cdot(N_t\cdot I\cdot\sqrt{2})\cdot\sin(m\cdot\ell\cdot y)\cdot dy \quad \text{(A6.3b)}$$

for η–odd

where N_t = number of series turns per layer per slot,

I = rms phase current,

η = slots per pole per phase,

and W = width of the current ribbon.

The negative sign inside the sign of integration in the expressions for K'_m corresponds to the direction of current shown in Figure A6.1. Simplification of the expressions for K'_m for both even and odd values of η gives

$$K'_m = -\frac{4\eta}{\pi \cdot m \cdot W} \cdot (N_t \cdot I \cdot \sqrt{2}) \cdot \sin\left(\frac{m\pi}{2}\right) \times \sin\left(\frac{m\pi}{2} \cdot \frac{W}{\tau}\right) \cdot \left[\frac{\sin((m\pi/2) \cdot (\eta \cdot \lambda/\tau))}{\eta \cdot \sin((m\pi/2) \cdot (\lambda/\tau))}\right] \tag{A6.4a}$$

If the stator conductors are represented by current filament instead of current ribbon, Equation A6.4a takes the following limiting form:

$$K'_m = -\frac{2\eta}{\tau} \cdot (N_t \cdot I \cdot \sqrt{2}) \cdot \sin\left(\frac{m\pi}{2}\right) \cdot \left[\frac{\sin((m\pi/2) \cdot (\eta \cdot \lambda/\tau))}{\eta \cdot \sin((m\pi/2) \cdot (\lambda/\tau))}\right] \tag{A6.4b}$$

The stator current sheet simulating both layers of phase R in region 2 can thus be given by

$$K_{xS2}^{R} = K_{xT2}^{R} + K_{xB2}^{R}$$

Hence, from Equations A6.4a and A6.4b, we get

$$K_{xS2}^{R} = \sum_{m-odd}^{\infty} K'_m \cdot 2 \cdot \cos\left(\frac{m\pi}{2} \cdot \frac{\eta \cdot \lambda}{\tau}\right) \cdot \sin\left(\frac{m\pi}{\tau} \cdot y\right) \cdot \sin(\omega t) \tag{A6.4c}$$

The current sheets simulating the other phases can be obtained from this equation by introducing the requisite phase shifts in y and t.

A6.2 End Windings: Regions 1 and 3

The parts of the stator winding located in regions 1 and 3 (Figure A6.1) are referred to as end windings. The current sheets simulating the end windings can be obtained by noting that the axial component of the surface current density is continuous at $x = \pm L'/2$ and that the divergence of current density in these regions is zero for both top- and bottom-layer windings separately. Thus, in view of Equations A6.4a and A6.4b we have, for region 1

$$K_{xT1}^{R} = \sum_{m-odd}^{\infty} K_m' \cdot \sin\left[\frac{m\pi}{\tau} \cdot \left\{y + \frac{\eta \cdot \lambda}{2} - \left(x + \frac{L'}{2}\right) \cdot \frac{\tau - \eta \cdot \lambda}{L'' - L'}\right\}\right] \cdot \sin(\omega t) \quad \text{(A6.5a)}$$

$$K_{yT1}^{R} = \sum_{m-odd}^{\infty} K_m' \cdot \frac{\tau - \eta \cdot \lambda}{L'' - L'} \times \sin\left[\frac{m\pi}{\tau} \cdot \left\{y + \frac{\eta \cdot \lambda}{2} - \left(x + \frac{L'}{2}\right) \cdot \frac{\tau - \eta \cdot \lambda}{L'' - L'}\right\}\right] \cdot \sin(\omega t) \quad \text{(A6.5b)}$$

$$K_{xB1}^{R} = \sum_{m-odd}^{\infty} K_m' \cdot \sin\left[\frac{m\pi}{\tau} \cdot \left\{y - \frac{\eta \cdot \lambda}{2} + \left(x + \frac{L'}{2}\right) \cdot \frac{\tau - \eta \cdot \lambda}{L'' - L'}\right\}\right] \cdot \sin(\omega t) \quad \text{(A6.5c)}$$

$$K_{yB1}^{R} = -\sum_{m-odd}^{\infty} K_m' \cdot \frac{\tau - \eta \cdot \lambda}{L'' - L'} \times \sin\left[\frac{m\pi}{\tau} \cdot \left\{y - \frac{\eta \cdot \lambda}{2} + \left(x + \frac{L'}{2}\right) \cdot \frac{\tau - \eta \cdot \lambda}{L'' - L'}\right\}\right] \cdot \sin(\omega t) \quad \text{(A6.5d)}$$

where K_{xT1}^{R}, K_{yT1}^{R} are components of the current sheet simulating the top layer of phase R, and K_{xB1}^{R}, K_{yB1}^{R} are components of the current sheet simulating the bottom layer of the same phase. Hence, the components of the resultant current sheet for both the layers of phase R in region 1 are

$$K_{xS1}^{R} = K_{xT1}^{R} + K_{xB1}^{R}$$

Hence, from Equations A6.5a and A6.5c, we get

$$K_{xS1}^{R} = \sum_{m-odd}^{\infty} K_m' \cdot 2 \times \cos\left[\frac{m\pi}{\tau} \cdot \left\{\frac{\eta \cdot \lambda}{2} - \left(x + \frac{L'}{2}\right) \cdot \frac{\tau - \eta \cdot \lambda}{L'' - L'}\right\}\right] \cdot \sin\left(\frac{m\pi}{\tau} \cdot y\right) \cdot \sin(\omega t) \quad \text{(A6.6a)}$$

and $K_{yS1}^{R} = K_{yT1}^{R} + K_{yB1}^{R}$

Thus, from Equations A6.5b and A6.5d, we get

$$K_{yS1}^{R} = \sum_{m-odd}^{\infty} K_m' \cdot \frac{\tau - \eta \cdot \lambda}{L'' - L'} \cdot 2 \cdot \sin\left[\frac{m\pi}{\tau} \cdot \left\{\frac{\eta \cdot \lambda}{2} - \left(x + \frac{L'}{2}\right) \cdot \frac{\tau - \eta \cdot \lambda}{L'' - L'}\right\}\right]$$
$$\times \cos\left(\frac{m\pi}{\tau} \cdot y\right) \cdot \sin(\omega t) \qquad \text{(A6.6b)}$$

Similarly, in view of Equations A6.1 and A6.2, we have

$$K_{xT3}^{R} = \sum_{m-odd}^{\infty} K_m' \cdot \sin\left[\frac{m\pi}{\tau} \cdot \left\{y + \frac{\eta \cdot \lambda}{2} + \left(x - \frac{L'}{2}\right) \cdot \frac{\tau - \eta \cdot \lambda}{L'' - L'}\right\}\right] \cdot \sin(\omega t) \qquad \text{(A6.7a)}$$

$$K_{yT3}^{R} = -\sum_{m-odd}^{\infty} K_m' \cdot \frac{\tau - \eta \cdot \lambda}{L'' - L'} \cdot \sin\left[\frac{m\pi}{\tau} \cdot \left\{y + \frac{\eta \cdot \lambda}{2} + \left(x - \frac{L'}{2}\right) \cdot \frac{\tau - \eta \cdot \lambda}{L'' - L'}\right\}\right] \cdot \sin(\omega t)$$
$$\text{(A6.7b)}$$

$$K_{xB3}^{R} = \sum_{m-odd}^{\infty} K_m' \cdot \sin\left[\frac{m\pi}{\tau} \cdot \left\{y - \frac{\eta \cdot \lambda}{2} - \left(x - \frac{L'}{2}\right) \cdot \frac{\tau - \eta \cdot \lambda}{L'' - L'}\right\}\right] \cdot \sin(\omega t) \qquad \text{(A6.7c)}$$

$$K_{yB3}^{R} = \sum_{m-odd}^{\infty} K_m' \cdot \frac{\tau - \eta \cdot \lambda}{L'' - L'} \cdot \sin\left[\frac{m\pi}{\tau} \cdot \left\{y - \frac{\eta \cdot \lambda}{2} - \left(x - \frac{L'}{2}\right) \cdot \frac{\tau - \eta \cdot \lambda}{L'' - L'}\right\}\right] \cdot \sin(\omega t)$$
$$\text{(A6.7d)}$$

where K_{xT3}^{R}, K_{yT3}^{R} are components of the current sheet simulating the top layer of phase R, and K_{xB3}^{R}, K_{yB3}^{R} are components of the current sheet simulating the bottom layer of the same phase. Hence, the components of the resultant current sheet for both the layers of phase R in region 1 are

$$K_{xS3}^{R} = K_{xT3}^{R} + K_{xB3}^{R}$$

Thus,

$$K_{xS3}^{R} = \sum_{m-odd}^{\infty} K_m' \cdot 2 \cdot \cos\left[\frac{m\pi}{\tau} \cdot \left\{\frac{\eta \cdot \lambda}{2} + \left(x - \frac{L'}{2}\right) \cdot \frac{\tau - \eta \cdot \lambda}{L'' - L'}\right\}\right]$$
$$\times \sin\left(\frac{m\pi}{\tau} \cdot y\right) \cdot \sin(\omega t) \qquad \text{(A6.8a)}$$

and

$$K_{yS3}^R = K_{yT3}^R + K_{yB3}^R$$

Giving,

$$K_{yS3}^R = -\sum_{m-odd}^{\infty} K'_m \cdot \frac{\tau - \eta \cdot \lambda}{L'' - L'} \cdot 2$$

$$\times \sin\left[\frac{m\pi}{\tau} \cdot \left\{\frac{\eta \cdot \lambda}{2} + \left(x - \frac{L'}{2}\right) \cdot \frac{\tau - \eta \cdot \lambda}{L'' - L'}\right\}\right] \cdot \cos\left(\frac{m\pi}{\tau} \cdot y\right) \cdot \sin(\omega t)$$

(A6.8b)

Equations A6.4, A6.5, A6.6a, A6.6b, A6.8a and A6.8b define the stator current sheet for phase R in various regions. The current sheet simulating the other phases in these regions can be obtained by introducing the requisite phase shifts in y and t.

A6.3 Consolidated Expressions

The stator current sheet simulating one phase of the stator winding has been obtained in a piece-wise manner. Current sheets obtained individually in the last section can be defined in an integrated form for the entire machine length. If $L_o(\geq L'')$ is the inner distance between the axially located end covers, the stator current sheet can be given by the following half-range Fourier series:

$$K_{xS}^R = \sum_{p-odd}^{\infty} K_p^R(y,t) \cdot \cos\left(\frac{p\pi}{L_o} \cdot x\right) \quad \text{over} - L_o/2 \geq x \geq L_o/2 \tag{A6.9}$$

where

$$K_p^R(y,t) = \frac{4}{L_o} \cdot \left[\int_0^{L'/2} K_{xS2}^R \cdot \cos\left(\frac{p\pi}{L_o} \cdot x\right) \cdot dx + \int_{L'/2}^{L''/2} K_{xS3}^R \cdot \cos\left(\frac{p\pi}{L_o} \cdot x\right) \cdot dx\right] \tag{A6.9a}$$

Thus, in view of Equations A6.5 and A6.8a, we have

$$K_p^R(y,t) = \frac{8}{L_o} \cdot \sin(\omega t) \cdot \sum_{m-odd}^{\infty} K'_m \cdot \sin\left(\frac{m\pi}{\tau} \cdot y\right) \cdot \int_0^{L'/2} \cos\left(\frac{m\pi}{2} \cdot \frac{\eta \cdot \lambda}{\tau}\right) \cdot \cos\left(\frac{p\pi}{L_o} \cdot x\right)$$

$$\times dx + \int_{L'/2}^{L''/2} \cos\left[\frac{m\pi}{\tau} \cdot \left\{\frac{\eta \cdot \lambda}{2} + \left(x - \frac{L'}{2}\right) \cdot \frac{\tau - \eta \cdot \lambda}{L'' - L'}\right\}\right] \cdot \cos\left(\frac{p\pi}{L_o} \cdot x\right) \cdot dx$$

On simplifying the above equation, we get

$$K_p^R(y,t) = \sum_{m-odd}^{\infty} K_{mp} \cdot \sin\left(\frac{m\pi}{\tau} \cdot y\right) \cdot \sin(\omega t) \tag{A6.10}$$

where

$$K_{mp} = K'_m \cdot \frac{8}{\pi} \cdot \frac{\begin{bmatrix} \cos\left(\frac{m\pi}{2} \cdot \lambda_o\right) \cdot \sin\left(\frac{p\pi}{2} \cdot \lambda_1\right) \cdot \frac{m^2}{p} \cdot (1-\lambda_o)^2 \\ + \sin\left(\frac{m\pi}{2}\right) \cdot \cos\left(\frac{p\pi}{2} \cdot \lambda_2 \cdot\right) \cdot m \cdot (1-\lambda_o) \cdot (\lambda_2 - \lambda_1) \\ - \sin\left(\frac{m\pi}{2} \cdot \lambda_o\right) \cdot \cos\left(\frac{p\pi}{2} \cdot \lambda_1\right) \cdot m \cdot (1-\lambda_o) \cdot (\lambda_2 - \lambda_1) \end{bmatrix}}{\left[m^2 \cdot (1-\lambda_o)^2 - p^2 \cdot (\lambda_2 - \lambda_1)^2\right]}, \tag{A6.11}$$

$$\lambda_o = \frac{\eta \cdot \lambda}{\tau} \tag{A6.11a}$$

$$\lambda_1 = L'/L_o \tag{A6.11b}$$

and

$$\lambda_2 = L''/L_o \tag{A6.11c}$$

Thus, from Equations A6.9 and A6.10, we have

$$K_{xS}^R = \sum_{p-odd}^{\infty}\left[\sum_{m-odd}^{\infty} K_{mp} \cdot \sin\left(\frac{m\pi}{\tau} \cdot y\right) \cdot \sin(\omega t)\right] \cdot \cos\left(\frac{p\pi}{L_o} \cdot x\right) \tag{A6.12a}$$

Similarly, the peripheral component of the stator current sheet for phase R can be given by

$$K_{yS}^R = -\frac{\tau}{L_o} \cdot \sum_{p-odd}^{\infty}\left[\sum_{m-odd}^{\infty} K_{mp} \cdot \left(\frac{p}{m}\right) \cdot \cos\left(\frac{m\pi}{\tau} \cdot y\right) \cdot \sin(\omega t)\right] \cdot \sin\left(\frac{p\pi}{L_o} \cdot x\right) \tag{A6.12b}$$

The surface current sheets simulating the other phases of the stator winding can be obtained by introducing the requisite phase shifts in y and t in

Equations A6.12 and A6.13. Therefore, for a three-phase winding, the current sheets simulating Y and B phases

$$K_{xS}^{Y} = \sum_{p-odd}^{\infty}\left[\sum_{m-odd}^{\infty} K_{mp} \cdot \sin\left\{\frac{m\pi}{\tau}\cdot\left(y-\frac{2}{3}\cdot\tau\right)\right\}\cdot\sin\left(\omega t-\frac{2}{3}\cdot\pi\right)\right]\cdot\cos\left(\frac{p\pi}{L_o}\cdot x\right) \tag{A6.13a}$$

$$K_{yS}^{Y} = -\frac{\tau}{L_o}\cdot\sum_{p-odd}^{\infty}\left[\sum_{m-odd}^{\infty} K_{mp}\cdot\left(\frac{p}{m}\right)\cdot\cos\left\{\frac{m\pi}{\tau}\cdot\left(y-\frac{2}{3}\cdot\tau\right)\right\}\cdot\sin\left(\omega t-\frac{2}{3}\cdot\pi\right)\right] \times \sin\left(\frac{p\pi}{L_o}\cdot x\right) \tag{A6.13b}$$

$$K_{xS}^{B} = \sum_{p-odd}^{\infty}\left[\sum_{m-odd}^{\infty} K_{mp} \cdot \sin\left\{\frac{m\pi}{\tau}\cdot\left(y-\frac{4}{3}\cdot\tau\right)\right\}\cdot\sin\left(\omega t-\frac{4}{3}\cdot\pi\right)\right]\cdot\cos\left(\frac{p\pi}{L_o}\cdot x\right) \tag{A6.14a}$$

$$K_{yS}^{B} = -\frac{\tau}{L_o}\cdot\sum_{p-odd}^{\infty}\left[\sum_{m-odd}^{\infty} K_{mp}\cdot\left(\frac{p}{m}\right)\cdot\cos\left\{\frac{m\pi}{\tau}\cdot\left(y-\frac{4}{3}\cdot\tau\right)\right\}\cdot\sin\left(\omega t-\frac{2}{3}\cdot\pi\right)\right] \times \sin\left(\frac{p\pi}{L_o}\cdot x\right) \tag{A6.14b}$$

The components of the resultant stator current sheet due to all the three phases are therefore given by

$$K_{xS} = K_{xS}^{R} + K_{xS}^{Y} + K_{xS}^{B}$$

$$= \sum_{p-odd}^{\infty}\sum_{m-odd}^{\infty} K_{mp}\cdot\left[\sin\left(\frac{m\pi}{\tau}\cdot y\right)\cdot\sin(\omega t)\cdot\left\{1-\cos\left(\frac{m\pi 2}{3}\right)\right\} + \cos\left(\frac{m\pi}{\tau}\cdot y\right)\cdot\cos(\omega t)\cdot\sqrt{3}\cdot\sin\left(\frac{m\pi 2}{3}\right)\right]\cdot\cos\left(\frac{p\pi}{L_o}\cdot x\right) \tag{A6.15a}$$

and

$$K_{yS} = K_{yS}^{R} + K_{yS}^{Y} + K_{yS}^{B}$$

$$= -\frac{\tau}{L_o} \cdot \sum_{p-odd}^{\infty} \sum_{m-odd}^{\infty} K_{mp} \cdot \left(\frac{p}{m}\right) \cdot \left[\cos\left(\frac{m\pi}{\tau} \cdot y\right) \cdot \sin(\omega t) \cdot \left\{1 - \cos\left(\frac{m\pi 2}{3}\right)\right\}\right.$$

$$\left. -\sin\left(\frac{m\pi}{\tau} \cdot y\right) \cdot \cos(\omega t) \cdot \sqrt{3} \cdot \sin\left(\frac{m\pi 2}{3}\right)\right] \cdot \sin\left(\frac{p\pi}{L_o} \cdot x\right) \qquad \text{(A6.15b)}$$

The above equations show that, in a current sheet simulating, a balanced three-phase, integral-slot winding and triplen peripheral harmonics are absent. The orders of nontriplen odd harmonics M are given as follows:

$$M = \frac{6 \cdot n - 3 + (-1)^n}{2} \qquad \text{(A6.16)}$$

where $n = 1, 2, 3, 4, \ldots$.

Therefore, putting

$$\ell = \pi/\tau \qquad \text{(A6.17)}$$

Equations A6.15a and A6.15b can be rewritten as

$$K_{xS} = \mathcal{R}e \sum_{p-odd}^{\infty} \left[\sum_{n=1}^{\infty} \frac{3}{2} \cdot K_{Mp} \cdot (-1)^{n-1} \cdot \exp \cdot j\{M \cdot \ell \cdot (-1)^{n-1} \cdot y - \omega t\}\right] \cdot \cos\left(\frac{p\pi}{L_o} \cdot x\right)$$

(A6.17a)

$$K_{yS} = \mathcal{R}e \sum_{p-odd}^{\infty} \left[-\sum_{n=1}^{\infty} \frac{3}{2} \cdot K_{Mp} \cdot \left(j \cdot \frac{\tau}{L_o} \cdot \frac{p}{M}\right) \cdot \exp \cdot j\{M \cdot \ell \cdot (-1)^{n-1} \cdot y - \omega t\}\right]$$

$$\times \sin\left(\frac{p\pi}{L_o} \cdot x\right)$$

(A6.17b)

If, for the peripheral distribution, only fundamental is considered, these equations reduce to

$$K_{xS} = \mathcal{R}e \sum_{p-odd}^{\infty} \left[\frac{3}{2} \cdot K_{1p} \cdot \exp \cdot j\{M \cdot \ell \cdot y - \omega t\}\right] \cdot \cos\left(\frac{p\pi}{L_o} \cdot x\right) \qquad \text{(A6.18a)}$$

$$K_{yS} = \mathcal{R}e \sum_{p-odd}^{\infty} \left[-\frac{3}{2} \cdot K_{1p} \cdot \left(j \cdot \frac{\tau}{L_o} \cdot p\right) \cdot \exp \cdot j\{M \cdot \ell \cdot y - \omega t\}\right] \cdot \sin\left(\frac{p\pi}{L_o} \cdot x\right)$$

(A6.18b)

Note: The treatment presented above is restricted to three-phase stator winding. The current sheet simulation of a general polyphase stator winding may be seen in Mukerji, S. K., Linear electromagnetic field analysis of the solid rotor induction machine, PhD thesis, Department of Electrical Engineering, Indian Institute of Technology, Bombay, pp. 308–321, June 1967.

Index

A

Absolute error, 57, 63; *see also* Approximation theorem
  Laplacian of, 58
  as mean square error, 64
Air-gap field, 111, 275, 276; *see also* Laplacian fields
  χ-plane to w-plane, 114–116
  contour in w-plane, 113
  current-carrying conductor, 112
  Schwarz–Christoffel transformation, 112
  in solid-rotor machine, 199
  z-plane to w-plane, 112–114
Air-gap permeance, 92, 186–189; *see also* Rectangular double-slotting
  Carter's coefficient, 93
  effect, 184
  estimation of, 338
  flux reduction, 93
  gap permeance, 93
  gap-extension factor, 93
  net flux, 92
  potential distribution, 92
  variations of, 94
Air-gap potential, 241
ALGLIB, 323
Ampere's law, 3, 12; *see also* Maxwell's equations
  experimental law, 13
Analytical techniques, 7–8; *see also* Numerical techniques
Angular frequency, 176
  of ac, 137, 140
  and slip frequency, 186
Anisotropic homogeneous region, 6, 171, 327
Anisotropic media, 73
  approximation theorem problems, 73
  field theory for, 6
  2D fields in, 172–175, 326–327
Aperiodic field distributions, 75, 94, 347
Aperiodical field distribution modelling, 94; *see also* Laplacian fields
  arbitrary orientation of two teeth, 99–101
  potential distribution in air gap, 94–95
  tooth-opposite-slot orientation, 97–98
  tooth-opposite-tooth orientation, 94
  unknown function evaluation, 95–97, 98–99, 102–107
Approximation theorem, 4, 57; *see also* Field theorems; Theorems
  absolute error, 57
  base value, 62–63
  cumulative error, 58, 68
  error for point on bounding surface, 58
  error function, 58
  exercise in, 73
  for Laplacian field, 57–63
  for Maxwell's equations, 68–73
  relative error, 59
  for vector magnetic potential, 63–68
Arbitrary constants, 8, 136, 183, 235, 270–271
  complex, 276
  determination, 194–197
  evaluation, 77–81, 82–86, 88–92, 121–124, 395–398
Arbitrary functions, 8
Arbitrary orientation of tooth-slot configuration, 75, 86–88
Arbitrary orientation of two teeth, 99–101; *see also* Integrations leading to simultaneous algebraic equations
  evaluation of unknown functions, 102–107
  integrals in first integration of LHS, 381–382

Arbitrary orientation of two teeth (*Continued*)
  integrals in fourth integration of RHS, 386–389
  integrals in integration of equations, 389–393
  integrals in second integration of RHS, 382–385
  integrals in third integration of LHS, 385–386
Armature winding
  boundary conditions, 121–124
  magnetic field intensity, 118–120
  magnetic field near, 116–117, 338
  power dissipation, 49
  simulation of, 6
  surface current density, 117–118
Asynchronous operation of ideal hysteresis machine, 285
Axial component of eddy current density, 250
  distribution, 234
  of rotor, 233
  on rotor surface, 252
Axial component of rotor magnetic field, 248
Axial current sheet, 399
Axioms, 28
Axis of skewed rotor slots, 185

B

Base value, 62–63; *see also* Approximation theorem
Bessel equation of zero order, 158–159
Bessel functions, 159
Biot–Savart's law, 3, 29; *see also* Field equations
Boundary conditions, 7; *see also* Approximation theorem; Uniqueness theorem
  field distribution in air gap, 208
  laminated-rotor polyphase induction machines, 194
  permeability for iron and, 121
  polyphase excitation, 134
  at stator air-gap surface, 241–242
Boundary surface, 32–33
Braking torque, 200, 286

C

Cage rotor induction machines, 175–183
Carter's coefficient, 93; *see also* Air-gap permeance
  exercise on, 124
Case studies, *see* Field theory of hysteresis machines; Leakage inductance of transformers; Single-phase induction motors with composite poles; Slot leakage inductance; Transient fields due to impact excitations
Cauchy's principal value of integral, 365
χ plane, 108
  to w-plane, 110–111, 114–116
Circular conductors; *see also* Eddy currents
  carrying ac current, 160
  current density along conductor axis, 161
  distribution of current density in, 159
  skin effects, 159–160
Classical depth of penetration, 22
Closed surfaces, 33
Complex arbitrary constant, 276
Complex conductivity, 22, 190, 326
  components, 172
Complex permeability, 50
Complex permittivity, 50
Complex Poynting vector, 281
Composite poles of single-phase induction motors, 286, 346
Computational errors, 314
Conducting plate with surface current sheets, 299
Conducting region, 233
  damped wave equations, 25
  eddy current and, 54, 127, 163, 341
  magnetic field in, 300, 341
  rotor core, 232, 233, 344
Conduction current, 13; *see also* Maxwell's equations
  density, 13
  energy dissipation, 27
Conductivity, 128, 140, 144, 172, 183
  complex, 22, 172
  as constant scalar quantity, 14
  for solid rotor iron, 287, 346

Constitutive equations, 49
Continuity equation, 25–26; *see also* Field equations
Contour in w-plane, 113
Convection currents, 168
Core
  cylindrical core with surface current sheet, 157
  with equilateral triangular cross-section, 145
  with hexagonal cross-sections, 149–152
  with octagonal cross-section, 153
  with octagonal cross-sections, 152–157
  with right angled triangular cross-section, 140
  solid rectangular, 138
  with triangular cross-sections, 140, 144–149
Coulomb's gauge condition, 19; *see also* Field equations
Coulomb's law, 3, 29; *see also* Field equations
Coupling fields, 3
Cuba software, 323
Cubature code, 323
Cuboid examples, 33–37
Cumulative error, 58; *see also* Approximation theorem
Curl
  absolute error, 64
  eddy current density by, 340
  equations in phasor form, 68
  of gradient, 18
  of magnetic field intensity, 118
  of vector, 18
Current, 3
  -carrying conductor, 112, 262
  -carrying excitation winding, 298
  discrete, 13
  filament, 201
  -free homogeneous region, 18
  ribbon, 399, 401
  sheet simulation, 399
  source density, 14
Current density
  along conductor axis, 161
  distribution, 159
  eddy, 136
  surface, 134
Current impact excitation, 298, 300; *see also* Transient fields due to impact excitations
  electromagnetic field, 301, 303–305
  field expressions, 302
  magnetic field, 300
  surface resistivity, 300

D

Damped wave equations, 16, 25, 299, 300
Damping, 16, 303
Dark region, 33
Density of excitation current, 291
Depth of penetration, 22; *see also* Field equations
Diamond coil, 399
Dielectric reluctance motors, 56
Diffusion equation, *see* Eddy currents—equation
Discrete charges, 12
Discrete conductors, 231
Discrete currents, 13
Discretisation error, 314
Displacement current density, 13; *see also* Maxwell's equations
Displacement vector, 39
Distributed charge, 12
Distribution factor, 223
Divergence, 28
  of curl of vector, 18
  of gradient, 19, 20
  theorem, 13, 32, 44, 46, 58, 64
Domain of machines, 2
Domain of numerical analysis, 314

E

Eddy currents, 127; *see also* Circular conductors; Field equations; Laminated rectangular core; Polyphase excitation; Solid rotor induction machines
  Bessel equation of zero order, 158–159
  in circular cores, 131, 157–159
  cylindrical core with surface current sheet, 157

Eddy currents (*Continued*)
density, 130, 136, 278
density components, 163–168
density phase angle, 250, 251
equation, 3, 16, 135
in equilateral triangular cross-section core, 145
exercise, 168–169
in hexagonal cross-section core, 149–152
in large plates, 132
loss, 130–131, 252–257, 278–279, 282
machines, 127
in octagonal cross-section core, 152–157
phenomena, 5
in rectangular cross-section core, 137–139
in regular polygonal cross-section core, 143–144
in right angled triangular cross-section core, 140
single-phase excitation, 133–134
solid rectangular core, 138
square cross-section, 142
surface current densities, 133, 157
target zone, 146
torch function, 144
in triangular cross-section core, 139–143, 144–149
Eddy currents in laminated rectangular cores, 161
alternating current, 162
arbitrary constants evaluation, 395–398
eddy current density components, 163–168
surface current density, 162
Eigenvalue problems, 316
Electric field
components, 192
intensity components, 178–179
Electric flux; *see also* Maxwell's equations
cutting, 17
density, 13, 197
Electric lines of force, 11; *see also* Maxwell's equations
Electrical machines; *see also* Eddy currents; Hysteresis machine action; Induction machine action
factors affecting performance, 75
problem of slotting, 4
rotating, 2
slots, 75; *see also* Rectangular double-slotting
Electrical motors, 49; *see also* Poynting theorem
Electromagnetic field, 1
laws, 3
Electromagnetic phenomena, 28; *see also* Maxwell's equations
Electromechanical energy conversion, 49; *see also* Poynting theorem
Electromotive-force (emf), 4, 12, 16; *see also* Maxwell's equations
emf, *see* Electromotive-force (emf)
End-effects, 6, 231
Energy conversion devices, 3
Equation types, 323; *see also* Hysteresis machine field theory; Integrations leading to simultaneous algebraic equations; Relations involving simple summations; Relations without summation or integration; Simultaneous linear algebraic equations; Transient fields in plates
evaluated integrals, 323
expression category, 323–324
Equipotential surfaces, 108
Error function, 58, 60, 65, 70; *see also* Approximation theorem
Evaluation of unknown functions, 95–97, 98–99, 102–107
Extrapolation, 313

F

Faraday's law of electromagnetic induction, 12; *see also* Maxwell's equations
FDM, *see* Finite difference method (FDM)
Field
in anisotropic rotor region, 189–194
approach, 1–2

Induction machines, 2
  axial end regions of, 116
  with skewed rotor slots, 184–186
Induction-cum-hysteresis generator, 286
Induction-cum-hysteresis motor, 285
Infinite series, 80, 81, 85, 86, 91, 92
Inhomogeneous laminated core region, 171
Integral transform, 365
Integrations leading to simultaneous algebraic equations, 347–348; *see also* Equation types; Tooth-opposite-slot orientation; Tooth-opposite-tooth orientation
  arbitrary orientation of teeth, 358–363
Interpolation, 313
Iterative methods, 314, 315

L

Lagrange multiplier, 316
Laminated core, 161, 168
Laminated iron cores, 6
Laminated rectangular core, 162; *see also* Eddy currents
  eddy current density components, 163–168
  eddy current loss in, 161
  surface current density, 162
  time-varying magnetic field, 163
Laminated-rotor polyphase induction machines, 171
  air-gap field, 186–189
  arbitrary constant, 183
  arbitrary constant determination, 194–197
  axis of skewed rotor slots, 185
  boundary conditions, 194
  complex conductivity 172, 190
  electric field components, 178–179, 192
  electric flux density component, 197
  exercise, 197
  fields in anisotropic rotor region, 189–194
  idealized machine, 175
  induction machines with skewed rotor slots, 184–186
  magnetic field intensity components, 190
  rotor induction machines, 175–183
  rotor parameters, 183–184
  skew angle, 184
  stator current sheet, 176
  stator surface current density, 185
  2D fields in anisotropic media, 172–175
Laplacian fields, 75; *see also* Air-gap field; Fringing flux; Magnetic field; Aperiodical field distribution modelling; Rectangular double-slotting
  exercise, 124–125
  leakage inductance, 116
Laplacian operator, 19
Large-scale electromagnetic phenomena, 28; *see also* Maxwell's equations
Leakage inductance, 116
Leakage inductance of transformers, 265; *see also* Field theory of hysteresis machines; Single-phase induction motors with composite poles; Slot leakage inductance; Transient fields due to impact excitations
  arbitrary constants, 270–271
  constant vector potential surfaces, 268
  current density distribution, 266–267
  exercise, 310
  flux linkages of, 272
  leakage inductance, 271–273
  magnetic flux density, 269
  physical configuration, 266
  sandwich winding, 266
  vector magnetic potential, 267–268
Left-hand side (LHS) 14
Lienard–Wiechert fields, 29; *see also* Field equations
Linear equations, 315
Linear programming, 316
Lorentz's force, 55–56
Lost flux, 93
Low voltage (LV), 266
LV, *see* Low voltage (LV)

M

Magnetic cores, 127; *see also* Eddy currents
Magnetic field, 116; *see also* Laplacian fields; Magnetic field intensity
  arbitrary constant evaluation, 121–124
  boundary conditions, 121
  distributions, 233
  field expressionselection, 121
  intensity, 233, 235, 239, 240, 241
  leakage inductance, 116
  near armature winding, 116, 338
  in plate, 137
  source, 232, 233
  surface current density, 117–118
  tangential components, 236, 239
Magnetic field intensity, 118, 127, 233, 239, 240; *see also* Magnetic field
  components, 190, 241
  in region 1, 118–119
  in region 2, 119
  in region 3, 119–120
  in region 4, 120
Magnetic field phase angle, 248, 249
Magnetic flux; *see also* Maxwell's equations
  cutting, 17
  lines, 11
Magnetic permeance, 76; *see also* Air-gap permeance
Magnetic poles, 12
Magnetic potential in current-free region, 236, 239, 240; *see also* Scalar electric and magnetic potentials
Magnetic potential in shaft and stator frame, 236
Magneto-motive-force (mmf), 4, 12–13
Matrix decomposition techniques, 315
Maxwell's equations, 3, 11; *see also* Field equations
  approximation theorem for, 68–73
  axioms, 28
  for fields in moving media, 16–17
  in integral form, 11–13
  large-scale electromagnetic phenomena, 28
  magnetic flux lines, 11
  in point form, 13–14
  total charge inside surface, 12
Mean square error, 59, 70
Mechanical power, 282, 283
MEMS, *see* Micro-electro-mechanical systems (MEMS)
Micro-electro-mechanical systems (MEMS), 3
mmf, *see* Magneto-motive-force (mmf)
Motional emf, 16
Multidimensional integrals, 322

N

Negative mechanical power, 286
Net flux over tooth-pitch, 92
Newton–Cote's formulae, 318
Nonlinear equations, 315
Non-square matrices, 315
Numerical analysis, 313; *see also* Numerical computation
  algebraic equation, 315
  coefficients, 315
  computational errors, 314
  constant term, 315
  discretisation error, 314
  eigenvalue problems, 316
  extrapolation, 313
  FDM, 317
  FEM, 317
  Horner scheme, 314–315
  interpolation, 313
  iterative methods, 315
  linear equations, 315–316
  matrix decomposition techniques, 315
  nonlinear equations, 315
  numerical stability, 314
  optimisation problem, 316
  regression, 313–314
  round-off error, 314
  truncation errors, 314
Numerical computation, 313; *see also* Equation types; Numerical analysis; Numerical integration
Numerical integration, 318; *see also* Equation types
  abscissas and weight factors, 320, 321
  Gauss–Hermite quadrature, 318, 319–320

Gauss–Laguerre quadrature, 318, 320
integration over bounded and unbounded intervals, 318–319
multidimensional integrals, 322
numerical integration software, 322–323
parabolic rule, 318–319
trapezoidal rule, 318
variable change, 321–322
Numerical quadrature, *see* Numerical integration
Numerical stability, 314
Numerical techniques, 7
analytical techniques, 7–8
FEM, 7

P

Periodic fields, 21
Phase angle
of eddy current density, 250, 251
of magnetic field, 248, 249
Poissons' summation formula, 91
Poisson's equation, 3
Polyphase excitation, 134–137; *see also* Eddy currents
arbitrary constant, 136
boundary conditions, 134
eddy current density, 136
eddy current equation, 135
magnetic field in plate, 137
surface current densities, 134
Polyphase induction machines, 5, 171
laminated iron cores, 6
simulation of armature winding, 6
unlaminated iron cores, 6
Positive-definite matrix, 315
Potential distribution, 32–33
in air gap, 92, 94–95
Power
components, 282–283
density, 282
flow components, 54–55
frequency, 133
Poynting theorem, 4, 49; *see also* Field theorems; Theorems
constitutive equations, 49
electrical motors, 49
electromechanical energy conversion, 49
force components, 55–57
power flow components, 54–55
Poynting vector, 50, 281
velocity vector, 53
Poynting vector, 50, 281
Presentation problems, 29
Pseudo torch function, 38, 39

Q

Quantum mechanics, 28
Quasi-Monte Carlo methods, 322

R

Radial component
of eddy current density, 251
of rotor magnetic field, 249
Rectangular double-slotting, 76; *see also* Air-gap permeance; Laplacian fields
arbitrary constant evaluation, 77–81, 82–86, 88–92
arbitrary orientation of tooth and slot, 86–88
magnetic permeance, 76
Poissons' summation formula, 91
tooth-opposite-slot orientation, 81–82
tooth-opposite-tooth orientation, 76–77
Rectangular region examples, 37–40
Regression, 313
Regular field plane, 110; *see also* Fringing flux
Relations involving simple summations, 329; *see also* Equation types; Transient fields in plates
eddy currents in core, 329–330
leakage inductance of transformers, 330
slot leakage inductance, 330
Relations without summation or integration, 324; *see also* Equation types
conductor air-gap field, 324–325
eddy currents, 325, 326
flux line, 324
fringing flux, 324

Relations without summation or integration (*Continued*)
  induction machine, 327
  Kelvin's functions, 326
  magnetic and electric field intensity evaluation, 327
  Maxwell's equation expressions, 327
  poly-phase excitation, 326
  rotor machine, 327–328
  Schwarz–Christoffel transformation, 324
  single-phase excitation, 326
  2D fields in anisotropic media, 326–327
Relative error, 59; *see also* Approximation theorem
Relaxation time, 26–28; *see also* Field equations
Retarded potentials, 23–25; *see also* Field equations
RHS, *see* Right-hand side (RHS)
Right-hand side (RHS), 15
rms, *see* Root mean square (rms)
Root mean square (rms), 59, 144
Rotor
  hysteresis, 284
  input power, 259
  parameters, 183–184
  power input, 131–132
Rotor and stator length effect, 248; *see also* Three-dimensional fields in solid-rotor induction machines
  eddy current density axial component, 252
  magnetic field peripheral component, 251
Round-off errors, 314

S

Sandwich winding, 266
Scalar and vector potentials, 29; *see also* Field equations
Scalar electric and magnetic potentials, 17–18; *see also* Magnetic potential in current-free region
Schwarz–Christoffel transformation, 108, 112; *see also* Air-gap field; Fringing flux
Scilab, 323
Self-excited hysteresis and induction generators, 286
Shaded pole-type single-phase induction motors, 286
Signum function, 365
Simpson's rule, 318
  parabolic rule, 318–319
  trapezoidal rule, 318
Simulation of armature winding, 6
Simultaneous linear algebraic equations, 331; *see also* Equation types
  air-gap permeance, 338
  distribution in different region, 334–336
  eddy currents in cores, 339–341
  magnetic field near armature winding, 338
  potential distribution evaluation, 331–334
  potential distributions in arbitrary tooth orientation, 336
  potential distributions in tooth-opposite-slot orientation, 334
  potential estimation, 336–338
  single-phase induction motors, 346
  3D fields in solid rotor induction machines, 344–347
  tooth-ripple harmonics, 341–343
Single-phase excitation, 133–134
Single-phase induction motors with composite poles, 286; *see also* Field theory of hysteresis machines; Leakage inductance of transformers; Tooth-ripple harmonics; Transient fields due to impact excitations
  excitation current density, 291
  field distribution, 288
  field expression evaluation, 296–298
  idealised machine, 287–288
  magnetic field intensity, 287, 288, 289, 292, 293, 294, 295
  simplifying assumptions, 287
  stator air-gap, 288
  vector magnetic potential, 292
Singular value decomposition, 316
Singular value problems, *see* Eigenvalue problems

Skew angle, 184
Skewed rotor slots, 184–186
Skewed stator slots, 399
Skin effects, 159–160
Slip
  frequency, 186
  -independent force, 284
  -power relation, 259, 280–281, 283
  -torque characteristics, 200, 201
Slot, 75; *see also* Rectangular double-slotting
  equipotential surfaces, 92
  pitch, 87, 93, 202, 400
  problem, 4
  width, 76, 94
Slot leakage inductance, 261; *see also* Leakage inductance of transformers; Single-phase induction motors with composite poles; Transient fields due to impact excitations
  current density distribution, 261–262
  current-carrying conductor, 262
  flux density, 263
  flux density components, 263
  inductance, 264–265
  leakage inductance, 265
  physical configuration, 261
  total flux linkage, 265
  vector magnetic potential, 263–264
Solid rotor induction machines, 127, 200–203; *see also* Eddy currents; Field distribution in air gap; Field distribution in solid rotor; Field distribution in stator slots; Tooth-ripple phenomena
  components of, 128
  conductors over pole pitch, 202
  current-carrying slot and empty slots, 203
  eddy current density, 130
  eddy current loss, 130–131, 228
  field components, 129–130
  force density, 131, 229
  machine performances, 227
  magnetic field intensity, 127
  mechanical power developed, 131, 229–230
  model, 231, 232
  phases in stator slots, 202
  rotor input power, 131–132, 230–231
  with three-phase double-layer stator winding, 201
  torque, 231
  2D model, 128
Space-phase angle, 285
Speed–torque curve, 284, 285
Squirrel cage induction machine, 5, 6, 171
Stable operation of hysteresis generator, 286
Statistical average, 29
Stator, 2
  air-gap, 288
  current sheet, 176, 275
  surface current density, 185
Stator poles, 130, 252, 344
  exercise, 310
  solid and laminated parts of, 287
Stator slots
  conductors in, 202
  equipotential and flux lines, 111
  exercise, 197
  field distribution in, 203
  magnetic field intensity components, 204, 342
  magneto-static potential in, 87
  opening, 399
  scalar magnetic potential in, 77, 81
  skewing of, 116
  three-phase winding in, 201
  winding harmonics, 199, 200
Stator winding, 399
  double-layer, 400
  end winding, 401–404
  integrated current sheets, 404–408
  main winding, 399–401
Stray losses, 199
Super synchronous speed, 284, 285, 286
Supply frequency, 117, 128, 207, 325, 342
Surface
  current density, 3, 116, 117–118, 133, 157, 162
  current distribution, 242
  resistivity, 300
Synchronous generator, 286

T

Tangential components of magnetic field, 236, 239
Target zone, 146
Teeth saturation, 2
Theorems, 31; *see also* Approximation theorem; Poynting theorem; Uniqueness theorem
  exercise, 73
  Helmholtz theorem, 45–48
Three-dimensional fields in solid-rotor induction machines, 231; *see also* Finite machine length effect; Rotor and stator length effect
  air-gap potential, 241
  arbitrary constants, 235
  boundary conditions at stator air-gap surface, 241–242
  conducting region, 233
  current sheet, 231
  eddy current density axial component, 233, 234
  eddy current loss, 252–257
  end-effects, 231
  exercise, 259
  force density, 257–258
  integration, 243–248
  magnetic field distributions, 233
  magnetic field intensity, 233, 235, 239, 240, 241
  magnetic field source, 232, 233
  magnetic field tangential components, 236, 239
  magnetic potential, 236, 239, 240
  mechanical power developed, 258–259
  performance parameters, 248
  rotor core, 232
  rotor input power, 259
  slip-power relation, 259
  solid-rotor induction machine model, 231, 232
  surface current distribution, 242
Time varying coordinates, 52
Time varying electromagnetic fields, 16, 163
  eddy currents, 127, 132
  steady-state sinusoidal, 49
Time-phase angle, 284
Tolerance limit, 57, 60, 73
Tooth
  -pitch, 92, 93
  -ripple harmonics, 199, 341; *see also* Single-phase induction motors with composite poles
  width, 76, 107, 331, 334, 336, 347
Tooth-opposite-slot orientation, 81–82, 97–98, 352–358; *see also* Integrations leading to simultaneous algebraic equations
  integrals involved, 375–378, 378–380
  unknown function evaluation, 98–99
Tooth-opposite-tooth orientation, 76–77, 94, 348–352; *see also* Integrations leading to simultaneous algebraic equations
  evaluation of unknown functions, 95–97
  integrals involved, 367–370, 371–373
Tooth-ripple phenomena, 199; *see also* Solid rotor induction machines
  slip/torque characteristics, 200, 201
Torch function, 142, 144
Torque, 3, 4, 5, 75, 200, 201, 231
  calculation of starting, 288
  exercise, 197, 310
  load, 284, 285
  net, 200
  rotor hysteresis impact, 283
  speed–torque curve, 284–286
Transformation
  from χ plane to w plane, 110–111, 114–116
  from z plane to w plane, 108–110, 112–114
Transformer, 2
  emf., 16
Transient fields due to impact excitations, 298; *see also* Current impact excitation; Field theory of hysteresis machines; Leakage inductance

of transformers; Single-phase induction motors with composite poles; Voltage impact excitation
conducting plate with surface current sheets, 299
current-carrying excitation winding, 298
damped wave equations, 300
exercises, 310
transients, 298
Transient fields in plates, 330; *see also* Relations involving simple summations
current impact excitation, 330
voltage impact excitation, 331
Truncated series, 85, 91
Truncation errors, 314
Two dimensional field
fields in anisotropic media, 172–175
problems, 20
Type 1 current impact excitation, 298
Type 1 transients, 298
Type 1 voltage impact excitation, 298
Type 2 current impact excitation, 298
Type 2 transients, 298
Type 2 voltage impact excitation, 298, 305

U

Unbounded regions, 45
Uniqueness theorem, 4, 31; *see also* Field theorems; Theorems
difference potential, 32
example of cuboid, 33–37
example of rectangular region, 37–40
for Laplace and Poisson equations, 31
for Maxwell's equations, 43–45
potential distribution, 32–33
for vector magnetic potentials, 41–43
Unlaminated iron cores, 6
Unlaminated rotor polyphase induction machines, 199

V

Values of functions, 314, 362
Vector magnetic potential, 18–21
Velocity of light, 16
in free space, 176, 177, 186
Velocity vector, 53
Voltage impact excitation, 298, 305; *see also* Transient fields due to impact excitations
electromagnetic field, 306, 308–310
transient current density, 306
voltage drop, 308
Volumetric current density, 3

W

Wave equation, 3
damped, 16, 25, 300
in free space, 303, 308
Well-conditioned problem, 314
Winding harmonics, 199
Wound rotor induction machines, 175–183
w-plane, 108
χ plane to, 114–116
contour in, 113
positive and negative parts, 110
from *z*-plane to, 109, 112–114

Z

z plane, 109, 115
to *w* plane, 108–110, 112–114, 324, 325